Data Analysis Using Regression and Multilevel/Hierarchical Models

Data Analysis Using Regression and Multilevel/Hierarchical Models is a comprehensive manual for the applied researcher who wants to perform data analysis using linear and nonlinear regression and multilevel models. The book introduces and demonstrates a wide variety of models, at the same time instructing the reader in how to fit these models using freely available software packages. The book illustrates the concepts by working through scores of real data examples that have arisen in the authors' own applied research, with programming code provided for each one. Topics covered include causal inference, including regression, poststratification, matching, regression discontinuity, and instrumental variables, as well as multilevel logistic regression and missing-data imputation. Practical tips regarding building, fitting, and understanding are provided throughout.

Andrew Gelman is Professor of Statistics and Professor of Political Science at Columbia University. He has published more than 150 articles in statistical theory, methods, and computation and in applications areas including decision analysis, survey sampling, political science, public health, and policy. His other books are *Bayesian Data Analysis* (1995, second edition 2003) and *Teaching Statistics: A Bag of Tricks* (2002).

Jennifer Hill is Assistant Professor of Public Affairs in the Department of International and Public Affairs at Columbia University. She has coauthored articles that have appeared in the *Journal of the American Statistical Association, American Political Science Review, American Journal of Public Health, Developmental Psychology*, the *Economic Journal*, and the *Journal of Policy Analysis and Management*, among others.

Analytical Methods for Social Research

Analytical Methods for Social Research presents texts on empirical and formal methods for the social sciences. Volumes in the series address both the theoretical underpinnings of analytical techniques and their application in social research. Some series volumes are broad in scope, cutting across a number of disciplines. Others focus mainly on methodological applications within specific fields such as political science, sociology, demography, and public health. The series serves a mix of students and researchers in the social sciences and statistics.

Series Editors:

R. Michael Alvarez, *California Institute of Technology*
Nathaniel L. Beck, *New York University*
Lawrence L. Wu, *New York University*

Other Titles in the Series:

Event History Modeling: A Guide for Social Scientists, by Janet M. Box-Steffensmeier
   and Bradford S. Jones
Ecological Inference: New Methodological Strategies, edited by Gary King, Ori Rosen,
   and Martin A. Tanner
Spatial Models of Parliamentary Voting, by Keith T. Poole
Essential Mathematics for Political and Social Research, by Jeff Gill
Political Game Theory: An Introduction, by Nolan McCarty and Adam Meirowitz

Data Analysis Using Regression and Multilevel/Hierarchical Models

ANDREW GELMAN
Columbia University

JENNIFER HILL
Columbia University

CAMBRIDGE
UNIVERSITY PRESS

CAMBRIDGE UNIVERSITY PRESS
Cambridge, New York, Melbourne, Madrid, Cape Town, Singapore, São Paulo, Delhi

Cambridge University Press
32 Avenue of the Americas, New York, NY 10013-2473, USA

www.cambridge.org
Information on this title: www.cambridge.org/9780521686891

First published 2007
Reprinted with corrections 2007
11th printing 2009

Printed in the United States of America

A catalog record for this publication is available from the British Library.

Library of Congress Cataloging in Publication Data
Gelman, Andrew.
Data analysis using regression and multilevel/hierarchical models / Andrew Gelman.
Jennifer Hill.
p.   cm. – (Analytical methods for social research)
Includes bibliographical references.
ISBN 0-521-86706-1 (hardcover) – ISBN 0-521-68689-X (pbk.)
1. Regression analysis.   2. Multilevel modes (Statistics).   1. Hill, Jennifer, 1969–
II. Title.   III. Series.
HA31.3.G45   2006
519.5'36–dc22   2006040566

ISBN 978-0-521-86706-1 hardback
ISBN 978-0-521-68689-1 paperback

For Zacky and for Audrey

Contents

List of examples

Preface

Aim of this book

This book originated as lecture notes for a course in regression and multilevel modeling, offered by the statistics department at Columbia University and attended by graduate students and postdoctoral researchers in social sciences (political science, economics, psychology, education, business, social work, and public health) and statistics. The prerequisite is statistics up to and including an introduction to multiple regression.

Advanced mathematics is not assumed—it is important to understand the linear model in regression, but it is not necessary to follow the matrix algebra in the derivation of least squares computations. It is useful to be familiar with exponents and logarithms, especially when working with generalized linear models.

After completing Part 1 of this book, you should be able to fit classical linear and generalized linear regression models—and do more with these models than simply look at their coefficients and their statistical significance. Applied goals include causal inference, prediction, comparison, and data description. After completing Part 2, you should be able to fit regression models for multilevel data. Part 3 takes you from data collection, through model understanding (looking at a table of estimated coefficients is usually not enough), to model checking and missing data. The appendixes include some reference materials on key tips, statistical graphics, and software for model fitting.

What you should be able to do after reading this book and working through the examples

This text is structured through models and examples, with the intention that after each chapter you should have certain skills in fitting, understanding, and displaying models:

- *Part 1A:* Fit, understand, and graph classical regressions and generalized linear models.

  - *Chapter 3:* Fit linear regressions and be able to interpret and display estimated coefficients.
  - *Chapter 4:* Build linear regression models by transforming and combining variables.
  - *Chapter 5:* Fit, understand, and display logistic regression models for binary data.
  - *Chapter 6:* Fit, understand, and display generalized linear models, including Poisson regression with overdispersion and ordered logit and probit models.

- *Part 1B:* Use regression to learn about quantities of substantive interest (not just regression coefficients).

  - *Chapter 7:* Simulate probability models and uncertainty about inferences and predictions.

- *Chapter 8:* Check model fits using fake-data simulation and predictive simulation.

- *Chapter 9:* Understand assumptions underlying causal inference. Set up regressions for causal inference and understand the challenges that arise.

- *Chapter 10:* Understand the assumptions underlying propensity score matching, instrumental variables, and other techniques to perform causal inference when simple regression is not enough. Be able to use these when appropriate.

- *Part 2A:* Understand and graph multilevel models.

  - *Chapter 11:* Understand multilevel data structures and models as generalizations of classical regression.

  - *Chapter 12:* Understand and graph simple varying-intercept regressions and interpret as partial-pooling estimates.

  - *Chapter 13:* Understand and graph multilevel linear models with varying intercepts and slopes, non-nested structures, and other complications.

  - *Chapter 14:* Understand and graph multilevel logistic models.

  - *Chapter 15:* Understand and graph multilevel overdispersed Poisson, ordered logit and probit, and other generalized linear models.

- *Part 2B:* Fit multilevel models using the software packages R and Bugs.

  - *Chapter 16:* Fit varying-intercept regressions and understand the basics of Bugs. Check your programming using fake-data simulation.

  - *Chapter 17:* Use Bugs to fit various models from Part 2A.

  - *Chapter 18:* Understand Bayesian inference as a generalization of least squares and maximum likelihood. Use the Gibbs sampler to fit multilevel models.

  - *Chapter 19:* Use redundant parameterizations to speed the convergence of the Gibbs sampler.

- *Part 3:*

  - *Chapter 20:* Perform sample size and power calculations for classical and hierarchical models: standard-error formulas for basic calculations and fake-data simulation for harder problems.

  - *Chapter 21:* Calculate and understand contrasts, explained variance, partial pooling coefficients, and other summaries of fitted multilevel models.

  - *Chapter 22:* Use the ideas of analysis of variance to summarize fitted multilevel models; use multilevel models to perform analysis of variance.

  - *Chapter 23:* Use multilevel models in causal inference.

  - *Chapter 24:* Check the fit of models using predictive simulation.

  - *Chapter 25:* Use regression to impute missing data in multivariate datasets.

In summary, you should be able to fit, graph, and understand classical and multilevel linear and generalized linear models and to use these model fits to make predictions and inferences about quantities of interest, including causal treatment effects.

Data for the examples and homework assignments and other resources for teaching and learning

The website www.stat.columbia.edu/~gelman/arm/ contains datasets used in the examples and homework problems of the book, as well as sample computer code. The website also includes some tips for teaching regression and multilevel modeling through class participation rather than lecturing. We plan to update these tips based on feedback from instructors and students; please send your comments and suggestions to gelman@stat.columbia.edu.

Outline of a course

When teaching a course based on this book, we recommend starting with a self-contained review of linear regression, logistic regression, and generalized linear models, focusing not on the mathematics but on understanding these methods and implementing them in a reasonable way. This is also a convenient way to introduce the statistical language R, which we use throughout for modeling, computation, and graphics. One thing that will probably be new to the reader is the use of random simulations to summarize inferences and predictions.

We then introduce multilevel models in the simplest case of nested linear models, fitting in the Bayesian modeling language Bugs and examining the results in R. Key concepts covered at this point are partial pooling, variance components, prior distributions, identifiability, and the interpretation of regression coefficients at different levels of the hierarchy. We follow with non-nested models, multilevel logistic regression, and other multilevel generalized linear models.

Next we detail the steps of fitting models in Bugs and give practical tips for reparameterizing a model to make it converge faster and additional tips on debugging. We also present a brief review of Bayesian inference and computation. Once the student is able to fit multilevel models, we move in the final weeks of the class to the final part of the book, which covers more advanced issues in data collection, model understanding, and model checking.

As we show throughout, multilevel modeling fits into a view of statistics that unifies substantive modeling with accurate data fitting, and graphical methods are crucial both for seeing unanticipated features in the data and for understanding the implications of fitted models.

Acknowledgments

We thank the many students and colleagues who have helped us understand and implement these ideas. Most important have been Jouni Kerman, David Park, and Joe Bafumi for years of suggestions throughout this project, and for many insights into how to present this material to students.

In addition, we thank Hal Stern and Gary King for discussions on the structure of this book; Chuanhai Liu, Xiao-Li Meng, Zaiying Huang, John Boscardin, Jouni Kerman, Alan Zaslavsky, David Dunson, Maria Grazia Pittau, Aleks Jakulin, and Yu-Sung Su for discussions about multilevel modeling and statistical computation; Iven Van Mechelen and Hans Berkhof for discussions about model checking; Iain Pardoe for discussions of average predictive effects and other summaries of regression models; Matt Salganik and Wendy McKelvey for suggestions on the presentation of sample size calculations; T. E. Raghunathan, Donald Rubin, Rajeev Dehejia, Michael Sobel, Guido Imbens, Samantha Cook, Ben Hansen, Dylan Small, and Ed Vytlacil for concepts of missing-data modeling and causal inference; Eric

Loken for help in understanding identifiability in item-response models; Niall Bolger, Agustin Calatroni, John Carlin, Rafael Guerrero-Preston, Oliver Kuss, Reid Landes, Eduardo Leoni, and Dan Rabinowitz for code in Stata, SAS, and SPSS; Hans Skaug for code in AD Model Builder; Uwe Ligges, Sibylle Sturtz, Douglas Bates, Peter Dalgaard, Martyn Plummer, and Ravi Varadhan for help with multilevel modeling and general advice on R; and the students in Statistics / Political Science 4330 at Columbia for their invaluable feedback throughout.

Collaborators on specific examples mentioned in this book include Phillip Price on the home radon study; Tom Little, David Park, Joe Bafumi, and Noah Kaplan on the models of opinion polls and political ideal points; Jane Waldfogel, Jeanne Brooks-Gunn, and Wen Han for the mothers and children's intelligence data; Lex van Geen and Alex Pfaff on the arsenic in Bangladesh; Gary King on election forecasting; Jeffrey Fagan and Alex Kiss on the study of police stops; Tian Zheng and Matt Salganik on the social network analysis; John Carlin for the data on mesquite bushes and the adolescent-smoking study; Alessandra Casella and Tom Palfrey for the storable-votes study; Rahul Dodhia for the flight simulator example; Boris Shor, Joe Bafumi, and David Park on the voting and income study; Alan Edelman for the internet connections data; Donald Rubin for the Electric Company and educational-testing examples; Jeanne Brooks-Gunn and Jane Waldfogel for the mother and child IQ scores example and Infant Health and Development Program data; Nabila El-Bassel for the risky behavior data; Lenna Nepomnyaschy for the child support example; Howard Wainer with the Advanced Placement study; Iain Pardoe for the prison-sentencing example; James Liebman, Jeffrey Fagan, Valerie West, and Yves Chretien for the death-penalty study; Marcia Meyers, Julien Teitler, Irv Garfinkel, Marilyn Sinkowicz, and Sandra Garcia with the Social Indicators Study; Wendy McKelvey for the cockroach and rodent examples; Stephen Arpadi for the zinc and HIV study; Eric Verhoogen and Jan von der Goltz for the Progresa data; and Iven van Mechelen, Yuri Goegebeur, and Francis Tuerlincx on the stochastic learning models. These applied projects motivated many of the methodological ideas presented here, for example the display and interpretation of varying-intercept, varying-slope models from the analysis of income and voting (see Section 14.2), the constraints in the model of senators' ideal points (see Section 14.3), and the difficulties with two-level interactions as revealed by the radon study (see Section 21.7). Much of the work in Section 5.7 and Chapter 21 on summarizing regression models was done in collaboration with Iain Pardoe.

Many errors were found and improvements suggested by Brad Carlin, John Carlin, Samantha Cook, Caroline Rosenthal Gelman, Kosuke Imai, Jonathan Katz, Uwe Ligges, Wendy McKelvey, Jong-Hee Park, Martyn Plummer, Phillip Price, Song Qian, Giuseppe Ragusa, Dylan Small, Elizabeth Stuart, Sibylle Sturtz, Alex Tabarrok, and Shravan Vasishth. Brian MacDonald's copyediting has saved us from much embarrassment, and we also thank Yu-Sung Su for typesetting help, Sarah Ryu for assistance with indexing, and Ed Parsons and his colleagues at Cambridge University Press for their help in putting this book together. We especially thank Bob O'Hara and Gregor Gorjanc for incredibly detailed and useful comments on the nearly completed manuscript.

We also thank the developers of free software, especially R (for statistical computation and graphics) and Bugs (for Bayesian modeling), and also Emacs and LaTex (used in the writing of this book). We thank Columbia University for its collaborative environment for research and teaching, and the U.S. National Science Foundation for financial support. Above all, we thank our families for their love and support during the writing of this book.

## CHAPTER 1

Why?

1.1 What is multilevel regression modeling?

Consider an educational study with data from students in many schools, predicting in each school the students' grades y on a standardized test given their scores on a pre-test x and other information. A separate regression model can be fit within each school, and the parameters from these schools can themselves be modeled as depending on school characteristics (such as the socioeconomic status of the school's neighborhood, whether the school is public or private, and so on). The student-level regression and the school-level regression here are the two levels of a *multilevel model.*

In this example, a multilevel model can be expressed in (at least) three equivalent ways as a student-level regression:

- A model in which the coefficients vary by school (thus, instead of a model such as $y = \alpha + \beta x + \text{error}$, we have $y = \alpha_j + \beta_j x + \text{error}$, where the subscripts j index schools),

- A model with more than one variance component (student-level and school-level variation),

- A regression with many predictors, including an indicator variable for each school in the data.

More generally, we consider a multilevel model to be a regression (a linear or generalized linear model) in which the parameters—the regression coefficients—are given a probability model. This second-level model has parameters of its own—the *hyperparameters* of the model—which are also estimated from data.

The two key parts of a multilevel model are varying coefficients, and a model for those varying coefficients (which can itself include group-level predictors). Classical regression can sometimes accommodate varying coefficients by using indicator variables. The feature that distinguishes multilevel models from classical regression is in the modeling of the variation between groups.

Models for regression coefficients

To give a preview of our notation, we write the regression equations for two multilevel models. To keep notation simple, we assume just one student-level predictor x (for example, a pre-test score) and one school-level predictor u (for example, average parents' incomes).

Varying-intercept model. First we write the model in which the regressions have the same slope in each of the schools, and only the intercepts vary. We use the notation i for individual students and $j[i]$ for the school j containing student i:[1]

$$
\begin{aligned}
y_i &= \alpha_{j[i]} + \beta x_i + \epsilon_i, \quad \text{for students } i = 1, \ldots, n \\
\alpha_j &= a + b u_j + \eta_j, \quad \text{for schools } j = 1, \ldots, J.
\end{aligned} \tag{1.1}
$$

Here, x_i and u_j represent predictors at the student and school levels, respectively, and ϵ_i and η_j are independent error terms at each of the two levels. The model can be written in several other equivalent ways, as we discuss in Section 12.5.

The number of "data points" J (here, schools) in the higher-level regression is typically much less than n, the sample size of the lower-level model (for students in this example).

Varying-intercept, varying-slope model. More complicated is the model where intercepts and slopes both can vary by school:

$$
\begin{aligned}
y_i &= \alpha_{j[i]} + \beta_{j[i]} x_i + \epsilon_i, \quad \text{for students } i = 1, \ldots, n \\
\alpha_j &= a_0 + b_0 u_j + \eta_{j1}, \quad \text{for schools } j = 1, \ldots, J \\
\beta_j &= a_1 + b_1 u_j + \eta_{j2}, \quad \text{for schools } j = 1, \ldots, J.
\end{aligned}
$$

Compared to model (1.1), this has twice as many vectors of varying coefficients (α, β), twice as many vectors of second-level coefficients (a, b), and potentially correlated second-level errors η_1, η_2. We will be able to handle these complications.

Labels

"Multilevel" or "hierarchical." Multilevel models are also called *hierarchical*, for two different reasons: first, from the structure of the data (for example, students clustered within schools); and second, from the model itself, which has its own hierarchy, with the parameters of the within-school regressions at the bottom, controlled by the hyperparameters of the upper-level model.

Later we shall consider non-nested models—for example, individual observations that are nested within states and years. Neither "state" nor "year" is above the other in a hierarchical sense. In this sort of example, we can consider individuals, states, and years to be three different levels without the requirement of a full ordering or hierarchy. More complex structures, such as three-level nesting (for example, students within schools within school districts) are also easy to handle within the general multilevel framework.

Why we avoid the term "random effects." Multilevel models are often known as random-effects or mixed-effects models. The regression coefficients that are being modeled are called *random effects*, in the sense that they are considered random outcomes of a process identified with the model that is predicting them. In contrast, *fixed effects* correspond either to parameters that do not vary (for example, fitting the same regression line for each of the schools) or to parameters that vary but are not modeled themselves (for example, fitting a least squares regression model with various predictors, including indicators for the schools). A *mixed-effects* model includes both fixed and random effects; for example, in model (1.1), the varying intercepts α_j have a group-level model, but β is fixed and does not vary by group.

---

[1] The model can also be written as $y_{ij} = \alpha_j + \beta x_{ij} + \epsilon_{ij}$, where y_{ij} is the measurement from student i in school j. We prefer using the single sequence i to index all students (and $j[i]$ to label schools) because this fits in better with our multilevel modeling framework with data and models at the individual and group levels. The data are y_i because they can exist without reference to the groupings, and we prefer to include information about the groupings as numerical data— that is, the index variable $j[i]$—rather than through reordering the data through subscripting. We discuss the structure of the data and models further in Chapter 11.

Fixed effects can be viewed as special cases of random effects, in which the higher-level variance (in model (1.1), this would be σ_α^2) is set to 0 or ∞. Hence, in our framework, all regression parameters are "random," and the term "multilevel" is all-encompassing. As we discuss on page 245, we find the terms "fixed," "random," and "mixed" effects to be confusing and often misleading, and so we avoid their use.

1.2 Some examples from our own research

Multilevel modeling can be applied to just about any problem. Just to give a feel of the ways it can be used, we give here a few examples from our applied work.

Combining information for local decisions: home radon measurement and remediation

Radon is a carcinogen—a naturally occurring radioactive gas whose decay products are also radioactive—known to cause lung cancer in high concentrations and estimated to cause several thousand lung cancer deaths per year in the United States. The distribution of radon levels in U.S. homes varies greatly, with some houses having dangerously high concentrations. In order to identify the areas with high radon exposures, the Environmental Protection Agency coordinated radon measurements in a random sample of more than 80,000 houses throughout the country.

To simplify the problem somewhat, our goal in analyzing these data was to estimate the distribution of radon levels in each of the approximately 3000 counties in the United States, so that homeowners could make decisions about measuring or remediating the radon in their houses based on the best available knowledge of local conditions. For the purpose of this analysis, the data were structured hierarchically: houses within counties. If we were to analyze multiple measurements within houses, there would be a three-level hierarchy of measurements, houses, and counties.

In performing the analysis, we had an important predictor—the floor on which the measurement was taken, either basement or first floor; radon comes from underground and can enter more easily when a house is built into the ground. We also had an important county-level predictor—a measurement of soil uranium that was available at the county level. We fit a model of the form (1.1), where y_i is the logarithm of the radon measurement in house i, x is the floor of the measurement (that is, 0 for basement and 1 for first floor), and u is the uranium measurement at the county level. The errors ϵ_i in the first line of (1.1) represent "within-county variation," which in this case includes measurement error, natural variation in radon levels within a house over time, and variation between houses (beyond what is explained by the floor of measurement). The errors η_j in the second line represent variation between counties, beyond what is explained by the county-level uranium predictor.

The hierarchical model allows us to fit a regression model to the individual measurements while accounting for systematic unexplained variation among the 3000 counties. We return to this example in Chapter 12.

Modeling correlations: forecasting presidential elections

It is of practical interest to politicians and theoretical interest to political scientists that the outcomes of elections can be forecast with reasonable accuracy given information available months ahead of time. To understand this better, we set up a model to forecast presidential elections. Our predicted outcomes were the Democratic Party's share of the two-party vote in each state in each of the 11 elections from 1948 through 1988, yielding 511 data points (the analysis excluded states that were won by third parties), and we had various predictors, including the performance of the Democrats in the previous election, measures of state-level and national economic trends, and national opinion polls up to two months before the election.

We set up our forecasting model two months before the 1992 presidential election and used it to make predictions for the 50 states. Predictions obtained using classical regression are reasonable, but when the model is evaluated historically (fitting to all but one election and then using the model to predict that election, then repeating this for the different past elections), the associated predictive intervals turn out to be too narrow: that is, the predictions are not as accurate as claimed by the model. Fewer than 50% of the predictions fall in the 50% predictive intervals, and fewer than 95% are inside the 95% intervals. The problem is that the 511 original data points are *structured*, and the state-level errors are *correlated*. It is overly optimistic to say that we have 511 independent data points.

Instead, we model

$$y_i = \beta_0 + X_{i1}\beta_1 + X_{i2}\beta_2 + \cdots + X_{ik}\beta_k + \eta_{t[i]} + \delta_{r[i],t[i]} + \epsilon_i, \text{ for } i = 1, \ldots, n, \quad (1.2)$$

where $t[i]$ is a indicator for time (election year), and $r[i]$ is an indicator for the region of the country (Northeast, Midwest, South, or West), and $n = 511$ is the number of state-years used to fit the model. For each election year, η_t is a nationwide error and the $\delta_{r,t}$'s are four independent regional errors.

The error terms must then be given distributions. As usual, the default is the normal distribution, which for this model we express as

$$\eta_t \quad \sim \quad N(0, \sigma_\eta^2), \text{ for } t = 1, \ldots, 11$$
$$\delta_{r,t} \quad \sim \quad N(0, \sigma_\delta^2), \text{ for } r = 1, \ldots, 4; \, t = 1, \ldots, 11$$
$$\epsilon_i \quad \sim \quad N(0, \sigma_\epsilon^2), \text{ for } i = 1, \ldots, 511. \quad (1.3)$$

In the multilevel model, all the parameters $\beta, \sigma_\eta, \sigma_\delta, \sigma_\epsilon$ are estimated from the data.

We can then make a prediction by simulating the election outcome in the 50 states in the next election year, $t = 12$:

$$y_i = \beta_0 + X_{i1}\beta_1 + X_{i2}\beta_2 + \cdots + X_{ik}\beta_k + \eta_{12} + \delta_{r[i],12} + \epsilon_i, \text{ for } i = n+1, \ldots, n+50.$$

To define the predictive distribution of these 50 outcomes, we need the point predictors $X_i\beta = \beta_0 + X_{i1}\beta_1 + X_{i2}\beta_2 + \cdots + X_{ik}\beta_k$ and the state-level errors ϵ as before, but we also need a new national error η_{12} and four new regional errors $\delta_{r,12}$, which we simulate from the distributions (1.3). The variation from these gives a more realistic statement of prediction uncertainties.

Small-area estimation: state-level opinions from national polls

In a micro-level version of election forecasting, it is possible to predict the political opinions of individual voters given demographic information and where they live. Here the data sources are opinion polls rather than elections.

For example, we analyzed the data from seven CBS News polls from the 10 days immediately preceding the 1988 U.S. presidential election. For each survey respondent i, we label $y_i = 1$ if he or she preferred George Bush (the Republican candidate), 0 if he or she preferred Michael Dukakis (the Democrat). We excluded respondents who preferred others or had no opinion, leaving a sample size n of about 6000. We then fit the model,

$$\Pr(y_i = 1) = \text{logit}^{-1}(X_i\beta),$$

where X included 85 predictors:

- A constant term
- An indicator for "female"
- An indicator for "black"
- An indicator for "female and black"
- 4 indicators for age categories (18–29, 30–44, 45–64, and 65+)
- 4 indicators for education categories (less than high school, high school, some college, college graduate)
- 16 indicators for age × education
- 51 indicators for states (including the District of Columbia)
- 5 indicators for regions (Northeast, Midwest, South, West, and D.C.)
- The Republican share of the vote for president in the state in the previous election.

In classical regression, it would be unwise to fit this many predictors because the estimates will be unreliable, especially for small states. In addition, it would be necessary to leave predictors out of each batch of indicators (the 4 age categories, the 4 education categories, the 16 age × education interactions, the 51 states, and the 5 regions) to avoid collinearity.

With a multilevel model, the coefficients for each batch of indicators are fit to a probability distribution, and it is possible to include all the predictors in the model. We return to this example in Section 14.1.

Social science modeling: police stops by ethnic group with variation across precincts

There have been complaints in New York City and elsewhere that the police harass members of ethnic minority groups. In 1999 the New York State Attorney General's Office instigated a study of the New York City police department's "stop and frisk" policy: the lawful practice of "temporarily detaining, questioning, and, at times, searching civilians on the street." The police have a policy of keeping records on every stop and frisk, and this information was collated for all stops (about 175,000 in total) over a 15-month period in 1998–1999. We analyzed these data to see to what extent different ethnic groups were stopped by the police. We focused on blacks (African Americans), hispanics (Latinos), and whites (European Americans). We excluded others (about 4% of the stops) because of sensitivity to ambiguities in classifications. The ethnic categories were as recorded by the police making the stops.

It was found that blacks and hispanics represented 50% and 33% of the stops, respectively, despite constituting only 26% and 24%, respectively, of the population of the city. An arguably more relevant baseline comparison, however, is to the number of crimes committed by members of each ethnic group. Data on actual crimes are not available, of course, so as a proxy we used the number of arrests within New York City in 1997 as recorded by the Division of Criminal Justice Services (DCJS) of New York State. We used these numbers to represent the frequency of crimes that the police might suspect were committed by members of each group. When compared in that way, the ratio of stops to previous DCJS arrests was 1.24 for whites, 1.53 for blacks, and 1.72 for hispanics—the minority groups still appeared to be stopped disproportionately often.

These ratios are suspect too, however, because they average over the whole city. Suppose the police make more stops in high-crime areas but treat the different ethnic groups equally within any locality. Then the citywide ratios could show strong differences between ethnic groups even if stops are entirely determined by location rather than ethnicity. In order to separate these two kinds of predictors, we performed a multilevel analysis using the city's 75 precincts. For each ethnic group $e = 1, 2, 3$ and precinct $p = 1, \ldots, 75$, we model the number of stops y_{ep} using an overdispersed Poisson regression. The exponentiated coefficients from this model represent relative rates of stops compared to arrests for the different ethnic groups, after controlling for precinct. We return to this example in Section 15.1.

1.3 Motivations for multilevel modeling

Multilevel models can be used for a variety of inferential goals including causal inference, prediction, and descriptive modeling.

Learning about treatment effects that vary

One of the basic goals of regression analysis is estimating treatment effects—how does y change when some x is varied, with all other inputs held constant? In many applications, it is not an overall effect of x that is of interest, but how this effect varies in the population. In classical statistics we can study this variation using *interactions*: for example, a particular educational innovation may be more effective for girls than for boys, or more effective for students who expressed more interest in school in a pre-test measurement.

Multilevel models also allow us to study effects that vary by group, for example an intervention that is more effective in some schools than others (perhaps because of unmeasured school-level factors such as teacher morale). In classical regression, estimates of varying effects can be noisy, especially when there are few observations per group; multilevel modeling allows us to estimate these interactions to the extent supported by the data.

Using all the data to perform inferences for groups with small sample size

A related problem arises when we are trying to estimate some group-level quantity, perhaps a local treatment effect or maybe simply a group-level average (as in the small-area estimation example on page 4). Classical estimation just using the local information can be essentially useless if the sample size is small in the group. At the other extreme, a classical regression ignoring group indicators can be misleading in ignoring group-level variation. Multilevel modeling allows the estimation of group averages and group-level effects, compromising between the overly noisy within-group estimate and the oversimplified regression estimate that ignores group indicators.

Prediction

Regression models are commonly used for predicting outcomes for new cases. But what if the data vary by group? Then we can make predictions for new units in existing groups or in new groups. The latter is difficult to do in classical regression:

if a model ignores group effects, it will tend to understate the error in predictions for new groups. But a classical regression that includes group effects does not have any automatic way of getting predictions for a new group.

A natural attack on the problem is a two-stage regression, first including group indicators and then fitting a regression of estimated group effects on group-level predictors. One can then forecast for a new group, with the group effect predicted from the group-level model, and then the observations predicted from the unit-level model. However, if sample sizes are small in some groups, it can be difficult or even impossible to fit such a two-stage model classically, and fully accounting for the uncertainty at both levels leads directly to a multilevel model.

Analysis of structured data

Some datasets are collected with an inherent multilevel structure, for example, students within schools, patients within hospitals, or data from cluster sampling. Statistical theory—whether sampling-theory or Bayesian—says that inference should include the factors used in the design of data collection. As we shall see, multilevel modeling is a direct way to include indicators for clusters at all levels of a design, without being overwhelmed with the problems of overfitting that arise from applying least squares or maximum likelihood to problems with large numbers of parameters.

More efficient inference for regression parameters

Data often arrive with multilevel structure (students within schools and grades, laboratory assays on plates, elections in districts within states, and so forth). Even simple cross-sectional data (for example, a random sample survey of 1000 Americans) can typically be placed within a larger multilevel context (for example, an annual series of such surveys). The traditional alternatives to multilevel modeling are *complete pooling*, in which differences between groups are ignored, and *no pooling*, in which data from different sources are analyzed separately. As we shall discuss in detail throughout the book, both these approaches have problems: no pooling ignores information and can give unacceptably variable inferences, and complete pooling suppresses variation that can be important or even the main goal of a study. The extreme alternatives can in fact be useful as preliminary estimates, but ultimately we prefer the *partial pooling* that comes out of a multilevel analysis.

Including predictors at two different levels

In the radon example described in Section 1.2, we have outcome measurements at the individual level and predictors at the individual and county levels. How can this information be put together? One possibility is simply to run a classical regression with predictors at both levels. But this does not correct for differences between counties *beyond* what is included in the predictors. Another approach would be to augment this model with indicators (dummy variables) for the counties. But in a classical regression it is not possible to include county-level indicators as well along with county-level predictors—the predictors would become collinear (see the end of Section 4.5 for a discussion of collinearity and nonidentifiability in this context).

Another approach is to fit the model with county indicators but without the county-level predictors, and then to fit a second model. This is possible but limited because it relies on the classical regression estimates of the coefficients for those county-level indicators—and if the data are sparse within counties, these estimates won't be very good. Another possibility in the classical framework would be to fit separate models in each group, but this is not possible unless the sample size is large in each group. The multilevel model provides a coherent model that simultaneously incorporates both individual- and group-level models.

Getting the right standard error: accurately accounting for uncertainty in prediction and estimation

Another motivation for multilevel modeling is for predictions, for example, when forecasting state-by-state outcomes of U.S. presidential elections, as described in Section 1.2. To get an accurate measure of predictive uncertainty, one must account for correlation of the outcome between states in a given election year. Multilevel modeling is a convenient way to do this.

For certain kinds of predictions, multilevel models are essential. For example, consider a model of test scores for students within schools. In classical regression, school-level variability might be modeled by including an indicator variable for each school. In this framework though, it is impossible to make a prediction for a new student in a new school, because there would not be an indicator for this new school in the model. This prediction problem is handled seamlessly using multilevel models.

1.4 Distinctive features of this book

The topics and methods covered in this book overlap with many other textbooks on regression, multilevel modeling, and applied statistics. We differ from most other books in these areas in the following ways:

- We present methods and software that allow the reader to fit complicated, linear or nonlinear, nested or non-nested models. We emphasize the use of the statistical software packages R and Bugs and provide code for many examples as well as methods such as redundant parameterization that speed computation and lead to new modeling ideas.

- We include a wide range of examples, almost all from our own applied research. The statistical methods are thus motivated in the best way, as successful practical tools.

- Most books define regression in terms of matrix operations. We avoid much of this matrix algebra for the simple reason that it is now done automatically by computers. We are more interested in understanding the "forward," or predictive, matrix multiplication $X\beta$ than the more complicated inferential formula $(X^tX)^{-1}X^ty$. The latter computation and its generalizations are important but can be done out of sight of the user. For details of the underlying matrix algebra, we refer readers to the regression textbooks listed in Section 3.8.

- We try as much as possible to display regression results graphically rather than through tables. Here we apply ideas such as those presented in the books by Ramsey and Schafer (2001) for classical regression and Kreft and De Leeuw (1998) for multilevel models. We consider graphical display of model estimates to be not just a useful teaching method but also a necessary tool in applied research.

Statistical texts commonly recommend graphical displays for model diagnostics. These can be very useful, and we refer readers to texts such as Cook and Weisberg

(1999) for more on this topic—but here we are emphasizing graphical displays of the fitted models themselves. It is our experience that, even when a model fits data well, we have difficulty understanding it if all we do is look at tables of regression coefficients.

- We consider multilevel modeling as generally applicable to structured data, not limited to clustered data, panel data, or nested designs. For example, in a random-digit-dialed survey of the United States, one can, and should, use multilevel models if one is interested in estimating differences among states or demographic subgroups—even if no multilevel structure is in the survey design.

Ultimately, you have to learn these methods by doing it yourself, and this chapter is intended to make things easier by recounting stories about how we learned this by doing it ourselves. But we warn you ahead of time that we include more of our successes than our failures.

Costs and benefits of our approach

Doing statistics as described in this book is not easy. The difficulties are not mathematical but rather conceptual and computational. For classical regressions and generalized linear models, the actual fitting is easy (as illustrated in Part 1), but programming effort is still required to graph the results relevantly and to simulate predictions and replicated data. When we move to multilevel modeling, the fitting itself gets much more complicated (see Part 2B), and displaying and checking the models require correspondingly more work. Our emphasis on R and Bugs means that an initial effort is required simply to learn and use the software. Also, compared to usual treatments of multilevel models, we describe a wider variety of modeling options for the researcher so that more decisions will need to be made.

A simpler alternative is to use classical regression and generalized linear modeling where possible—this can be done in R or, essentially equivalently, in Stata, SAS, SPSS, and various other software—and then, when multilevel modeling is really needed, to use functions that adapt classical regression to handle simple multilevel models. Such functions, which can be run with only a little more effort than simple regression fitting, exist in many standard statistical packages.

Compared to these easier-to-use programs, our approach has several advantages:

- We can fit a greater variety of models. The modular structure of Bugs allows us to add complexity where needed to fit data and study patterns of interest.
- By working with simulations (rather than simply point estimates of parameters), we can directly capture inferential uncertainty and propagate it into predictions (as discussed in Chapter 7 and applied throughout the book). We can directly obtain inference for quantities other than regression coefficients and variance parameters.
- R gives us flexibility to display inferences and data flexibly.

We recognize, however, that other software and approaches may be useful too, either as starting points or to check results. Section C.4 describes briefly how to fit multilevel models in several other popular statistical software packages.

1.5 Computing

We perform computer analyses using the freely available software R and Bugs. Appendix C gives instructions on obtaining and using these programs. Here we outline how these programs fit into our overall strategy for data analysis.

Our general approach to statistical computing

In any statistical analysis, we like to be able to directly manipulate the data, model, and inferences. We just about never know the right thing to do ahead of time, so we have to spend much of our effort examining and cleaning the data, fitting many different models, summarizing the inferences from the models in different ways, and then going back and figuring how to expand the model to allow new data to be included in the analysis.

It is important, then, to be able to select subsets of the data, to graph whatever aspect of the data might be of interest, and to be able to compute numerical summaries and fit simple models easily. All this can be done within R—you will have to put some initial effort into learning the language, but it will pay off later.

You will almost always need to try many different models for any problem: not just different subsets of predictor variables as in linear regression, and not just minor changes such as fitting a logit or probit model, but entirely different formulations of the model—different ways of relating observed inputs to outcomes. This is especially true when using new and unfamiliar tools such as multilevel models. In Bugs, we can easily alter the internal structure of the models we are fitting, in a way that cannot easily be done with other statistical software.

Finally, our analyses are almost never simply summarized by a set of parameter estimates and standard errors. As we illustrate throughout, we need to look carefully at our inferences to see if they make sense and to understand the operation of the model, and we usually need to postprocess the parameter estimates to get predictions or generalizations to new settings. These inference manipulations are similar to data manipulations, and we do them in R to have maximum flexibility.

Model fitting in Part 1

Part 1 of this book uses the R software for three general tasks: (1) fitting classical linear and generalized linear models, (2) graphing data and estimated models, and (3) using simulation to propagate uncertainty in inferences and predictions (see Sections 7.1–7.2 for more on this).

Model fitting in Parts 2 and 3

When we move to multilevel modeling, we begin by fitting directly in R; however, for more complicated models we move to Bugs, which has a general language for writing statistical models. We call Bugs from R and continue to use R for preprocessing of data, graphical display of data and inferences, and simulation-based prediction and model checking.

R and S

Our favorite all-around statistics software is R, which is a free open-source version of S, a program developed in the 1970s and 1980s at Bell Laboratories. S is also available commercially as S-Plus. We shall refer to R throughout, but other versions of S generally do the same things.

R is excellent for graphics, classical statistical modeling (most relevant here are the `lm()` and `glm()` functions for linear and generalized linear models), and various nonparametric methods. As we discuss in Part 2, the `lmer()` function provides quick fits in R for many multilevel models. Other packages such as `MCMCpack` exist to fit specific classes of models in R, and other such programs are in development.

Beyond the specific models that can be fit by these packages, R is fully pro-
grammable and can thus fit any model, if enough programming is done. It is possible to link R to Fortran or C to write faster programs. R also can choke on large
datasets (which is one reason we automatically "thin" large Bugs outputs before
reading into R; see Section 16.9).

Bugs

Bugs (an acronym for *Bayesian Inference using Gibbs Sampling*) is a program developed by statisticians at the Medical Research Council in Cambridge, England.
As of this writing, the most powerful versions available are WinBugs 1.4 and Open-
Bugs. In this book, when we say "Bugs," we are referring to WinBugs 1.4; however,
the code should also work (perhaps with some modification) under OpenBugs or
future implementations.

The Bugs modeling language has a modular form that allows the user to put
together all sorts of Bayesian models, including most of the multilevel models currently fit in social science applications. The two volumes of online examples in Bugs
give some indication of the possibilities—in fact, it is common practice to write a
Bugs script by starting with an example with similar features and then altering it
step by step to fit the particular problem at hand.

The key advantage of Bugs is its generality in setting up models; its main disadvantage is that it is slow and can get stuck with large datasets. These problems can
be somewhat reduced in practice by randomly sampling from the full data to create
a smaller dataset for preliminary modeling and debugging, saving the full data until
you are clear on what model you want to fit. (This is simply a computational trick
and should not be confused with *cross-validation*, a statistical method in which a
procedure is applied to a subset of the data and then checked using the rest of
the data.) Bugs does not always use the most efficient simulation algorithms, and
currently its most powerful version runs only in Windows, which in practice reduces
the ability to implement long computations in time-share with other processes.

When fitting complicated models, we set up the data in R, fit models in Bugs,
then go back to R for further statistical analysis using the fitted models.

Some models cannot be fit in Bugs. For these we illustrate in Section 15.3 a
new R package under development called Umacs (universal Markov chain sampler).
Umacs is less automatic than Bugs and requires more knowledge of the algebra of
Bayesian inference.

Other software

Some statistical software has been designed specifically for fitting multilevel models, notably MLWin and HLM. It is also possible to fit some multilevel models in
R, Stata, SAS, and other general-purpose statistical software, but without the flexibility of modeling in Bugs. The models allowed by these programs are less general
than available in Bugs; however, they are generally faster and can handle larger
datasets. We discuss these packages further in Section C.4.

Data and code for examples

Data and computer code for the examples and exercises in the book can be downloaded at the website www.stat.columbia.edu/~gelman/arm/, which also includes
other supporting materials for this book.

## CHAPTER 2

Concepts and methods from basic probability and statistics

Simple methods from introductory statistics have three important roles in regression and multilevel modeling. First, simple probability distributions are the building blocks for elaborate models. Second, multilevel models are generalizations of classical complete-pooling and no-pooling estimates, and so it is important to understand where these classical estimates come from. Third, it is often useful in practice to construct quick confidence intervals and hypothesis tests for small parts of a problem—before fitting an elaborate model, or in understanding the output from such a model.

This chapter provides a quick review of some of these methods.

2.1 Probability distributions

A probability distribution corresponds to an urn with a potentially infinite number of balls inside. When a ball is drawn at random, the "random variable" is what is written on this ball.

Areas of application of probability distributions include:

- Distributions of data (for example, heights of men, heights of women, heights of adults), for which we use the notation y_i, $i = 1, \ldots, n$.

- Distributions of parameter values, for which we use the notation θ_j, $j = 1, \ldots, J$, or other Greek letters such as α, β, γ. We shall see many of these with the multilevel models in Part 2 of the book. For now, consider a regression model (for example, predicting students' grades from pre-test scores) fit separately in each of several schools. The coefficients of the separate regressions can be modeled as following a distribution, which can be estimated from data.

- Distributions of error terms, which we write as ϵ_i, $i = 1, \ldots, n$—or, for group-level errors, η_j, $j = 1, \ldots, J$.

A "distribution" is how we describe a set of objects that are not identified, or when the identification gives no information. For example, the heights of a set of unnamed persons have a distribution, as contrasted with the heights of a particular set of your friends.

The basic way that distributions are used in statistical modeling is to start by fitting a distribution to data y, then get predictors X and model y given X with errors ϵ. Further information in X can change the distribution of the ϵ's (typically, by reducing their variance). Distributions are often thought of as data summaries, but in the regression context they are more commonly applied to ϵ's.

Normal distribution; means and variances

The Central Limit Theorem of probability states that the sum of many small independent random variables will be a random variable with an approximate normal

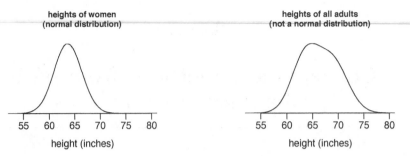

Figure 2.1 *(a) Heights of women (which approximately follow a normal distribution, as predicted from the Central Limit Theorem), and (b) heights of all adults in the United States (which have the form of a mixture of two normal distributions, one for each sex).*

distribution. If we write this summation of independent components as $z = \sum_{i=1}^{n} z_i$, then the mean and variance of z are the sums of the means and variances of the z_i's: $\mu_z = \sum_{i=1}^{n} \mu_{z_i}$ and $\sigma_z = \sqrt{\sum_{i=1}^{n} \sigma_{z_i}^2}$. We write this as $z \sim N(\mu_z, \sigma_z^2)$.

The Central Limit Theorem holds in practice—that is, $\sum_{i=1}^{n} z_i$ actually follows an approximate normal distribution—if the individual $\sigma_{z_i}^2$'s are small compared to the total variance σ_z^2.

For example, the heights of women in the United States follow an approximate normal distribution. The Central Limit Theorem applies here because height is affected by many small additive factors. In contrast, the distribution of heights of all adults in the United States is not so close to normality. The Central Limit Theorem does not apply here because there is a single large factor—sex—that represents much of the total variation. See Figure 2.1.

Linear transformations. Linearly transformed normal distributions are still normal. For example, if y are men's heights in inches (with mean 69.1 and standard deviation 2.9), then $2.54y$ are their heights in centimeters (with mean $2.54 \cdot 69 = 175$ and standard deviation $2.54 \cdot 2.9 = 7.4$).

For an example of a slightly more complicated calculation, suppose we take independent samples of 100 men and 100 women and compute the difference between the average heights of the men and the average heights of the women. This difference will be normally distributed with mean $69.1 - 63.7 = 5.4$ and standard deviation $\sqrt{2.9^2/100 + 2.7^2/100} = 0.4$ (see Exercise 2.4).

Means and variances of sums of correlated random variables. If x and y are random variables with means μ_x, μ_y, standard deviations σ_x, σ_y, and correlation ρ, then $x + y$ has mean $\mu_x + \mu_y$ and standard deviation $\sqrt{\sigma_x^2 + \sigma_y^2 + 2\rho\sigma_x\sigma_y}$. More generally, the weighted sum $ax + by$ has mean $a\mu_x + b\mu_y$, and its standard deviation is $\sqrt{a^2\sigma_x^2 + b^2\sigma_y^2 + 2ab\rho\sigma_x\sigma_y}$. From this we can derive, for example, that $x - y$ has mean $\mu_x - \mu_y$ and standard deviation $\sqrt{\sigma_x^2 + \sigma_y^2 - 2\rho\sigma_x\sigma_y}$.

Estimated regression coefficients. Estimated regression coefficients are themselves linear combinations of data (formally, the estimate $(X^tX)^{-1}X^ty$ is a linear combination of the data values y), and so the Central Limit Theorem again applies, in this case implying that, for large samples, estimated regression coefficients are approximately normally distributed. Similar arguments apply to estimates from logistic regression and other generalized linear models, and for maximum likelihood

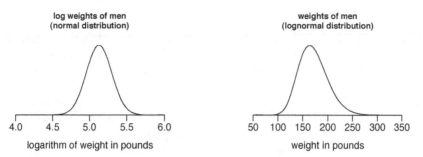

Figure 2.2 *Weights of men (which approximately follow a lognormal distribution, as pre-dicted from the Central Limit Theorem from combining several small multiplicative fac-tors), plotted on the logarithmic and original scales.*

estimation in general (see Section 18.1), for well-behaved models with large sample sizes.

Multivariate normal distribution

More generally, a random vector $z = (z_1, \ldots, z_K)$ with a K-dimensional *multivari-ate normal distribution* with a vector mean μ and a covariance matrix Σ is written as $z \sim N(\mu, \Sigma)$. The diagonal elements of Σ are the variances of the K individual random variables z_k; thus, we can write $z_k \sim N(\mu_k, \Sigma_{kk})$. The off-diagonal elements of Σ are the covariances between different elements of z, defined so that the cor-relation between z_j and z_k is $\Sigma_{jk}/\sqrt{\Sigma_{jj}\Sigma_{kk}}$. The multivariate normal distribution sometimes arises when modeling data, but in this book we encounter it in models for vectors of regression coefficients.

Approximate normal distribution of regression coefficients and other parameter es-timates. The least squares estimate of a vector of linear regression coefficients β is $\hat{\beta} = (X^t X)^{-1} X^t y$ (see Section 3.4), which, when viewed as a function of data y (considering the predictors X as constants), is a linear combination of the data. Using the Central Limit Theorem, it can be shown that the distribution of $\hat{\beta}$ will be approximately multivariate normal if the sample size is large. We describe in Chapter 7 how we use this distribution to summarize uncertainty in regression inferences.

Lognormal distribution

It is often helpful to model all-positive random variables on the logarithmic scale. For example, the logarithms of men's weights (in pounds) have an approximate normal distribution with mean 5.13 and standard deviation 0.17. Figure 2.2 shows the distributions of log weights and weights among men in the United States. The exponential of the mean and standard deviations of log weights are called the *geo-metric mean* and *geometric standard deviation* of the weights; in this example, they are 169 pounds and 1.18, respectively. When working with this *lognormal distri-bution*, we sometimes want to compute the mean and standard deviation on the original scale; these are $\exp(\mu + \frac{1}{2}\sigma^2)$ and $\exp(\mu + \frac{1}{2}\sigma^2)\sqrt{\exp(\sigma^2) - 1}$, respectively. For the men's weights example, these come to 171 pounds and 29 pounds.

Binomial distribution

If you take 20 shots in basketball, and each has 0.3 probability of succeeding, and if these shots are independent of each other (that is, success in one shot does not increase or decrease the probability of success in any other shot), then the number of shots that succeed is said to have a *binomial distribution* with $n = 20$ and $p = 0.3$, for which we use the notation $y \sim \text{Binomial}(n, p)$. As can be seen even in this simple example, the binomial model is typically only an approximation with real data, where in multiple trials, the probability p of success can vary, and for which outcomes can be correlated. Nonetheless, the binomial model is a useful starting point for modeling such data. And in some settings—most notably, independent sampling with Yes/No responses—the binomial model generally is appropriate, or very close to appropriate.

Poisson distribution

The *Poisson distribution* is used for count data such as the number of cases of cancer in a county, or the number of hits to a website during a particular hour, or the number of persons named Michael whom you know:

- If a county has a population of 100,000, and the average rate of a particular cancer is 45.2 per million persons per year, then the number of cancers in this county could be modeled as Poisson with expectation 4.52.

- If hits are coming at random, with an average rate of 380 per hour, then the number of hits in any particular hour could be modeled as Poisson with expectation 380.

- If you know approximately 1000 persons, and 1% of all persons in the population are named Michael, and you are as likely to know Michaels as anyone else, then the number of Michaels you know could be modeled as Poisson with expectation 10.

As with the binomial distribution, the Poisson model is almost always an idealization, with the first example ignoring systematic differences among counties, the second ignoring clustering or burstiness of the hits, and the third ignoring factors such as sex and age that distinguish Michaels, on average, from the general population.

Again, however, the Poisson distribution is a starting point—as long as its fit to data is checked. The model can be expanded to account for "overdispersion" in data, as we discuss in the context of Figure 2.5 on page 21.

2.2 Statistical inference

Sampling and measurement error models

Statistical inference is used to learn from incomplete or imperfect data. There are two standard paradigms for inference:

- In the *sampling model*, we are interested in learning some characteristics of a population (for example, the mean and standard deviation of the heights of all women in the United States), which we must estimate from a sample, or subset, of that population.

- In the *measurement error model*, we are interested in learning aspects of some underlying pattern or law (for example, the parameters a and b in the model

$y = a + bx$), but the data are measured with error (most simply, $y = a + bx + \epsilon$, although one can also consider models with measurement error in x).

These two paradigms are different: the sampling model makes no reference to measurements, and the measurement model can apply even when complete data are observed. In practice, however, we often combine the two approaches when creating a statistical model.

For example, consider a regression model predicting students' grades from pretest scores and other background variables. There is typically a sampling aspect to such a study, which is performed on some set of students with the goal of generalizing to a larger population. The model also includes measurement error, at least implicitly, because a student's test score is only an imperfect measure of his or her abilities.

This book follows the usual approach of setting up regression models in the measurement-error framework ($y = a + bx + \epsilon$), with the sampling interpretation implicit in that the errors $\epsilon_i, \ldots, \epsilon_n$ can be considered as a random sample from a distribution (for example, $N(0, \sigma^2)$) that represents a hypothetical "superpopulation." We consider these issues in more detail in Chapter 21; at this point, we raise this issue only to clarify the connection between probability distributions (which are typically modeled as draws from an urn, or distribution, as described at the beginning of Section 2.1) and the measurement error models used in regression.

Parameters and estimation

The goal of statistical inference for the sorts of *parametric models* that we use is to estimate underlying parameters and summarize our uncertainty in these estimates. We discuss inference more formally in Chapter 18; here it is enough to say that we typically understand a fitted model by plugging in estimates of its parameters, and then we consider the uncertainty in the parameter estimates when assessing how much we actually have learned from a given dataset.

Standard errors

The standard error is the standard deviation of the parameter estimate and gives us a sense of our uncertainty about a parameter and can be used in constructing confidence intervals, as we discuss in the next section. When estimating the mean of an infinite population, given a simple random sample of size n, the standard error is $\sigma/\sqrt{n}$, where σ is the standard deviation of the measurements in the population.

Standard errors for proportions

Consider a survey of size n with y Yes responses and $n - y$ No responses. The estimated proportion of the population who would answer Yes to this survey is $\hat{p} = y/n$, and the standard error of this estimate is $\sqrt{\hat{p}(1 - \hat{p})/n}$. This estimate and standard error are usually reasonable unless $y = 0$ or $n - y = 0$, in which case the resulting standard error estimate of zero is misleading.[1]

---

[1] A reasonable quick correction when y or $n-y$ is near zero is to use the estimate $\hat{p} = (y+1)/(n+2)$ with standard error $\sqrt{\hat{p}(1 - \hat{p})/n}$; see Agresti and Coull (1998).

2.3 Classical confidence intervals

Confidence intervals from the normal and t distributions

The usual 95% confidence interval for large samples based on the normal distribution is an estimate ± 2 standard errors. Also from the normal distribution, an estimate ± 1 standard error is a 68% interval, and an estimate $\pm 2/3$ of a standard error is a 50% interval. A 50% interval is particularly easy to interpret since the true value should be as likely to be inside as outside the interval. A 95% interval is about three times as wide as a 50% interval. The t distribution can be used to correct for uncertainty in the estimation of the standard error.

Continuous data. For example, suppose an object is weighed five times, with measurements $y = 35, 34, 38, 35, 37$, which have an average value of 35.8 and a standard deviation of 1.6. In R, we can create the 50% and 95% t intervals (based on 4 degrees of freedom) as follows:

R code
```
n <- length(y)
estimate <- mean(y)
se <- sd(y)/sqrt(n)
int.50 <- estimate + qt(c(.25,.75),n-1)*se
int.95 <- estimate + qt(c(.025,.975),n-1)*se
```

Proportions. Confidence intervals for proportions come directly from the standard-error formula. For example, if 700 persons in a random sample support the death penalty and 300 oppose it, then a 95% interval for the proportion of supporters in the population is simply $[0.7 \pm 2\sqrt{0.7 \cdot 0.3/1000}] = [0.67, 0.73]$ or, in R,

R code
```
estimate <- y/n
se <- sqrt (estimate*(1-estimate)/n)
int.95 <- estimate + qnorm(c(.025,.975))*se
```

Discrete data. For nonbinary discrete data, we can simply use the continuous formula for the standard error. For example, consider a hypothetical survey that asks 1000 randomly selected adults how many dogs they own, and suppose 600 have no dog, 300 have 1 dog, 50 have 2 dogs, 30 have 3 dogs, and 20 have 4 dogs. What is a 95% confidence interval for the average number of dogs in the population? If the data are not already specified in a file, we can quickly code the data vector R:

R code
```
y <- rep (c(0,1,2,3,4), c(600,300,50,30,20))
```

We can then continue by computing the mean, standard deviation, and standard error, as shown with continuous data above.

Comparisons, visual and numerical

Confidence intervals can often be compared visually, as in Figure 2.3, which displays 68% confidence intervals for the proportion of American adults supporting the death penalty (among those with an opinion on the question), from a series of Gallup polls. For an example of a formal comparison, consider a change in the estimated support for the death penalty from $80\% \pm 1.4\%$ to $74\% \pm 1.3\%$. The estimated difference is 6%, with a standard error of $\sqrt{(1.4\%)^2 + (1.3\%)^2} = 1.9\%$.

Linear transformations

To get confidence intervals for a linear transformed parameter, simply transform the intervals. For example, in the example on page 18, the 95% interval for the number

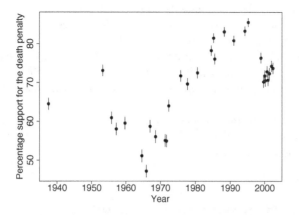

Figure 2.3 *Illustration of visual comparison of confidence intervals. Graph displays the proportion of respondents supporting the death penalty (estimates ±1 standard error—that is, 68% confidence intervals—under the simplifying assumption that each poll was a simple random sample of size 1000), from Gallup polls over time.*

of dogs per person is $[0.52, 0.62]$. Suppose this (hypothetical) random sample were taken in a city of 1 million adults. The confidence interval for the total number of pet dogs in the city is then $[520{,}000, 620{,}000]$.

Weighted averages

Confidence intervals for other derived quantities can be determined by appropriately combining the separate means and variances. For example, suppose that separate surveys conducted in France, Germany, Italy, and other countries yield estimates of $55\% \pm 2\%$, $61\% \pm 3\%$, $38\% \pm 3\%$, ..., for some opinion question. The estimated proportion for all adults in the European Union is $\frac{N_1}{N_{\text{tot}}}55\% + \frac{N_2}{N_{\text{tot}}}61\% + \frac{N_3}{N_{\text{tot}}}38\% + \cdots$, where $N_1, N_2, N_3, \ldots$ are the total number of adults in France, Germany, Italy, ..., and N_{tot} is the total number in the European Union. The standard error of this weighted average is $\sqrt{(\frac{N_1}{N_{\text{tot}}}2\%)^2 + (\frac{N_2}{N_{\text{tot}}}3\%)^2 + (\frac{N_3}{N_{\text{tot}}}3\%)^2 + \cdots}$.

Given N, p, se—the vectors of population sizes, estimated proportions of Yes responses, and standard errors—we can compute the weighted average and its 95% confidence interval in R:

```
w.avg <- sum(N*p)/sum(N)
se.w.avg <- sqrt (sum ((N*se/sum(N))^2))
int.95 <- w.avg + c(-2,2)*se.w.avg
```
R code

Using simulation to compute confidence intervals for ratios, logarithms, odds ratios, logits, and other functions of estimated parameters

For quantities more complicated than linear transformations, sums, and averages, we can compute standard errors and approximate confidence intervals using simulation. Section 7.2 discusses this in detail; here we illustrate with a quick example.

Consider a survey of 1100 persons, of whom 700 support the death penalty, 300 oppose, and 100 express no opinion. An estimate of the proportion in the population who support the death penalty, among those with an opinion, is 0.7, with a 95% confidence interval is $[0.67, 0.73]$ (see page 18).

Now suppose these 1000 respondents include 500 men and 500 women, and suppose that the death penalty was supported by 75% of the men in the sample and only 65% of the women. We would like to estimate the *ratio* of support for the death penalty among men to that among women. The estimate is easily seen to be $0.75/0.65 = 1.15$—men support it 15% more than women—but computing the standard error is more challenging. The most direct approach, which we recommend, uses simulation.

In R we create 10,000 simulation draws of the inference for men and for women, compute the ratio for each draw, and then determine a 95% interval based on the central 95% of these simulations:

R code
```
n.men <- 500
p.hat.men <- 0.75
se.men <- sqrt (p.hat.men*(1-p.hat.men)/n.men)

n.women <- 500
p.hat.women <- 0.65
se.women <- sqrt (p.hat.women*(1-p.hat.women)/n.women)

n.sims <- 10000
p.men <- rnorm (n.sims, p.hat.men, se.men)
p.women <- rnorm (n.sims, p.hat.women, se.women)
ratio <- p.men/p.women
int.95 <- quantile (ratio, c(.025,.975))
```

which yields a 95% interval of $[1.06, 1.25]$.

2.4 Classical hypothesis testing

The possible outcomes of a hypothesis test are "reject" or "not reject." It is never possible to "accept" a statistical hypothesis, only to find that the data are not sufficient to reject it.

Comparisons of parameters to fixed values and each other: interpreting confidence intervals as hypothesis tests

The hypothesis that a parameter equals zero (or any other fixed value) is directly tested by fitting the model that includes the parameter in question and examining its 95% interval. If the interval excludes zero (or the specified fixed value), then the hypothesis is rejected at the 5% level.

Testing whether two parameters are equal is equivalent to testing whether their difference equals zero. We do this by including both parameters in the model and then examining the 95% interval for their difference. As with inference for a single parameter, the confidence interval is commonly of more interest than the hypothesis test. For example, if support for the death penalty has decreased by $6\% \pm 2.1\%$, then the magnitude of this estimated difference is probably as important as that the change is statistically significantly different from zero.

The hypothesis of whether a parameter is positive is directly assessed via its confidence interval. If both ends of the 95% confidence interval exceed zero, then we are at least 95% sure (under the assumptions of the model) that the parameter is positive. Testing whether one parameter is greater than the other is equivalent to examining the confidence interval for their difference and testing for whether it is entirely positive.

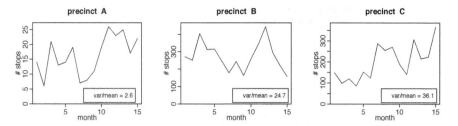

Figure 2.4 *Number of stops by the New York City police for each month over a 15-month period, for three different precincts (chosen to show different patterns in the data).*

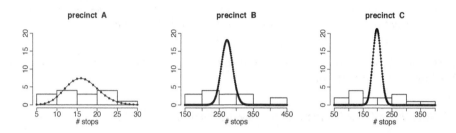

Figure 2.5 *Histograms of monthly counts of stops for the three precincts displayed in 2.4, with fitted Poisson distributions overlain. The data are much more variable than the fitted distributions, indicating overdispersion that is mild in precinct A and huge in precincts B and C.*

Testing for the existence of a variance component

We illustrate with the example of overdispersion in the binomial or Poisson model. For example, the police stop-and-frisk study (see Sections 1.2, 6.2, and 15.1) includes data from a 15-month period. We can examine the data within each precinct to see if the month-to-month variation is greater than would be expected by chance.

Figure 2.4 shows the number of police stops by month, in each of three different precincts. If the data in any precinct really came from a Poisson distribution, we would expect the variance among the counts, $\text{var}_{t=1}^{15} y_t$, to be approximately equal to their mean, $\text{avg}_{t=1}^{15} y_t$. The ratio of variance/mean is thus a measure of dispersion, with var/mean = 1 indicating that the Poisson model is appropriate, and var/mean > 1 indicating overdispersion (and var/mean < 1 indicating underdispersion, but in practice this is much less common). In this example, all three precincts are overdispersed, with variance/mean ratios well over 1.

To give a sense of what this overdispersion implies, Figure 2.5 plots histograms of the monthly counts in each precinct, with the best-fitting Poisson distributions superimposed. The observed counts are much more variable than the model in each case.

Underdispersion

Count data with variance less than the mean would indicate *underdispersion*, but this is rare in actual data. In the police example, underdispersion could possibly result from a "quota" policy in which officers are encouraged to make approximately the same number of stops each month. Figure 2.6 illustrates with hypothetical data in which the number of stops is constrained to be close to 50 each month. In this

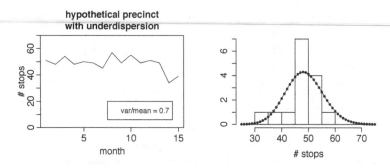

Figure 2.6 *(a) Time series and (b) histogram of number of stops by month for a hypothetical precinct with underdispersed counts. The theoretical Poisson distribution (with parameter set to the mean of the data) is overlain on the histogram.*

particular dataset, the mean is 49 and the variance is 34, and the underdispersion is clear in the histogram.

Multiple hypothesis testing and why we do not worry about it

A concern is sometimes expressed that if you test a large number of hypotheses, then you're bound to reject some. For example, with 100 different hypothesis tests, you would expect about 5 to be statistically significant at the 5% level—even if all the hypotheses were true. This concern is sometimes allayed by *multiple comparisons* procedures, which adjust significance levels to account for the multiplicity of tests.

From our data analysis perspective, however, we are not concerned about multiple comparisons. For one thing, we almost never expect any of our "point null hypotheses" (that is, hypotheses that a parameter equals zero, or that two parameters are equal) to be true, and so we are not particularly worried about the possibility of rejecting them too often. If we examine 100 parameters or comparisons, we expect about half the 50% intervals and about 5% of the 95% intervals to exclude the true values. There is no need to correct for the multiplicity of tests if we accept that they will be mistaken on occasion.

2.5 Problems with statistical significance

A common statistical error is to summarize comparisons by statistical significance and to draw a sharp distinction between significant and nonsignificant results. The approach of summarizing by statistical significance has two pitfalls, one that is obvious and one that is less well known.

First, statistical significance does not equal practical significance. For example, if the estimated predictive effect of height on earnings were $10 per inch with a standard error of $2, this would be statistically but not practically significant. Conversely, an estimate of $10,000 per inch with a standard error of $10,000 would not be statistically significant, but it has the possibility of being practically significant (and also the possibility of being zero; that is what "not statistically significant" means).

The second problem is that changes in statistical significance are not themselves significant. By this, we are not merely making the commonplace observation that any particular threshold is arbitrary—for example, only a small change is required to move an estimate from a 5.1% significance level to 4.9%, thus moving it into statistical significance. Rather, we are pointing out that even large changes in significance levels can correspond to small, nonsignificant changes in the underlying variables.

For example, consider two independent studies with effect estimates and standard errors of 25 ± 10 and 10 ± 10. The first study is statistically significant at the 1% level, and the second is not at all significant at 1 standard error away from zero. Thus it would be tempting to conclude that there is a large difference between the two studies. In fact, however, the difference is not even close to being statistically significant: the estimated difference is 15, with a standard error of $\sqrt{10^2 + 10^2} = 14$.

Section 21.8 gives a practical example of the pitfalls of using statistical significance to classify studies, along with a discussion of how these comparisons can be better summarized using a multilevel model.

2.6 55,000 residents desperately need your help!

We illustrate the application of basic statistical methods with a story. One day a couple of years ago, we received a fax, entitled HELP!, from a member of a residential organization:

> Last week we had an election for the Board of Directors. Many residents believe, as I do, that the election was rigged and what was supposed to be votes being cast by 5,553 of the 15,372 voting households is instead a fixed vote with fixed percentages being assigned to each and every candidate making it impossible to participate in an honest election.
>
> The unofficial election results I have faxed along with this letter represent the tallies. Tallies were given after 600 were counted. Then again at 1200, 2444, 3444, 4444, and final count at 5553.
>
> After close inspection we believe that there was nothing random about the count and tallies each time and that specific unnatural percentages or rigged percentages were being assigned to each and every candidate.
>
> Are we crazy? In a community this diverse and large, can candidates running on separate and opposite slates as well as independents receive similar vote percentage increases tally after tally, plus or minus three or four percent? Does this appear random to you? What do you think? HELP!

Figure 2.7 shows a subset of the data. These vote tallies were deemed suspicious because the proportion of the votes received by each candidate barely changed throughout the tallying. For example, Clotelia Smith's vote share never went below 34.6% or above 36.6%. How can we HELP these people and test their hypothesis?

We start by plotting the data: for each candidate, the proportion of vote received after 600, 1200, ... votes; see Figure 2.8. These graphs are difficult to interpret, however, since the data points are not in any sense independent: the vote at any time point includes all the votes that came before. We handle this problem by subtraction to obtain the number of votes for each candidate in the intervals between the vote tallies: the first 600 votes, the next 600, the next 1244, then next 1000, then next 1000, and the final 1109, with the total representing all 5553 votes.

Figure 2.9 displays the results. Even after taking differences, these graphs are fairly stable—but how does this variation compare to what would be expected if votes were actually coming in at random? We formulate this as a hypothesis test and carry it out in five steps:

1. *The null hypothesis* is that the voters are coming to the polls at random. The fax writer believed the data contradicted the null hypothesis; this is what we want to check.

2. *The test statistic* is some summary of the data used to check the hypothesis. Because the concern was that the votes were unexpectedly stable as the count

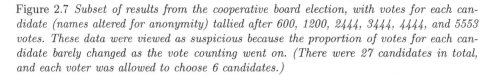

Clotelia Smith	208	416	867	1259	1610	2020
Earl Coppin	55	106	215	313	401	505
Clarissa Montes	133	250	505	716	902	1129
...	...	...	...	...	...	...

Figure 2.7 *Subset of results from the cooperative board election, with votes for each candidate (names altered for anonymity) tallied after 600, 1200, 2444, 3444, 4444, and 5553 votes. These data were viewed as suspicious because the proportion of votes for each candidate barely changed as the vote counting went on. (There were 27 candidates in total, and each voter was allowed to choose 6 candidates.)*

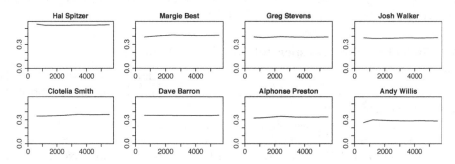

Figure 2.8 *Proportion of votes received by each candidate in the cooperative board election, after each stage of counting: 600, 1200, 2444, ..., 5553 votes. There were 27 candidates in total; for brevity we display just the leading 8 vote-getters here. The vote proportions appear to be extremely stable over time; this might be misleading, however, since the vote at any time point includes all the previous vote tallies. See Figure 2.9.*

proceeded, we define a test statistic to summarize that variability. For each candidate i, we label $y_{i1}, \ldots, y_{i6}$ to be the numbers of votes received by the candidates in each of the six recorded stages of the count. (For example, from Figure 2.7, the values of $y_{i1}, y_{i2}, \ldots, y_{i6}$ for Earl Coppin are $55, 51, \ldots, 104$.) We then compute $p_{it} = y_{it}/n_t$ for $t = 1, \ldots, 6$, the proportion of the votes received by candidate i at each stage. The test statistic for candidate i is then the sample standard deviation of these six values $p_{i1}, \ldots, p_{i6}$,

$$T_i = \text{sd}_{t=1}^6 p_{it},$$

a measure of the variation in his or her votes over time.

3. *The theoretical distribution of the test statistic if the null hypothesis were true.* Under the null hypothesis, the six subsets of the election are simply six different random samples of the voters, with a proportion π_i who would vote for candidate i. From the binomial distribution, the proportion p_{it} then has a mean of π_i and a variance of $\pi_i(1 - \pi_i)/n_t$. On average, the variance of the six p_{it}'s will equal the average of the six theoretical variances, and so the variance of the p_{it}'s—whose square root is our test statistic—should equal, on average, the theoretical value $\text{avg}_{t=1}^6 \pi_i(1 - \pi_i)/n_t$. The probabilities π_i are not known, so we follow standard practice and insert the empirical probabilities, p_i, so that the expected value of the test statistic, for each candidate i, is

$$T_i^{\text{theory}} = \sqrt{p_i(1 - p_i)\text{avg}_{t=1}^6 (1/n_t)}.$$

4. *Comparing the test statistic to its theoretical distribution.* Figure 2.10 plots the observed and theoretical values of the test statistic for each of the 27 candidates,

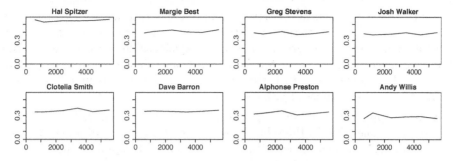

Figure 2.9 *Proportion of votes received by each of the 8 leading candidates in the cooperative board election, at each disjoint stage of voting: the first 600 votes, the next 600, the next 1244, then next 1000, then next 1000, and the final 1109, with the total representing all 5553 votes. The plots here and in Figure 2.8 have been put on a common scale which allows easy comparison of candidates, although at the cost of making it difficult to see details in the individual time series.*

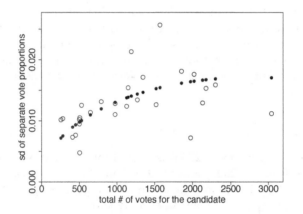

Figure 2.10 *The open circles show, for each of the 27 candidates in the cooperative board election, the standard deviation of the proportions of the vote received by the candidate in the first 600, next 600, next 1244, ..., and the final 1109 votes, plotted versus the total number of votes received by the candidate. The solid dots show the expected standard deviation of the separate vote proportions for each candidate, based on the binomial model that would be appropriate if voters were coming to the polls at random. The actual standard deviations appear consistent with the theoretical model.*

as a function of the total number of votes received by the candidate. The theoretical values follow a simple curve (which makes sense, since the total number of votes determines the empirical probabilities p_i, which determine T_i^{theory}), and the actual values appear to fit the theory fairly well, with some above and some below.

5. *Summary comparisons using χ^2 tests.* We can also express the hypothesis tests numerically. Under the null hypothesis, the probability of a candidate receiving votes is independent of the time of each vote, and thus the 2×6 table of votes including or excluding each candidate would be consistent with the model of independence. (See Figure 2.10 for an example.) We can then compute for each candidate a χ^2 statistic, $\sum_{j=1}^{2} \sum_{t=1}^{6} (\text{observed}_{jt} - \text{expected}_{jt})^2 / \text{expected}_{jt}$, and compare to a χ^2 distribution with $(6-1) \times (2-1) = 5$ degrees of freedom.

Unlike the usual application of χ^2 testing, in this case we are looking for un-expectedly *low* values of the χ^2 statistic (and thus p-values close to 1), which would indicate vote proportions that have suspiciously little variation over time. In fact, however, the χ^2 tests for the 27 candidates show no suspicious patterns: the p-values range from 0 to 1, with about 10% below 0.1, about 10% above 0.9, and no extreme p-values at either end.

Another approach would be to perform a χ^2 test on the entire 27×6 table of votes over time (that is, the table whose first row is the top row of the left table on Figure 2.7, then continues with the data from Earl Coppin, Clarissa Montes, and so forth). This test is somewhat suspect since it ignores that the votes come in batches (each voter can choose up to 6 candidates) but is a convenient summary test. The value of the χ^2 statistic is 115, which, when compared to a χ^2 distribution with $(27 - 1) \times (6 - 1) = 130$ degrees of freedom, has a p-value of 0.83—indicating slightly less variation than expected, but not statistically significant. That is, if the null hypothesis were true, we would not be particularly surprised to see a χ^2 statistic of 115.

We thus conclude that the intermediate vote tallies are consistent with random voting. As we explained to the writer of the fax, opinion polls of 1000 people are typically accurate to within 2%, and so, if voters really are arriving at random, it makes sense that batches of 1000 votes are highly stable. This does not rule out the possibility of fraud, but it shows that this aspect of the voting is consistent with the null hypothesis.

2.7 Bibliographic note

De Veaux, Velleman, and Bock (2006) is a good introductory statistics textbook, and Ramsey and Schafer (2001) and Snedecor and Cochran (1989) are also good sources for classical statistical methods. A quick summary of probability distributions appears in appendix A of Gelman et al. (2003).

Agresti and Coull (1998) consider the effectiveness of various quick methods of inference for binomial proportions. Gilovich, Vallone, and Tversky (1985) discuss the applicability of the binomial model to basketball shooting, along with psychological difficulties in interpreting binomial data.

See Browner and Newman (1987), Krantz (1999), and Gelman and Stern (2006) for further discussion and references on the problems with statistical significance.

The data on heights and weights of Americans come from Brainard and Burmaster (1992). The voting example in Section 2.6 comes from Gelman (2004c).

2.8 Exercises

The data for the assignments in this and other chapters are at www.stat.columbia.edu/~gelman/arm/examples/. See Appendix C for further details.

1. A test is graded from 0 to 50, with an average score of 35 and a standard deviation of 10. For comparison to other tests, it would be convenient to rescale to a mean of 100 and standard deviation of 15.

   (a) How can the scores be linearly transformed to have this new mean and standard deviation?

   (b) There is another linear transformation that also rescales the scores to have mean 100 and standard deviation 15. What is it, and why would you *not* want to use it for this purpose?

2. The following are the proportions of girl births in Vienna for each month in 1908 and 1909 (out of an average of 3900 births per month):

.4777 .4875 .4859 .4754 .4874 .4864 .4813 .4787 .4895 .4797 .4876 .4859
.4857 .4907 .5010 .4903 .4860 .4911 .4871 .4725 .4822 .4870 .4823 .4973

The data are in the folder `girls`. von Mises (1957) used these proportions to claim that the sex ratios were less variable than would be expected by chance.

(a) Compute the standard deviation of these proportions and compare to the standard deviation that would be expected if the sexes of babies were independently decided with a constant probability over the 24-month period.

(b) The actual and theoretical standard deviations from (a) differ, of course. Is this difference statistically significant? (Hint: under the randomness model, the actual variance should have a distribution with expected value equal to the theoretical variance, and proportional to a χ^2 with 23 degrees of freedom.)

3. Demonstration of the Central Limit Theorem: let $x = x_1 + \cdots + x_{20}$, the sum of 20 independent Uniform(0,1) random variables. In R, create 1000 simulations of x and plot their histogram. On the histogram, overlay a graph of the normal density function. Comment on any differences between the histogram and the curve.

4. Distribution of averages and differences: the heights of men in the United States are approximately normally distributed with mean 69.1 inches and standard deviation 2.9 inches. The heights of women are approximately normally distributed with mean 63.7 inches and standard deviation 2.7 inches. Let x be the average height of 100 randomly sampled men, and y be the average height of 100 randomly sampled women. In R, create 1000 simulations of $x - y$ and plot their histogram. Using the simulations, compute the mean and standard deviation of the distribution of $x - y$ and compare to their exact values.

5. Correlated random variables: suppose that the heights of husbands and wives have a correlation of 0.3. Let x and y be the heights of a married couple chosen at random. What are the mean and standard deviation of the average height, $(x + y)/2$?

Part 1A: Single-level regression

We start with an overview of classical linear regression and generalized linear models, focusing on practical issues of fitting, understanding, and graphical display. We also use this as an opportunity to introduce the statistical package R.

CHAPTER 3

Linear regression: the basics

Linear regression is a method that summarizes how the average values of a numerical *outcome* variable vary over subpopulations defined by linear functions of *predictors*. Introductory statistics and regression texts often focus on how regression can be used to represent relationships between variables, rather than as a comparison of average outcomes. By focusing on regression as a comparison of averages, we are being explicit about its limitations for defining these relationships causally, an issue to which we return in Chapter 9. Regression can be used to predict an outcome given a linear function of these predictors, and regression coefficients can be thought of as comparisons across predicted values or as comparisons among averages in the data.

3.1 One predictor

We begin by understanding the coefficients without worrying about issues of estimation and uncertainty. We shall fit a series of regressions predicting cognitive test scores of three- and four-year-old children given characteristics of their mothers, using data from a survey of adult American women and their children (a subsample from the National Longitudinal Survey of Youth).

For a binary predictor, the regression coefficient is the difference between the averages of the two groups

We start by modeling the children's test scores given an indicator for whether the mother graduated from high school (coded as 1) or not (coded as 0). The fitted model is

$$\text{kid.score} = 78 + 12 \cdot \text{mom.hs} + \text{error}, \tag{3.1}$$

but for now we focus on the deterministic part,

$$\widehat{\text{kid.score}} = 78 + 12 \cdot \text{mom.hs}, \tag{3.2}$$

where $\widehat{\text{kid.score}}$ denotes either predicted or expected test score given the `mom.hs` predictor.

This model summarizes the difference in average test scores between the children of mothers who completed high school and those with mothers who did not. Figure 3.1 displays how the regression line runs through the mean of each subpopulation.

The intercept, 78, is the average (or predicted) score for children whose mothers did not complete high school. To see this algebraically, consider that to obtain predicted scores for these children we would just plug 0 into this equation. To obtain average test scores for children (or the predicted score for a single child) whose mothers were high school graduates, we would just plug 1 into this equation to obtain $78 + 12 \cdot 1 = 91$.

The difference between these two subpopulation means is equal to the coefficient

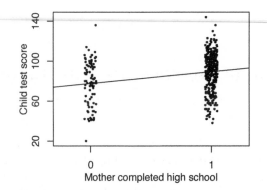

Figure 3.1 *Child's test score plotted versus an indicator for whether mother completed high school. Superimposed is the regression line, which runs through the average of each subpopulation defined by maternal education level. The indicator variable for high school completion has been* jittered; *that is, a random number has been added to each value so that the points do not lie on top of each other.*

on `mom.hs`. This coefficient tells us that children of mothers who have completed high school score 12 points higher on average than children of mothers who have not completed high school.

Regression with a continuous predictor

If we regress instead on a continuous predictor, mother's score on an IQ test, the fitted model is

$$\text{kid.score} = 26 + 0.6 \cdot \text{mom.iq} + \text{error}, \tag{3.3}$$

and is shown in Figure 3.2. We can think of the points on the line either as predicted test scores for children at each of several maternal IQ levels, or average test scores for subpopulations defined by these scores.

If we compare average child test scores for subpopulations that differ in maternal IQ by 1 point, we expect to see that the group with higher maternal IQ achieves 0.6 points more on average. Perhaps a more interesting comparison would be between groups of children whose mothers' IQ differed by 10 points—these children would be expected to have scores that differed by 6 points on average.

To understand the constant term in the regression we must consider a case with zero values of all the other predictors. In this example, the intercept of 26 reflects the predicted test scores for children whose mothers have IQ scores of zero. This is not the most helpful quantity—we don't observe any women with zero IQ. We will discuss a simple transformation in the next section that gives the intercept a more useful interpretation.

3.2 Multiple predictors

Regression coefficients are more complicated to interpret with multiple predictors because the interpretation for any given coefficient is, in part, contingent on the other variables in the model. Typical advice is to interpret each coefficient "with all the other predictors held constant." We illustrate with an example, followed by an elaboration in which the simple interpretation of regression coefficients does not work.

For instance, consider a linear regression predicting child test scores from mater-

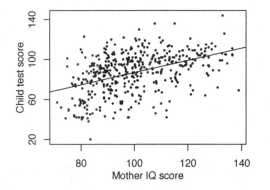

Figure 3.2 *Child's test score plotted versus maternal IQ with regression line superimposed. Each point on the line can be conceived of either as a predicted child test score for children with mothers who have the corresponding IQ, or as the average score for a subpopulation of children with mothers with that IQ.*

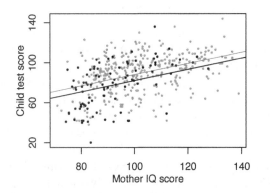

Figure 3.3 *Child's test score plotted versus maternal IQ. Light dots represent children whose mothers graduated from high school and dark dots represent children whose mothers did not graduate from high school. Superimposed are the regression lines from the regression of child's test score on maternal IQ and maternal high school indicator (the darker line for children whose mothers did not complete high school, the lighter line for children whose mothers did complete high school).*

nal education and maternal IQ. The fitted model is

$$\text{kid.score} = 26 + 6 \cdot \text{mom.hs} + 0.6 \cdot \text{mom.iq} + \text{error}, \qquad (3.4)$$

and is displayed in Figure 3.3. This model forces the slope of the regression of child's test score on mother's IQ score to be the same for each maternal education subgroup. The next section considers models in which the slopes of the two lines differ. First, however, we interpret the coefficients in model (3.4):

1. *The intercept.* If a child had a mother with an IQ of 0 and who did not complete high school (thus, mom.hs = 0), then we would predict this child's test score to be 26. This is not a useful prediction, since no mothers have IQs of 0.

2. *The coefficient of maternal high school completion.* Comparing children whose mothers have the same IQ, but who differed in whether they completed high school, the model predicts an expected difference of 6 in their test scores.

3. *The coefficient of maternal IQ.* Comparing children with the same value of mom.hs, but whose mothers differ by 1 point in IQ, we would expect to see

a difference of 0.6 points in the child's test score (equivalently, a difference of 10 in mothers' IQs corresponds to a difference of 6 points for their children).

It's not always possible to change one predictor while holding all others constant

We interpret the regression slopes as comparisons of individuals that differ in one predictor while being *at the same levels of the other predictors*. In some settings, one can also imagine manipulating the predictors to change some or hold others constant—but such an interpretation is not necessary. This becomes clearer when we consider situations in which it is logically impossible to change the value of one predictor while keeping the value of another constant. For example, if a model includes both IQ and IQ^2 as predictors, it does not make sense to consider changes in IQ with IQ^2 held constant. Or, as we discuss in the next section, if a model includes `mom.hs`, `mom.IQ`, and their interaction, `mom.hs * mom.IQ`, it is not meaningful to consider any of these three with the other two held constant.

Counterfactual and predictive interpretations

In the more general context of multiple linear regression, it pays to be more explicit about how we interpret coefficients in general. We distinguish between two interpretations of regression coefficients.

- The *predictive interpretation* considers how the outcome variable differs, on average, when comparing two groups of units that differ by 1 in the relevant predictor while being identical in all the other predictors. Under the linear model, the coefficient is the expected difference in y between these two units. This is the sort of interpretation we have described thus far.

- The *counterfactual interpretation* is expressed in terms of changes within individuals, rather than comparisons between individuals. Here, the coefficient is the expected change in y caused by adding 1 to the relevant predictor, while leaving all the other predictors in the model unchanged. For example, "changing maternal IQ from 100 to 101 would lead to an expected increase of 0.6 in child's test score." This sort of interpretation arises in causal inference.

Most introductory statistics and regression texts warn against the latter interpretation but then allow for similar interpretations such as "a change of 10 in maternal IQ is *associated* with a change of 6 points in child's score." Thus, the counterfactual interpretation is probably more familiar to you—and is sometimes easier to understand. However, as we discuss in detail in Chapter 9, the counterfactual interpretation can be inappropriate without making some strong assumptions.

3.3 Interactions

In model (3.4), the slope of the regression of child's test score on mother's IQ was forced to be equal across subgroups defined by mother's high school completion, but inspection of the data in Figure 3.3 suggests that the slopes differ substantially. A remedy for this is to include an *interaction* between `mom.hs` and `mom.iq`—that is, a new predictor which is defined as the product of these two variables. This allows the slope to vary across subgroups. The fitted model is

kid.score $= -11 + 51 \cdot$ mom.hs $+ 1.1 \cdot$ mom.iq $- 0.5 \cdot$ mom.hs $\cdot$ mom.iq $+$ error

and is displayed in Figure 3.4a, where we see the separate regression lines for each subgroup defined by maternal education.

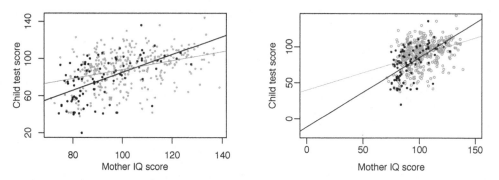

Figure 3.4 *(a) Regression lines of child's test score on mother's IQ with different symbols for children of mothers who completed high school (light circles) and those whose mothers did not complete high school (dark dots). The interaction allows for a different slope in each group, with light and dark lines corresponding to the light and dark points. (b) Same plot but with horizontal axis extended to zero to reveal the intercepts of the lines.*

Figure 3.4b shows the regression line and uncertainty on a scale with the x-axis extended to zero to display the intercepts—the points on the y-axis where the lines cross zero. This highlights the fact that not only is the value meaningless in terms of its interpretation, it is also so far out of the range of our data as to be highly unreliable as a subpopulation estimate.

Care must be taken in interpreting the coefficients in this model. We derive meaning from the coefficients (or, sometimes, functions of the coefficients) by examining average or predicted test scores within and across specific subgroups. Some coefficients are interpretable only for certain subgroups.

1. *The intercept* represents the predicted test scores for children whose mothers did not complete high school and had IQs of 0—not a meaningful scenario. (As we discuss in Sections 4.1–4.2, intercepts can be more interpretable if input variables are centered before including them as regression predictors.)

2. *The coefficient of* `mom.hs` can be conceived as the difference between the predicted test scores for children whose mothers did not complete high school and had IQs of 0, and children whose mothers did complete high school and had IQs of 0. You can see this by just plugging in the appropriate numbers and comparing the equations. Since it is implausible to imagine mothers with IQs of zero, this coefficient is not easily interpretable.

3. *The coefficient of* `mom.iq` can be thought of as the comparison of mean test scores across children whose mothers did not complete high school, but whose mothers differ by 1 point in IQ. This is the slope of the dark line in Figure 3.4.

4. *The coefficient on the interaction term* represents the *difference* in the slope for `mom.iq`, comparing children with mothers who did and did not complete high school: that is, the difference between the slopes of the light and dark lines in Figure 3.4.

An equivalent way to understand the model is to look at the separate regression lines for children of mothers who completed high school and those whose mothers

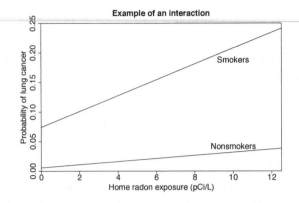

Figure 3.5 *Illustration of interactions between smoking and home radon level on the lifetime probability of lung cancer in men. The effects of radon are much more severe for smokers. The lines are estimated based on case-control studies; see Lin et al. (1999) for references.*

did not:

$$
\begin{aligned}
\text{no hs: kid.score} &= -11 + 51 \cdot 0 + 1.1 \cdot \text{mom.iq} - 0.5 \cdot 0 \cdot \text{mom.iq} \\
&= -11 + 1.1 \cdot \text{mom.iq} \\
\text{hs: kid.score} &= -11 + 51 \cdot 1 + 1.1 \cdot \text{mom.iq} - 0.5 \cdot 1 \cdot \text{mom.iq} \\
&= 40 + 0.6 \cdot \text{mom.iq}.
\end{aligned}
$$

The estimated slopes of 1.1 for children whose mothers did not complete high school and 0.6 for children of mothers who did are directly interpretable. The intercepts still suffer from the problem of only being interpretable at mothers' IQs of 0.

When should we look for interactions?

Interactions can be important. In practice, inputs that have large main effects also tend to have large interactions with other inputs (however, small main effects do not preclude the possibility of large interactions). For example, smoking has a huge effect on cancer. In epidemiologial studies of other carcinogens, it is crucial to adjust for smoking both as a main effect and as an interaction. Figure 3.5 illustrates with the example of home radon exposure: high levels of radon are associated with greater likelihood of cancer—but this difference is much greater for smokers than for nonsmokers.

Including interactions is a way to allow a model to be fit differently to different subsets of data. These two approaches are related, as we discuss later in the context of multilevel models.

Interpreting regression coefficients in the presence of interactions

Models with interactions can often be more easily interpreted if we first pre-process the data by centering each input variable about its mean or some other convenient reference point. We discuss this in Section 4.2 in the context of linear transformations.

3.4 Statistical inference

When illustrating specific examples, it helps to use descriptive variable names. In order to discuss more general theory and data manipulations, however, we shall adopt generic mathematical notation. This section introduces this notation and discusses the stochastic aspect of the model as well.

Units, outcome, predictors, and inputs

We refer to the individual data points as *units*—thus, the answer to the question, "What is the unit of analysis?" will be something like "persons" or "schools" or "congressional elections," *not* something like "pounds" or "miles." Multilevel models feature more than one set of units (for example, both persons and schools), as we discuss later on.

We refer to the X-variables in the regression as *predictors* or "predictor variables," and y as the *outcome* or "outcome variable." We do *not* use the terms "dependent" and "independent" variables, because we reserve those terms for their use in describing properties of probability distributions.

Finally, we use the term *inputs* for the information on the units that goes into the X-variables. Inputs are not the same as predictors. For example, consider the model that includes the interaction of maternal education and maternal IQ:

$$\text{kid.score} = 58 + 16 \cdot \text{mom.hs} + 0.5 \cdot \text{mom.iq} - 0.2 \cdot \text{mom.hs} \cdot \text{mom.iq} + \text{error.}$$

This regression has four *predictors*—maternal high school, maternal IQ, maternal high school × IQ, and the constant term—but only two *inputs*, maternal education and IQ.

Regression in vector-matrix notation

We follow the usual notation and label the outcome for the ith individual as y_i and the deterministic prediction as $X_i\beta = \beta_1 X_{i1} + \cdots + \beta_k X_{ik}$, indexing the persons in the data as $i = 1, \ldots, n = 1378$. In our most recent example, y_i is the ith child's test score, and there are $k = 4$ predictors in the vector X_i (the ith row of the matrix X): X_{i1}, a *constant term* that is defined to equal 1 for all persons; X_{i2}, the mother's high school completion status (coded as 0 or 1); X_{i3}, the mother's test score; and X_{i4}, the interaction between mother's test score and high school completion status. The vector β of coefficients has length $k = 4$ as well. The errors from the model are labeled as ϵ_i and assumed to follow a normal distribution with mean 0 and standard deviation σ, which we write as $N(0, \sigma^2)$. The parameter σ represents the variability with which the outcomes deviate from their predictions based on the model. We use the notation $\tilde{y}$ for unobserved data to be predicted from the model, given predictors $\tilde{X}$; see Figure 3.6.

Two ways of writing the model

The classical linear regression model can then be written mathematically as

$$\begin{aligned} y_i &= X_i\beta + \epsilon_i \\ &= \beta_1 X_{i1} + \cdots + \beta_k X_{ik} + \epsilon_i, \quad \text{for } i = 1, \ldots, n, \end{aligned}$$

where the errors ϵ_i have independent normal distributions with mean 0 and standard deviation σ.

1.4	1	0.69	−1	−0.69	0.5	2.6	0.31
1.8	1	1.85	1	1.85	1.94	2.71	3.18
0.3	1	3.83	1	3.83	2.23	2.53	3.81
1.5	1	0.5	−1	−0.5	1.85	2.5	1.73
2.0	1	2.29	−1	−2.29	2.99	3.26	2.51
2.3	1	1.62	1	1.62	0.51	0.77	1.01
0.2	1	2.29	−1	2.29	1.57	1.8	2.44
0.9	1	1.8	1		3.72	1.1	1.32
1.8	1	1.22	1	1.22	1.13	1.05	2.66
1.8	1	0.92	−1	−0.92	2.29	2.2	2.95
0.2	1	1.7	1	1.7	0.12	0.17	2.86
2.3	1	1.46	−1	−1.46	2.28	2.4	2.04
−0.3	1	4.3	1	4.3	2.3	1.87	0.48
0.4	1	3.64	−1	−3.64	1.9	1.13	0.51
1.5	1	2.27	1	2.27	0.47	3.04	3.12
?	1	1.63	−1	−1.63	0.84	2.35	1.25
	1	0.65	−1	−0.65	2.08	1.26	2.3
	1	1.83	−1	−1.83	1.84	1.58	2.99
?	1	2.58	1	2.58	2.03	1.8	1.39
?	1	0.07	−1	−0.07	2.1	2.32	1.27

Figure 3.6 *Notation for regression modeling. The model is fit to the observed outcomes y given predictors X. As described in the text, the model can then be applied to predict unobserved outcomes $\tilde{y}$ (indicated by small question marks), given predictors on new data $\tilde{X}$.*

An equivalent representation is

$$y_i \sim \mathrm{N}(X_i\beta, \sigma^2), \text{ for } i = 1, \ldots, n,$$

where X is an n by k matrix with i^{th} row X_i, or, using multivariate notation,

$$y \sim \mathrm{N}(X\beta, \sigma^2 I),$$

where y is a vector of length n, X is a $n \times k$ matrix of predictors, β is a column vector of length k, and I is the $n \times n$ identity matrix. Fitting the model (in any of its forms) using least squares yields estimates $\hat{\beta}$ and $\hat{\sigma}$.

Fitting and summarizing regressions in R

We can fit regressions using the `lm()` function in R. We illustrate with the model including mother's high school completion and IQ as predictors, for simplicity not adding the interaction for now. We shall label this model as `fit.3` as it is the third model fit in this chapter:

R code
```
fit.3 <- lm (kid.score ~ mom.hs + mom.iq)
display (fit.3)
```

(The spaces in the R code are not necessary, but we include them to make the code more readable.) The result is

R output
```
lm(formula = kid.score ~ mom.hs + mom.iq)
            coef.est coef.se
(Intercept)    25.7     5.9
mom.hs          5.9     2.2
```

```
mom.iq              0.6      0.1
  n = 434, k = 3
  residual sd = 18.1, R-Squared = 0.21
```

The `display()` function was written by us (see Section C.2 for details) to give a clean printout focusing on the most pertinent pieces of information: the coefficients and their standard errors, the sample size, number of predictors, residual standard deviation, and R^2.

In contrast, the default R option,

```
print (fit.3)
```
                                                                            R code

displays too little information, giving only the coefficient estimates with no standard errors and no information on the residual standard deviations:

```
Call:
lm(formula = kid.score ~ mom.hs + mom.iq)
```
                                                                            R code

```
Coefficients:
(Intercept)        mom.hs        mom.iq
   25.73154       5.95012       0.56391
```

Another option in R is the `summary()` function:

```
summary (fit.3)
```
                                                                            R code

but this produces a mass of barely digestible information displayed to many decimal places:

```
Call:
lm(formula = formula("kid.score ~ mom.hs + mom.iq"))
```
                                                                            R output

```
Residuals:
    Min     1Q  Median     3Q     Max
-52.873 -12.663   2.404  11.356  49.545

Coefficients:
             Estimate Std. Error t value Pr(>|t|)
(Intercept) 25.73154    5.87521   4.380 1.49e-05 ***
mom.hs       5.95012    2.21181   2.690 0.00742 **
mom.iq       0.56391    0.06057   9.309 < 2e-16 ***
---
Signif. codes:  0 '***' 0.001 '**' 0.01 '*' 0.05 '.' 0.1 ' ' 1

Residual standard error: 18.14 on 431 degrees of freedom
Multiple R-Squared: 0.2141,      Adjusted R-squared: 0.2105
F-statistic: 58.72 on 2 and 431 DF,  p-value: < 2.2e-16
```

We prefer our `display()` function, which consisely presents the most relevant information from the model fit.

Least squares estimate of the vector of regression coefficients, β

For the model $y = X\beta + \epsilon$, the least squares estimate is the $\hat{\beta}$ that minimizes the sum of squared errors, $\sum_{i=1}^{n}(y_i - X_i\hat{\beta})^2$, for the given data X, y. Intuitively, the least squares criterion seems useful because, if we are trying to predict an outcome using other variables, we want to do so in such a way as to minimize the error of our prediction.

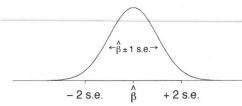

Figure 3.7 *Distribution representing uncertainty in an estimated regression coefficient. The range of this distribution corresponds to the possible values of β that are consistent with the data. When using this as an uncertainty distribution, we assign an approximate 68% chance that β will lie within 1 standard error of the point estimate, $\hat{\beta}$, and an approximate 95% chance that β will lie within 2 standard errors. Assuming the regression model is correct, it should happen only about 5% of the time that the estimate, $\hat{\beta}$, falls more than 2 standard errors away from the true β.*

The least squares estimate is also the maximum likelihood estimate if the errors ϵ_i are independent with equal variance and normally distributed (see Section 18.1). In any case, the least squares estimate can be expressed in matrix notation as $\hat{\beta} = (X^t X)^{-1} X^t y$. In practice, the computation is performed using various efficient matrix decompositions without ever fully computing $X^t X$ or inverting it. For our purposes, it is merely useful to realize that $\hat{\beta}$ is a linear function of the outcomes y.

Standard errors: uncertainty in the coefficient estimates

The estimates $\hat{\beta}$ come with standard errors, as displayed in the regression output. The standard errors represent estimation uncertainty. We can roughly say that coefficient estimates within 2 standard errors of $\hat{\beta}$ are consistent with the data. Figure 3.7 shows the normal distribution that approximately represents the range of possible values of β. For example, in the model on page 38, the coefficient of `mom.hs` has an estimate $\hat{\beta}$ of 5.9 and a standard error of 2.2; thus the data are roughly consistent with values of β in the range $[5.9 \pm 2 \cdot 2.2] = [1.5, 10.3]$. More precisely, one can account for the uncertainty in the standard errors themselves by using the t distribution with degrees of freedom set to the number of data points minus the number of estimated coefficients, but the normal approximation works fine when the degrees of freedom are more than 30 or so.

The uncertainty in the coefficient estimates will also be correlated (except in the special case of studies with balanced designs). All this information is encoded in the estimated covariance matrix $V_\beta \hat{\sigma}^2$, where $V_\beta = (X^t X)^{-1}$. The diagonal elements of $V_\beta \hat{\sigma}^2$ are the estimation variances of the individual components of β, and the off-diagonal elements represent covariances of estimation. Thus, for example, $\sqrt{V_{\beta\,11}}\,\hat{\sigma}$ is the standard error of $\hat{\beta}_1$, $\sqrt{V_{\beta\,22}}\,\hat{\sigma}$ is the standard error of $\hat{\beta}_2$, and $V_{\beta\,12}/\sqrt{V_{\beta\,11}V_{\beta\,22}}$ is the correlation of the estimates $\hat{\beta}_1, \hat{\beta}_2$.

We do not usually look at this covariance matrix; rather, we summarize inferences using the coefficient estimates and standard errors, and we use the covariance matrix for predictive simulations, as described in Section 7.2.

Residuals, r_i

The *residuals*, $r_i = y_i - X_i \hat{\beta}$, are the differences between the data and the fitted values. As a byproduct of the least squares estimation of β, the residuals r_i will be uncorrelated with all the predictors in the model. If the model includes a constant

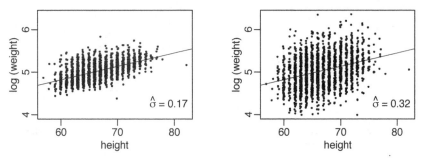

Figure 3.8 *Two hypothetical datasets with the same regression line, $y = a + bx$, but different values of the residual standard deviation, σ. The left plot shows actual data from a survey of adults; the right plot shows data with random noise added to y.*

term, then the residuals must be uncorrelated with a constant, which means they must have mean 0. This is a byproduct of how the model is estimated; it is *not* a regression assumption. We shall discuss later in the chapter how residuals can be used to diagnose problems with the model.

Residual standard deviation $\hat{\sigma}$ and explained variance R^2

The residual standard deviation, $\hat{\sigma} = \sqrt{\sum_{i=1}^{n} r_i^2/(n-k)}$, summarizes the scale of the residuals. For example, in the test scores example, $\hat{\sigma} = 18$, which tells us that the linear model can predict children's test scores to about an accuracy of 18 points. Said another way, we can think of this standard deviation as a measure of the average distance each observation falls from its prediction from the model.

The fit of the model can be summarized by $\hat{\sigma}$ (the smaller the residual variance, the better the fit) and by R^2, the fraction of variance "explained" by the model. The "unexplained" variance is $\hat{\sigma}^2$, and if we label s_y as the standard deviation of the data, then $R^2 = 1 - \hat{\sigma}^2/s_y^2$. In the test scores regression, R^2 is a perhaps disappointing 22%. (However, in a deeper sense, it is presumably a good thing that this regression has a low R^2—that is, that a child's achievement cannot be accurately predicted given only these maternal characteristics.)

The quantity $n - k$, the number of data points minus the number of estimated coefficients, is called the *degrees of freedom* for estimating the residual errors. In classical regression, k must be less than n—otherwise, the data could be fit perfectly, and it would not be possible to estimate the regression errors at all.

Difficulties in interpreting residual standard deviation and explained variance

As we make clear throughout the book, we are generally more interested in the "deterministic" part of the model, $y = X\beta$, than in the variation, ϵ. However, when we do look at the residual standard deviation, $\hat{\sigma}$, we are typically interested in it for its own sake—as a measure of the unexplained variation in the data—or because of its relevance to the precision of inferences about the regression coefficients β. (As discussed already, standard errors for β are proportional to σ.) Figure 3.8 illustrates two regressions with the same deterministic model, $y = a + bx$, but different values of σ.

Interpreting the proportion of explained variance, R^2, can be tricky because its numerator and denominator can be changed in different ways. Figure 3.9 illustrates with an example where the regression model is identical, but R^2 decreases because

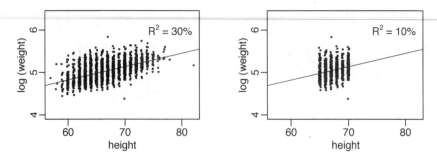

Figure 3.9 *Two hypothetical datasets with the same regression line, $y = a + bx$ and residual standard deviation, σ, but different values of the explained variance, R^2. The left plot shows actual data; the right plot shows data restricted to heights between 65 and 70 inches.*

the model is estimated on a subset of the data. (Going from the left to right plots in Figure 3.9, the residual standard deviation σ is unchanged but the standard deviation of the raw data, s_y, decreases when we restrict to this subset; thus, $R^2 = 1 - \hat{\sigma}^2/s_y^2$ declines.) Even though R^2 is much lower in the right plot, the model fits the data just as well as in the plot on the left.

Statistical significance

Roughly speaking, if a coefficient estimate is more than 2 standard errors away from zero, then it is called *statistically significant*. When an estimate is statistically significant, we are fairly sure that the sign ($+$ or $-$) of the estimate is stable, and not just an artifact of small sample size.

People sometimes think that if a coefficient estimate is not significant, then it should be excluded from the model. We disagree. It is fine to have nonsignificant coefficients in a model, as long as they make sense. We discuss this further in Section 4.6.

Uncertainty in the residual standard deviation

Under the model, the estimated residual variance, $\hat{\sigma}^2$, has a sampling distribution centered at the true value, σ^2, and proportional to a χ^2 distribution with $n - k$ degrees of freedom. We make use of this uncertainty in our predictive simulations, as described in Section 7.2.

3.5 Graphical displays of data and fitted model

Displaying a regression line as a function of one input variable

We displayed some aspects of our test scores model using plots of the data in Figures 3.1–3.3.

We can make a plot such as Figure 3.2 as follows:

R code
```
fit.2 <- lm (kid.score ~ mom.iq)
plot (mom.iq, kid.score, xlab="Mother IQ score", ylab="Child test score")
curve (coef(fit.2)[1] + coef(fit.2)[2]*x, add=TRUE)
```

The function `plot()` creates the scatterplot of observations, and `curve` superimposes the regression line using the saved coefficients from the `lm()` call (as extracted using the `coef()` function). The expression within `curve()` can also be written using matrix notation in R:

```
curve (cbind(1,x) %*% coef(fit.2), add=TRUE)
```
R code

Displaying two fitted regression lines

Model with no interaction.   For the model with two inputs, we can create a graph with two sets of points and two regression lines, as in Figure 3.3:

```
fit.3 <- lm (kid.score ~ mom.hs + mom.iq)
colors <- ifelse (mom.hs==1, "black", "gray")
plot (mom.iq, kid.score, xlab="Mother IQ score", ylab="Child test score",
  col=colors, pch=20)
curve (cbind (1, 1, x) %*% coef(fit.3), add=TRUE, col="black")
curve (cbind (1, 0, x) %*% coef(fit.3), add=TRUE, col="gray")
```
R code

Setting `pch=20` tells the `plot()` function to display the data using small dots, and the `col` option sets the colors of the points, which we have assigned to black or gray according to the value of `mom.hs`.[1] Finally, the calls to `curve()` superimpose the regression lines for the two groups defined by maternal high school completion.

Model with interaction.   We can set up the same sort of plot for the model with interactions, with the only difference being that the two lines have different slopes:

```
fit.4 <- lm (kid.score ~ mom.hs + mom.iq + mom.hs:mom.iq)
colors <- ifelse (mom.hs==1, "black", "gray")
plot (mom.iq, kid.score, xlab="Mother IQ score", ylab="Child test score",
  col=colors, pch=20)
curve (cbind (1, 1, x, 1*x) %*% coef(fit.4), add=TRUE, col="black")
curve (cbind (1, 0, x, 0*x) %*% coef(fit.4), add=TRUE, col="gray")
```
R code

The result is shown in Figure 3.4.

Displaying uncertainty in the fitted regression

As discussed in Section 7.2, we can use the `sim()` function in R to create simulations that represent our uncertainty in the estimated regression coefficients. Here we briefly describe how to use these simulations to display this inferential uncertainty. For simplicity we return to a model with just one predictor:

```
fit.2 <- lm (kid.score ~ mom.iq)
```
R code

yielding

```
            coef.est coef.se
(Intercept)    25.8      5.9
mom.iq          0.6      0.1
  n = 434, k = 2
  residual sd = 18.3, R-Squared = 0.2
```
R output

The following code creates Figure 3.10, which shows the fitted regression line along with several simulations representing uncertainty about the line:

---

[1] An alternative sequence of commands is
```
plot (mom.iq, kid.score, xlab="Mother IQ score", ylab="Child test score", type="n")
points (mom.iq[mom.hs==1], kid.score[mom.hs==1], pch=20, col="black")
points (mom.iq[mom.hs==0], kid.score[mom.hs==0], pch=20, col="gray")
```
Here, `plot()`, called with the `type="n"` option, sets up the axes but without plotting the points. Then each call to `points()` superimposes the observations for each group (defined by maternal high school completion) separately—each using a different symbol.

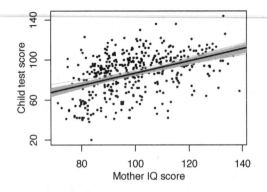

Figure 3.10 *Data and regression of child's test score on maternal IQ, with the solid line showing the fitted regression model and light lines indicating uncertainty in the fitted regression.*

R code
```
fit.2.sim <- sim (fit.2)
plot (mom.iq, kid.score, xlab="Mother IQ score", ylab="Child test score")
for (i in 1:10){
   curve (fit.2.sim$beta[i,1] + fit.2.sim$beta[i,2]*x, add=TRUE,col="gray")
}
curve (coef(fit.2)[1] + coef(fit.2)[2]*x, add=TRUE, col="black")
```

The for (i in i:10) loop allows us to display 10 different simulations.[2] Figure 3.10 also illustrates the uncertainty we have about *predictions* from our model. This uncertainty increases with greater departures from the mean of the predictor variable.

Displaying using one plot for each input variable

Now consider the regression model with the indicator for maternal high school completion included:

R code
```
fit.3 <- lm (kid.score ~ mom.hs + mom.iq)
```

We display this model in Figure 3.11 as two plots, one for each of the two input variables with the other held at its average value:

R code
```
beta.hat <- coef (fit.3)
beta.sim <- sim (fit.3)$beta
par (mfrow=c(1,2))

plot (mom.iq, kid.score, xlab="Mother IQ score", ylab="Child test score")
for (i in 1:10){
   curve (cbind (1, mean(mom.hs), x) %*% beta.sim[i,], lwd=.5,
      col="gray", add=TRUE)
}
curve (cbind (1, mean(mom.hs), x) %*% beta.hat, col="black", add=TRUE)

plot (mom.hs, kid.score, xlab="Mother completed high school",
```

---

[2] Another way to code this loop in R is to use the apply() function, for example,
```
Oneline <- function (beta) {curve (beta[1]+beta[2]*x, add=TRUE, col="gray")}
apply (fit.2.sim$beta, 1, Oneline)
```
Using apply() in this way is cleaner for experienced R users; the looped form as shown in the text is possibly easier for R novices to understand.

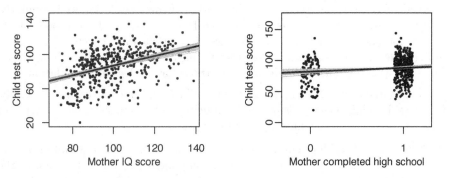

Figure 3.11 *Data and regression of child's test score on maternal IQ and high school completion, shown as a function of each of the two input variables (with light lines indicating uncertainty in the regressions). Values for high school completion have been jittered to make the points more distinct.*

```
    ylab="Child test score")
for (i in 1:10){
  curve (cbind (1, x, mean(mom.iq)) %*% beta.sim[i,], lwd=.5,
      col="gray", add=TRUE)
}
curve (cbind (1, x, mean(mom.iq)) %*% beta.hat, col="black", add=TRUE)
```

3.6 Assumptions and diagnostics

We now turn to the assumptions of the regression model, along with diagnostics that can be used to assess whether some of these assumptions are reasonable. Some of the most important assumptions, however, rely on the researcher's knowledge of the subject area and may not be directly testable from the available data alone.

Assumptions of the regression model

We list the assumptions of the regression model in *decreasing* order of importance.

1. *Validity.* Most importantly, the data you are analyzing should map to the research question you are trying to answer. This sounds obvious but is often overlooked or ignored because it can be inconvenient. Optimally, this means that the outcome measure should accurately reflect the phenomenon of interest, the model should include all relevant predictors, and the model should generalize to the cases to which it will be applied.

   For example, with regard to the outcome variable, a model of earnings will not necessarily tell you about patterns of total assets. A model of test scores will not necessarily tell you about child intelligence or cognitive development.

   Choosing inputs to a regression is often the most challenging step in the analysis. We are generally encouraged to include all "relevant" predictors, but in practice it can be difficult to determine which are necessary and how to interpret coefficients with large standard errors. Chapter 9 discusses the choice of inputs for regressions used in causal inference.

   A sample that is representative of all mothers and children may not be the most appropriate for making inferences about mothers and children who participate in the Temporary Assistance for Needy Families program. However, a carefully selected subsample may reflect the distribution of this population well. Similarly, results regarding diet and exercise obtained from a study performed on patients at risk for heart disease may not be generally applicable to generally healthy individuals. In this case assumptions would have to be made about how results for the at-risk population might relate to those for the healthy population.

Data used in empirical research rarely meet all (if any) of these criteria precisely. However, keeping these goals in mind can help you be precise about the types of questions you can and cannot answer reliably.

2. *Additivity and linearity.* The most important mathematical assumption of the regression model is that its deterministic component is a linear function of the separate predictors: $y = \beta_1 x_1 + \beta_2 x_2 + \cdots$.

If additivity is violated, it might make sense to transform the data (for example, if $y = abc$, then $\log y = \log a + \log b + \log c$) or to add interactions. If linearity is violated, perhaps a predictor should be put in as $1/x$ or $\log(x)$ instead of simply linearly. Or a more complicated relationship could be expressed by including both x and x^2 as predictors.

For example, it is common to include both **age** and **age**^{2} as regression predictors. In medical and public health examples, this allows a health measure to decline with higher ages, with the rate of decline becoming steeper as age increases. In political examples, including both **age** and **age**^{2} allows the possibility of increasing slopes with age and also U-shaped patterns if, for example, the young and old favor taxes more than the middle-aged.

In such analyses we usually prefer to include age as a categorical predictor, as discussed in Section 4.5. Another option is to use a nonlinear function such as a spline or other generalized additive model. In any case, the goal is to add predictors so that the linear and additive model is a reasonable approximation.

3. *Independence of errors.* The simple regression model assumes that the errors from the prediction line are independent. We will return to this issue in detail when discussing multilevel models.

4. *Equal variance of errors.* If the variance of the regression errors are unequal, estimation is more efficiently performed using weighted least squares, where each point is weighted inversely proportional to its variance (see Section 18.4). In most cases, however, this issue is minor. Unequal variance does not affect the most important aspect of a regression model, which is the form of the predictor $X\beta$.

5. *Normality of errors.* The regression assumption that is generally *least* important is that the errors are normally distributed. In fact, for the purpose of estimating the regression line (as compared to predicting individual data points), the assumption of normality is barely important at all. Thus, in contrast to many regression textbooks, we do *not* recommend diagnostics of the normality of regression residuals.

If the distribution of residuals is of interest, perhaps because of predictive goals, this should be distinguished from the distribution of the data, y. For example, consider a regression on a single discrete predictor, x, which takes on the values 0, 1, and 2, with one-third of the population in each category. Suppose the true regression line is $y = 0.2 + 0.5x$ with normally distributed errors with standard deviation 0.1. Then a graph of the data y will show three fairly sharp modes centered at 0.2, 0.7, and 1.2. Other examples of such mixture distributions arise in economics, when including both employed and unemployed people, or

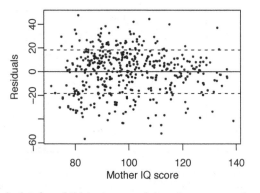

Figure 3.12 *Residual plot for child test score data when regressed on maternal IQ, with dotted lines showing ±1 standard-deviation bounds. The residuals show no striking patterns.*

the study of elections, when comparing districts with incumbent legislators of different parties.

Further assumptions are necessary if a regression coefficient is to be given a causal interpretation, as we discuss in Chapters 9 and 10.

Plotting residuals to reveal aspects of the data not captured by the model

A good way to diagnose violations of some of the assumptions just considered (importantly, linearity) is to plot the residuals r_i versus fitted values $X_i\hat{\beta}$ or simply individual predictors x_i; Figure 3.12 illustrates for the test scores example where child's test score is regressed simply on mother's IQ. The plot looks fine; there do not appear to be any strong patterns. In other settings, residual plots can reveal systematic problems with model fit, as is illustrated, for example, in Chapter 6.

3.7 Prediction and validation

Sometimes the goal of our model is to make predictions using new data. In the case of predictions of future time points, these data may eventually become available, allowing the researcher to see how well the model works for this purpose. Sometimes out-of-sample predictions are made for the explicit purpose of model checking, as we illustrate next.

Prediction

From model (3.4) on page 33, we would predict that a child of a mother who graduated from high school and with IQ of 100 would achieve a test score of $26 + 6 \cdot 1 + 0.6 \cdot 100 = 92$. If this equation represented the true model, rather than an estimated model, then we could use $\hat{\sigma} = 18$ as an estimate of the standard error for our prediction. Actually, the estimated error standard deviation is slightly higher than $\hat{\sigma}$, because of uncertainty in the estimate of the regression parameters—a complication that gives rise to those special prediction standard errors seen in most

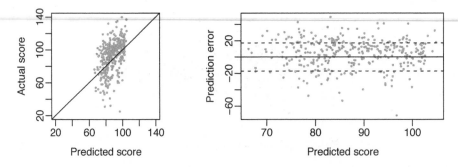

Figure 3.13 *Plots assessing how well the model fit to older children works in making predictions for younger children. The first panel compares predictions for younger children from a model against their actual values. The second panel compares residuals from these predictions against the predicted values.*

regression texts.[3] In R we can create a data frame for the new data and then use the `predict()` function. For example, the following code gives a point prediction and 95% predictive interval:

R code
```
x.new <- data.frame (mom.hs=1, mom.iq=100)
predict (fit.3, x.new, interval="prediction", level=0.95)
```

More generally, we can propagate predictive uncertainty using simulation, as explained in Section 7.2.

We use the notation $\tilde{y}_i$ for the outcome measured on a new data point and $\tilde{X}_i$ for the vector of predictors (in this example, $\tilde{X}_i = (1, 1, 100)$). The predicted value from the model is $\tilde{X}_i \hat{\beta}$, with a predictive standard error slightly higher than $\hat{\sigma}$. The normal distribution then implies that approximately 50% of the actual values should be within $\pm 0.67 \hat{\sigma}$ of the predictions, 68% should be within $\pm \hat{\sigma}$, and 95% within $\pm 2 \hat{\sigma}$.

We can similarly predict a vector of $\tilde{n}$ new outcomes, $\tilde{y}$, given a $\tilde{n} \times k$ matrix of predictors, $\tilde{X}$; see Figure 3.13.

External validation

The most fundamental way to test a model, in any scientific context, is to use it to make predictions and then compare to actual data.

Figure 3.13 illustrates with the test score data model, which was fit to data collected from 1986 and 1994 for children who were born before 1987. We apply the model to predict the outcomes of children born in 1987 or later (data collected from 1990 to 1998). This is not an ideal example for prediction because we would not necessarily expect the model for the older children to be appropriate for the younger children, even though tests for all children were taken at age 3 or 4. However, we can use it to demonstrate the methods for computing and evaluating predictions. We look at point predictions here and simulation-based predictions in Section 7.2.

The new data, $\tilde{y}$, are the outcomes for the 336 new children predicted from

---

[3] For example, in linear regression with one predictor, the "forecast standard error" around the prediction from a new data point with predictor value $\tilde{x}$ is

$$\hat{\sigma}_{\text{forecast}} = \hat{\sigma} \sqrt{1 + \frac{1}{n} + \frac{(\tilde{x} - \bar{x})^2}{\sum_{i=1}^{n} (x_i - \bar{x})^2}}.$$

mom.iq and mom.hs, using the model fit using the data from the older children. The first panel of Figure 3.13 plots actual values $\tilde{y}_i$ versus predicted values $\tilde{X}_i\hat{\beta}$, and the second panel plots residuals versus predicted values with dotted lines at $\pm\hat{\sigma}$ (approximate 68% error bounds; see Section 2.3). The error plot shows no obvious problems with applying the older-child model to the younger children, though from the scale we detect that the predictions have wide variability.

Even if we had detected clear problems with these predictions, this would not mean necessarily that there is anything wrong with the model as fit to the original dataset. However, we would need to understand it further before generalizing to other children.

3.8 Bibliographic note

Linear regression has been used for centuries in applications in the social and physical sciences; see Stigler (1986). Many introductory statistics texts have good discussions of simple linear regression, for example Moore and McCabe (1998) and De Veaux et al. (2006). Fox (2002) teaches R in the context of applied regression. In addition, the R website links to various useful free literature.

Carlin and Forbes (2004) provide an excellent introduction to the concepts of linear modeling and regression, and Pardoe (2006) is an introductory text focusing on business examples. For fuller treatments, Neter et al. (1996) and Weisberg provide accessible introductions to regression, and Ramsey and Schafer (2001) is a good complement, with a focus on issues such as model understanding, graphical display, and experimental design. Woolridge (2001) presents regression modeling from an econometric perspective. The R^2 summary of explained variance is analyzed by Wherry (1931); see also King (1986) for examples of common mistakes in reasoning with regression and Section 21.9 for more advanced references on R^2 and other methods for summarizing fitted models. Berk (2004) discusses the various assumptions implicit in regression analysis.

For more on children's test scores and maternal employment, see Hill et al. (2005). See Appendix B and Murrell (2005) for more on how to make the sorts of graphs shown in this chapter and throughout the book. The technique of jittering (used in Figure 3.1 and elsewhere in this book) comes from Chambers et al. (1983).

3.9 Exercises

1. The folder pyth contains outcome y and inputs x_1, x_2 for 40 data points, with a further 20 points with the inputs but no observed outcome. Save the file to your working directory and read it into R using the read.table() function.

   (a) Use R to fit a linear regression model predicting y from x_1, x_2, using the first 40 data points in the file. Summarize the inferences and check the fit of your model.

   (b) Display the estimated model graphically as in Figure 3.2.

   (c) Make a residual plot for this model. Do the assumptions appear to be met?

   (d) Make predictions for the remaining 20 data points in the file. How confident do you feel about these predictions?

   After doing this exercise, take a look at Gelman and Nolan (2002, section 9.4) to see where these data came from.

2. Suppose that, for a certain population, we can predict log earnings from log height as follows:

- A person who is 66 inches tall is predicted to have earnings of $30,000.
- Every increase of 1% in height corresponds to a predicted increase of 0.8% in earnings.
- The earnings of approximately 95% of people fall within a factor of 1.1 of predicted values.

(a) Give the equation of the regression line and the residual standard deviation of the regression.

(b) Suppose the standard deviation of log heights is 5% in this population. What, then, is the R^2 of the regression model described here?

3. In this exercise you will simulate two variables that are statistically independent of each other to see what happens when we run a regression of one on the other.

(a) First generate 1000 data points from a normal distribution with mean 0 and standard deviation 1 by typing `var1 <- rnorm(1000,0,1)` in R. Generate another variable in the same way (call it `var2`). Run a regression of one variable on the other. Is the slope coefficient statistically significant?

(b) Now run a simulation repeating this process 100 times. This can be done using a loop. From each simulation, save the z-score (the estimated coefficient of `var1` divided by its standard error). If the absolute value of the z-score exceeds 2, the estimate is statistically significant. Here is code to perform the simulation:[4]

R code

```
z.scores <- rep (NA, 100)
for (k in 1:100) {
  var1 <- rnorm (1000,0,1)
  var2 <- rnorm (1000,0,1)
  fit <- lm (var2 ~ var1)
  z.scores[k] <- coef(fit)[2]/se.coef(fit)[2]
}
```

How many of these 100 z-scores are statistically significant?

4. The `child.iq` folder contains a subset of the children and mother data discussed earlier in the chapter. You have access to children's test scores at age 3, mother's education, and the mother's age at the time she gave birth for a sample of 400 children. The data are a Stata file which you can read into R by saving in your working directory and then typing the following:

R code

```
library ("foreign")
iq.data <- read.dta ("child.iq.dta")
```

(a) Fit a regression of child test scores on mother's age, display the data and fitted model, check assumptions, and interpret the slope coefficient. When do you recommend mothers should give birth? What are you assuming in making these recommendations?

(b) Repeat this for a regression that further includes mother's education, interpreting both slope coefficients in this model. Have your conclusions about the timing of birth changed?

---

[4] We have initialized the vector of z-scores with missing values (NAs). Another approach is to start with `z.scores <- numeric(length=100)`, which would initialize with a vector of zeroes. In general, however, we prefer to initialize with NAs, because then when there is a bug in the code, it sometimes shows up as NAs in the final results, alerting us to the problem.

(c) Now create an indicator variable reflecting whether the mother has completed high school or not. Consider interactions between the high school completion and mother's age in family. Also, create a plot that shows the separate regression lines for each high school completion status group.

(d) Finally, fit a regression of child test scores on mother's age and education level for the first 200 children and use this model to predict test scores for the next 200. Graphically display comparisons of the predicted and actual scores for the final 200 children.

5. The folder `beauty` contains data from Hamermesh and Parker (2005) on student evaluations of instructors' beauty and teaching quality for several courses at the University of Texas. The teaching evaluations were conducted at the end of the semester, and the beauty judgments were made later, by six students who had not attended the classes and were not aware of the course evaluations.

(a) Run a regression using beauty (the variable `btystdave`) to predict course evaluations (`courseevaluation`), controlling for various other inputs. Display the fitted model graphically, and explaining the meaning of each of the coefficients, along with the residual standard deviation. Plot the residuals versus fitted values.

(b) Fit some other models, including beauty and also other input variables. Consider at least one model with interactions. For each model, state what the *predictors* are, and what the *inputs* are (see Section 2.1), and explain the meaning of each of its coefficients.

See also Felton, Mitchell, and Stinson (2003) for more on this topic.

## CHAPTER 4

Linear regression: before and after fitting the model

It is not always appropriate to fit a classical linear regression model using data in their raw form. As we discuss in Sections 4.1 and 4.4, linear and logarithmic transformations can sometimes help in the interpretation of the model. Nonlinear transformations of the data are sometimes necessary to more closely satisfy additivity and linearity assumptions, which in turn should improve the fit and predictive power of the model. Section 4.5 presents some other univariate transformations that are occasionally useful. We have already discussed interactions in Section 3.3, and in Section 4.6 we consider other techniques for combining input variables.

4.1 Linear transformations

Linear transformations do not affect the fit of a classical regression model, and they do not affect predictions: the changes in the inputs and the coefficients cancel in forming the predicted value $X\beta$.[1] However, well-chosen linear transformation can improve interpretability of coefficients and make a fitted model easier to understand. We saw in Chapter 3 how linear transformations can help with the interpretation of the intercept; this section provides examples involving the interpretation of the other coefficients in the model.

Scaling of predictors and regression coefficients. The regression coefficient β_j represents the average difference in y comparing units that differ by 1 unit on the jth predictor and are otherwise identical. In some cases, though, a difference of 1 unit on the x-scale is not the most relevant comparison. Consider, for example, a model fit to data we downloaded from a survey of adult Americans in 1994 that predicts their earnings (in dollars) given their height (in inches) and sex (coded as 1 for men and 2 for women):

$$\text{earnings} = -61000 + 1300 \cdot \text{height} + \text{error}, \qquad (4.1)$$

with a residual standard deviation of 19000. (A linear model is not really appropriate for these data, as we shall discuss soon, but we'll stick with the simple example for introducing the concept of linear transformations.)

Figure 4.1 shows the regression line and uncertainty on a scale with the x-axis extended to zero to display the intercept—the point on the y-axis where the line crosses zero. The estimated intercept of -61000 has little meaning since it corresponds to the predicted earnings for a person of zero height.

Now consider the following alternative forms of the model:

$$\text{earnings} = -61000 + 51 \cdot \text{height (in millimeters)} + \text{error}$$
$$\text{earnings} = -61000 + 81000000 \cdot \text{height (in miles)} + \text{error}.$$

How important is height? While \$51 does not seem to matter very much, \$81,000,000

---

[1] In contrast, in a multilevel model, linear transformations can change the fit of a model and its predictions, as we explain in Section 13.6.

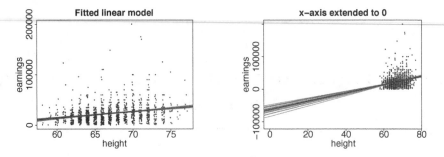

Figure 4.1 *Regression of earnings on height, earnings* $= -61000 + 1300 \cdot$ *height, with solid line showing the fitted regression model and light lines indicating uncertainty in the fitted regression. In the plot on the right, the x-scale is extended to zero to reveal the intercept of the regression line.*

is a lot. Yet, both these equations reflect the same underlying information. To understand these coefficients better, we need some sense of the variation in height in the population to which we plan to apply the model. One approach is to consider the standard deviation of heights in the data, which is 3.8 inches (or 97 millimeters, or 0.000061 miles). The expected difference in earnings corresponding to a 3.8-inch difference in height is $1300 \cdot 3.8 = \$51 \cdot 97 = \$81000000 \cdot 0.000061 = \$4900$, which is reasonably large but much smaller than the residual standard deviation of \$19000 unexplained by the regression.

Standardization using z-scores

Another way to scale the coefficients is to *standardize* the predictor by subtracting the mean and dividing by the standard deviation to yield a "z-score." In this example, `height` would be replaced by `z.height` $=$ (`height` $- 66.9$)/3.8, and the coefficient for `z.height` will be 4900. Then coefficients are interpreted in units of standard deviations with respect to the corresponding predictor just as they were, after the fact, in the previous example. In addition, standardizing predictors using z-scores will change our interpretation of the intercept to the mean of y when all predictor values are at their mean values.

We actually prefer to divide by 2 standard deviations to allow inferences to be more consistent with those for binary inputs, as we discuss in Section 4.2.

Standardization using reasonable scales

It is often useful to keep inputs on familiar scales such as inches, dollars, or years, but making convenient rescalings to aid in the interpretability of coefficients. For example, we might work with income/\$10000 or age/10.

For another example, in some surveys, party identification is on a 1–7 scale, from strong Republican to strong Democrat. The rescaled variable (PID $- 4$)/2, equals -1 for Republicans, 0 for moderates, and $+1$ for Democrats, and so the coefficient on this variable is directly interpretable.

4.2 Centering and standardizing, especially for models with interactions

Figure 4.1b illustrates the difficulty of interpreting the intercept term in a regression in a setting where it does not make sense to consider predictors set to zero. More generally, similar challenges arise in interpreting coefficients in models with interactions, as we saw in Section 3.3 with the following model:

```
lm(formula = kid.score ~ mom.hs + mom.iq + mom.hs:mom.iq)
              coef.est coef.se
(Intercept) ·    -11.5    13.8
mom.hs           51.3     15.3
mom.iq            1.1      0.2
mom.hs:mom.iq    -0.5      0.2
  n = 434, k = 4
  residual sd = 18.0, R-Squared = 0.23
```
R output

The coefficient on `mom.hs` is 51.3—does this mean that children with mothers who graduated from high school do, on average, 51.3 points better on their tests? No. The model includes an interaction, and 51.3 is the predicted difference for kids that differ in `mom.hs`, *among those with* `mom.iq` = 0. Since `mom.iq` is never even close to zero (see Figure 3.4 on page 35), the comparison at zero, and thus the coefficient of 51.3, is essentially meaningless.

Similarly, the coefficient of 1.1 for "main effect" of `mom.iq` is the slope for this variable, among those children for whom `mom.hs` = 0. This is less of a stretch (since `mom.hs` actually does equal zero for many of the cases in the data; see Figure 3.1 on page 32) but still can be somewhat misleading since `mom.hs` = 0 is at the edge of the data.

Centering by subtracting the mean of the data

We can simplify the interpretation of the regression model by first subtracting the mean of each input variable:

```
c.mom.hs <- mom.hs - mean(mom.hs)
c.mom.iq <- mom.iq - mean(mom.iq)
```
R code

The resulting regression is easier to interpret, with each main effect corresponding to a predictive difference with the other input at its average value:

```
lm(formula = kid.score ~ c.mom.hs + c.mom.iq + c.mom.hs:c.mom.iq)
                  coef.est coef.se
(Intercept)         87.6     0.9
c.mom.hs             2.8     2.4
c.mom.iq             0.6     0.1
c.mom.hs:c.mom.iq   -0.5     0.2
  n = 434, k = 4
  residual sd = 18.0, R-Squared = 0.23
```
R output

The residual standard deviation and R^2 do not change—linear transformation of the predictors does not affect the fit of a classical regression model—and the coefficient and standard error of the interaction do not change, but the main effects and the intercept move a lot and are now interpretable based on comparison to the mean of the data.

Using a conventional centering point

Another option is to center based on an understandable reference point, for example, the midpoint of the range for `mom.hs` and the population average IQ:

R code
```
c2.mom.hs <- mom.hs - 0.5
c2.mom.iq <- mom.iq - 100
```

In this parameterization, the coefficient of `c2.mom.hs` is the average predictive difference between a child with `mom.hs` = 1 and `mom.hs` = 0, for those children with `mom.iq` = 100. Similarly, the coefficient of `c2.mom.iq` corresponds to a comparison for the case `mom.hs` = 0.5, which includes no actual data but represents a midpoint of the range.

R output
```
lm(formula = kid.score ~ c2.mom.hs + c2.mom.iq + c2.mom.hs:c2.mom.iq)
                      coef.est coef.se
(Intercept)            86.8     1.2
c2.mom.hs               2.8     2.4
c2.mom.iq               0.7     0.1
c2.mom.hs:c2.mom.iq    -0.5     0.2
  n = 434, k = 4
  residual sd = 18.0, R-Squared = 0.23
```

Once again, the residual standard deviation, R^2, and coefficient for the interaction have not changed. The intercept and main effect have changed very little, because the points 0.5 and 100 happen to be close to the mean of `mom.hs` and `mom.iq` in the data.

Standardizing by subtracting the mean and dividing by 2 standard deviations

Centering helped us interpret the main effects in the regression, but it still leaves us with a scaling problem. The coefficient of `mom.hs` is much larger than that of `mom.iq`, but this is misleading, considering that we are comparing the complete change in one variable (mother completed high school or not) to a mere 1-point change in mother's IQ, which is not much at all (see Figure 3.4 on page 35).

A natural step is to scale the predictors by dividing by 2 standard deviations—we shall explain shortly why we use 2 rather than 1—so that a 1-unit change in the rescaled predictor corresponds to a change from 1 standard deviation below the mean, to 1 standard deviation above. Here are the rescaled predictors in the child testing example:

R code
```
z.mom.hs <- (mom.hs - mean(mom.hs))/(2*sd(mom.hs))
z.mom.iq <- (mom.iq - mean(mom.iq))/(2*sd(mom.iq))
```

We can now interpret all the coefficients on a roughly common scale (except for the intercept, which now corresponds to the average predicted outcome with all inputs at their mean):

R output
```
lm(formula = kid.score ~ z.mom.hs + z.mom.iq + z.mom.hs:z.mom.iq)
                    coef.est coef.se
(Intercept)          87.6     0.9
z.mom.hs              2.3     2.0
z.mom.iq             17.7     1.8
z.mom.hs:z.mom.iq   -11.9     4.0
  n = 434, k = 4
  residual sd = 18.0, R-Squared = 0.23
```

Why scale by 2 standard deviations?

We divide by 2 standard deviations rather than 1 to maintain coherence when considering binary input variables. To see this, consider the simplest binary x variable which takes on the values 0 and 1, each with probability 0.5. The standard deviation of x is then $\sqrt{0.5 \cdot 0.5} = 0.5$, and so the standardized variable, $(x - \mu_x)/(2\sigma_x)$, takes on the values ± 0.5, and its coefficient reflects comparisons between $x = 0$ and $x = 1$. In contrast, if we had divided by 1 standard deviation, the rescaled variable takes on the values ± 1, and its coefficient corresponds to half the difference between the two possible values of x. This identity is close to precise for binary inputs even when the frequencies are not exactly equal, since $\sqrt{p(1-p)} \approx 0.5$ when p is not too far from 0.5.

In a complicated regression with many predictors, it can make sense to leave binary inputs as is, and linearly transform continuous inputs, possibly by scaling using the standard deviation. In this case, dividing by 2 standard deviations ensures a rough comparability in the coefficients. In our children's testing example, the predictive difference corresponding to 2 standard deviations of mother's IQ is clearly much higher than the comparison of mothers with and without a high school education.

Multiplying each regression coefficient by 2 standard deviations of its predictor

For models with no interactions, a procedure that is equivalent to centering and rescaling is to leave the regression predictors as is, and then create rescaled regression coefficients by multiplying each β by two times the standard deviation of its corresponding x. This gives a sense of the importance of each variable, controlling for all the others in the linear model. As noted, scaling by 2 (rather than 1) standard deviations allows these scaled coefficients to be comparable to unscaled coefficients for binary predictors.

4.3 Correlation and "regression to the mean"

Consider a regression with a single predictor (in addition to the constant term); thus, $y = a + bx + \text{error}$. If both x and y are standardized—that is, if they are defined as x <- (x-mean(x))/sd(x) and y <- (y-mean(y))/sd(y)—then the regression intercept is zero and the slope is simply the correlation between x and y. Thus, the slope of a regression of two standardized variables must always be between -1 and 1, or, to put it another way, if a regression slope is more than 1 or less than -1, the variance of y must exceed that of x. In general, the slope of a regression with one predictor is $b = \rho \sigma_y / \sigma_x$, where ρ is the correlation between the two variables and σ_x and σ_y are the standard deviations of x and y.

The principal components line and the regression line

Some of the confusing aspects of regression can be understood in the simple case of standardized variables. Figure 4.2 shows a simulated-data example of standardized variables with correlation (and thus regression slope) 0.5. The left plot shows the *principal component line*, which goes closest through the cloud of points, in the sense of minimizing the sum of squared Euclidean distances between the points and the line. The principal component line in this case is simply $y = x$.

The right plot in Figure 4.2 shows the *regression line*, which minimizes the sum of the squares of the *vertical* distances between the points and the line—it is the

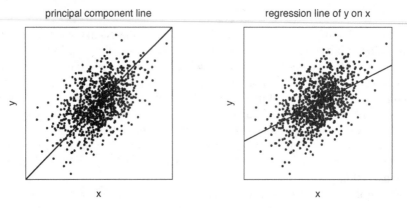

Figure 4.2 *Data simulated from a bivariate normal distribution with correlation 0.5. The regression line, which represents the best prediction of y given x, has half the slope of the principal component line, which goes closest through the cloud of points.*

familiar least squares line, $y = \hat{a} + \hat{b}x$, with $\hat{a}, \hat{b}$ chosen to minimize $\sum_{i=1}^{n}(y_i - (\hat{a} + \hat{b}x_i))^2$. In this case, $\hat{a} = 0$ and $\hat{b} = 0.5$; the regression line thus has slope 0.5.

When given this sort of scatterplot (without any lines superimposed) and asked to draw the regression line of y on x, students tend to draw the principal component line shown in Figure 4.2a. However, for the goal of predicting y from x, or for estimating the average of y for any given value of x, the regression line is in fact better—even if it does not appear so at first.

The superiority of the regression line for estimating the average of y given x can be seen from a careful study of Figure 4.2. For example, consider the points at the extreme left of either graph. They all lie above the principal components line but are roughly half below and half above the regression line. Thus, the principal component line underpredicts y for low values of x. Similarly, a careful study of the right side of each graph shows that the principal component line overpredicts y for high values of x. In contrast, the regression line again gives unbiased predictions, in the sense of going through the average value of y given x.

Regression to the mean

Recall that when x and y are standardized (that is, placed on a common scale, as in Figure 4.2), the regression line always has slope less than 1. Thus, when x is 1 standard deviations above the mean, the predicted value of y is somewhere between 0 and 1 standard deviations above the mean. This phenomenon in linear models—that y is predicted to be closer to the mean (in standard-deviation units) than x—is called *regression to the mean* and occurs in many vivid contexts.

For example, if a woman is 10 inches taller than the average for her sex, and the correlation of mothers' and (adult) sons' heights is 0.5, then her son's predicted height is 5 inches taller than the average for men. He is expected to be taller than average, but not so much taller—thus a "regression" (in the nonstatistical sense) to the average.

A similar calculation can be performed for any pair of variables that are not perfectly correlated. For example, let x_i and y_i be the number of games won by baseball team i in two successive seasons. They will not be correlated 100%; thus, we would expect the teams that did the best in season 1 (that is, with highest values of x) to do not as well in season 2 (that is, with values of y that are closer to the average for all the teams). Similarly, we would expect a team with a poor record in season 1 to improve in season 2.

A naive interpretation of regression to the mean is that heights, or baseball records, or other variable phenomena necessarily become more and more "average" over time. This view is mistaken because it ignores the error in the regression predicting y from x. For any data point x_i, the point prediction for its y_i will be regressed toward the mean, but the actual y_i that is observed will not be exactly where it is predicted. Some points end up falling closer to the mean and some fall further. This can be seen in Figure 4.2b.

4.4 Logarithmic transformations

When additivity and linearity (see Section 3.6) are not reasonable assumptions, a nonlinear transformation can sometimes remedy the situation. It commonly makes sense to take the logarithm of outcomes that are all-positive. For outcome variables, this becomes clear when we think about making predictions on the original scale. The regression model imposes no constraints that would force these predictions to be positive as well. However, if we take the logarithm of the variable, run the model, make predictions on the log scale, and then transform back (by exponentiating), the resulting predictions are necessarily positive because for any real a, $\exp(a) > 0$.

Perhaps more importantly, a linear model on the logarithmic scale corresponds to a multiplicative model on the original scale. Consider the linear regression model

$$\log y_i = b_0 + b_1 X_{i1} + b_2 X_{i2} + \cdots + \epsilon_i$$

Exponentiating both sides yields

$$
\begin{aligned}
y_i &= e^{b_0 + b_1 X_{i1} + b_2 X_{i2} + \cdots + \epsilon_i} \\
&= B_0 \cdot B_1^{X_{i1}} \cdot B_2^{X_{i2}} \cdots E_i
\end{aligned}
$$

where $B_0 = e^{b_0}$, $B_1 = e^{b_1}$, $B_2 = e^{b_2}$, ... are exponentiated regression coefficients (and thus are positive), and $E_i = e^{\epsilon_i}$ is the exponentiated error term (also positive). On the scale of the original data y_i, the predictors $X_{i1}, X_{i2}, \ldots$ come in multiplicatively.

Height and earnings example

We illustrate logarithmic regression by considering models predicting earnings from height. Expression (4.1) on page 53 shows a linear regression of earnings on height. However, it really makes more sense to model earnings on the logarithmic scale (our model here excludes those people who reported zero earnings). We can fit a regression to log earnings and then take the exponential to get predictions on the original scale.

Direct interpretation of small coefficients on the log scale. We take the logarithm of earnings and regress on height,

```
log.earn <- log (earn)                                    R code
earn.logmodel.1 <- lm (log.earn ~ height)
display (earn.logmodel.1)
```

yielding the following estimate:

```
lm(formula = log.earn ~ height)                           R output
            coef.est coef.se
(Intercept)     5.74    0.45
```

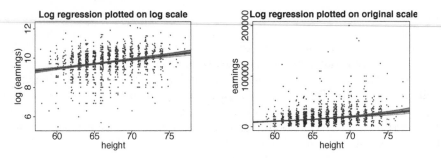

Figure 4.3 *Plot of regression of earnings on height, with solid line showing the fitted log regression model, log(earnings) = 5.78 + 0.06 · height, plotted on the logarithmic and un-transformed scales. Compare to the linear model (Figure 4.1a).*
scatterplot!data and regression lines superimposed

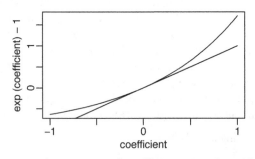

Figure 4.4 *Interpretation of exponentiated coefficients in a logarithmic regression model as relative difference (curved upper line), and the approximation exp(x) = 1 + x, which is valid for small coefficients x (straight line).*

```
height          0.06    0.01
  n = 1192, k = 2
  residual sd = 0.89, R-Squared = 0.06
```

The estimated coefficient $\beta_1 = 0.06$ implies that a difference of 1 inch in height corresponds to an expected positive difference of 0.06 in log(earnings), so that earnings are multiplied by exp(0.06). But exp(0.06) ≈ 1.06 (more precisely, it is 1.062). Thus, a difference of 1 in the predictor corresponds to an expected positive difference of about 6% in the outcome variable. Similarly, if β_1 were −0.06, then a positive difference of 1 inch of height would correspond to an expected *negative* difference of about 6% in earnings.

This correspondence does grow weaker as the magnitude of the coefficient increases. Figure 4.4 displays the deterioration of the correspondence as the coefficient size increases. The plot is restricted to coefficients in the range $(-1, 1)$ because, on the log scale, regression coefficients are typically (though not always) less than 1. A coefficient of 1 on the log scale implies that a change of one unit in the predictor is associated with a change of exp(1) = 2.7 in the outcome, and if predictors are parameterized in a reasonable way, it is unusual to see effects of this magnitude.

Why we use natural log rather than log-base-10

We prefer natural logs (that is, logarithms base e) because, as described above, coefficients on the natural-log scale are directly interpretable as approximate proportional differences: with a coefficient of 0.06, a difference of 1 in x corresponds to an approximate 6% difference in y, and so forth.[2]

Another approach is to take logarithms base 10, which we write as $\log_{10}$. The connection between the two different scales is that $\log_{10}(x) = \log(x)/\log(10) = \log(x)/2.30$. The advantage of $\log_{10}$ is that the predicted values themselves are easier to interpret; for example, when considering the earnings regressions, $\log_{10}(10,000) = 4$ and $\log_{10}(100,000) = 5$, and with some experience we can also quickly read off intermediate values—for example, if $\log_{10}(\text{earnings}) = 4.5$, then earnings $\approx 30,000$.

The disadvantage of $\log_{10}$ is that the resulting coefficients are harder to interpret. For example, if we define

```
log10.earn <- log10 (earn)
```
R code

the regression on height looks like

```
lm(formula = log10.earn ~ height)
            coef.est coef.se
(Intercept) 2.493    0.197
height      0.026    0.003
  n = 1187, k = 2
  residual sd = 0.388, R-Squared = 0.06
```
R output

The coefficient of 0.026 tells us that a difference of 1 inch in height corresponds to a difference of 0.026 in $\log_{10}(\text{earnings})$; that is, a multiplicative difference of $10^{0.026} = 1.062$. This is the same 6% change as before, but it cannot be seen by simply looking at the coefficient as could be done on the natural-log scale.

Building a regression model on the log scale

Adding another predictor. Each inch of height corresponds to a 6% increase in earnings—that seems like a lot! But men are mostly taller than women and also tend to have higher earnings. Perhaps the 6% predictive difference can be "explained" by differences between the sexes. Do taller people earn more, on average, than shorter people of the same sex? We can answer this question by including sex into the regression model—in this case, a predictor called `male` that equals 1 for men and 0 for women:

```
lm(formula = log.earn ~ height + male)
            coef.est coef.se
(Intercept)   8.15    0.60
height        0.02    0.01
male          0.42    0.07
  n = 1192, k = 3
  residual sd = 0.88, R-Squared = 0.09
```
R output

After controlling for sex, an inch of height corresponds to estimated predictive difference of 2%: under this model, two persons of the same sex but differing by 1 inch in height will differ, on average, by 2% in earnings. The predictive comparison of sex, however, is huge: comparing a man and a woman of the same height, the man's earnings are $\exp(0.42) = 1.52$ times the woman's; that is, 52% more. (We cannot simply convert the 0.42 to 42% because this coefficient is not so close to zero; see Figure 4.4.)

---

[2] Natural log is sometimes written as "ln" or "$\log_e$" but we simply write "log" since this is our default.

Naming inputs. Incidentally, we named this new input variable `male` so that it could be immediately interpreted. Had we named it `sex`, for example, we would always have to go back to the coding to check whether 0 and 1 referred to men and women, or vice versa.[3]

Checking statistical significance. The difference between the sexes is huge and well known, but the height comparison is interesting too—a 2% difference, for earnings of $50,000, comes to a nontrivial $1000 per inch. To judge statistical significance, we can check to see if the estimated coefficient is more than 2 standard errors from zero. In this case, with an estimate of 0.02 and standard error of 0.01, we would need to display to three decimal places to be sure (using the `digits` option in the `display()` function):

R output
```
lm(formula = log.earn ~ height + male)
            coef.est coef.se
(Intercept)    8.153   0.603
height         0.021   0.009
male           0.423   0.072
  n = 1192, k = 3
  residual sd = 0.88, R-Squared = 0.09
```

The coefficient for height indeed is statistically significant. Another way to check significance is to directly compute the 95% confidence interval based on the inferential simulations, as we discuss in Section 7.2.

Residual standard deviation and R^2. Finally, the regression model has a residual standard deviation of 0.88, implying that approximately 68% of log earnings will be within 0.88 of the predicted value. On the original scale, approximately 68% of earnings will be within a factor of $\exp(0.88) = 2.4$ of the prediction. For example, a 70-inch person has predicted earnings of $8.153 + 0.021 \cdot 70 = 9.623$, with a predictive standard deviation of approximately 0.88. Thus, there is an approximate 68% chance that this person has log earnings in the range $[9.623 \pm 0.88] = [8.74, 10.50]$, which corresponds to earnings in the range $[\exp(8.74), \exp(10.50)] = [6000, 36000]$. This very wide range tells us that the regression model does not predict earnings well—it is not very impressive to have a prediction that can be wrong by a factor of 2.4—and this is also reflected in the R^2, which is only 0.09, indicating that only 9% of the variance in the data is explained by the regression model. This low R^2 manifests itself graphically in Figure 4.3, where the range of the regression predictions is clearly much narrower than the range of the data.

Including an interaction. We now consider a model with an interaction between height and sex, so that the predictive comparison for height can differ for men and women:

R code
```
earn.logmodel.3 <- lm (log.earn ~ height + male + height:male)
```

which yields

R output
```
            coef.est coef.se
(Intercept)    8.388   0.844
height         0.017   0.013
male          -0.079   1.258
height:male    0.007   0.019
  n = 1192, k = 4
  residual sd = 0.88, R-Squared = 0.09
```

---

[3] Another approach would be to consider `sex` variable as a factor with two named levels, `male` and `female`; see page 68. Our point here is that, if the variable is coded numerically, it is convenient to give it the name `male` corresponding to the coding of 1.

That is,

$$\log(\text{earnings}) = 8.4 + 0.017 \cdot \text{height} - 0.079 \cdot \text{male} + 0.007 \cdot \text{height} \cdot \text{male}. \quad (4.2)$$

We shall interpret each of the four coefficients in this model.

- The *intercept* is the predicted log earnings if `height` and `male` both equal zero. Because heights are never close to zero, the intercept has no direct interpretation.

- The coefficient for `height` is the predicted difference in log earnings corresponding to a 1-inch difference in height, if `male` equals zero. Thus, the estimated predictive difference per inch of height is 1.7% for women. The estimate is less than 2 standard errors from zero, indicating that the data are consistent with a zero or negative predictive difference also.

- The coefficient for `male` is the predicted difference in log earnings between women and men, if `height` equals 0. Heights are never close to zero, and so the coefficient for `male` has no direct interpretation in this model. (We have already encountered this problem; for example, consider the difference between the intercepts of the two lines in Figure 3.4b on page 35.)

- The coefficient for `height:male` is the difference in slopes of the lines predicting log earnings on height, comparing men to women. Thus, an inch of height corresponds to 0.7% more of an increase in earnings among men than among women, and the estimated predictive difference per inch of height among men is 1.7% + 0.7% = 2.4%.

The interaction coefficient is not statistically significant, but it is plausible that the correlation between height and earnings is stronger for men and women, and so we keep it in the model, following general principles we discuss more fully in Section 4.6.

Linear transformation to make coefficients more interpretable. We can make the parameters in the interaction model clearer to interpret by rescaling the height predictor to have a mean of 0 and standard deviation 1:

```
z.height <- (height - mean(height))/sd(height)                                R code
```

For these data, `mean(height)` and `sd(height)` are 66.9 inches and 3.8 inches, respectively. Fitting the model to `z.height`, `male`, and their interaction yields

```
lm(formula = log.earn ~ z.height + male + z.height:male)                      R output
              coef.est coef.se
(Intercept)       9.53    0.05
z.height          0.07    0.05
male              0.42    0.07
z.height:male     0.03    0.07
  n = 1192, k = 4
  residual sd = 0.88, R-Squared = 0.09
```

We can now interpret all four of the coefficients:

- The *intercept* is the predicted log earnings if `z.height` and `male` both equal zero. Thus, a 66.9-inch tall woman is predicted to have log earnings of 9.53, and thus earnings of $\exp(9.53) = 14000$.

- The coefficient for `z.height` is the predicted difference in log earnings corresponding to a 1 standard-deviation difference in height, if `male` equals zero. Thus, the estimated predictive difference for a 3.8-inch increase in height is 7% for women.

- The coefficient for `male` is the predicted difference in log earnings between women and men, if `z.height` equals 0. Thus, a 66.9-inch man is predicted to have log earnings that are 0.42 higher than that of a 66.9-inch woman. This corresponds to a ratio of $\exp(0.42) = 1.52$, so the man is predicted to have 52% higher earnings than the woman.

- The coefficient for `z.height:male` is the difference in slopes between the predictive differences for height among women and men. Thus, a 3.8-inch difference of height corresponds to 3% more of an increase in earnings for men than for women, and the estimated predictive comparison among men is $7\% + 3\% = 10\%$.

One might also consider centering the predictor for sex, but here it is easy enough to interpret `male` = 0, which corresponds to the baseline category (in this case, women).

Further difficulties in interpretation

For a glimpse into yet another difficulty in interpreting regression coefficients, consider the simpler log earnings regression without the interaction term. The predictive interpretation of the height coefficient is simple enough: comparing two adults of the same sex, the taller person will be expected to earn 2% more per inch of height (see the model on page 61). This seems to be a reasonable comparison.

For the coefficient for sex, we would say: comparing two adults of the same height but different sex, the man will be expected to earn 52% more. But is this a relevant comparison? For example, if we are comparing a 66-inch woman to a 66-inch man, then we are comparing a tall woman to a short man. So, in some sense, they do not differ only in sex. Perhaps a more reasonable comparison would be of an "average woman" to an "average man."

The ultimate solution to this sort of problem must depend on why the model is being fit in the first place. For now we shall focus on the technical issues of fitting reasonable models to data. We return to issues of interpretation in Chapters 9 and 10.

Log-log model: transforming the input and outcome variables

If the log transformation is applied to an input variable as well as the outcome, the coefficient can be interpreted as the expected proportional change in y per proportional change in x. For example:

R output

```
lm(formula = log.earn ~ log.height + male)
              coef.est coef.se
(Intercept)     3.62     2.60
log.height      1.41     0.62
male            0.42     0.07
  n = 1192, k = 3
  residual sd = 0.88, R-Squared = 0.09
```

For each 1% difference in height, the predicted difference in earnings is 1.41%. The other input, `male`, is categorical so it does not make sense to take its logarithm.

In economics, the coefficient in a log-log model is sometimes called an "elasticity"; see Exercise 4.6 for an example.

Taking logarithms even when not necessary

If a variable has a narrow dynamic range (that is, if the ratio between the high and low values is close to 1), then it will not make much of a difference in fit if the regression is on the logarithmic or the original scale. For example, the standard deviation of log.height in our survey data is 0.06, meaning that heights in the data vary by only approximately a factor of 6%.

In such a situation, it might seem to make sense to stay on the original scale for reasons of simplicity. However, the logarithmic transformation can make sense even here, because coefficients are often more easily understood on the log scale. The choice of scale comes down to interpretability: whether it is easier to understand the model as proportional increase in earnings per inch, or per proportional increase in height.

For an input with a larger amount of relative variation (for example, heights of children, or weights of animals), it would make sense to work with its logarithm immediately, both as an aid in interpretation and likely an improvement in fit too.

4.5 Other transformations

Square root transformations

The square root is sometimes useful for compressing high values more mildly than is done by the logarithm. Consider again our height and earnings example.

Fitting a linear model to the raw, untransformed scale seemed inappropriate. Expressed in a different way than before, we would expect the differences between people earning nothing versus those earning $10,000 to be far greater than the differences between people earning, say, $80,000 versus $90,000. But under the linear model, these are all equal increments (as in model (4.1)), where an extra inch is worth $1300 more in earnings at all levels.

On the other hand, the log transformation seems too severe with these data. With logarithms, the differences between populations earning $5000 versus $10,000 is equivalent to the differences between those earning $40,000 versus those earning $80,000. On the square root scale, however, the differences between the 0 earnings and $10,000 earnings groups are about the same as comparisons between $10,000 and $40,000 or between $40,000 and $90,000. (These move from 0 to 100, 200, and 300 on the square root scale.) See Chapter 25 for more on this example.

Unfortunately, models on the square root scale lack the clean interpretation of the original-scale and log-transformed models. For one thing, large negative predictions on this scale get squared and become large positive values on the original scale, thus introducing a nonmonotonicity in the model. We are more likely to use the square root model for prediction than with models whose coefficients we want to understand.

Idiosyncratic transformations

Sometimes it is useful to develop transformations tailored for specific problems. For example, with the original height-earnings data it would have not been possible to simply take the logarithm of earnings as many observations had zero values. Instead, a model can be constructed in two steps: (1) model the probability that earnings exceed zero (for example, using a logistic regression; see Chapter 5); (2) fit a linear regression, conditional on earnings being positive, which is what we did

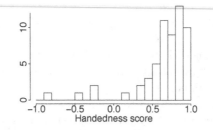

Figure 4.5 *Histogram of handedness scores of a sample of students. Scores range from −1 (completely left-handed) to +1 (completely right-handed) and are based on the responses to ten questions such as "Which hand do you write with?" and "Which hand do you use to hold a spoon?" The continuous range of responses shows the limitations of treating handedness as a dichotomous variable. From Gelman and Nolan (2002).*

in the example above. One could also model total income, but economists are often interested in modeling earnings alone.

In any case, plots and simulation should definitely be used to summarize inferences, since the coefficients of the two parts of the model combine nonlinearly in their joint prediction of earnings. We discuss this sort of model further in Sections 6.7 and 7.4.

What sort of transformed scale would be appropriate for a variable such as "assets" that can be negative, positive, or zero? One possibility is a discrete coding that compresses the high range, for example, 0 for assets in the range [−$100, $100], 1 for assets between $100 and $1000, 2 for assets between $1000 and $10,000, and so forth, and −1 for assets between −$100 and −$10,000, and so forth. Such a mapping could be expressed more fully as a continuous transformation, but for explanatory purposes it can be convenient to use a discrete scale.

Using continuous rather than discrete predictors

Many variables that appear binary or discrete can usefully be viewed as continuous. For example, rather than define "handedness" as −1 for left-handers and +1 for right-handers, one can use a standard ten-question handedness scale that gives an essentially continuous scale from −1 to 1 (see Figure 4.5).

We avoid discretizing continuous variables (except as a way of simplifying a complicated transformation, as described previously, or to model nonlinearity, as described later). A common mistake is to take a numerical measure and replace it with a binary "pass/fail" score. For example, suppose we tried to predict election winners, rather than continuous votes. Such a model would not work well, as it would discard much of the information in the data (for example, the distinction between a candidate receiving 51% or 65% of the vote). The model would be "wasting its effort" in the hopeless task of predicting the winner in very close cases. Even if our only goal is to predict the winners, we are better off predicting continuous vote shares and then transforming them into predictions about winners, as in our example with congressional elections in Section 7.3.

Using discrete rather than continuous predictors

In some cases, however, it is appropriate to discretize a continuous variable if a simple monotonic or quadratic relation does not seem appropriate. For example, in modeling political preferences, it can make sense to include age with four indicator variables: 18–29, 29–44, 45–64, and 65+, to allow for different sorts of generational patterns. Furthermore, variables that assign numbers to categories that are ordered but for which the gaps between neighboring categories are not always equivalent are often good candidates for discretization.

As an example, Chapter 3 described models for children's test scores given information about their mothers. Another input variable that can be used in these models is maternal employment, which is defined on a four-point ordered scale:

- mom.work = 1: mother did not work in first three years of child's life
- mom.work = 2: mother worked in second or third year of child's life
- mom.work = 3: mother worked part-time in first year of child's life
- mom.work = 4: mother worked full-time in first year of child's life.

Fitting a simple model using discrete predictors yields

```
lm(formula = kid.score ~ as.factor(mom.work), data = kid.iq)          R output
                       coef.est coef.se
(Intercept)              82.0     2.3
as.factor(mom.work)2      3.8     3.1
as.factor(mom.work)3     11.5     3.6
as.factor(mom.work)4      5.2     2.7
  n = 434, k = 4
  residual sd = 20.2, R-Squared = 0.02
```

This parameterization of the model allows for different averages for the children of mothers corresponding to each category of maternal employment. The "baseline" category (mom.work = 1) corresponds to children whose mothers do not go back to work at all in the first three years after the child is born; the average test score for these children is estimated by the intercept, 82.0. The average test scores for the children in the other categories is found by adding the corresponding coefficient to this baseline average. This parameterization allows us to see that the children of mothers who work part-time in the first year after the child is born achieve the highest average test scores, $82.0 + 11.5$. These families also tend to be the most advantaged in terms of many other sociodemographic characteristics as well, so a causal interpretation is not warranted.

Index and indicator variables

Index variables divide a population into categories. For example:

- male = 1 for males and 0 for females
- age = 1 for ages 18–29, 2 for ages 30–44, 3 for ages 45–64, 4 for ages 65+
- state = 1 for Alabama, ..., 50 for Wyoming
- county indexes for the 3082 counties in the United States.

Indicator variables are 0/1 predictors based on index variables. For example:

- sex.1 = 1 for females and 0 otherwise
  sex.2 = 1 for males and 0 otherwise
- age.1 = 1 for ages 18–29 and 0 otherwise
  age.2 = 1 for ages 30–44 and 0 otherwise
  age.3 = 1 for ages 45–64 and 0 otherwise
  age.4 = 1 for ages 65+ and 0 otherwise

- 50 indicators for `state`
- 3082 indicators for `county`.

As demonstrated in the previous section, including these variables as regression predictors allows for different means for the populations corresponding to each of the categories delineated by the variable.

When to use index or indicator variables.   When an input has only two levels, we prefer to code it with a single variable and name it appropriately; for example, as discussed earlier with the earnings example, the name `male` is more descriptive than `sex.1` and `sex.2`.

R also allows variables to be included as *factors* with named *levels*; for example, `sex` would have the levels `male` and `female`. In this book, however, we restrict ourselves to numerically defined variables, which is convenient for mathematical notation and also when setting up models in Bugs.

When an input has multiple levels, we prefer to create an index variable (thus, for example, `age`, which can take on the levels 1, 2, 3, 4), which can then be given indicators if necessary. As discussed in Chapter 11, multilevel modeling offers a general approach to such categorical predictors.

Identifiability

A model is said to be *nonidentifiable* if it contains parameters that cannot be estimated uniquely—or, to put it another way, that have standard errors of infinity. The offending parameters are called *nonidentified*. The most familiar and important example of nonidentifiability arises from collinearity of regression predictors. A set of predictors is collinear if there is a linear combination of them that equals 0 for all the data.

If an index variable takes on J values, then there are J associated indicator variables. A classical regression can include only $J-1$ of any set of indicators—if all J were included, they would be collinear with the constant term. (You could include a full set of J by excluding the constant term, but then the same problem would arise if you wanted to include a new set of indicators. For example, you could not include both of the sex categories and all four of the age categories. It is simpler just to keep the constant term and all but one of each set of indicators.)

For each index variable, the indicator that is excluded from the regression is known as the default, reference, or baseline condition because it is the implied category if all the $J-1$ indicators are set to zero. The default in R is to set the first level of a factor as the reference condition; other options include using the last level as baseline, selecting the baseline, and constraining the coefficients to sum to zero. There is some discussion in the regression literature on how best to set reference conditions, but we will not worry about it, because in multilevel models we can include all J indicator variables at once.

In practice, you will know that a regression is nonidentified because your computer program will give an error or return "NA" for a coefficient estimate (or it will be dropped by the program from the analysis and nothing will be reported except that it has been removed).

4.6 Building regression models for prediction

A model must be created before it can be fit and checked, and yet we put "model building" near the end of this chapter. Why? It is best to have a theoretical model laid out before any data analyses begin. But in practical data analysis it is usually easiest to start with a simple model and then build in additional complexity, taking care to check for problems along the way.

There are typically many reasonable ways in which a model can be constructed. Models may differ depending on the inferential goals or the way the data were collected. Key choices include how the input variables should be combined in creating predictors, and which predictors should be included in the model. In classical regression, these are huge issues, because if you include too many predictors in a model, the parameter estimates become so variable as to be useless. Some of these issues are less important in multilevel regression but they certainly do not disappear completely.

This section focuses on the problem of building models for prediction. Building models that can yield causal inferences is a related but separate topic that is addressed in Chapters 9 and 10.

General principles

Our general principles for building regression models for prediction are as follows:

1. Include all input variables that, for substantive reasons, might be expected to be important in predicting the outcome.

2. It is not always necessary to include these inputs as separate predictors—for example, sometimes several inputs can be averaged or summed to create a "total score" that can be used as a single predictor in the model.

3. For inputs that have large effects, consider including their interactions as well.

4. We suggest the following strategy for decisions regarding whether to exclude a variable from a prediction model based on expected sign and statistical significance (typically measured at the 5% level; that is, a coefficient is "statistically significant" if its estimate is more than 2 standard errors from zero):

   (a) If a predictor is not statistically significant and has the expected sign, it is generally fine to keep it in. It may not help predictions dramatically but is also probably not hurting them.

   (b) If a predictor is not statistically significant and does not have the expected sign (for example, incumbency having a negative effect on vote share), consider removing it from the model (that is, setting its coefficient to zero).

   (c) If a predictor *is* statistically significant and does not have the expected sign, then think hard if it makes sense. (For example, perhaps this is a country such as India in which incumbents are generally unpopular; see Linden, 2006.) Try to gather data on potential lurking variables and include them in the analysis.

   (d) If a predictor is statistically significant and has the expected sign, then by all means keep it in the model.

These strategies do not completely solve our problems but they help keep us from making mistakes such as discarding important information. They are predicated on having thought hard about these relationships *before* fitting the model. It's always easier to justify a coefficient's sign after the fact than to think hard ahead of time about what we expect. On the other hand, an explanation that is determined after running the model can still be valid. We should be able to adjust our theories in light of new information.

Example: predicting the yields of mesquite bushes

We illustrate some ideas of model checking with a real-data example that is nonetheless somewhat artificial in being presented in isolation from its applied context. Partly because this example is not a "success story" and our results are inconclusive, it represents the sort of analysis a student might perform in exploring a new dataset.

Data were collected in order to develop a method of estimating the total production (biomass) of mesquite leaves using easily measured parameters of the plant, before actual harvesting takes place. Two separate sets of measurements were taken, one on a group of 26 mesquite bushes and the other on a different group of 20 mesquite bushes measured at a different time of year. All the data were obtained in the same geographical location (ranch), but neither constituted a strictly random sample.

The outcome variable is the total weight (in grams) of photosynthetic material as derived from actual harvesting of the bush. The input variables are:

diam1:	diameter of the canopy (the leafy area of the bush) in meters, measured along the longer axis of the bush
diam2:	canopy diameter measured along the shorter axis
canopy.height:	height of the canopy
total.height:	total height of the bush
density:	plant unit density (# of primary stems per plant unit)
group:	group of measurements (0 for the first group, 1 for the second group)

It is reasonable to predict the leaf weight using some sort of regression model. Many formulations are possible. The simplest approach is to regress `weight` on all of the predictors, yielding the estimates:

R output

```
lm(formula = weight ~ diam1 + diam2 + canopy.height + total.height +
       density + group, data = mesquite)
                   coef.est coef.se
(Intercept)         -729     147
diam1                190     113
diam2                371     124
canopy.height        356     210
total.height        -102     186
density              131      34
group               -363     100
  n = 46, k = 7
  residual sd = 269, R-Squared = 0.85
```

To get a sense of the importance of each predictor, it is useful to know the range of each variable:

R output

	min	q25	median	q75	max	IQR
diam1	0.8	1.4	2.0	2.5	5.2	1.1
diam2	0.4	1.0	1.5	1.9	4.0	0.9
canopy.height	0.5	0.9	1.1	1.3	2.5	0.4
total.height	0.6	1.2	1.5	1.7	3.0	0.5
density	1.0	1.0	1.0	2.0	9.0	1.0
group	0.0	0.0	0.0	1.0	1.0	1.0
weight	60	220	360	690	4050	470

"IQR" in the last column refers to the *interquartile range*—the difference between the 75th and 25th percentile points of each variable.

But perhaps it is more reasonable to fit on the logarithmic scale, so that effects are multiplicative rather than additive:

```
lm(formula = log(weight) ~ log(diam1) + log(diam2) + log(canopy.height) +     R output
    log(total.height) + log(density) + group, data = mesquite)
                        coef.est coef.se   IQR of predictor
(Intercept)               5.35     0.17    --
log(diam1)                0.39     0.28    0.6
log(diam2)                1.15     0.21    0.6
log(canopy.height)        0.37     0.28    0.4
log(total.height)         0.39     0.31    0.4
log(density)              0.11     0.12    0.3
group                    -0.58     0.13    1.0
  n = 46, k = 7
  residual sd = 0.33, R-Squared = 0.89
```

Instead of, "each meter difference in canopy height is associated with an additional 356 grams of leaf weight," we have, "a difference of $x\%$ in canopy height is associated with an (approximate) positive difference of $0.37x\%$ in leaf weight" (evaluated at the same levels of all other variables across comparisons).

So far we have been throwing all the predictors directly into the model. A more "minimalist" approach is to try to come up with a simple model that makes sense. Thinking geometrically, we can predict leaf weight from the volume of the leaf canopy, which we shall roughly approximate as

$$\text{canopy.volume} = \text{diam1} \cdot \text{diam2} \cdot \text{canopy.height}.$$

This model is an oversimplification: the leaves are mostly on the surface of a bush, not in its interior, and so some measure of surface area is perhaps more appropriate. We shall return to this point shortly.

It still makes sense to work on the logarithmic scale:

```
lm(formula = log(weight) ~ log(canopy.volume))                              R output
                      coef.est coef.se
(Intercept)             5.17     0.08
log(canopy.volume)      0.72     0.05
  n = 46, k = 2
  residual sd = 0.41, R-Squared = 0.80
```

Thus, leaf weight is approximately proportional to `canopy.volume` to the 0.72 power. It is perhaps surprising that this power is not closer to 1. The usual explanation for this is that there is variation in `canopy.volume` that is unrelated to the weight of the leaves, and this tends to *attenuate* the regression coefficient—that is, to decrease its absolute value from the "natural" value of 1 to something lower. Similarly, regressions of "after" versus "before" typically have slopes of less than 1. (For another example, Section 7.3 has an example of forecasting congressional elections in which the vote in the previous election has a coefficient of only 0.58.)

The regression with only `canopy.volume` is satisfyingly simple, with an impressive R-squared of 80%. However, the predictions are still much worse than the model with all the predictors. Perhaps we should go back and put in the other predictors. We shall define:

$$\text{canopy.area} \;=\; \text{diam1} \cdot \text{diam2}$$
$$\text{canopy.shape} \;=\; \text{diam1}/\text{diam2}.$$

The set (canopy.volume, canopy.area, canopy.shape) is then just a different parameterization of the three canopy dimensions. Including them all in the model yields:

R output

```
lm(formula = log(weight) ~ log(canopy.volume) + log(canopy.area) +
       log(canopy.shape) + log(total.height) + log(density) + group)
                        coef.est coef.se
(Intercept)                 5.35    0.17
log(canopy.volume)          0.37    0.28
log(canopy.area)            0.40    0.29
log(canopy.shape)          -0.38    0.23
log(total.height)           0.39    0.31
log(density)                0.11    0.12
group                      -0.58    0.13
   n = 46, k = 7
   residual sd = 0.33, R-Squared = 0.89
```

This fit is identical to that of the earlier log-scale model (just a linear transformation of the predictors), but to us these coefficient estimates are more directly interpretable:

- Canopy volume and area are both positively associated with weight. Neither is statistically significant, but we keep them in because they both make sense: (1) a larger-volume canopy should have more leaves, and (2) conditional on volume, a canopy with larger cross-sectional area should have more exposure to the sun.

- The negative coefficient of canopy.shape implies that bushes that are more circular in cross section have more leaf weight (after controlling for volume and area). It is not clear whether we should "believe" this. The coefficient is not statistically significant; we could keep this predictor in the model or leave it out.

- Total height is positively associated with weight, which could make sense if the bushes are planted close together—taller bushes get more sun. The coefficient is not statistically significant, but it seems to make sense to "believe" it and leave it in.

- It is not clear how to interpret the coefficient for density. Since it is not statistically significant, maybe we can exclude it.

- For whatever reason, the coefficient for group is large and statistically significant, so we must keep it in. It would be a good idea to learn how the two groups differ so that a more relevant measurement could be included for which group is a proxy.

This leaves us with a model such as

R output

```
lm(formula = log(weight) ~ log(canopy.volume) + log(canopy.area) +
       group)
                        coef.est coef.se
(Intercept)                 5.22    0.09
log(canopy.volume)          0.61    0.19
log(canopy.area)            0.29    0.24
group                      -0.53    0.12
   n = 46, k = 4
   residual sd = 0.34, R-Squared = 0.87
```

or

R output

```
lm(formula = log(weight) ~ log(canopy.volume) + log(canopy.area) +
       log(canopy.shape) + log(total.height) + group)
                        coef.est coef.se
(Intercept)                 5.31    0.16
log(canopy.volume)          0.38    0.28
```

```
log(canopy.area)        0.41      0.29
log(canopy.shape)      -0.32      0.22
log(total.height)       0.42      0.31
group                  -0.54      0.12
  n = 46, k = 6
  residual sd = 0.33, R-Squared = 0.88
```

We want to include both volume and area in the model, since for geometrical reasons we expect both to be positively predictive of leaf volume. It would also make sense to look at some residual plots to look for any patterns in the data beyond what has been fitted by the model.

Finally, it would seem like a good idea to include interactions of **group** with the other predictors. Unfortunately, with only 46 data points, it turns out to be impossible to estimate these interactions accurately: none of them are statistically significant.

To conclude this example: we have had some success in transforming the outcome and input variables to obtain a reasonable predictive model. However, we do not have any clean way of choosing among the models (or combining them). We also do not have any easy way of choosing between the linear and log-transformation models, or bridging the gap between them. For this problem, the log model seems to make much more sense, but we would also like a data-based reason to prefer it, if it is indeed preferable.

4.7 Fitting a series of regressions

It is common to fit a regression model repeatedly, either for different datasets or to subsets of an existing dataset. For example, one could estimate the relation between height and earnings using surveys from several years, or from several countries, or within different regions or states within the United States.

As discussed in Part 2 of this book, multilevel modeling is a way to estimate a regression repeatedly, partially pooling information from the different fits. Here we consider the more informal procedure of estimating the regression separately—with no pooling between years or groups—and then displaying all these estimates together, which can be considered as an informal precursor to multilevel modeling.[4]

Predicting party identification

Political scientists have long been interested in party identification and its changes over time. We illustrate here with a series of cross-sectional regressions modeling party identification given political ideology and demographic variables.

We use the National Election Study, which asks about party identification on a 1–7 scale (1 = strong Democrat, 2 = Democrat, 3 = weak Democrat, 4 = independent, ..., 7 = strong Republican), which we treat as a continuous variable. We include the following predictors: political ideology (1 = strong liberal, 2 = liberal, ..., 7 = strong conservative), ethnicity (0 = white, 1 = black, 0.5 = other), age (as categories: 18–29, 30–44, 45–64, and 65+ years, with the lowest age category as a baseline), education (1 = no high school, 2 = high school graduate, 3 = some college, 4 =

---

[4] The method of repeated modeling, followed by time-series plots of estimates, is sometimes called the "secret weapon" because it is so easy and powerful but yet is rarely used as a data-analytic tool. We suspect that one reason for its rarity of use is that, once one acknowledges the time-series structure of a dataset, it is natural to want to take the next step and model that directly. In practice, however, there is a broad range of problems for which a cross-sectional analysis is informative, and for which a time-series display is appropriate to give a sense of trends.

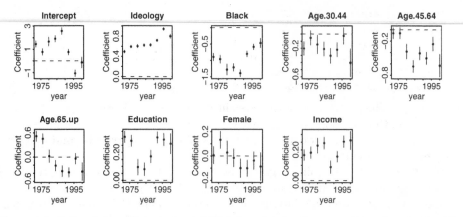

Figure 4.6 *Estimated coefficients (and 50% intervals) for the regression of party identifi-cation on political ideology, ethnicity, and other predictors, as fit separately to poll data from each presidential election campaign from 1976 through 2000. The plots are on differ-ent scales, with the input variables ordered roughly in declining order of the magnitudes of their coefficients. The set of plots illustrates the display of inferences from a series of regressions.*

college graduate), sex (0=male, 1=female), and income (1=0–16th percentile, 2= 17–33rd percentile, 3=34–67th percentile, 4=68–95th percentile, 5=96–100th per-centile).

Figure 4.6 shows the estimated coefficients tracked over time. Ideology and ethnic-ity are the most important,[5] and they remain fairly stable over time. The predictive differences for age and sex change fairly dramatically during the thirty-year period.

4.8 Bibliographic note

For additional reading on transformations, see Atkinson (1985), Mosteller and Tukey (1977), Box and Cox (1964), and Carroll and Ruppert (1981). Bring (1994) has a thorough discussion on standardizing regression coefficients; see also Blalock (1961) and Greenland, Schlessman, and Criqui (1986). Gelman (2007) discusses scaling inputs by dividing by two standard deviations. Harrell (2001) discusses strategies for regression modeling.

For more on the earnings and height example, see Persico, Postlewaite, and Sil-verman (2004) and Gelman and Nolan (2002). For more on the handedness example, see Gelman and Nolan (2002, sections 2.5 and 3.3.2). The historical background of regression to the mean is covered by Stigler (1986), and its connections to multilevel modeling are discussed by Stigler (1983).

The mesquite bushes example in Section 4.6 comes from an exam problem from the 1980s; we have not been able to track down the original data. For more on the ideology example in Section 4.7, see Bafumi (2005).

4.9 Exercises

1. Logarithmic transformation and regression: consider the following regression:

$$\log(\text{weight}) = -3.5 + 2.0 \log(\text{height}) + \text{error},$$

---

[5] Ideology is on a seven-point scale, so that its coefficients must be multiplied by 4 to get the expected change when comparing a liberal (ideology=2) to a conservative (ideology=6).

with errors that have standard deviation 0.25. Weights are in pounds and heights are in inches.

(a) Fill in the blanks: approximately 68% of the persons will have weights within a factor of __ and __ of their predicted values from the regression.

(b) Draw the regression line and scatterplot of log(weight) versus log(height) that make sense and are consistent with the fitted model. Be sure to label the axes of your graph.

2. The folder `earnings` has data from the Work, Family, and Well-Being Survey (Ross, 1990). Pull out the data on earnings, sex, height, and weight.

(a) In R, check the dataset and clean any unusually coded data.

(b) Fit a linear regression model predicting earnings from height. What transformation should you perform in order to interpret the intercept from this model as average earnings for people with average height?

(c) Fit some regression models with the goal of predicting earnings from some combination of sex, height, and weight. Be sure to try various transformations and interactions that might make sense. Choose your preferred model and justify.

(d) Interpret all model coefficients.

3. Plotting linear and nonlinear regressions: we downloaded data with weight (in pounds) and age (in years) from a random sample of American adults. We first created new variables: age10 = age/10 and age10.sq = $(age/10)^2$, and indicators age18.29, age30.44, age45.64, and age65up for four age categories. We then fit some regressions, with the following results:

```
lm(formula = weight ~ age10)                                          R output
               coef.est coef.se
(Intercept)     161.0     7.3
age10             2.6     1.6
  n = 2009, k = 2
  residual sd = 119.7, R-Squared = 0.00

lm(formula = weight ~ age10 + age10.sq)
               coef.est coef.se
(Intercept)      96.2    19.3
age10            33.6     8.7
age10.sq         -3.2     0.9
  n = 2009, k = 3
  residual sd = 119.3, R-Squared = 0.01

lm(formula = weight ~ age30.44 + age45.64 + age65up)
               coef.est coef.se
(Intercept)     157.2     5.4
age30.44TRUE     19.1     7.0
age45.64TRUE     27.2     7.6
age65upTRUE       8.5     8.7
  n = 2009, k = 4
  residual sd = 119.4, R-Squared = 0.01
```

(a) On a graph of weights versus age (that is, weight on y-axis, age on x-axis), draw the fitted regression line from the first model.

(b) On the same graph, draw the fitted regression line from the second model.

(c) On another graph with the same axes and scale, draw the fitted regression line from the third model. (It will be discontinuous.)

4. Logarithmic transformations: the folder **pollution** contains mortality rates and various environmental factors from 60 U.S. metropolitan areas (see McDonald and Schwing, 1973). For this exercise we shall model mortality rate given nitric oxides, sulfur dioxide, and hydrocarbons as inputs. This model is an extreme oversimplification as it combines all sources of mortality and does not adjust for crucial factors such as age and smoking. We use it to illustrate log transformations in regression.

(a) Create a scatterplot of mortality rate versus level of nitric oxides. Do you think linear regression will fit these data well? Fit the regression and evaluate a residual plot from the regression.

(b) Find an appropriate transformation that will result in data more appropriate for linear regression. Fit a regression to the transformed data and evaluate the new residual plot.

(c) Interpret the slope coefficient from the model you chose in (b).

(d) Now fit a model predicting mortality rate using levels of nitric oxides, sulfur dioxide, and hydrocarbons as inputs. Use appropriate transformations when helpful. Plot the fitted regression model and interpret the coefficients.

(e) Cross-validate: fit the model you chose above to the first half of the data and then predict for the second half. (You used all the data to construct the model in (d), so this is not really cross-validation, but it gives a sense of how the steps of cross-validation can be implemented.)

5. Special-purpose transformations: for a study of congressional elections, you would like a measure of the relative amount of money raised by each of the two major-party candidates in each district. Suppose that you know the amount of money raised by each candidate; label these dollar values D_i and R_i. You would like to combine these into a single variable that can be included as an input variable into a model predicting vote share for the Democrats.

(a) Discuss the advantages and disadvantages of the following measures:
   - The simple difference, $D_i - R_i$
   - The ratio, D_i/R_i
   - The difference on the logarithmic scale, $\log D_i - \log R_i$
   - The relative proportion, $D_i/(D_i + R_i)$.

(b) Propose an idiosyncratic transformation (as in the example on page 65) and discuss the advantages and disadvantages of using it as a regression input.

6. An economist runs a regression examining the relations between the average price of cigarettes, P, and the quantity purchased, Q, across a large sample of counties in the United States, assuming the following functional form, $\log Q = \alpha + \beta \log P$. Suppose the estimate for β is 0.3. Interpret this coefficient.

7. Sequence of regressions: find a regression problem that is of interest to you and can be performed repeatedly (for example, data from several years, or for several countries). Perform a separate analysis for each year, or country, and display the estimates in a plot as in Figure 4.6 on page 74.

8. Return to the teaching evaluations data from Exercise 3.5. Fit regression models predicting evaluations given many of the inputs in the dataset. Consider interactions, combinations of predictors, and transformations, as appropriate. Consider

several models, discuss in detail the final model that you choose, and also explain why you chose it rather than the others you had considered.

## CHAPTER 5

Logistic regression

Logistic regression is the standard way to model binary outcomes (that is, data y_i that take on the values 0 or 1). Section 5.1 introduces logistic regression in a simple example with one predictor, then for most of the rest of the chapter we work through an extended example with multiple predictors and interactions.

5.1 Logistic regression with a single predictor

Example: modeling political preference given income

Conservative parties generally receive more support among voters with higher incomes. We illustrate classical logistic regression with a simple analysis of this pattern from the National Election Study in 1992. For each respondent i in this poll, we label $y_i = 1$ if he or she preferred George Bush (the Republican candidate for president) or 0 if he or she preferred Bill Clinton (the Democratic candidate), for now excluding respondents who preferred Ross Perot or other candidates, or had no opinion. We predict preferences given the respondent's income level, which is characterized on a five-point scale.[1]

The data are shown as (jittered) dots in Figure 5.1, along with the fitted *logistic regression* line, a curve that is constrained to lie between 0 and 1. We interpret the line as the probability that $y = 1$ given x—in mathematical notation, $\Pr(y = 1|x)$.

We fit and display the logistic regression using the following R function calls:

```
fit.1 <- glm (vote ~ income, family=binomial(link="logit"))
display (fit.1)
```
R code

to yield

```
            coef.est coef.se
(Intercept)   -1.40    0.19
income         0.33    0.06
  n = 1179,  k = 2
  residual deviance = 1556.9, null deviance = 1591.2 (difference = 34.3)
```
R output

The fitted model is $\Pr(y_i = 1) = \text{logit}^{-1}(-1.40 + 0.33 \cdot \text{income})$. We shall define this model mathematically and then return to discuss its interpretation.

The logistic regression model

It would not make sense to fit the continuous linear regression model, $X\beta + \text{error}$, to data y that take on the values 0 and 1. Instead, we model the probability that $y = 1$,

$$\Pr(y_i = 1) = \text{logit}^{-1}(X_i\beta), \tag{5.1}$$

under the assumption that the outcomes y_i are independent given these probabilities. We refer to $X\beta$ as the *linear predictor*.

[1] See Section 4.7 for details on the income categories and other variables measured in this survey.

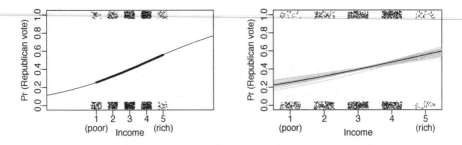

Figure 5.1 *Logistic regression estimating the probability of supporting George Bush in the 1992 presidential election, as a function of discretized income level. Survey data are indicated by jittered dots. In this example little is revealed by these jittered points, but we want to emphasize here that the data and fitted model can be put on a common scale. (a) Fitted logistic regression: the thick line indicates the curve in the range of the data; the thinner lines at the end show how the logistic curve approaches 0 and 1 in the limits. (b) In the range of the data, the solid line shows the best-fit logistic regression, and the light lines show uncertainty in the fit.*

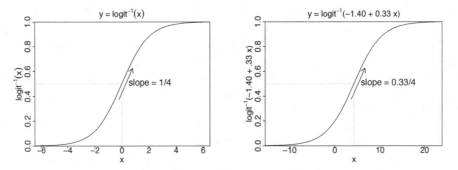

Figure 5.2 *(a) Inverse-logit function $logit^{-1}(x)$: the transformation from linear predictors to probabilities that is used in logistic regression. (b) An example of the predicted probabilities from a logistic regression model: $y = logit^{-1}(-1.40 + 0.33x)$. The shape of the curve is the same, but its location and scale have changed; compare the x-axes on the two graphs. For each curve, the dotted line shows where the predicted probability is 0.5: in graph (a), this is at $logit(0.5) = 0$; in graph (b), the halfway point is where $-1.40 + 0.33x = 0$, which is $x = 1.40/0.33 = 4.2$.*
The slope of the curve at the halfway point is the logistic regression coefficient divided by 4, thus 1/4 for $y = logit^{-1}(x)$ and 0.33/4 for $y = logit^{-1}(-1.40 + 0.33x)$. The slope of the logistic regression curve is steepest at this halfway point.

The function $logit^{-1}(x) = \frac{e^x}{1+e^x}$ transforms continuous values to the range $(0, 1)$, which is necessary, since probabilities must be between 0 and 1. This is illustrated for the election example in Figure 5.1 and more theoretically in Figure 5.2.

Equivalently, model (5.1) can be written

$$\Pr(y_i = 1) = p_i$$
$$\text{logit}(p_i) = X_i\beta, \tag{5.2}$$

where $\text{logit}(x) = \log(x/(1-x))$ is a function mapping the range $(0, 1)$ to the range $(-\infty, \infty)$. We prefer to work with logit^{-1} because it is natural to focus on the mapping from the linear predictor to the probabilities, rather than the reverse. However, you will need to understand formulation (5.2) to follow the literature and also when fitting logistic models in Bugs.

The inverse-logistic function is curved, and so the expected difference in y corresponding to a fixed difference in x is not a constant. As can be seen in Figure 5.2, the steepest change occurs at the middle of the curve. For example:

- logit$(0.5) = 0$, and logit$(0.6) = 0.4$. Here, adding 0.4 on the logit scale corresponds to a change from 50% to 60% on the probability scale.

- logit$(0.9) = 2.2$, and logit$(0.93) = 2.6$. Here, adding 0.4 on the logit scale corresponds to a change from 90% to 93% on the probability scale.

Similarly, adding 0.4 at the low end of the scale moves a probability from 7% to 10%. In general, any particular change on the logit scale is compressed at the ends of the probability scale, which is needed to keep probabilities bounded between 0 and 1.

5.2 Interpreting the logistic regression coefficients

Coefficients in logistic regression can be challenging to interpret because of the nonlinearity just noted. We shall try to generalize the procedure for understanding coefficients one at a time, as was done for linear regression in Chapter 3. We illustrate with the model, $\text{Pr(Bush support)} = \text{logit}^{-1}(-1.40 + 0.33 \cdot \text{income})$. Figure 5.1 shows the story, but we would also like numerical summaries. We present some simple approaches here and return in Section 5.7 to more comprehensive numerical summaries.

Evaluation at and near the mean of the data

The curve of the logistic function requires us to choose where to evaluate changes, if we want to interpret on the probability scale. The mean of the input variables in the data is often a useful starting point.

- As with linear regression, the *intercept* can only be interpreted assuming zero values for the other predictors. When zero is not interesting or not even in the model (as in the voting example, where income is on a 1–5 scale), the intercept must be evaluated at some other point. For example, we can evaluate Pr(Bush support) at the central income category and get $\text{logit}^{-1}(-1.40 + 0.33 \cdot 3) = 0.40$.

  Or we can evaluate Pr(Bush support) at the mean of respondents' incomes: $\text{logit}^{-1}(-1.40 + 0.33 \cdot \bar{x})$; in R we code this as[2]

  ```
  invlogit (-1.40 + 0.33*mean(income))
  ```
  R code

  or, more generally,

  ```
  invlogit (coef(fit.1)[1] + coef(fit.1)[2]*mean(income))
  ```
  R code

  For this dataset, $\bar{x} = 3.1$, yielding Pr(Bush support) $= 0.40$ at this central point.

- A difference of 1 in income (on this 1–5 scale) corresponds to a positive difference of 0.33 in the logit probability of supporting Bush. There are two convenient ways to summarize this directly in terms of probabilities.

  - We can evaluate how the probability differs with a unit difference in x near the central value. Since $\bar{x} = 3.1$ in this example, we can evaluate the logistic regression function at $x = 3$ and $x = 2$; the difference in $\text{Pr}(y = 1)$ corresponding to adding 1 to x is $\text{logit}^{-1}(-1.40+0.33\cdot3) - \text{logit}^{-1}(-1.40+0.33\cdot2) = 0.08$.

---

[2] We are using a function we have written, `invlogit <- function (x) {1/(1+exp(-x))}`.

A difference of 1 in income category corresponds to a positive difference of 8% in the probability of supporting Bush.

- Rather than consider a discrete change in x, we can compute the derivative of the logistic curve at the central value, in this case $\bar{x} = 3.1$. Differentiating the function $\text{logit}^{-1}(\alpha + \beta x)$ with respect to x yields $\beta e^{\alpha + \beta x}/(1 + e^{\alpha + \beta x})^2$. The value of the linear predictor at the central value of $\bar{x} = 3.1$ is $-1.40 + 0.33 \cdot 3.1 = -0.39$, and the slope of the curve—the "change" in $\Pr(y = 1)$ per small unit of "change" in x—at this point is $0.33 e^{-0.39}/(1 + e^{-0.39})^2 = 0.13$.

- For this example, the difference on the probability scale is the same value of 0.13 (to one decimal place); this is typical but in some cases where a unit difference is large, the differencing and the derivative can give slightly different answers. They will always be the same sign, however.

The "divide by 4 rule"

The logistic curve is steepest at its center, at which point $\alpha + \beta x = 0$ so that $\text{logit}^{-1}(\alpha + \beta x) = 0.5$ (see Figure 5.2). The slope of the curve—the derivative of the logistic function—is maximized at this point and attains the value $\beta e^0/(1 + e^0)^2 = \beta/4$. Thus, $\beta/4$ is the maximum difference in $\Pr(y = 1)$ corresponding to a unit difference in x.

As a rule of convenience, we can take logistic regression coefficients (other than the constant term) and divide them by 4 to get an upper bound of the predictive difference corresponding to a unit difference in x. This upper bound is a reasonable approximation near the midpoint of the logistic curve, where probabilities are close to 0.5.

For example, in the model $\Pr(\text{Bush support}) = \text{logit}^{-1}(-1.40 + 0.33 \cdot \text{income})$, we can divide 0.33/4 to get 0.08: a difference of 1 in income category corresponds to no more than an 8% positive difference in the probability of supporting Bush. Because the data in this case actually lie near the 50% point (see Figure 5.1), this "divide by 4" approximation turns out to be close to 0.13, the derivative evaluated at the central point of the data.

Interpretation of coefficients as odds ratios

Another way to interpret logistic regression coefficients is in terms of *odds ratios*. If two outcomes have the probabilities $(p, 1-p)$, then $p/(1-p)$ is called the *odds*. An odds of 1 is equivalent to a probability of 0.5—that is, equally likely outcomes. Odds of 0.5 or 2.0 represent probabilities of $(1/3, 2/3)$. The ratio of two odds—thus, $(p_1/(1-p_1))/(p_2/(1-p_2))$—is called an odds ratio. Thus, an odds ratio of 2 corresponds to a change from $p = 0.33$ to $p = 0.5$, or a change from $p = 0.5$ to $p = 0.67$.

An advantage of working with odds ratios (instead of probabilities) is that it is possible to keep scaling up odds ratios indefinitely without running into the boundary points of 0 and 1. For example, going from an odds of 2 to an odds of 4 increases the probability from 2/3 to 4/5; doubling the odds again increases the probability to 8/9, and so forth.

Exponentiated logistic regression coefficients can be interpreted as odds ratios. For simplicity, we illustrate with a model with one predictor, so that

$$\log\left(\frac{\Pr(y = 1|x)}{\Pr(y = 0|x)}\right) = \alpha + \beta x. \tag{5.3}$$

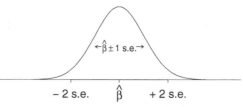

Figure 5.3 *Distribution representing uncertainty in an estimated regression coefficient (repeated from page 40). The range of this distribution corresponds to the possible values of β that are consistent with the data. When using this as an uncertainty distribution, we assign an approximate 68% chance that β will lie within 1 standard error of the point estimate, $\hat{\beta}$, and an approximate 95% chance that β will lie within 2 standard errors. Assuming the regression model is correct, it should happen only about 5% of the time that the estimate, $\hat{\beta}$, falls more than 2 standard errors away from the true β.*

Adding 1 to x (that is, changing x to $x+1$ in (5.3)) has the effect of adding β to both sides of the equation. Exponentiating both sides, the odds are then multiplied by e^{β}. For example, if $\beta = 0.2$, then a unit difference in x corresponds to a multiplicative change of $e^{0.2} = 1.22$ in the odds (for example, changing the odds from 1 to 1.22, or changing p from 0.5 to 0.55).

We find that the concept of odds can be somewhat difficult to understand, and odds ratios are even more obscure. Therefore we prefer to interpret coefficients on the original scale of the data when possible, for example, saying that adding 0.2 on the logit scale corresponds to a change in probability from $\text{logit}^{-1}(0)$ to $\text{logit}^{-1}(0.2)$.

Inference

Coefficient estimates and standard errors.   The coefficients in classical logistic regression are estimated using maximum likelihood, a procedure that can often work well for models with few predictors fit to reasonably large samples (but see Section 5.8 for a potential problem).

As with the linear model, the standard errors represent estimation uncertainty. We can roughly say that coefficient estimates within 2 standard errors of $\hat{\beta}$ are consistent with the data. Figure 5.3 shows the normal distribution that approximately represents the range of possible values of β. For the voting example, the coefficient of income has an estimate $\hat{\beta}$ of 0.33 and a standard error of 0.06; thus the data are roughly consistent with values of β in the range $[0.33 \pm 2 \cdot 0.06] = [0.21, 0.45]$.

Statistical significance.   As with linear regression, a coefficient is considered "statistically significant" if it is at least 2 standard errors away from zero. In the voting example, the coefficient of income is statistically significant and positive, meaning that we can be fairly certain that, in the population represented by this survey, positive differences in income generally correspond to positive (not negative) differences in the probability of supporting Bush for president.

Also as with linear regression, we usually do *not* try to interpret the statistical significance of the intercept. The sign of an intercept is not generally of any interest, and so it is usually meaningless to compare it to zero or worry about whether it is statistically significantly different from zero.

Finally, when considering multiple inputs, we follow the same principles as with linear regression when deciding when and how to include and combine inputs in a model, as discussed in Section 4.6.

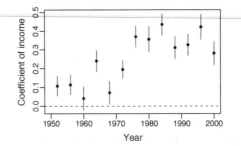

Figure 5.4 *Coefficient of income (on a 1–5 scale) with ±1 standard-error bounds in logistic regressions predicting Republican preference for president, as estimated separately from surveys in the second half of the twentieth century. The pattern of richer voters supporting Republicans has increased since 1970. The data used in the estimate for 1992 appear in Figure 5.1.*

Predictions. Logistic regression predictions are probabilistic, so for each unobserved future data point $\tilde{y}_i$, there is a predictive probability,

$$\tilde{p}_i = \Pr(\tilde{y}_i = 1) = \text{logit}^{-1}(\tilde{X}_i\beta),$$

rather than a point prediction. For example, for a voter not in the survey with income level 5 (recall the 5-point scale in Figure 5.1), the predicted *probability* of supporting Bush is $\Pr(\tilde{y}_i = 1) = \text{logit}^{-1}(-1.40 + 0.33 \cdot 5) = 0.55$. We do not say that our prediction for the *outcome* is 0.55, since the outcome $\tilde{y}_i$—support for Bush or not—itself will be 0 or 1.

Fitting and displaying the model in R

After fitting the logistic regression using the **glm** function (see page 79), we can graph the data and fitted line (see Figure 5.1a) as follows:

R code
```
plot (income, vote)
curve (invlogit (coef(fit.1)[1] + coef(fit.1)[2]*x), add=TRUE)
```

(The R code we actually use to make the figure has more steps so as to display axis labels, jitter the points, adjust line thickness, and so forth.) Figure 5.1b has dotted lines representing uncertainty in the coefficients; we display these by adding the following to the plotting commands:

R code
```
sim.1 <- sim (fit.1)
for (j in 1:10){
  curve (invlogit (sim.1$beta[j,1] + sim.1$beta[j,2]*x),
    col="gray", lwd=.5, add=TRUE)}
```

We demonstrate further use of the **sim** function in Chapter 7.

Displaying the results of several logistic regressions

We can display estimates from a series of logistic regressions in a single graph, just as was done in Section 4.7 for linear regression coefficients. Figure 5.4 illustrates with the estimate ±1 standard error for the coefficient for income on presidential preference, fit to National Election Studies pre-election polls from 1952 through 2000. Higher income has consistently been predictive of Republican support, but the connection has become stronger over the years.

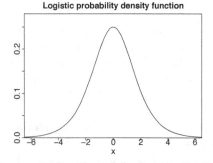

Figure 5.5 *The probability density function of the logistic distribution, which is used for the error term in the latent-data formulation (5.4) of logistic regression. The logistic curve in Figure 5.2a is the cumulative distribution function of this density. The maximum of the density is 0.25, which corresponds to the maximum slope of 0.25 in the inverse-logit function of Figure 5.2a.*

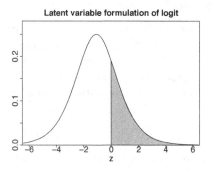

Figure 5.6 *The probability density function of the latent variable z_i in model (5.4) if the linear predictor, $X_i\beta$, has the value -1.07. The shaded area indicates the probability that $z_i > 0$, so that $y_i = 1$ in the logistic regression.*

5.3 Latent-data formulation

We can interpret logistic regression directly—as a nonlinear model for the probability of a "success" or "yes" response given some predictors—and also indirectly, using what are called unobserved or *latent* variables. In this formulation, each discrete outcome y_i is associated with a continuous, unobserved outcome z_i, defined as follows:

$$y_i = \begin{cases} 1 & \text{if } z_i > 0 \\ 0 & \text{if } z_i < 0 \end{cases}$$

$$z_i = X_i\beta + \epsilon_i, \tag{5.4}$$

with independent errors ϵ_i that have the *logistic* probability distribution. The logistic distribution is shown in Figure 5.5 and is defined so that

$$\Pr(\epsilon_i < x) = \text{logit}^{-1}(x) \text{ for all } x.$$

Thus, $\Pr(y_i = 1) = \Pr(z_i > 0) = \Pr(\epsilon_i > -X_i\beta) = \text{logit}^{-1}(X_i\beta)$, and so models (5.1) and (5.4) are equivalent.

Figure 5.6 illustrates for an observation i with income level $x_i = 1$ (that is, a person in the lowest income category), whose linear predictor, $X_i\beta$, thus has the value $-1.40 + 0.33 \cdot 1 = -1.07$. The curve illustrates the distribution of the latent variable z_i, and the shaded area corresponds to the probability that $z_i > 0$, so that $y_i = 1$. In this example, $\Pr(y_i = 1) = \text{logit}^{-1}(-1.07) = 0.26$.

Interpretation of the latent variables

Latent variables are a computational trick but they can also be interpreted substantively. For example, in the pre-election survey, $y_i = 1$ for Bush supporters and 0 for Clinton supporters. The unobserved continuous z_i can be interpreted as the respondent's "utility" or preference for Bush, compared to Clinton: the sign of the utility tells us which candidate is preferred, and its magnitude reveals the strength of the preference.

Only the sign of z_i, not its magnitude, can be determined directly from binary data. However, we can learn more about the z_i's given the logistic regression predictors. In addition, in some settings direct information is available about the z_i's; for example, a survey can ask "feeling thermometer" questions such as, "Rate your feelings about George Bush on a 1–10 scale, with 1 being the most negative and 10 being the most positive."

Nonidentifiability of the latent variance parameter

The logistic probability density function in Figure 5.5 appears bell-shaped, much like the normal density that is used for errors in linear regression. In fact, the logistic distribution is very close to the normal distribution with mean 0 and standard deviation 1.6—an identity that we discuss further on page 118 in the context of "probit regression." For now, we merely note that the logistic model (5.4) for the latent variable z is closely approximated by the normal regression model,

$$z_i = X_i\beta + \epsilon_i, \quad \epsilon_i \sim \mathrm{N}(0, \sigma^2), \tag{5.5}$$

with $\sigma = 1.6$. This then raises the question, why not estimate σ?

We cannot estimate the parameter σ in model (5.5) because it is not identified when considered jointly with the regression parameter β. If all the elements of β are multiplied by a positive constant and σ is also multiplied by that constant, then the model does not change. For example, suppose we fit the model

$$z_i = -1.40 + 0.33x_i + \epsilon_i, \quad \epsilon_i \sim \mathrm{N}(0, 1.6^2).$$

This is equivalent to the model

$$z_i = -14.0 + 3.3x_i + \epsilon_i, \quad \epsilon_i \sim \mathrm{N}(0, 16^2),$$

or

$$z_i = -140 + 33x_i + \epsilon_i, \quad \epsilon_i \sim \mathrm{N}(0, 160^2).$$

As we move from each of these models to the next, z is multiplied by 10, but the *sign* of z does not change. Thus all the models have the same implications for the observed data y: for each model, $\Pr(y_i = 1) \approx \mathrm{logit}^{-1}(-1.40 + 0.33x_i)$ (only approximate because the logistic distribution is not exactly normal).

Thus, model (5.5) has an essential indeterminacy when fit to binary data, and it is standard to resolve this by setting the variance parameter σ to a fixed value, for example 1.6, which is essentially equivalent to the unit logistic distribution.

5.4 Building a logistic regression model: wells in Bangladesh

We illustrate the steps of building, understanding, and checking the fit of a logistic regression model using an example from economics (or perhaps it is psychology, or public health): modeling the decisions of households in Bangladesh about whether to change their source of drinking water.

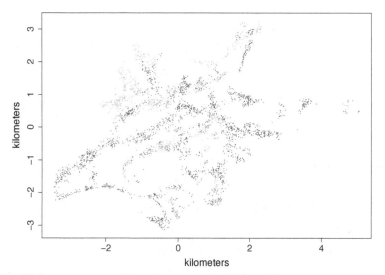

Figure 5.7 *Wells in an area of Araihazar upazila, Bangladesh. Light and dark dots represent wells with arsenic greater than and less than the safety standard of 0.5 (in units of hundreds of micrograms per liter). (The wells are located where people live. The empty areas between the wells are mostly cropland.) Safe and unsafe wells are intermingled in most of the area, which suggests that users of unsafe wells can switch to nearby safe wells.*

Background

Many of the wells used for drinking water in Bangladesh and other South Asian countries are contaminated with natural arsenic, affecting an estimated 100 million people. Arsenic is a cumulative poison, and exposure increases the risk of cancer and other diseases, with risks estimated to be proportional to exposure.

Any locality can include wells with a range of arsenic levels, as can be seen from the map in Figure 5.7 of all the wells in a collection of villages in a small area of Bangladesh. The bad news is that even if your neighbor's well is safe, it does not mean that yours is safe. However, the corresponding good news is that, if your well has a high arsenic level, you can probably find a safe well nearby to get your water from—if you are willing to walk the distance and your neighbor is willing to share. (The amount of water needed for drinking is low enough that adding users to a well would not exhaust its capacity, and the surface water in this area is subject to contamination by microbes, hence the desire to use water from deep wells.)

In the area shown in Figure 5.7, a research team from the United States and Bangladesh measured all the wells and labeled them with their arsenic level as well as a characterization as "safe" (below 0.5 in units of hundreds of micrograms per liter, the Bangladesh standard for arsenic in drinking water) or "unsafe" (above 0.5). People with unsafe wells were encouraged to switch to nearby private or community wells or to new wells of their own construction.

A few years later, the researchers returned to find out who had switched wells. We shall perform a logistic regression analysis to understand the factors predictive of well switching among the users of unsafe wells. In the notation of the previous section, our outcome variable is

$$y_i = \begin{cases} 1 & \text{if household } i \text{ switched to a new well} \\ 0 & \text{if household } i \text{ continued using its own well.} \end{cases}$$

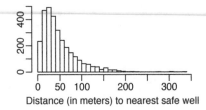

Figure 5.8 *Histogram of distance to the nearest safe well, for each of the unsafe wells in the Araihazar dataset (see Figure 5.7).*

We consider the following inputs:

- A constant term

- The distance (in meters) to the closest known safe well

- The arsenic level of respondent's well

- Whether any members of the household are active in community organizations

- The education level of the head of household.

We shall first fit the model just using distance to nearest well and then put in arsenic concentration, organizational membership, and education.

Logistic regression with just one predictor

We fit the logistic regression in R:

R code
```
fit.1 <- glm (switch ~ dist, family=binomial(link="logit"))
```

Displaying this yields

R output
```
glm(formula = switch ~ dist, family=binomial(link="logit"))
            coef.est coef.se
(Intercept)   0.6060  0.0603
dist         -0.0062  0.0010
  n = 3020, k = 2
    residual deviance = 4076.2, null deviance = 4118.1 (difference = 41.9)
```

The coefficient for `dist` is −0.0062, which seems low, but this is misleading since distance is measured in meters, so this coefficient corresponds to the difference between, say, a house that is 90 meters away from the nearest safe well and a house that is 91 meters away.

Figure 5.8 shows the distribution of `dist` in the data. It seems more reasonable to rescale distance in 100-meter units:

R code
```
dist100 <- dist/100
```

and refitting the logistic regression yields

R output
```
glm(formula = switch ~ dist100, family=binomial(link="logit"))
            coef.est coef.se
(Intercept)   0.61    0.06
dist100      -0.62    0.10
  n = 3020, k = 2
    residual deviance = 4076.2, null deviance = 4118.1 (difference = 41.9)
```

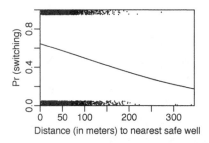

Figure 5.9 *Graphical expression of the fitted logistic regression,* $\Pr(\textit{switching wells}) = logit^{-1}(0.61 - 0.62 \cdot \texttt{dist100})$, *with (jittered) data overlain. The predictor* `dist100` *is* `dist`/*100: distance to the nearest safe well in 100-meter units.*

Graphing the fitted model

In preparing to plot the data, we first create a function to jitter the binary outcome while keeping the points between 0 and 1:

```
jitter.binary <- function(a, jitt=.05){                              R code
  ifelse (a==0, runif (length(a), 0, jitt), runif (length(a), 1-jitt, 1))
}
```

We can then graph the data and fitted model:[3]

```
switch.jitter <- jitter.binary (switch)                              R code
plot (dist, switch.jitter)
curve (invlogit (coef(fit.1)[1] + coef(fit.1)[2]*x), add=TRUE)
```

The result is displayed in Figure 5.9. The probability of switching is about 60% for people who live near a safe well, declining to about 20% for people who live more than 300 meters from any safe well. This makes sense: the probability of switching is higher for people who live closer to a safe well.

Interpreting the logistic regression coefficients

We can interpret the coefficient estimates using evaluations of the inverse-logit function and its derivative, as in the example of Section 5.1. Our model here is

$$\Pr(\text{switch}) = \text{logit}^{-1}(0.61 - 0.62 \cdot \text{dist100}).$$

1. The constant term can be interpreted when `dist100` = 0, in which case the probability of switching is $\text{logit}^{-1}(0.61) = 0.65$. Thus, the model estimates a 65% probability of switching if you live right next to an existing safe well.

2. We can evaluate the predictive difference with respect to `dist100` by computing the derivative at the average value of `dist100` in the dataset, which is 0.48 (that is, 48 meters; see Figure 5.8). The value of the linear predictor here is $0.61 - 0.62 \cdot 0.48 = 0.31$, and so the slope of the curve at this point is $-0.62e^{0.31}/(1+e^{0.31})^2 = -0.15$. Thus, adding 1 to `dist100`—that is, adding 100 meters to the distance to the nearest safe well—corresponds to a negative difference in the probability of switching of about 15%.

---

[3] Another display option, which would more clearly show the differences between households that did and did not switch, would be to overlay separate histograms of `dist` for the switchers and nonswitchers.

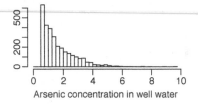

Figure 5.10 *Histogram of arsenic levels in unsafe wells (those exceeding 0.5) in the measured area of Araihazar, Bangladesh (see Figure 5.7).*

3. More quickly, the "divide by 4 rule" gives us $-0.62/4 = -0.15$. This comes out the same, to two decimal places, as was calculated using the derivative because the curve passes through the 50% point right in the middle of the data (see Figure 5.9).

In addition to interpreting its magnitude, we can look at the statistical significance of the coefficient for distance. The slope is estimated well, with a standard error of only 0.10, which is tiny compared to the coefficient estimate of -0.62. The approximate 95% interval is $[-0.82, -0.42]$, which is clearly statistically significantly different from zero.

Adding a second input variable

We now extend the well-switching example by adding the arsenic level of the existing well as a regression input. At the levels present in the Bangladesh drinking water, the health risks from arsenic are roughly proportional to exposure, and so we would expect switching to be more likely from wells with high arsenic levels. Figure 5.10 shows the arsenic levels of the unsafe wells before switching.

R code

```
fit.3 <- glm (switch ~ dist100 + arsenic, family=binomial(link="logit"))
```

which, when displayed, yields

R output

```
                coef.est coef.se
(Intercept)       0.00     0.08
dist100          -0.90     0.10
arsenic           0.46     0.04
  n = 3020, k = 3
  residual deviance = 3930.7, null deviance = 4118.1 (difference = 187.4)
```

Thus, comparing two wells with the same arsenic level, every 100 meters in distance to the nearest safe well corresponds to a *negative* difference of 0.90 in the logit probability of switching. Similarly, a difference of 1 in arsenic concentration corresponds to a 0.46 *positive* difference in the logit probability of switching. Both coefficients are statistically significant, each being more than 2 standard errors away from zero. And both their signs make sense: switching is easier if there is a nearby safe well, and if a household's existing well has a high arsenic level, there should be more motivation to switch.

For a quick interpretation, we divide the coefficients by 4: thus, 100 meters more in distance corresponds to an approximately 22% lower probability of switching, and 1 unit more in arsenic concentration corresponds to an approximately 11% positive difference in switching probability.

Comparing these two coefficients, it would at first seem that distance is a more important factor than arsenic level in determining the probability of switching.

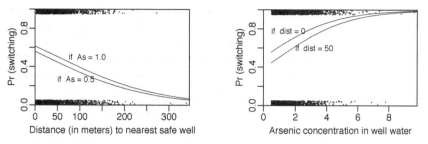

Figure 5.11 *Fitted logistic regression of probability of switching from an unsafe well as a function of two variables, plotted (a) as a function of distance to nearest safe well and (b) as a function of arsenic level of existing well. For each plot, the other input variable is held constant at different representative values.*

Such a statement is misleading, however, because in our data `dist100` shows less variation than `arsenic`: the standard deviation of distances to the nearest well is 0.38 (in units of 100 meters), whereas arsenic levels have a standard deviation of 1.10 on the scale used here. Thus, the logistic regression coefficients corresponding to 1-standard-deviation differences are $-0.90 \cdot 0.38 = -0.34$ for distance and $0.46 \cdot 1.10 = 0.51$ for arsenic level. Dividing by 4 yields the quick summary estimate of a 1-standard-deviation difference in distance or arsenic level corresponding to an 8% negative difference or a 13% positive difference, respectively, in Pr(switch).

Comparing the coefficient estimates when adding a predictor

The coefficient for `dist100` changes from -0.62 in the original model to 0.90 when arsenic level is added to the model. This change occurs because wells that are far from the nearest safe well are also likely to be particularly high in arsenic.

Graphing the fitted model with two predictors

The most natural way to graph the regression of y on two predictors might be as a three-dimensional surface, with the vertical axis showing $\Pr(y = 1)$ as a function of predictors plotted on the two horizontal axes.

However, we find such graphs hard to read, so instead we make separate plots as a function of each of the two variables; see Figure 5.11. As with the lines in Figure 3.4, we can plot the focus input variable on the x-axis and use multiple lines to show the fit for different values of the other input. To produce Figure 5.11a, we first plot the (jittered) data points, forcing zero to be included in the x-range of the plot because it is a natural baseline comparison for distance:

```
plot (dist, switch.jitter, xlim=c(0,max(dist)))
```
R code

We next add the fitted curves:

```
curve (invlogit (cbind (1, x/100,  .5) %*% coef(fit.3)), add=TRUE)
curve (invlogit (cbind (1, x/100, 1.0) %*% coef(fit.3)), add=TRUE)
```
R code

We need to divide x by 100 here because the plot is in the scale of meters but the model is defined in terms of $dist100 = dist/100$.

The object created by `cbind(1,x/100,.5)` is an $n \times 3$ matrix constructed from a column of 1's, the vector `x` (used internally by the `curve` function), and a vector of .5's. In constructing the matrix, R automatically expands the scalars 1 and .5 to the length of the vector `x`. For the two lines, we pick arsenic levels of 0.5

and 1.0 because 0.5 is the minimum value of arsenic concentration (since we are only studying users of unsafe wells), and a difference of 0.5 represents a reasonable comparison, given the distribution of arsenic levels in the data (see Figure 5.10).

Similar commands are used to make Figure 5.11b, showing the probability of switching as a function of arsenic concentration with distance held constant:

R code
```
plot (arsenic, switch.jitter, xlim=c(0,max(arsenic)))
curve (invlogit (cbind (1, 0, x) %*% coef(fit.3)), add=TRUE)
curve (invlogit (cbind (1,.5, x) %*% coef(fit.3)), add=TRUE)
```

5.5 Logistic regression with interactions

We continue our modeling by adding the interaction between the two inputs:

R code
```
fit.4 <- glm (switch ~ dist100 + arsenic + dist100:arsenic,
   family=binomial(link="logit"))
display (fit.4)
```

which yields

R output

	coef.est	coef.se
(Intercept)	-0.15	0.12
dist100	-0.58	0.21
arsenic	0.56	0.07
dist100:arsenic	-0.18	0.10

n = 3020, k = 4
residual deviance = 3927.6, null deviance = 4118.1 (difference = 190.5)

To understand the numbers in the table, we use the following tricks:

- Evaluating predictions and interactions at the mean of the data, which have average values of 0.48 for dist100 and 1.66 for arsenic (that is, a mean distance of 48 meters to the nearest safe well, and a mean arsenic level of 1.66 among the unsafe wells).

- Dividing by 4 to get approximate predictive differences on the probability scale.

We now interpret each regression coefficient in turn.

- *Constant term:* $\text{logit}^{-1}(-0.15) = 0.47$ is the estimated probability of switching, if the distance to the nearest safe well is 0 and the arsenic level of the current well is 0. This is an impossible condition (since arsenic levels all exceed 0.5 in our set of unsafe wells), so we do not try to interpret the constant term. Instead, we can evaluate the prediction at the average values of dist100 = 0.48 and arsenic = 1.66, where the probability of switching is $\text{logit}^{-1}(-0.15 - 0.58 \cdot 0.48 + 0.56 \cdot 1.66 - 0.18 \cdot 0.48 \cdot 1.66) = 0.59$.

- *Coefficient for distance:* this corresponds to comparing two wells that differ by 1 in dist100, if the arsenic level is 0 for both wells. Once again, we should not try to interpret this.

  Instead, we can look at the average value, arsenic = 1.66, where distance has a coefficient of $-0.58 - 0.18 \cdot 1.66 = -0.88$ on the logit scale. To quickly interpret this on the probability scale, we divide by 4: $-0.88/4 = -0.22$. Thus, at the mean level of arsenic in the data, each 100 meters of distance corresponds to an approximate 22% *negative* difference in probability of switching.

- *Coefficient for arsenic:* this corresponds to comparing two wells that differ by 1 in arsenic, if the distance to the nearest safe well is 0 for both.

  Instead, we evaluate the comparison at the average value for distance, dist100 =

0.48, where arsenic has a coefficient of $0.56 - 0.18 \cdot 0.48 = 0.47$ on the logit scale. To quickly interpret this on the probability scale, we divide by 4: $0.47/4 = 0.12$. Thus, at the mean level of distance in the data, each additional unit of arsenic corresponds to an approximate 12% *positive* difference in probability of switching.

- *Coefficient for the interaction term:* this can be interpreted in two ways. Looking from one direction, for each additional unit of arsenic, the value -0.18 is added to the coefficient for distance. We have already seen that the coefficient for distance is -0.88 at the average level of arsenic, and so we can understand the interaction as saying that the importance of distance as a predictor increases for households with higher existing arsenic levels.

  Looking at it the other way, for each additional 100 meters of distance to the nearest well, the value -0.18 is added to the coefficient for arsenic. We have already seen that the coefficient for distance is 0.47 at the average distance to nearest safe well, and so we can understand the interaction as saying that the importance of arsenic as a predictor decreases for households that are farther from existing safe wells.

Centering the input variables

As discussed earlier in the context of linear regression, before fitting interactions it makes sense to center the input variables so that we can more easily interpret the coefficients. The centered inputs are:

```
c.dist100 <- dist100 - mean(dist100)
c.arsenic <- arsenic - mean(arsenic)
```
R code

We do not fully standardize these—that is, we do not scale by their standard deviations—because it is convenient to be able to consider known differences on the original scales of the data (100-meter distances and arsenic-concentration units).

Refitting the interaction model using the centered inputs

We can refit the model using the centered input variables, which will make the coefficients much easier to interpret:

```
fit.5 <- glm (switch ~ c.dist100 + c.arsenic + c.dist100:c.arsenic,
    family=binomial(link="logit"))
```
R code

We center the *inputs*, not the *predictors*. Hence, we do not center the interaction (dist100*arsenic); rather, we include the interaction of the two centered input variables. Displaying fit.5 yields

```
                   coef.est coef.se
(Intercept)           0.35    0.04
c.dist100            -0.88    0.10
c.arsenic             0.47    0.04
c.dist100:c.arsenic  -0.18    0.10
  n = 3020, k = 4
  residual deviance = 3927.6, null deviance = 4118.1 (difference = 190.5)
```
R output

Interpreting the inferences on this new scale:

- *Constant term:* $\text{logit}^{-1}(0.35) = 0.59$ is the estimated probability of switching, if c.dist100 = c.arsenic = 0, that is, if distance to nearest safe well and arsenic level are at their averages in the data. (We obtained this same calculation, but with more effort, with our earlier model with uncentered inputs.)

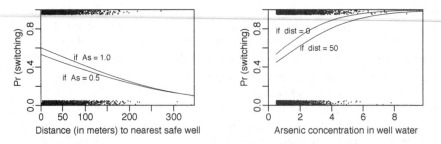

Figure 5.12 *Fitted logistic regression of probability of switching from an unsafe well as a function of distance to nearest safe well and arsenic level of existing well, for the model with interactions. Compare to the no-interaction model in Figure 5.11.*

- *Coefficient for distance:* this is the coefficient for distance (on the logit scale) if arsenic level is at its average value. To quickly interpret this on the probability scale, we divide by 4: $-0.88/4 = -0.22$. Thus, at the mean level of arsenic in the data, each 100 meters of distance corresponds to an approximate 22% *negative* difference in probability of switching.

- *Coefficient for arsenic:* this is the coefficient for arsenic level if distance to nearest safe well is at its average value. To quickly interpret this on the probability scale, we divide by 4: $0.47/4 = 0.12$. Thus, at the mean level of distance in the data, each additional unit of arsenic corresponds to an approximate 12% *positive* difference in probability of switching.

- *Coefficient for the interaction term:* this is unchanged by centering and has the same interpretation as before.

The predictions for new observations are unchanged. The linear centering of the predictors changes the interpretations of the coefficients but does not change the underlying model.

Statistical significance of the interaction

As can be seen from the regression table on the previous page, `c.dist100:c.arsenic` has an estimated coefficient of -0.18 with a standard error of 0.10. The estimate is not quite 2 standard errors away from zero and so is not quite statistically significant. However, the negative sign makes sense—it is plausible that arsenic level becomes a less important predictor for households that are farther from the nearest safe well, and the magnitude of the association is also plausible. So we keep the interaction in, following our general rules for regression coefficients and statistical significance, as given in Section 4.6.

Graphing the model with interactions

The clearest way to visualize the interaction model is to plot the regression curves as a function for each picture. The result is shown in Figure 5.12, the first graph of which we make in R as follows (with similar commands for the other graph):

R code
```
plot (dist, switch.jitter, xlim=c(0,max(dist)))
curve (invlogit (cbind(1,x/100, .5, .5*x/100) %*% coef(fit.4)), add=TRUE)
curve (invlogit (cbind(1,x/100,1.0,1.0*x/100) %*% coef(fit.4)), add=TRUE)
```

As Figure 5.12 makes clear, the interaction is not large in the range of most of the data. The largest pattern that shows up is in Figure 5.12a, where the two lines intersect at around 300 meters. This graph shows evidence that the differences in switching associated with differences in arsenic level are large if you are close to a safe well, but with a diminishing effect if you are far from any safe well. This interaction makes some sense; however, there is some uncertainty in the size of the interaction (from the earlier regression table, an estimate of -0.18 with a standard error of 0.10), and as Figure 5.12a shows, there are only a few data points in the area where the interaction makes much of a difference.

The interaction also appears in Figure 5.12b, this time in a plot of probability of switching as a function of arsenic concentration, at two different levels of distance.

Adding social predictors

Are well users more likely to switch if they have community connections or more education? To see, we add two inputs:

- assoc = 1 if a household member is in any community organization

- educ = years of education of the well user.

We actually work with educ4 = educ/4, for the usual reasons of making its regression coefficient more interpretable—it now represents the predictive difference of adding four years of education.[4]

```
glm(formula = switch ~ c.dist100 + c.arsenic + c.dist100:c.arsenic +          R output
      assoc + educ4, family=binomial(link="logit"))
                        coef.est coef.se
(Intercept)              0.20     0.07
c.dist100               -0.88     0.11
c.arsenic                0.48     0.04
c.dist100:c.arsenic     -0.16     0.10
assoc                   -0.12     0.08
educ4                    0.17     0.04
    n = 3020, k = 6
    residual deviance = 3905.4, null deviance = 4118.1 (difference = 212.7)
```

For households with unsafe wells, belonging to a community association surprisingly is *not* predictive of switching, after controlling for the other factors in the model. However, persons with higher education are more likely to switch: the crude estimated difference is $0.17/4 = 0.04$, or a 4% positive difference in switching probability when comparing households that differ by 4 years of education.[5]

The coefficient for education makes sense and is statistically significant, so we keep it in the model. The coefficient for association does not make sense and is not statistically significant, so we remove it. (See Section 4.6 for a fuller discussion of including or excluding regression predictors.) We are left with

```
glm(formula = switch ~ c.dist100 + c.arsenic + c.dist100:c.arsenic +          R output
      educ4, family = binomial(link = "logit"))
                    coef.est coef.se
(Intercept)          0.15     0.06
```

[4] The levels of education among the 3000 respondents varied from 0 to 17 years, with nearly a third having zero. We repeated our analysis with a discrete recoding of the education variable (0 = 0 years, 1 = 1–8 years, 2 = 9–12 years, 3 = 12+ years), and our results were essentially unchanged.

[5] Throughout this example, we have referred to "coefficients" and "differences," rather than to "effects" and "changes," because the observational nature of the data makes it difficult to directly interpret the regression model causally. We continue causal inference more carefully in Chapter 9, briefly discussing the arsenic problem at the end of Section 9.8.

```
c.dist100              -0.87      0.11
c.arsenic               0.48      0.04
c.dist100:c.arsenic    -0.16      0.10
educ4                   0.17      0.04
   n = 3020, k = 5
   residual deviance = 3907.9, null deviance = 4118.1 (difference = 210.2)
```

Adding further interactions

When inputs have large main effects, it is our general practice to include their interactions as well. We first create a centered education variable:

R code
```
c.educ4 <- educ4 - mean(educ4)
```

and then fit a new model interacting it with distance to nearest safe well and arsenic level of the existing well:

R output
```
glm(formula=switch~c.dist100 + c.arsenic + c.educ4 + c.dist100:c.arsenic +
    c.dist100:c.educ4 + c.arsenic:c.educ4, family=binomial(link="logit"))
                        coef.est coef.se
(Intercept)             0.36      0.04
c.dist100              -0.90      0.11
c.arsenic               0.49      0.04
c.educ4                 0.18      0.04
c.dist100:c.arsenic    -0.12      0.10
c.dist100:c.educ4       0.32      0.11
c.arsenic:c.educ4       0.07      0.04
   n = 3020, k = 7
   residual deviance = 3891.7, null deviance = 4118.1 (difference = 226.4)
```

We can interpret these new interactions by understanding how education modifies the predictive difference corresponding to distance and arsenic.

- *Interaction of distance and education:* a difference of 4 years of education corresponds to a difference of 0.32 in the coefficient for `dist100`. As we have already seen, `dist100` has a negative coefficient on average; thus positive changes in education *reduce* distance's negative association. This makes sense: people with more education probably have other resources so that walking an extra distance to get water is not such a burden.

- *Interaction of arsenic and education:* a difference of 4 years of education corresponds to a difference of 0.07 in the coefficient for `arsenic`. As we have already seen, `arsenic` has a positive coefficient on average; thus increasing education *increases* arsenic's positive association. This makes sense: people with more education could be more informed about the risks of arsenic and thus more sensitive to increasing arsenic levels (or, conversely, less in a hurry to switch from wells with arsenic levels that are relatively low).

As before, centering allows us to interpret the main effects as coefficients when other inputs are held at their average values in the data.

Standardizing predictors

We should think seriously about standardizing all predictors as a default option when fitting models with interactions. The struggles with `dist100` and `educ4` in this example suggest that standardization—by subtracting the mean from each of the continuous input variables and dividing by 2 standard deviations, as suggested near the end of Section 4.2—might be the simplest approach.

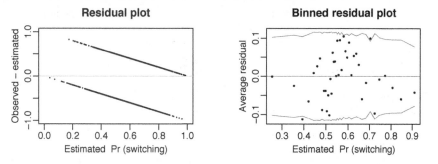

Figure 5.13 *(a) Residual plot and (b) binned residual plot for the well-switching model shown on page 96. The strong patterns in the raw residual plot arise from the discreteness of the data and inspire us to use the binned residual plot instead. The bins are not equally spaced; rather, each bin has an equal number of data points. The light lines in the binned residual plot indicate theoretical 95% error bounds.*

5.6 Evaluating, checking, and comparing fitted logistic regressions

Residuals and binned residuals

We can define residuals for logistic regression, as with linear regression, as observed minus expected values:

$$\text{residual}_i = y_i - \text{E}(y_i|X_i) = y_i - \text{logit}^{-1}(X_i\beta).$$

The data y_i are discrete and so are the residuals. For example, if $\text{logit}^{-1}(X_i\beta) = 0.7$, then $\text{residual}_i = -0.7$ or $+0.3$, depending on whether $y_i = 0$ or 1. As a result, plots of raw residuals from logistic regression are generally not useful. For example, Figure 5.13a plots residuals versus fitted values for the well-switching regression.

Instead, we plot *binned residuals* by dividing the data into categories (bins) based on their fitted values, and then plotting the average residual versus the average fitted value for each bin. The result appears in Figure 5.13b; here we divided the data into 40 bins of equal size.[6] The dotted lines (computed as $2\sqrt{p(1-p)/n}$, where n is the number of points per bin, $3020/40 = 75$ in this case) indicate ± 2 standard-error bounds, within which one would expect about 95% of the binned residuals to fall, if the model were actually true. One of the 40 binned residuals in Figure 5.13b falls outside the bounds, which is not a surprise, and no dramatic pattern appears.

Plotting binned residuals versus inputs of interest

We can also look at residuals in a more structured way by binning and plotting them with respect to individual input variables or combinations of inputs. For example, in the well-switching example, Figure 5.14a displays the average residual in each bin as defined by distance to the nearest safe well, and Figure 5.14b shows average residuals, binned by arsenic levels.

This latter plot shows a disturbing pattern, with an extreme negative residual in the first three bins: people with wells in the lowest bin (which turns out to correspond to arsenic levels between 0.51 and 0.53) are about 20% less likely to

---

[6] There is typically some arbitrariness in choosing the number of bins: we want each bin to contain enough points so that the averaged residuals are not too noisy, but it helps to have many bins so as to see more local patterns in the residuals. For this example, 40 bins seemed to give sufficient resolution, while still having enough points per bin. Another approach would be to apply a nonparametric smoothing procedure such as lowess (Cleveland, 1979) to the residuals.

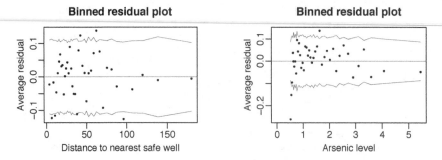

Figure 5.14 *Plots of residuals for the well-switching model, binned and plotted versus (a) distance to nearest well and (b) arsenic level. The dotted lines in the binned residual plot indicate theoretical 95% error bounds that would be appropriate if the model were true. The second plot shows a problem with the model in the lowest bins of arsenic levels.*

switch than is predicted by the model: the average predicted probability of switching for these users is 49%, but actually only 32% of them switched. There is also a slight pattern in the residuals as a whole, with positive residuals (on average) in the middle of the range of arsenic and negative residuals at the high end.

Considering a log transformation

To experienced regression modelers, a rising and then falling pattern of residuals such as in Figure 5.14b is a signal to consider taking the logarithm of the predictor on the x axis—in this case, arsenic level. Another option would be to add a quadratic term to the regression; however, since arsenic is an all-positive variable, it makes sense to consider its logarithm. We do not, however, model distance on the log scale, since the residual plot, as shown in Figure 5.13a, indicates a good fit of the linear model.

We define

R code
```
log.arsenic <- log(arsenic)
c.log.arsenic <- log.arsenic - mean (log.arsenic)
```

and then fit the same model as before, using `log.arsenic` in place of `arsenic`:

R output
```
glm(formula = switch ~ c.dist100 + c.log.arsenic + c.educ4 +
        c.dist100:c.log.arsenic + c.dist100:c.educ4 + c.log.arsenic:c.educ4,
        family = binomial(link = "logit"))
                              coef.est coef.se
(Intercept)                     0.35    0.04
c.dist100                      -0.98    0.11
c.log.arsenic                   0.90    0.07
c.educ4                         0.18    0.04
c.dist100:c.log.arsenic        -0.16    0.19
c.dist100:c.educ4               0.34    0.11
c.log.arsenic:c.educ4           0.06    0.07
    n = 3020, k = 7
    residual deviance = 3863.1, null deviance = 4118.1 (difference = 255)
```

This is qualitatively similar to the model on the original scale: the interactions have the same sign as before, and the signs of the main effects are also unchanged.

Figure 5.15a shows the predicted probability of switching as a function of arsenic level. Compared to the model in which arsenic was included as a linear predictor

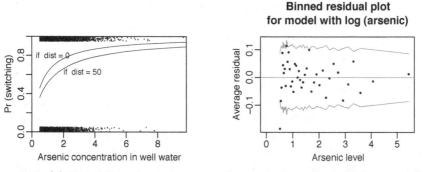

Figure 5.15 *(a) Probability of switching as a function of arsenic level (at two different values of* **dist** *and with education held constant at its average value), for the model that includes arsenic on the logarithmic scale. Compared to Figure 5.11b (the corresponding plot with arsenic level included as a linear predictor), the model looks similar, but with a steeper slope at the low end of the curve and a more gradual slope at the high end.*
(b) Average residuals for this model, binned by arsenic level. Compared to Figure 5.14b, the residual plot still shows problems at the lowest arsenic levels but otherwise looks cleaner.

(see Figure 5.11b on page 91), the curves are compressed at the left and stretched out at the right.

Figure 5.15b displays the residuals for the log model, again binned by arsenic level. Compared to the earlier model, the residuals look better but there is still a problem at the very low end. Users of wells with arsenic levels just above 0.50 are less likely to switch than predicted by the model. At this point, we do not know if this can be explained psychologically (measurements just over the threshold do not seem so bad), through measurement error (perhaps some of the wells we have recorded as 0.51 or 0.52 were measured before or after and found to have arsenic levels below 0.5), or for some other reason.

Error rate and comparison to the null model

The *error rate* is defined as the proportion of cases for which the deterministic prediction—guessing $y_i = 1$ if logit$^{-1}(X_i\beta) > 0.5$ and guessing $y_i = 0$ if logit$^{-1}(X_i\beta) < 0.5$—is wrong. In R, we could write:

```
error.rate <- mean ((predicted>0.5 & y==0) | (predicted<.5 & y==1))
```
R code

The error rate should always be less than $1/2$ (otherwise we could simply set all the β's to 0 and get a better-fitting model), but in many cases we would expect it to be much lower. We can compare it to the error rate of the *null model,* which is simply to assign the same probability to each y_i. This is simply logistic regression with only a constant term, and the estimated probability will simply be the proportion of 1's in the data, or $p = \sum_{i=1}^{n} y_i/n$ (recalling that each $y_i = 0$ or 1). The error rate of the null model is then p or $1-p$, whichever is lower.

For example, in the well-switching example, the null model has an error rate of 42% (58% of the respondents are switchers and 42% are not, thus the model with no predictors gives each person a 58% chance of switching, which corresponds to a point prediction of switching for each person, and that guess will be wrong 42% of the time). Our final logistic regression model (as calculated in R as shown) has an error rate of 36%. The model correctly predicts the behavior of 64% of the respondents.

The error rate is not a perfect summary of model misfit, because it does not distinguish between predictions of 0.6 and 0.9, for example. But, as with R^2 for the linear model, it is easy to interpret and is often helpful in understanding the model fit. An error rate equal to the null rate is terrible, and the best possible error rate is zero. Thus, the well-switching model is not particularly impressive with an error rate of 38%, a mere 4% better than simply guessing that all people will switch.

This low error rate does not mean the model is useless—as the plots showed, the fitted model is highly predictive of the probability of switching. But most of the data are close to the mean level of the inputs (distances of less than 100 meters to the nearest safe well, and arsenic levels between 0.5 and 1.0), and so for most of the data, the simple mean prediction, Pr(switch)=0.58, works well. The model is informative near the extremes, but relatively few data points are out there and so the overall predictive accuracy of the model is not high.

Deviance

For logistic regressions and other discrete-data models, it does not quite make sense to calculate residual standard deviation and R^2, for pretty much the same reason that the models are not simply fit by least squares—the squared error is not the mathematically optimal measure of model error. Instead, it is standard to use *deviance*, a statistical summary of model fit, defined for logistic regression and other generalized linear models to be an analogy to residual standard deviation.

For now, you should know the following properties of deviance:

- Deviance is a measure of error; lower deviance means better fit to data.

- If a predictor that is simply random noise is added to a model, we expect deviance to decrease by 1, on average.

- When an informative predictor is added to a model, we expect deviance to decrease by more than 1. When k predictors are added to a model, we expect deviance to decrease by more than k.

For classical (non-multilevel) models, the deviance is defined as -2 times the logarithm of the likelihood function (up to an arbitrary additive constant, since we are always comparing deviances, never evaluating them on their own).

For example, in the first model fit to the well-switching example, the display on page 88 reports that the "null deviance" is 4118.1 and the "residual deviance" is 4076.2. The null deviance corresponds to the null model, with just the constant term. Thus, by adding `dist` as a predictor in the model, the deviance has decreased by 41.9. This is much more than the expected decrease of 1 if the predictor were noise, so it has clearly improved the fit.

The next fitted model uses `dist100 = dist/100` as a predictor instead. The deviance stays at 4076.2, because linear transformations have no effect on predictions in classical regression models. (We shall see, however, that linear transformations can make a difference in multilevel models.)

We then add the `arsenic` predictor, and the deviance decreases to 3930.7, a drop of 145.5—once again, much more than the expected decrease of 1 if the new predictor were noise, so it has clearly improved the fit.

The following model including the interaction between `dist` and `arsenic` has a residual deviance of 3927.6, a decrease of 3.1 from the previous model, only a bit more than the expected decrease of 1 if the new predictor were noise. This decrease in deviance is not statistically significant (we can see this because the coefficient for the added predictor is less than 2 standard errors from zero) but, as discussed in Section 5.5, we keep the interaction in the model because it makes sense in the applied context.

Adding the social predictors `assoc` and `educ` to the regression decreases the deviance to 3905.4, implying better prediction than all the previous models. Removing `assoc` increases the deviance only a small amount, to 3907.9. Adding interactions of education with distance and arsenic level reduces the deviance by quite a bit more, to 3891.7.

Transforming arsenic on to the log scale—that is, removing `arsenic` from the model and replacing it with `log.arsenic`, takes the deviance down to 3863.1, another large improvement.

For multilevel models, deviance is generalized to the deviance information criterion (DIC), as described in Section 24.3.

5.7 Average predictive comparisons on the probability scale

As illustrated, for example, by Figure 5.11 on page 91, logistic regressions are nonlinear on the probability scale—that is, a specified difference in one of the x variables does *not* correspond to a constant difference in $\Pr(y = 1)$. As a result, logistic regression coefficients cannot directly be interpreted on the scale of the data. Logistic regressions are inherently more difficult than linear regressions to interpret.

Graphs such as Figure 5.11 are useful, but for models with many predictors, or where graphing is inconvenient, it is helpful to have a summary, comparable to the linear regression coefficient, which gives the expected, or average, difference in $\Pr(y=1)$ corresponding to a unit difference in each of the input variables.

Example: well switching in Bangladesh

For a model with nonlinearity or interactions, or both, this *average predictive comparison* depends on the values of the input variables, as we shall illustrate with the well-switching example. To keep the presentation clean at this point, we shall work with a simple no-interaction model,

```
fit.10 <- glm (switch ~ dist100 + arsenic + educ4,
    family=binomial(link="logit"))
```

R code

which yields

```
            coef.est coef.se
(Intercept)   -0.21    0.09
dist100       -0.90    0.10
arsenic        0.47    0.04
educ4          0.17    0.04
  n = 3020, k = 4
  residual deviance = 3910.4, null deviance = 4118.1 (difference = 207.7)
```

R output

giving the probability of switching as a function of distance to the nearest well (in 100-meter units), arsenic level, and education (in 4-year units).

Average predictive difference in probability of switching, comparing households that are next to, or 100 meters from, the nearest safe well. Let us compare two households—one with `dist100` $= 0$ and one with `dist100` $= 1$—but identical in the other input variables, `arsenic` and `educ4`. The *predictive difference* in probability of switching between these two households is

$$\delta(\text{arsenic}, \text{educ4}) = \text{logit}^{-1}(-0.21 - 0.90 \cdot 1 + 0.47 \cdot \text{arsenic} + 0.17 \cdot \text{educ4}) -$$
$$\text{logit}^{-1}(-0.21 - 0.90 \cdot 0 + 0.47 \cdot \text{arsenic} + 0.17 \cdot \text{educ4}). \quad (5.6)$$

We write δ as a function of `arsenic` and `educ4` to emphasize that it depends on the levels of these other variables.

We average the predictive differences over the n households in the data to obtain:

$$\text{average predictive difference:} \ = \frac{1}{n} \sum_{i=1}^{n} \delta(\text{arsenic}_i, \text{educ4}_i). \quad (5.7)$$

In R:

R code
```
b <- coef (fit.10)
hi <- 1
lo <- 0
delta <- invlogit (b[1] + b[2]*hi + b[3]*arsenic + b[4]*educ4) -
         invlogit (b[1] + b[2]*lo + b[3]*arsenic + b[4]*educ4)
print (mean(delta))
```

The result is -0.20, implying that, on average in the data, households that are 100 meters from the nearest safe well are 20% less likely to switch, compared to househoulds that are right next to the nearest safe well, at the same arsenic and education levels.

Average predictive difference in probability of switching, comparing households with existing arsenic levels of 0.5 and 1.0. We can similarly compute the predictive difference, and average predictive difference, comparing households at two different arsenic levels, assuming equality in distance to nearest safe well and education levels. We choose `arsenic` $= 0.5$ and 1.0 as comparison points because 0.5 is the lowest unsafe level, 1.0 is twice that, and this comparison captures much of the range of the data (see Figure 5.10 on page 90). Here is the computation:

R code
```
hi <- 1.0
lo <- 0.5
delta <- invlogit (b[1] + b[2]*dist100 + b[3]*hi + b[4]*educ4) -
         invlogit (b[1] + b[2]*dist100 + b[3]*lo + b[4]*educ4)
print (mean(delta))
```

The result is 0.06—so this comparison corresponds to a 6% difference in probability of switching.

Average predictive difference in probability of switching, comparing householders with 0 and 12 years of education. Similarly, we can compute an average predictive difference of the probability of switching for householders with 0 compared to 12 years of education (that is, comparing `educ4` $= 0$ to `educ4` $= 3$):

R code
```
hi <- 3
lo <- 0
delta <- invlogit (b[1]+b[2]*dist100+b[3]*arsenic+b[4]*hi) -
         invlogit (b[1]+b[2]*dist100+b[3]*arsenic+b[4]*lo)
print (mean(delta))
```

which comes to 0.12.

Average predictive comparisons in the presence of interactions

We can perform similar calculations for models with interactions. For example, consider the average predictive difference, comparing `dist` $= 0$ to `dist` $= 100$, for the model that includes a distance $\times$ arsenic interaction:

```
fit.11 <- glm (switch ~ dist100 + arsenic + educ4 + dist100:arsenic,
    family=binomial(link="logit"))
```
R code

which, when displayed, yields

```
                coef.est coef.se
(Intercept)       -0.35    0.13
dist100           -0.60    0.21
arsenic            0.56    0.07
educ4              0.17    0.04
dist100:arsenic   -0.16    0.10
  n = 3020, k = 5
  residual deviance = 3907.9, null deviance = 4118.1 (difference = 210.2)
```
R output

Here is the R code for computing the average predictive difference comparing dist1 $= 1$ to dist1 $= 0$:

```
b <- coef (fit.11)
hi <- 1
lo <- 0
delta <- invlogit (b[1] + b[2]*hi + b[3]*arsenic + b[4]*educ4 +
                b[5]*hi*arsenic) -
        invlogit (b[1] + b[2]*lo + b[3]*arsenic + b[4]*educ4 +
                b[5]*lo*arsenic)
print (mean(delta))
```
R code

which comes to -0.19.

General notation for predictive comparisons

Considering each input one at a time, we use the notation u for the *input of interest* and v for the vector of all other inputs. Suppose we are considering comparisons of $u = u^{(1)}$ to $u = u^{(0)}$ with all other inputs held constant (for example, we have considered the comparison of households that are 0 meters or 100 meters from the nearest safe well). The *predictive difference* in probabilities between two cases, differing only in u, is

$$\delta(u^{(\text{hi})}, u^{(\text{lo})}, v, \beta) \ = \ \Pr(y{=}1|u^{(\text{hi})}, v, \beta) - \Pr(y{=}1|u^{(\text{lo})}, v, \beta), \qquad (5.8)$$

where the vertical bar in these expressions is read "conditional on" (for example, the probability that $y = 1$ given $u^{(\text{hi})}$, v, and β).

The average predictive difference then averages over the n points in the dataset used to fit the logistic regression:

$$\Delta(u^{(\text{hi})}, u^{(\text{lo})}) = \frac{1}{n} \sum_{i=1}^{n} \delta(u^{(\text{hi})}, u^{(\text{lo})}, v_i, \beta), \qquad (5.9)$$

where v_i represents the vector of other inputs (in our example, arsenic and education levels) for data point i. These expressions generalize formulas (5.6) and (5.7).

For models with interactions, the predictive difference formula (5.8) must be computed carefully, with awareness of where each input enters into the regression model. The distinction between input variables (in this case, distance, arsenic, and

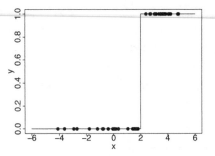

Figure 5.16 *Example of data for which a logistic regression model is nonidentifiable. The outcome y equals 0 for all data below $x = 2$ and 1 for all data above $x = 2$, hence the best-fit logistic regression line is $y = logit^{-1}(\infty(x - 2))$, which has an infinite slope at $x = 2$.*

education) and predictors (constant term, distance, arsenic, education, and distance × arsenic) is crucial. We discuss average predictive comparisons further in Section 21.4.

5.8 Identifiability and separation

There are two reasons that a logistic regression can be nonidentified (that is, have parameters that cannot be estimated from the available data and model, as discussed in Section 4.5 in the context of linear regression):

1. As with linear regression, if predictors are collinear, then estimation of the linear predictor, $X\beta$, does not allow separate estimation of the individual parameters β. We can handle this kind of nonidentifiability in the same way that we would proceed for linear regression, as described in Section 4.5.

2. A completely separate identifiability problem, called *separation*, can arise from the discreteness of the data.

   • If a predictor x_j is completely aligned with the outcome, so that $y = 1$ for all the cases where x_j exceeds some threshold T, and $y = 0$ for all cases where $x_j < T$, then the best estimate for the coefficient β_j is ∞. Figure 5.16 shows an example. Exercise 5.11 gives an example with a binary predictor.

   • Conversely, if $y = 1$ for all cases where $x_j < T$, and $y = 0$ for all cases where $x_j > T$, then $\hat{\beta}_j$ will be $-\infty$.

   • More generally, this problem will occur if any linear combination of predictors is perfectly aligned with the outcome. For example, suppose that $7x_1 + x_2 - 3x_3$ is completely positively aligned with the data, with $y = 1$ if and only if this linear combination of predictors exceeds some threshold. Then the linear combination $7\hat{\beta}_1 + \hat{\beta}_2 - 3\hat{\beta}_3$ will be estimated at ∞, which will cause at least one of the three coefficients $\beta_1, \beta_2, \beta_3$ to be estimated at ∞ or $-\infty$.

One way to handle separation is using a Bayesian or penalized-likelihood approach (implemented for R in our `bayesglm()` function in the `arm` package and also with the `brlr()` function in the `brlr` package) that provides a small amount of information on all the regression coefficients, including those that are not identified from the data alone. (See Chapter 18 for more on Bayesian inference.)

5.9 Bibliographic note

According to Cramer (2003, chapter 9), logistic regression was introduced for binary data in the mid-twentieth century and has become increasingly popular as computational improvements have allowed it to become a routine data-analytic tool.

For more on income and voting in presidential elections, see Gelman, Shor, et al. (2005). The example of drinking water in Bangladesh is described further by van Geen et al. (2003) and Gelman, Trevisani, et al. (2004).

Binned residual plots and related tools for checking the fit of logistic regressions are discussed by Landwehr, Pregibon, and Shoemaker (1984), Gelman, Goegebeur, et al. (2000), Pardoe and Cook (2002), and Pardoe (2004).

Deviance is discussed by McCullagh and Nelder (1989); related ideas include the Akaike (1973) information criterion (AIC), C_p (Mallows, 1973), and the deviance information criterion (DIC; Spiegelhalter et al., 2002). See also Fox (2002) for an applied overview and Gelman et al. (2003, sections 6.7–6.8) for a Bayesian perspective.

Nonidentifiability of logistic regression and separation in discrete data are discussed by Albert and Anderson (1984), Lesaffre and Albert (1989), Heinze and Schemper (2003), as well as in the book by Agresti (2002). Firth (1993), Zorn (2005), and Gelman et al. (2007) present Bayesian solutions.

5.10 Exercises

1. The folder **nes** contains the survey data of presidential preference and income for the 1992 election analyzed in Section 5.1, along with other variables including sex, ethnicity, education, party identification, and political ideology.

    (a) Fit a logistic regression predicting support for Bush given all these inputs. Consider how to include these as regression predictors and also consider possible interactions.

    (b) Evaluate and compare the different models you have fit. Consider coefficient estimates and standard errors, residual plots, and deviances.

    (c) For your chosen model, discuss and compare the importance of each input variable in the prediction.

2. Without using a computer, sketch the following logistic regression lines:

    (a) $\Pr(y = 1) = \text{logit}^{-1}(x)$

    (b) $\Pr(y = 1) = \text{logit}^{-1}(2 + x)$

    (c) $\Pr(y = 1) = \text{logit}^{-1}(2x)$

    (d) $\Pr(y = 1) = \text{logit}^{-1}(2 + 2x)$

    (e) $\Pr(y = 1) = \text{logit}^{-1}(-2x)$

3. You are interested in how well the combined earnings of the parents in a child's family predicts high school graduation. You are told that the probability a child graduates from high school is 27% for children whose parents earn no income and is 88% for children whose parents earn $60,000. Determine the logistic regression model that is consistent with this information. (For simplicity you may want to assume that income is measured in units of $10,000).

4. Perform a logistic regression for a problem of interest to you. This can be from a research project, a previous class, or data you download. Choose one variable

of interest to be the outcome, which will take on the values 0 and 1 (since you are doing logistic regression).

(a) Analyze the data in R. Use the `display()` function to summarize the results.

(b) Fit several different versions of your model. Try including different predictors, interactions, and transformations of the inputs.

(c) Choose one particular formulation of the model and do the following:

    i. Describe how each input affects $\Pr(y=1)$ in the fitted model. You must consider the estimated coefficient, the range of the input values, and the nonlinear inverse-logit function.

    ii. What is the error rate of the fitted model? What is the error rate of the null model?

    iii. Look at the deviance of the fitted and null models. Does the improvement in fit seem to be real?

    iv. Use the model to make predictions for some test cases of interest.

5. In a class of 50 students, a logistic regression is performed of course grade (pass or fail) on midterm exam score (continuous values with mean 60 and standard deviation 15). The fitted model is $\Pr(\text{pass}) = \text{logit}^{-1}(-24 + 0.4x)$.

(a) Graph the fitted model. Also on this graph put a scatterplot of hypothetical data consistent with the information given.

(b) Suppose the midterm scores were transformed to have a mean of 0 and standard deviation of 1. What would be the equation of the logistic regression using these transformed scores as a predictor?

(c) Create a new predictor that is pure noise (for example, in R you can create `newpred <- rnorm (n,0,1)`). Add it to your model. How much does the deviance decrease?

6. Latent-data formulation of the logistic model: take the model $\Pr(y = 1) = \text{logit}^{-1}(1 + 2x_1 + 3x_2)$ and consider a person for whom $x_1 = 1$ and $x_2 = 0.5$. Sketch the distribution of the latent data for this person. Figure out the probability that $y=1$ for the person and shade the corresponding area on your graph.

7. Limitations of logistic regression: consider a dataset with $n = 20$ points, a single predictor x that takes on the values $1, \ldots, 20$, and binary data y. Construct data values $y_1, \ldots, y_{20}$ that are inconsistent with any logistic regression on x. Fit a logistic regression to these data, plot the data and fitted curve, and explain why you can say that the model does not fit the data.

8. Building a logistic regression model: the folder `rodents` contains data on rodents in a sample of New York City apartments.

(a) Build a logistic regression model to predict the presence of rodents (the variable `rodent2` in the dataset) given indicators for the ethnic groups (`race`). Combine categories as appropriate. Discuss the estimated coefficients in the model.

(b) Add to your model some other potentially relevant predictors describing the apartment, building, and community district. Build your model using the general principles explained in Section 4.6. Discuss the coefficients for the ethnicity indicators in your model.

9. Graphing logistic regressions: the well-switching data described in Section 5.4 are in the folder `arsenic`.

(a) Fit a logistic regression for the probability of switching using log (distance to nearest safe well) as a predictor.

(b) Make a graph similar to Figure 5.9 displaying Pr(switch) as a function of distance to nearest safe well, along with the data.

(c) Make a residual plot and binned residual plot as in Figure 5.13.

(d) Compute the error rate of the fitted model and compare to the error rate of the null model.

(e) Create indicator variables corresponding to dist < 100, 100 ≤ dist < 200, and dist > 200. Fit a logistic regression for Pr(switch) using these indicators. With this new model, repeat the computations and graphs for part (a) of this exercise.

10. Model building and comparison: continue with the well-switching data described in the previous exercise.

(a) Fit a logistic regression for the probability of switching using, as predictors, distance, log(arsenic), and their interaction. Interpret the estimated coefficients and their standard errors.

(b) Make graphs as in Figure 5.12 to show the relation between probability of switching, distance, and arsenic level.

(c) Following the procedure described in Section 5.7, compute the average predictive differences corresponding to:

  i. A comparison of dist = 0 to dist = 100, with arsenic held constant.
  ii. A comparison of dist = 100 to dist = 200, with arsenic held constant.
  iii. A comparison of arsenic = 0.5 to arsenic = 1.0, with dist held constant.
  iv. A comparison of arsenic = 1.0 to arsenic = 2.0, with dist held constant.

  Discuss these results.

11. Identifiability: the folder nes has data from the National Election Studies that were used in Section 5.1 to model vote preferences given income. When we try to fit a similar model using ethnicity as a predictor, we run into a problem. Here are fits from 1960, 1964, 1968, and 1972:

```
glm(formula = vote ~ female + black + income,                         R output
    family=binomial(link="logit"), subset=(year==1960))
            coef.est coef.se
(Intercept) -0.14     0.23
female       0.24     0.14
black       -1.03     0.36
income       0.03     0.06

glm(formula = vote ~ female + black + income,
    family=binomial(link="logit"), subset=(year==1964))
            coef.est coef.se
(Intercept) -1.15     0.22
female      -0.09     0.14
black      -16.83   420.40
income       0.19     0.06

glm(formula = vote ~ female + black + income,
    family=binomial(link="logit"), subset=(year==1968))
            coef.est coef.se
(Intercept)  0.47     0.24
```

```
female      -0.01     0.15
black       -3.64     0.59
income      -0.03     0.07

glm(formula = vote ~ female + black + income,
   family=binomial(link="logit"), subset=(year==1972))
                 coef.est coef.se
(Intercept)  0.67      0.18
female      -0.25      0.12
black       -2.63      0.27
income       0.09      0.05
```

What happened with the coefficient of black in 1964? Take a look at the data and figure out where this extreme estimate came from. What can be done to fit the model in 1964?

## CHAPTER 6

Generalized linear models

6.1 Introduction

Generalized linear modeling is a framework for statistical analysis that includes linear and logistic regression as special cases. Linear regression directly predicts continuous data y from a *linear predictor* $X\beta = \beta_0 + X_1\beta_1 + \cdots + X_k\beta_k$. Logistic regression predicts $\Pr(y = 1)$ for binary data from a linear predictor with an inverse-logit transformation. A generalized linear model involves:

1. A data vector $y = (y_1, \ldots, y_n)$

2. Predictors X and coefficients β, forming a linear predictor $X\beta$

3. A *link function* g, yielding a vector of transformed data $\hat{y} = g^{-1}(X\beta)$ that are used to model the data

4. A data distribution, $p(y|\hat{y})$

5. Possibly other parameters, such as variances, overdispersions, and cutpoints, involved in the predictors, link function, and data distribution.

The options in a generalized linear model are the transformation g and the data distribution p.

- In *linear regression*, the transformation is the identity (that is, $g(u) \equiv u$) and the data distribution is normal, with standard deviation σ estimated from data.

- In *logistic regression*, the transformation is the inverse-logit, $g^{-1}(u) = \text{logit}^{-1}(u)$ (see Figure 5.2a on page 80) and the data distribution is defined by the probability for binary data: $\Pr(y\,{=}\,1) = \hat{y}$.

This chapter discusses several other classes of generalized linear model, which we list here for convenience:

- The *Poisson* model (Section 6.2) is used for count data; that is, where each data point y_i can equal 0, 1, 2, ... The usual transformation g used here is the logarithmic, so that $g(u) = \exp(u)$ transforms a continuous linear predictor $X_i\beta$ to a positive $\hat{y}_i$. The data distribution is Poisson.

  It is usually a good idea to add a parameter to this model to capture *overdispersion*, that is, variation in the data beyond what would be predicted from the Poisson distribution alone.

- The *logistic-binomial* model (Section 6.3) is used in settings where each data point y_i represents the number of successes in some number n_i of tries. (This n_i, the number of tries for data point i, is not the same as n, the number of data points.) In this model, the transformation g is the inverse-logit and the data distribution is binomial.

  As with Poisson regression, the binomial model is typically improved by the inclusion of an overdispersion parameter.

- The *probit* model (Section 6.4) is the same as logistic regression but with the logit function replaced by the normal cumulative distribution, or equivalently with the normal distribution instead of the logistic in the latent-data errors.

- *Multinomial* logit and probit models (Section 6.5) are extensions of logistic and probit regressions for categorical data with more than two options, for example survey responses such as Strongly Agree, Agree, Indifferent, Disagree, Strongly Disagree. These models use the logit or probit transformation and the multinomial distribution and require additional parameters to model the multiple possibilities of the data.

  Multinomial models are further classified as *ordered* (for example, Strongly Agree, ..., Strongly Disagree) or *unordered* (for example, Vanilla, Chocolate, Strawberry, Other).

- *Robust* regression models (Section 6.6) replace the usual normal or logistic models by other distributions[1] (usually the so-called Student-t family of models) that allow occasional extreme values.

This chapter briefly goes through many of these models, with an example of overdispersed Poisson regression in Section 6.2 and an ordered logistic example in Section 6.5. Finally, in Section 6.8 we discuss the connections between generalized linear models and behavioral models of choice that are used in psychology and economics, using as an example the logistic regression for well switching in Bangladesh. The chapter is not intended to be a comprehensive overview of generalized linear models; rather, we want to give a sense of the variety of regression models that can be appropriate for different data structures that we have seen in applications.

Fitting generalized linear models in R

Because of the variety of options involved, generalized linear modeling can be more complicated than linear and logistic regressions. The starting point in R is the `glm()` function, which we have already used extensively for logistic regression in Chapter 5 and is a generalization of the linear-modeling function `lm()`. We can use `glm()` directly to fit logistic-binomial, probit, and Poisson regressions, among others, and to correct for overdispersion where appropriate, with `bayesglm()` available when there is separation or the model is otherwise nonidentified. Ordered logit and probit regressions can be fit using the `polr()` and `bayespolr()` functions, unordered probit models can be fit using the `mnp` package, and t models can be fit using the `hett` package in R. (See Appendix C for information on these and other R packages.) Beyond this, most of these models and various generalizations can be fit in Bugs, as we discuss in Part 2B of this book in the context of multilevel modeling.

6.2 Poisson regression, exposure, and overdispersion

The Poisson distribution is used to model variation in count data (that is, data that can equal $0, 1, 2, \ldots$). After a brief introduction, we illustrate in detail with the example of New York City police stops that we introduced in Section 1.2.

Traffic accidents

In the Poisson model, each unit i corresponds to a setting (typically a spatial location or a time interval) in which y_i events are observed. For example, i could

---

[1] In the statistical literature, generalized linear models have been defined using exponential-family models, a particular class of data distributions that excludes, for example, the t distribution. For our purposes, however, we use the term "generalized linear model" to apply to any model with a linear predictor, link function, and data distribution, not restricting to exponential-family models.

index street intersections in a city and y_i could be the number of traffic accidents at intersection i in a given year.

As with linear and logistic regression, the variation in y can be explained with linear predictors X. In the traffic accidents example, these predictors could include: a constant term, a measure of the average speed of traffic near the intersection, and an indicator for whether the intersection has a traffic signal. The basic Poisson regression model has the form

$$y_i \sim \text{Poisson}(\theta_i). \tag{6.1}$$

The parameter θ_i must be positive, so it makes sense to fit a linear regression on the logarithmic scale:

$$\theta_i = \exp(X_i\beta). \tag{6.2}$$

Interpreting Poisson regression coefficients

The coefficients β can be exponentiated and treated as multiplicative effects. For example, suppose the traffic accident model is

$$y_i \sim \text{Poisson}(\exp(2.8 + 0.012X_{i1} - 0.20X_{i2})),$$

where X_{i1} is average speed (in miles per hour, or mph) on the nearby streets and $X_{i2} = 1$ if the intersection has a traffic signal or 0 otherwise. We can then interpret each coefficient as follows:

- The constant term gives the intercept of the regression, that is, the prediction if $X_{i1} = 0$ and $X_{i2} = 0$. Since this is not possible (no street will have an average speed of 0), we will not try to interpret the constant term.
- The coefficient of X_{i1} is the expected difference in y (on the logarithmic scale) for each additional mph of traffic speed. Thus, the expected multiplicative increase is $e^{0.012} = 1.012$, or a 1.2% positive difference in the rate of traffic accidents per mph. Since traffic speeds vary by tens of mph, it would actually make sense to define X_{i1} as speed in tens of mph, in which case its coefficient would be 0.12, corresponding to a 12% increase (more precisely, $e^{0.12} = 1.127$: a 12.7% increase) in accident rate per ten mph.
- The coefficient of X_{i2} tells us that the predictive difference of having a traffic signal can be found be multiplying the accident rate by $\exp(-0.20) = 0.82$ yielding a reduction of 18%.

As with regression models in general, each coefficient is interpreted as a comparison in which one predictor differs by one unit while all the other predictors remain at the same level, which is not necessarily the most appropriate assumption when extending the model to new settings. For example, installing traffic signals in all the intersections in the city would *not* necessarily be expected to reduce accidents by 18%.

Poisson regression with an exposure input

In most applications of Poisson regression, the counts can be interpreted relative to some baseline or "exposure," for example, the number of vehicles that travel through the intersection. In the general Poisson regression model, we think of y_i as the number of cases in a process with rate θ_i and exposure u_i.

$$y_i \sim \text{Poisson}(u_i\theta_i), \tag{6.3}$$

where, as before, $\theta_i = \exp(X_i\beta)$. The logarithm of the exposure, $\log(u_i)$, is called the *offset* in generalized linear model terminology.

The regression coefficients β summarize the associations between the predictors and θ_i (in our example, the rate of traffic accidents per vehicle).

Including log(exposure) as a predictor in the Poisson regression. Putting the logarithm of the exposure into the model as an offset, as in model (6.3), is equivalent to including it as a regression predictor, but with its coefficient fixed to the value 1. Another option is to include it as a predictor and let its coefficient be estimated from the data. In some settings, this makes sense in that it can allow the data to be fit better; in other settings, it is simpler to just keep it as an offset so that the estimated rate θ has a more direct interpretation.

Differences between the binomial and Poisson models

The Poisson model is similar to the binomial model for count data (see Section 6.3) but is applied in slightly different situations:

- If each data point y_i can be interpreted as the number of "successes" out of n_i trials, then it is standard to use the binomial/logistic model (as described in Section 6.3) or its overdispersed generalization.

- If each data point y_i does not have a natural limit—it is not based on a number of independent trials—then it is standard to use the Poisson/logarithmic regression model (as described here) or its overdispersed generalization.

Example: police stops by ethnic group

For the analysis of police stops:

- The units i are precincts and ethnic groups ($i = 1, \ldots, n = 3 \times 75$).

- The outcome y_i is the number of stops of members of that ethnic group in that precinct.

- The exposure u_i is the number of arrests by people of that ethnic group in that precinct in the previous year as recorded by the Department of Criminal Justice Services (DCJS).

- The inputs are the precinct and ethnicity indexes.

- The predictors are a constant, 74 precinct indicators (for example, precincts 2–75, with precinct 1 as the baseline), and 2 ethnicity indicators (for example, for hispanics and whites, with blacks as the baseline).

We illustrate model fitting in three steps. First, we fit a model with the offset and a constant term alone:

R output
```
glm(formula = stops ~ 1, family=poisson, offset=log(arrests))
            coef.est coef.se
(Intercept)    -3.4     0.0
  n = 225, k = 1
  residual deviance = 44877, null deviance = 44877 (difference = 0)
```

Next, we add ethnicity indicators:

R output
```
glm(formula = stops ~ factor(eth), family=poisson,
      offset=log(arrests))
            coef.est coef.se
(Intercept)   -3.30    0.00
```

```
factor(eth)2        0.06      0.01
factor(eth)3       -0.18      0.01
   n = 225, k = 3
   residual deviance = 44133, null deviance = 44877 (difference = 744.1)
```

The two ethnicity coefficients are highly statistically significant, and the deviance has decreased by 744, much more than the 2 that would be expected if ethnicity had no explanatory power in the model. Compared to the baseline category 1 (blacks), we see that category 2 (hispanics) has 6% more stops, and category 3 (whites) has 18% fewer stops, in proportion to DCJS arrest rates.

Now we add the 75 precincts:

```
glm(formula = stops ~ factor(eth) + factor(precinct), family=poisson,          R output
    offset=log(arrests))
                        coef.est coef.se
(Intercept)               -4.03    0.05
factor(eth)2               0.00    0.01
factor(eth)3              -0.42    0.01
factor(precinct)2         -0.06    0.07
factor(precinct)3          0.54    0.06
    . . .
factor(precinct)75         1.41    0.08
   n = 225, k = 77
   residual deviance = 2828.6, null deviance = 44877 (difference = 42048.4)
   overdispersion parameter = 18.2
```

The decrease in the deviance, from 44,000 to 2800, is huge—much larger than the decrease of 74 that would be expected if the precinct factor were random noise. After controlling for precincts, the ethnicity coefficients have changed a bit—blacks and hispanics (categories 1 and 2) have approximately the same rate of being stopped, and whites (category 3) have about a 42% lower chance than minorities of being stopped—all in comparison to the DCJS arrest rates, which are used as a baseline.[2]

Thus, controlling for precinct actually increases the difference between whites and minorities in the rate of stops. We explore this issue further in Section 15.1.

We can also look at the precinct coefficients in the regression—for example, the stop rates (per DCJS arrest) after controlling for ethnicity, are approximately 6% lower in precinct 2, $\exp(0.54) = 1.72$ times as high in precinct 3, ..., and $\exp(1.41) = 4.09$ times as high in precinct 75, as compared to the baseline precinct 1.

The exposure input

In this example, stops by police are compared to the number of arrests in the previous year, so that the coefficient for the "hispanic" or "white" indicator will be greater than 1 if the people in that group are stopped disproportionately to their rates of arrest, as compared to blacks. Similarly, the coefficients for the indicators for precincts 2–75 will exceed 1 for those precincts where stops are more frequent than in precinct 1, as compared to their arrest rates in the previous year.

In Section 15.1 we shall consider another possible analysis that uses population, rather than previous year's arrests, as the exposure.

---

[2] More precisely, the exponentiated coefficient for whites is $\exp(-0.42) = 0.66$, so their chance of being stopped is actually 34% lower—the approximation $\exp(-\beta) \approx 1 - \beta$ is accurate only when β is close to 0.

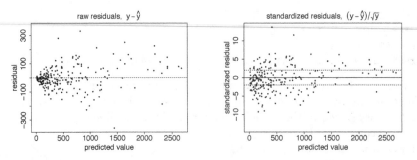

Figure 6.1 *Testing for overdispersion in a Poisson regression model: (a) residuals versus predicted values, (b) standardized residuals versus predicted values. As expected from the model, the variance of the residuals increases as predicted values increase. The standardized residuals should have mean 0 and standard deviation 1 (hence the lines at ±2 indicating approximate 95% error bounds). The variance of the standardized residuals is much greater than 1, indicating a large amount of overdispersion.*

Overdispersion

Poisson regressions do not supply an independent variance parameter σ, and as a result can be overdispersed and usually are, a point we considered briefly on page 21 and pursue further here in a regression context. Under the Poisson distribution the variance equals the mean—that is, the standard deviation equals the square root of the mean. In the model (6.3), $E(y_i) = u_i\theta_i$ and $sd(y_i) = \sqrt{u_i\theta_i}$. We define the standardized residuals:

$$
\begin{aligned}
z_i &= \frac{y_i - \hat{y}_i}{sd(\hat{y}_i)} \\
&= \frac{y_i - u_i\hat{\theta}_i}{\sqrt{u_i\hat{\theta}_i}},
\end{aligned}
\tag{6.4}
$$

where $\hat{\theta}_i = e^{X_i\hat{\theta}}$. If the Poisson model is true, then the z_i's should be approximately independent (not exactly independent, since the same estimate $\hat{\beta}$ is used in computing all of them), each with mean 0 and standard deviation 1. If there is overdispersion, however, we would expect the z_i's to be larger, in absolute value, reflecting the extra variation beyond what is predicted under the Poisson model.

We can test for overdispersion in classical Poisson regression by computing the sum of squares of the n standardized residuals, $\sum_{i=1}^{n} z_i^2$, and comparing this to the χ_{n-k}^2 distribution, which is what we would expect under the model (using $n-k$ rather than n degrees of freedom to account for the estimation of k regression coefficients). The χ_{n-k}^2 distribution has average value $n-k$, and so the ratio,

$$
\text{estimated overdispersion} = \frac{1}{n-k} \sum_{i=1}^{n} z_i^2,
\tag{6.5}
$$

is a summary of the overdispersion in the data compared to the fitted model.

For example, the classical Poisson regression for the police stops has $n = 225$ data points and $k = 77$ linear predictors. Figure 6.1 plots the residuals $y_i - \hat{y}_i$ and standardized residuals $z_i = (y_i - \hat{y}_i)/sd(\hat{y}_i)$, as a function of predicted values from the Poisson regression model. As expected from the Poisson model, the variance of the residuals increases as the predicted values increase, and the variance of the standardized residuals is approximately constant. However, the standardized residuals have a variance much greater than 1, indicating serious overdispersion.

To program the overdispersion test in R:

```
yhat <- predict (glm.police, type="response")
z <- (stops-yhat)/sqrt(yhat)
cat ("overdispersion ratio is ", sum(z^2)/(n-k), "\n")
cat ("p-value of overdispersion test is ", pchisq (sum(z^2), n-k), "\n")
```
R code

The sum of squared standardized residuals is $\sum_{i=1}^{n} z_i^2 = 2700$, compared to an expected value of $n-k = 148$. The estimated overdispersion factor is $2700/148 = 18.2$, and the p-value is 1, indicating that the probability is essentially zero that a random variable from a χ_{148}^2 distribution would be as large as 2700. In summary, the police stops data are overdispersed by a factor of 18, which is huge—even an overdispersion factor of 2 would be considered large—and also statistically significant.

Adjusting inferences for overdispersion

In this example, the basic correction for overdispersion is to multiply all regression standard errors by $\sqrt{18.2} = 4.3$. Luckily, it turns out that our main inferences are not seriously affected. The parameter of primary interest is α_3—the log of the rate of stops for whites compared to blacks—which is estimated at -0.42 ± 0.01 before (see the regression display on page 113) and now becomes -0.42 ± 0.04. Transforming back to the original scale, whites are stopped at an estimated 66% of the rate of blacks, with an approximate 50% interval of $e^{-0.42 \pm (2/3)0.04} = [0.64, 0.67]$ and an approximate 95% interval of $e^{-0.42 \pm 2 \cdot 0.04} = [0.61, 0.71]$.

Fitting the overdispersed-Poisson or negative-binomial model

More simply, we can fit an overdispersed model using the `quasipoisson` family:

```
glm(formula = stops ~ factor(eth) + factor(precinct), family=quasipoisson,
        offset=log(arrests))
                      coef.est coef.se
(Intercept)            -4.03    0.21
factor(eth)2            0.00    0.03
factor(eth)3          -0.42    0.04
factor(precinct)2    -0.06    0.30
factor(precinct)3     0.54    0.24
 . . .
factor(precinct)75    1.41    0.33
  n = 225, k = 77
  residual deviance = 2828.6, null deviance = 44877 (difference = 42048.4)
  overdispersion parameter = 18.2
```
R output

We write this model as

$$y_i \sim \text{overdispersed Poisson} \left(u_i \exp(X_i\beta), \omega\right),$$

where ω is the overdispersion parameter (estimated at 18.2 in this case). Strictly speaking, "overdispersed Poisson" is not a single model but rather describes any count-data model for which the variance of the data is ω times the mean, reducing to the Poisson if $\omega = 1$.

A specific model commonly used in this scenario is the so-called negative-binomial distribution:

$$y_i \sim \text{Negative-binomial} \left(\text{mean} = u_i \exp(X_i\beta), \text{overdispersion} = \omega\right).$$

Unfortunately, the negative-binomial distribution is conventionally expressed not based on its mean and overdispersion but rather in terms of parameters a and b, where the mean of the distribution is a/b and the overdispersion is $1 + 1/b$. One must check the parameterization when fitting such models, and it can be helpful to double-check by simulating datasets from the fitted model and checking that they look like the actual data (see Section 8.3).

We return to the police stops example, correcting for overdispersion using a multilevel model, in Section 15.1.

6.3 Logistic-binomial model

Chapter 5 discussed logistic regression for binary (Yes/No or 0/1) data. The logistic model can also be used for count data, using the binomial distribution (see page 16) to model the number of "successess" out of a specified number of possibilities, with the probability of success being fit to a logistic regression.

The binomial model for count data, applied to death sentences

We illustrate binomial logistic regression in the context of a study of the proportion of death penalty verdicts that were overturned, in each of 34 states in the 23 years, 1973–1995. The units of this analysis are the $34 \times 23 = 784$ state-years (actually, we only have $n = 450$ state-years in our analysis, since different states have restarted the death penalty at different times since 1973). For each state-year i, we label n_i as the number of death sentences in that state in that year and y_i as the number of these verdicts that were later overturned by higher courts. Our model has the form

$$
\begin{aligned}
y_i &\sim \text{Binomial}(n_i, p_i) \\
p_i &= \text{logit}^{-1}(X_i \beta),
\end{aligned}
\tag{6.6}
$$

where X is a matrix of predictors. To start, we use

- A constant term

- 33 indicators for states

- A time trend for years (that is, a variable that equals 1 for 1973, 2 for 1974, 3 for 1975, and so on).

This model could also be written as

$$
\begin{aligned}
y_{st} &\sim \text{Binomial}(n_{st}, p_{st}) \\
p_{st} &= \text{logit}^{-1}(\mu + \alpha_s + \beta t),
\end{aligned}
$$

with subscripts s for state and t for time (that is, year-1972). We prefer the form (6.6) because of its greater generality. But it is useful to be able to go back and forth between the two formulations.

Overdispersion

When logistic regression is applied to count data, it is possible—in fact, usual—for the data to have more variation than is explained by the model. This overdispersion problem arises because the logistic regression model does not have a variance parameter σ.

More specifically, if data y have a binomial distribution with parameters n and p, then the mean of y is np and the standard deviation of y is $\sqrt{np(1-p)}$. As in model (6.4), we define the standardized residual for each data point i as

$$z_i = \frac{y_i - \hat{y}_i}{\text{sd}(\hat{y}_i)}$$

$$= \frac{y_i - n_i\hat{p}_i}{\sqrt{n_i\hat{p}_i(1 - \hat{p}_i)}}, \tag{6.7}$$

where $p_i = \text{logit}^{-1}(X_i\hat{\beta})$. If the binomial model is true, then the z_i's should be approximately independent, each with mean 0 and standard deviation 1.

As with the Poisson model, we can then compute the estimated overdispersion $\frac{1}{n-k}\sum_{i=1}^{n} z_i^2$ (see model (6.5) on page 114) and formally test for overdispersion by comparing $\sum_{i=1}^{n} z_i^2$ to a χ^2_{n-k} distribution. (The n here represents the number of data points and is unrelated to the notation n_i in models (6.6) and (6.7) referring to the number of cases in state-year i.)

In practice, overdispersion happens almost all the time that logistic regression (or Poisson regression, as discussed in Section 6.2) is applied to count data. In the more general family of distributions known as overdispersed models, the standard deviation can have the form $\sqrt{\omega np(1-p)}$, where $\omega > 1$ is known as the *overdispersion parameter*. The overdispersed model reduces to binomial logistic regression when $\omega = 1$.

Adjusting inferences for overdispersion

As with Poisson regression, a simple correction for overdispersion is to multiply the standard errors of all the coefficient estimates by the square root of the estimated overdispersion (6.5). Without this adjustment, the confidence intervals would be too narrow, and inferences would be overconfident.

Overdispersed binomial regressions can be fit in R using the `glm()` function with the `quasibinomial(link="logit")` family. A corresponding distribution is the beta-binomial.

Binary-data model as a special case of the count-data model

Logistic regression for binary data as in Chapter 5 is a special case of the binomial form (6.6) with $n_i \equiv 1$ for all i. Overdispersion at the level of the individual data points cannot occur in the binary model, which is why we did not introduce overdispersed models in Chapter 5.

Count-data model as a special case of the binary-data model

Conversely, the binomial model (6.6) can be expressed in the binary-data form (5.1) by considering each of the n_i cases as a separate data point. The sample size of this expanded regression is $\sum_i n_i$, and the data points are 0's and 1's: each unit i corresponds to y_i ones and $n_i - y_i$ zeroes. Finally, the X matrix is expanded to have $\sum_i n_i$ rows, where the ith row of the original X matrix becomes n_i identical rows in the expanded matrix. In this parameterization, overdispersion could be included in a multilevel model by creating an index variable for the original measurements (in the death penalty example, taking on the values $1, \ldots, 450$) and including a varying coefficient or error term at this level.

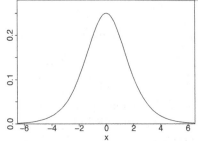

Figure 6.2 *Normal density function with mean 0 and standard deviation 1.6. For most practical purposes, this is indistinguishable from the logistic density (Figure 5.5 on page 85). Thus we can interpret coefficients in probit models as logistic regression coefficients divided by 1.6.*

6.4 Probit regression: normally distributed latent data

The *probit* model is the same as the logit, except it replaces the logistic by the normal distribution (see Figure 5.5). We can write the model directly as

$$\Pr(y_i = 1) = \Phi(X_i\beta),$$

where Φ is the normal cumulative distribution function. In the latent-data formulation,

$$
\begin{aligned}
y_i &= \begin{cases} 1 & \text{if } z_i > 0 \\ 0 & \text{if } z_i < 0 \end{cases} \\
z_i &= X_i\beta + \epsilon_i \\
\epsilon_i &\sim \text{N}(0, 1),
\end{aligned}
\tag{6.8}
$$

that is, a normal distribution for the latent errors with mean 0 and standard deviation 1.

More generally, the model can have an error variance, so that the last line of (6.8) is replaced by

$$\epsilon_i \sim \text{N}(0, \sigma^2),$$

but then σ is nonidentified, because the model is unchanged if we multiply σ by some constant c and then multiply the vector β by c also. Hence we need some restriction on the parameters, and the standard approach is to fix $\sigma = 1$ as in (6.8).

Probit or logit?

As is shown in Figure 6.2 (compare to Figure 5.5 on page 85), the probit model is close to the logit with the residual standard deviation of ϵ set to 1 rather than 1.6. As a result, coefficients in a probit regression are typically close to logistic regression coefficients divided by 1.6. For example, here is the probit version of the logistic regression model on page 88 for well switching:

R output
```
glm(formula = switch ~ dist100, family=binomial(link="probit"))
             coef.est coef.se
(Intercept)    0.38     0.04
dist100       -0.39     0.06
  n = 3020, k = 2
  residual deviance = 4076.3, null deviance = 4118.1 (difference = 41.8)
```

For the examples we have seen, the choice of logit or probit model is a matter of taste or convenience, for example, in interpreting the latent normal errors of probit models. When we see probit regression coefficients, we can simply multiply them by 1.6 to obtain the equivalent logistic coefficients. For example, the model we have just fit, $\Pr(y = 1) = \Phi(0.38 - 0.39x)$, is essentially equivalent to the logistic model $\Pr(y = 1) = \text{logit}^{-1}(1.6(0.38 - 0.39x)) = \text{logit}^{-1}(0.61 - 0.62x)$, which indeed is the logit model estimated on page 88.

6.5 Ordered and unordered categorical regression

Logistic and probit regression can be extended to multiple categories, which can be ordered or unordered. Examples of ordered categorical outcomes include Democrat, Independent, Republican; Yes, Maybe, No; Always, Frequently, Often, Rarely, Never. Examples of unordered categorical outcomes include Liberal, Labor, Conservative; Football, Basketball, Baseball, Hockey; Train, Bus, Automobile, Walk; White, Black, Hispanic, Asian, Other. We discuss ordered categories first, including an extended example, and then briefly discuss regression models for unordered categorical variables.

The ordered multinomial logit model

Consider a categorical outcome y that can take on the values $1, 2, \ldots, K$. The ordered logistic model can be written in two equivalent ways. First we express it as a series of logistic regressions:

$$
\begin{aligned}
\Pr(y > 1) &= \text{logit}^{-1}(X\beta) \\
\Pr(y > 2) &= \text{logit}^{-1}(X\beta - c_2) \\
\Pr(y > 3) &= \text{logit}^{-1}(X\beta - c_3) \\
&\cdots \\
\Pr(y > K-1) &= \text{logit}^{-1}(X\beta - c_{K-1}).
\end{aligned}
\tag{6.9}
$$

The parameters c_k (which are called thresholds or *cutpoints*, for reasons which we shall explain shortly) are constrained to increase: $0 = c_1 < c_2 < \cdots < c_{K-1}$, because the probabilities in (6.9) are strictly decreasing (assuming that all K outcomes have nonzero probabilities of occurring). Since c_1 is defined to be 0, the model with K categories has $K-2$ free parameters c_k in addition to β. This makes sense since $K=2$ for the usual logistic regression, for which only β needs to be estimated.

The cutpoints $c_2, \ldots, c_{K-1}$ can be estimated using maximum likelihood, simultaneously with the coefficients β. For some datasets, however, the parameters can be nonidentified, as with logistic regression for binary data (see Section 5.8).

The expressions in (6.9) can be subtracted to get the probabilities of individual outcomes:

$$
\begin{aligned}
\Pr(y = k) &= \Pr(y > k-1) - \Pr(y > k) \\
&= \text{logit}^{-1}(X\beta - c_{k-1}) - \text{logit}^{-1}(X\beta - c_k).
\end{aligned}
$$

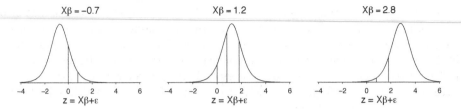

Figure 6.3 *Illustration of cutpoints in an ordered categorical logistic model. In this example, there are $K = 4$ categories and the cutpoints are $c_1 = 0, c_2 = 0.8, c_3 = 1.8$. The three graphs illustrate the distribution of the latent outcome z corresponding to three different values of the linear predictor, $X\beta$. For each, the cutpoints show where the outcome y will equal 1, 2, 3, or 4.*

Latent variable interpretation with cutpoints

The ordered categorical model is easiest to understand by generalizing the latent variable formulation (5.4) to K categories:

$$
y_i = \begin{cases}
1 & \text{if } z_i < 0 \\
2 & \text{if } z_i \in (0, c_2) \\
3 & \text{if } z_i \in (c_2, c_3) \\
& \dots \\
K-1 & \text{if } z_i \in (c_{K-2}, c_{K-1}) \\
K & \text{if } z_i > c_{K-1}
\end{cases}
$$
$$
z_i = X_i\beta + \epsilon_i, \tag{6.10}
$$

with independent errors ϵ_i that have the logistic distribution, as in (5.4).

Figure 6.3 illustrates the latent variable model and shows how the distance between any two adjacent cutpoints c_{k-1}, c_k affects the probability that $y = k$. We can also see that if the linear predictor $X\beta$ is high enough, y will almost certainly take on the highest possible value, and if $X\beta$ is low enough, y will almost certainly equal the lowest possible value.

Example: storable votes

We illustrate ordered categorical data analysis with a study from experimental economics, on the topic of "storable votes." This example is somewhat complicated, and illustrates both the use and potential limitations of the ordered logistic model. In the experiment under study, college students were recruited to play a series of voting games. In each game, a set of k players vote on two issues, with the twist being that each player is given a total of 4 votes. On the first issue, a player has the choice of casting 1, 2, or 3 votes, with the remaining votes cast on the second issue. The winning side of each issue is decided by majority vote, at which point the players on the winning side each get positive payoffs, which are drawn from a uniform distribution on the interval $[1, 100]$.

To increase their expected payoffs, players should follow a strategy of casting more votes for issues where their potential payoffs are higher. The way this experiment is conducted, the players are told the distribution of possible payoffs, and they are told their potential payoff for each issue just before the vote. Thus, in making the choice of how many votes to cast in the first issue, each player knows his or her potential payoff for that vote only. Then, the players are told their potential payoffs for the second vote, but no choice is involved at this point since they will automatically

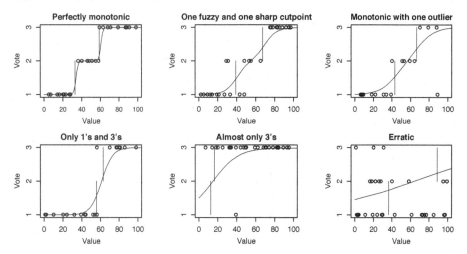

Figure 6.4 *Data from some example individuals in the storable votes study. Vertical lines show estimated cutpoints, and curves show expected responses as estimated using ordered logistic regressions. The two left graphs show data that fit the model reasonably well; the others fit the model in some ways but not perfectly.*

spend all their remaining votes. Players' strategies can thus be summarized as their choices of initial votes, $y = 1, 2,$ or 3, given their potential payoff, x.

Figure 6.4 graphs the responses from six of the hundred or so students in the experiment, with these six chosen to represent several different patterns of data. We were not surprised to see that responses were generally monotonic—that is, students tend to spend more votes when their potential payoff is higher—but it was interesting to see the variety of approximately monotonic strategies that were chosen.

As is apparent in Figure 6.4, most individuals' behaviors can be summarized by three parameters—the cutpoint between votes of 1 and 2, the cutpoint between 2 and 3, and the fuzziness of these divisions. The two cutpoints characterize the chosen monotone strategy, and the sharpness of the divisions indicates the consistency with which the strategy is followed.

Three parameterizations of the ordered logistic model. It is convenient to model the responses using an ordered logit, using a parameterization slightly different from that of model (6.10) to match up with our understanding of the monotone strategies. The model is

$$y_i = \begin{cases} 1 & \text{if } z_i < c_{1.5} \\ 2 & \text{if } z_i \in (c_{1.5}, c_{2.5}) \\ 3 & \text{if } z_i > c_{2.5} \end{cases}$$

$$z_i \sim \text{logistic}(x_i, \sigma^2). \tag{6.11}$$

In this model, the cutpoints $c_{1.5}$ and $c_{2.5}$ are on the 1–100 scale of the data x, and the scale σ of the errors ϵ corresponds to the fuzziness of the cutpoints.

This model has the same number of parameters as the conventional parameterization (6.10)—two regression coefficients have disappeared, while one additional free cutpoint and an error variance have been added. Here is model (6.10) with

$K = 3$ categories and one predictor x,

$$y_i = \begin{cases} 1 & \text{if } z_i < 0 \\ 2 & \text{if } z_i \in (0, c_2) \\ 3 & \text{if } z_i > c_2 \end{cases}$$

$$z_i = \alpha + \beta x + \epsilon_i, \tag{6.12}$$

with independent errors $\epsilon_i \sim \text{logistic}(0, 1)$.

Yet another version of the model keeps the two distinct cutpoints but removes the constant term, α; thus,

$$y_i = \begin{cases} 1 & \text{if } z_i < c_{1|2} \\ 2 & \text{if } z_i \in (0, c_{2|3}) \\ 3 & \text{if } z_i > c_{2|3} \end{cases}$$

$$z_i = \beta x + \epsilon_i, \tag{6.13}$$

with independent errors $\epsilon_i \sim \text{logistic}(0, 1)$.

The three models are in fact equivalent, with z_i/β in (6.13) and $(z_i - \alpha)/\beta$ in (6.12) corresponding to z_i in (6.11) and the parameters matching up as follows:

Model (6.11)	Model (6.12)	Model (6.13)
$c_{1.5}$	$-\alpha/\beta$	$-c_{1|2}/\beta$
$c_{2.5}$	$(c_2 - \alpha)/\beta$	$-c_{2|3}/\beta$
σ	$1/\beta$	$1/\beta$

We prefer parameterization (6.11) because we can directly interpret $c_{1.5}$ and $c_{2.5}$ as thresholds on the scale of the input x, and σ corresponds to the gradualness of the transitions from 1's to 2's and from 2's to 3's. It is sometimes convenient, however, to fit the model using the standard parameterizations (6.12) and (6.13), and so it is helpful to be able to go back and forth between the models.

Fitting the model in R. We can fit ordered logit (or probit) models using `bayespolr()`, which is adapted from the `polr` ("proportional odds logistic regression") function from the `MASS` package in R.[3] We illustrate with data from one of the persons in the storable votes study:

R code

```
fit.1 <- bayespolr (factor(y) ~ x)
display (fit.1)
```

which yields

R output

```
       coef.est coef.se
x      0.10     0.04
1|2 3.46        1.53
2|3 7.03        2.44
  n = 20, k = 3 (including 2 intercepts)
  residual deviance = 32.2, null deviance is not computed by polr
```

From the output we can see this has fitted a model of the form (6.13), with estimates $\hat{\beta} = 0.10$, $\hat{c}_{1|2} = 3.46$ and $\hat{c}_{2|3} = 7.03$. Transforming to model (6.11) using the table of the three models, we get $\hat{c}_{1.5} = 3.46/0.10 = 34.6$, $\hat{c}_{2.5} = 7.03/0.10 = 70.3$, and $\hat{\sigma} = 1/0.10 = 10$.

---

[3] Compared to `polr()`, the `bayespolr()` function adds some prior information to allow the model to be fit in the presence of separation or other nonidentifiability.

Displaying the fitted model. Figure 6.4 shows the cutpoints $c_{1.5}, c_{2.5}$ and expected votes $E(y)$ as a function of x, as estimated from the data from each of several students. From the model (6.11), the expected votes can be written as

$$
\begin{aligned}
E(y|x) &= 1 \cdot \Pr(y = 1|x) + 2 \cdot \Pr(y = 2|x) + 3 \cdot \Pr(y = 3|x) \\
&= 1 \cdot \left(1 - \text{logit}^{-1} \left(\frac{x - c_{1.5}}{\sigma} \right) \right) + \\
&\quad + 2 \cdot \left(\text{logit}^{-1} \left(\frac{x - c_{1.5}}{\sigma} \right) - \text{logit}^{-1} \left(\frac{x - c_{2.5}}{\sigma} \right) \right) + \\
&\quad + 3 \cdot \text{logit}^{-1} \left(\frac{x - c_{2.5}}{\sigma} \right),
\end{aligned}
\tag{6.14}
$$

where $\text{logit}^{-1}(x) = e^x/(1+e^x)$ is the logistic curve displayed in Figure 5.2a on page 80. Expression (6.14) looks complicated but is easy to program as a function in R:

```
expected <- function (x, c1.5, c2.5, sigma){                    R code
  p1.5 <- invlogit ((x-c1.5)/sigma)
  p2.5 <- invlogit ((x-c2.5)/sigma)
  return ((1*(1-p1.5) + 2*(p1.5-p2.5) + 3*p2.5))
}
```

The data, cutpoints, and curves in Figure 6.4 can then be plotted as follows:

```
plot (x, y, xlim=c(0,100), ylim=c(1,3), xlab="Value", ylab="Vote")    R code
lines (rep (c1.5, 2), c(1,2))
lines (rep (c2.5, 2), c(2,3))
curve (expected (x, c1.5, c2.5, sigma), add=TRUE)
```

Having displayed these estimates for individuals, the next step is to study the distribution of the parameters in the population, to understand the range of strategies applied by the students. In this context, the data have a multilevel structure—30 observations for each of several students—and we pursue this example further in Section 15.2 in the chapter on multilevel generalized linear models.

Alternative approaches to modeling ordered categorical data

Ordered categorical data can be modeled in several ways, including:

- Ordered logit model with $K-1$ cutpoint parameters, as we have just illustrated.
- The same model in probit form.
- Simple linear regression (possibly preceded by a simple transformation of the outcome values). This can be a good idea if the number of categories is large and if they can be considered equally spaced. This presupposes that a reasonable range of the categories is actually used. For example, if ratings are on a 1 to 10 scale, but in practice always equal 9 or 10, then a linear model probably will not work well.
- Separate logistic regressions—that is, a logistic regression model for $y = 1$ versus $y = 2, \ldots, K$; then, if $y \geq 2$, a logistic regression for $y = 2$ versus $y = 3, \ldots, K$; and so on up to a model, if $y \geq K-1$ for $y = K-1$ versus $y = K$. Or this can be set up using the probit model. Separate logistic (or probit) regressions have the advantage of more flexibility in fitting data but the disadvantage of losing the simple latent-variable interpretation of the cutpoint model we have described.
- Finally, robit regression, which we discuss in Section 6.6, is a competitor to logistic regression that accounts for occasional aberrant data such as the outlier in the upper-right plot of Figure 6.4.

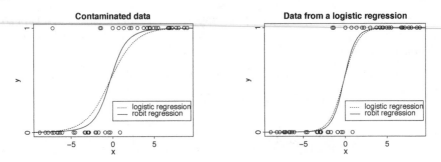

Figure 6.5 *Hypothetical data to be fitted using logistic regression: (a) a dataset with an "outlier" (the unexpected y = 1 value near the upper left); (b) data simulated from a logistic regression model, with no outliers. In each plot, the dotted and solid lines show the fitted logit and robit regressions, respectively. In each case, the robit line is steeper—especially for the contaminated data—because it effectively downweights the influence of points that do not appear to fit the model.*

Unordered categorical regression

As discussed at the beginning of Section 6.5, it is sometimes appropriate to model discrete outcomes as unordered. An example that arose in our research was the well-switching problem. As described in Section 5.4, households with unsafe wells had the option to switch to safer wells. But the actual alternatives are more complicated and can be summarized as: (0) do nothing, (1) switch to an existing private well, (2) switch to an existing community well, (3) install a new well yourself. If these are coded as 0, 1, 2, 3, then we can model $\Pr(y \geq 1), \Pr(y \geq 2 | y \geq 1), \Pr(y = 3 | y \geq 2)$. Although the four options could be considered to be ordered in some way, it does not make sense to apply the ordered multinomial logit or probit model, since different factors likely influence the three different decisions. Rather, it makes more sense to fit separate logit (or probit) models to each of the three components of the decision: (a) do you switch or do nothing? (b) if you switch, do you switch to an existing well or build a new well yourself? (c) if you switch to an existing well, is it a private or community well? More about this important category of model can be found in the references at the end of this chapter.

6.6 Robust regression using the t model

The t distribution instead of the normal

When a regression model can have occasional very large errors, it is generally more appropriate to use a Student-t rather than normal distribution for the errors. The basic form of the regression is unchanged—$y = X\beta + \epsilon$—but with a different distribution for the ϵ's and thus a slightly different method for estimating β (see the discussion of maximum likelihood estimation in Chapter 18) and a different distribution for predictions. Regressions estimated using the t model are said to be *robust* in that the coefficient estimates are less influenced by individual outlying data points. Regressions with t errors can be fit using the `tlm()` function in the `hett` package in R.

Robit instead of logit or probit

Logistic regression (and the essentially equivalent probit regression) are flexible and convenient for modeling binary data, but they can run into problems with outliers. Outliers are usually thought of as extreme observations, but in the context of discrete data, an "outlier" is more of an *unexpected* observation. Figure 6.5a illustrates, with data simulated from a logistic regression, with an extreme point switched from 0 to 1. In the context of the logistic model, an observation of $y = 1$ for this value of x would be extremely unlikely, but in real data this sort of "misclassification" can definitely occur. Hence this graph represents the sort of data to which we might fit a logistic regression, even though this model is not exactly appropriate.

For another illustration of a logistic regression with an aberrant data point, see the upper-right plot in Figure 6.4. That is an example with three outcomes; for simplicity, we restrict our attention here to binary outcomes.

Logistic regression can be conveniently "robustified" by generalizing the latent-data formulation (5.4):

$$y_i = \begin{cases} 1 & \text{if } z_i > 0 \\ 0 & \text{if } z_i < 0 \end{cases}$$
$$z_i = X_i\beta + \epsilon_i,$$

to give the latent errors ϵ a t distribution:

$$\epsilon_i \sim t_\nu\left(0, \frac{\nu - 2}{\nu}\right), \tag{6.15}$$

with the degrees-of-freedom parameter $\nu > 2$ estimated from the data and the t distribution scaled so that its standard deviation equals 1.

The t model for the ϵ_i's allows the occasional unexpected prediction—a positive value of z for a highly negative value of the linear predictor $X\beta$, or vice versa. Figure 6.5a illustrates with the simulated "contaminated" dataset: the solid line shows $\Pr(y = 1)$ as a function of the x for the fitted robit regression, and it is quite a bit steeper than the fitted logistic model. The t distribution effectively downweights the discordant data point so that the model better fits the main part of the data.

Figure 6.5b shows what happens with data that actually come from a logistic model: here, the robit model is close to the logit, which makes sense since it does not find discrepancies.

Mathematically, the robit model can be considered as a generalization of probit and an approximate generalization of logit. Probit corresponds to the degrees of freedom $\nu = \infty$, and logit is very close to the robit model with $\nu = 7$.

6.7 Building more complex generalized linear models

The models we have considered so far can handle many regression problems in practice. For continuous data we start with linear regression with normal errors, consider appropriate transformations and interactions as discussed in Chapter 4, and switch to a t error model for data with occasional large errors. For binary data we use logit, probit, or perhaps robit, again transforming input variables and considering residual plots as discussed in Chapter 5. For count data, the starting points are the overdispersed binomial and Poisson distributions, and for discrete outcomes with more than two categories we can fit ordered or unordered multinomial logit or probit regression. Here we briefly describe some situations where it is helpful to consider other models.

Mixed discrete/continuous data

Earnings is an example of an outcome variable with both discrete and continuous aspects. In our earnings and height regressions in Chapter 4, we preprocessed the data by removing all respondents with zero earnings. In general, however, it can be appropriate to model a variable such as earnings in two steps: first a logistic regression for $\Pr(y > 0)$, then a linear regression on $\log(y)$, conditional on $y > 0$. Predictions for such a model then must be done in two steps, most conveniently using simulation (see Chapter 7).

When modeling an outcome in several steps, programming effort is sometimes required to convert inferences on to the original scale of the data. For example, in a two-step model for predicting earnings given height and sex, we first use a logistic regression to predict whether earnings are positive:

R code
```
earn.pos <- ifelse (earnings>0, 1, 0)
fit.1a <- glm (earn.pos ~ height + male, family=binomial(link="logit"))
```

yielding the fit

R output
```
              coef.est coef.se
(Intercept)    -3.85    2.07
height          0.08    0.03
male            1.70    0.32
  n = 1374, k = 3
  residual deviance = 988.3, null deviance = 1093.2 (difference = 104.9)
```

We then fit a linear regression to the logarithms of positive earnings:

R code
```
log.earn <- log(earnings)
fit.1b <- lm (log.earn ~ height + male, subset = earnings>0)
```

yielding the fit

R output
```
              coef.est coef.se
(Intercept)     8.12    0.60
height          0.02    0.01
male            0.42    0.07
  n = 1187, k = 3
  residual sd = 0.88, R-Squared = 0.09
```

Thus, for example, a 66-inch-tall woman has a probability $\text{logit}^{-1}(-3.85 + 0.08 \cdot 66 + 1.70 \cdot 0) = 0.81$, or an 81% chance, of having positive earnings. If her earnings are positive, their predicted value is $\exp(8.12 + 0.02 \cdot 66 + 0.42 \cdot 0) = 12600$. Combining these gives a mixture of a spike at 0 and a lognormal distribution, which is most easily manipulated using simulations, as we discuss in Sections 7.4 and 25.4.

Latent-data models. Another way to model mixed data is through latent data, for example positing an "underlying" income level z_i—the income that person i would have if he or she were employed—that is observed only if $y_i > 0$. *Tobit regression* is one such model that is popular in econometrics.

Cockroaches and the zero-inflated Poisson model

The binomial and Poisson models, and their overdispersed generalizations, all can be expressed in terms of an underlying continuous probability or rate of occurrence of an event. Sometimes, however, the underlying rate itself has discrete aspects.

For example, in a study of cockroach infestation in city apartments, each apartment i was set up with traps for several days. We label u_i as the number of trap-days and y_i as the number of cockroaches trapped. With a goal of predicting cockroach infestation given predictors X (including income and ethnicity of the apartment dwellers, indicators for neighborhood, and measures of quality of the apartment), we would start with the model

$$y_i \sim \text{overdispersed Poisson} (u_i e^{X_i \beta}, \omega). \tag{6.16}$$

It is possible, however, for the data to have more zeroes (that is, apartments i with cockroach counts $y_i = 0$) than predicted by this model.[3] A natural explanation is that some apartments have truly a zero (or very near-zero) rate of cockroaches, whereas others simply have zero counts from the discreteness of the data. The *zero-inflated* model places (6.16) into a mixture model:

$$y_i \begin{cases} = 0, \text{ if } S_i = 0 \\ \sim \text{overdispersed Poisson} (u_i e^{X_i \beta}, \omega), \text{ if } S_i = 1. \end{cases}$$

Here, S_i is an indicator of whether apartment i has any cockroaches at all, and it could be modeled using logistic regression:

$$\Pr(S_i = 1) = \text{logit}^{-1}(X_i \gamma),$$

where γ is a new set of regression coefficients for this part of the model. Estimating this two-stage model is not simple—the S_i's are not observed and so one cannot directly estimate γ; and we do not know which zero observations correspond to $S_i = 0$ and which correspond to outcomes of the Poisson distribution, so we cannot directly estimate β. Some R functions have been written to fit such models and they can also be fit using Bugs.

Other models

The basic choices of linear, logistic, and Poisson models, along with mixtures of these models and their overdispersed, robust, and multinomial generalizations, can handle many regression problems. However, other distributional forms have been used for specific sorts of data; these include exponential, gamma, and Weibull models for waiting-time data, and hazard models for survival data. More generally, nonparametric models including generalized additive models, neural networks, and many others have been developed for going beyond the generalized linear modeling framework by allowing data-fitted nonlinear relations between inputs and the data.

6.8 Constructive choice models

So far we have considered regression modeling as a descriptive tool for studying how an outcome can be predicted given some input variables. A completely different approach, sometimes applicable to choice data such as in the examples in Chapters 5 and 6 on logistic regression and generalized linear models, is to model the decisions as a balancing of goals or utilities.

We demonstrate this idea using the example of well switching in Bangladesh (see Section 5.4). How can we understand the relation between distance, arsenic level, and the decision to switch? It makes sense that people with higher arsenic levels would be more likely to switch, but what coefficient values should we expect? Should the relation be on the log or linear scale? The actual health risk is believed

---

[3] In our actual example, the overdispersed Poisson model did a reasonable job predicting the number of zeroes; see page 161. But in other similar datasets the zero-inflated model can both make sense and fit data well, hence our presentation here.

to be linear in arsenic concentration; does that mean that a logarithmic model is inappropriate? Such questions can be addressed using a model for individual decisions.

To set up a *choice model*, we must specify a *value function*, which represents the strength of preference for one decision over the other—in this case, the preference for switching as compared to not switching. The value function is scaled so that zero represents indifference, positive values correspond to a preference for switching, and negative values result in not switching. This model is thus similar to the latent-data interpretation of logistic regression (see page 85); and in fact that model is a special case, as we shall see here.

Logistic or probit regression as a choice model in one dimension

There are simple one-dimensional choice models that reduce to probit or logit regression with a single predictor, as we illustrate with the model of switching given distance to nearest well. From page 88, the logistic regression is

R output
```
glm(formula = switch ~ dist100, family=binomial(link="logit"))
            coef.est coef.se
(Intercept)     0.61    0.06
dist100        -0.62    0.10
  n = 3020, k = 2
  residual deviance = 4076.2, null deviance = 4118.1 (difference = 41.9)
```

Now let us think about it from first principles as a decision problem. For household i, define

- a_i = the benefit of switching from an unsafe to a safe well

- $b_i + c_i x_i$ = the cost of switching to a new well a distance x_i away.

We are assuming a utility theory in which the benefit (in reduced risk of disease) can be expressed on the same scale as the cost (the inconvenience of no longer using one's own well, plus the additional effort—proportional to distance—required to carry the water).

Logit model. Under the utility model, household i will switch if $a_i > b_i + c_i x_i$. However, we do not have direct measurements of the a_i's, b_i's, and c_i's. All we can learn from the data is the probability of switching as a function of x_i; that is,

$$\Pr(\text{switch}) = \Pr(y_i = 1) = \Pr(a_i > b_i + c_i x_i), \qquad (6.17)$$

treating a_i, b_i, c_i as random variables whose distribution is determined by the (unknown) values of these parameters in the population.

Expression (6.17) can be written as

$$\Pr(y_i = 1) = \Pr\left(\frac{a_i - b_i}{c_i} > x_i\right),$$

a re-expression that is useful in that it puts all the random variables in the same place and reveals that the population relation between y and x depends on the distribution of $(a - b)/c$ in the population.

For convenience, label $d_i = (a_i - b_i)/c_i$: the net benefit of switching to a neighboring well, divided by the cost per distance traveled to a new well. If d_i has a logistic distribution in the population, and if d is independent of x, then $\Pr(y = 1)$ will have the form of a logistic regression on x, as we shall show here.

If d_i has a logistic distribution with center μ and scale σ, then $d_i = \mu + \sigma \epsilon_i$,

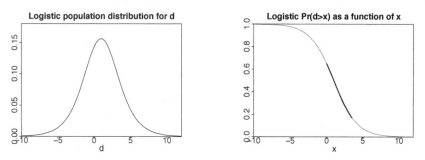

Figure 6.6 *(a) Hypothesized logistic distribution of $d_i = (a_i - b_i)/c_i$ in the population and (b) corresponding logistic regression curve of the probability of switching given distance. These both correspond to the model, $\Pr(y_i = 1) = \Pr(d_i > x_i) = \text{logit}^{-1}(0.61 - 0.62x)$. The dark part of the curve in (b) corresponds to the range of x (distance in 100-meter units) in the well-switching data; see Figure 5.9 on page 89.*

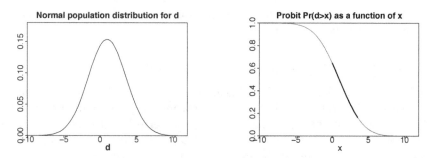

Figure 6.7 *(a) Hypothesized normal distribution of $d_i = (a_i - b_i)/c_i$ with mean 0.98 and standard deviation 2.6 and (b) corresponding probit regression curve of the probability of switching given distance. These both correspond to the model, $\Pr(y_i = 1) = \Pr(d_i > x_i) = \Phi(0.38 - 0.39x)$. Compare to Figure 6.6.*

where ϵ_i has the unit logistic density; see Figure 5.2 on page 80. Then

$$\Pr(\text{switch}) = \Pr(d_i > x) = \Pr\left(\frac{d_i - \mu}{\sigma} > \frac{x - \mu}{\sigma}\right)$$

$$= \text{logit}^{-1}\left(\frac{\mu - x}{\sigma}\right) = \text{logit}^{-1}\left(\frac{\mu}{\sigma} - \frac{1}{\sigma}x\right),$$

which is simply a logistic regression with coefficients μ/σ and $-1/\sigma$. We can then fit the logistic regression and solve for μ and σ. For example, the well-switching model, $\Pr(y = 1) = \text{logit}^{-1}(0.61 - 0.62x)$, corresponds to $\mu/\sigma = 0.61$ and $-1/\sigma = -0.62$; thus $\sigma = 1/0.62 = 1.6$ and $\mu = 0.61/0.62 = 0.98$. Figure 6.6 shows the distribution of d, along with the curve of $\Pr(d > x)$ as a function of x.

Probit model. A similar model is obtained by starting with a normal distribution for the utility parameter: $d \sim \text{N}(\mu, \sigma^2)$. In this case,

$$\Pr(\text{switch}) = \Pr(d_i > x) = \Pr\left(\frac{d_i - \mu}{\sigma} > \frac{x - \mu}{\sigma}\right)$$

$$= \Phi\left(\frac{\mu - x}{\sigma}\right) = \Phi\left(\frac{\mu}{\sigma} - \frac{1}{\sigma}x\right),$$

which is simply a probit regression. The model $\Pr(y = 1) = \Phi(0.38 - 0.39x)$ corresponds to $\mu/\sigma = 0.38$ and $-1/\sigma = -0.39$; thus $\sigma = 1/0.39 = 2.6$ and

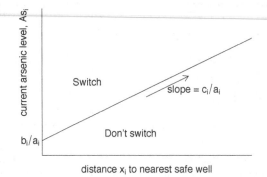

Figure 6.8 *Decision options for well switching given arsenic level of current well and distance to the nearest safe well, based on the decision rule: switch if $a_i \cdot (\text{As})_i > b_i + cx_i$.*

$\mu = 0.38/0.39 = 0.98$. Figure 6.7 shows this model, which is nearly identical to the logistic model shown in Figure 6.6.

Choice models, discrete data regressions, and latent data

Logistic regression and generalized linear models are usually set up as methods for estimating the probabilities of different outcomes y given predictors x. A fitted model represents an entire population, with the "error" in the model coming in through probabilities that are not simply 0 or 1 (hence, the gap between data points and fitted curves in graphs such as Figure 5.9 on page 89).

In contrast, choice models are defined at the level of the individual, as we can see in the well-switching example, where each household i has, along with its own data X_i, y_i, its own parameters a_i, b_i, c_i that determine its utility function and thus its decision of whether to switch.

Logistic or probit regression as a choice model in multiple dimensions

We can extend the well-switching model to multiple dimensions by considering the arsenic level of the current well as a factor in the decision.

- $a_i \cdot (\text{As})_i$ = the benefit of switching from an unsafe well with arsenic level As_i to a safe well. (It makes sense for the benefit to be proportional to the current arsenic level, because risk is believed to be essentially proportional to cumulative exposure to arsenic.)

- $b_i + c_i x_i$ = the cost of switching to a new well a distance x_i away.

Household i should then switch if $a_i \cdot (\text{As})_i > b_i + cx_i$—the decision thus depends on the household's arsenic level $(\text{As})_i$, its distance x_i to the nearest well, and its utility parameters a_i, b_i, c_i.

Figure 6.8 shows the decision space for an individual household, depending on its arsenic level and distance to the nearest safe well. Given a_i, b_i, c_i, the decision under this model is deterministic. However, a_i, b_i, c_i are not directly observable—all we see are the decisions ($y_i = 0$ or 1) for households, given their arsenic levels As_i and distances x_i to the nearest safe well.

Certain distributions of (a, b, c) in the population reduce to the fitted logistic regression, for example, if a_i and c_i are constants and b_i/a_i has a logistic distribution that is independent of $(\text{As})_i$ and x_i. More generally, choice models reduce to logistic regressions if the factors come in additively, with coefficients that do not vary in the population, and if there is a fixed cost (b_i in this example) that has a logistic distribution in the population.

Other distributions of (a, b, c) are possible. The corresponding models can be fit, treating these utility parameters as latent data. There is no easy way of fitting such models using `glm()` in R (except for the special cases that reduce to logit and probit), but they can be fit in Bugs (see Exercise 17.7).

Insights from decision models

A choice model can give us some insight even if we do not formally fit it. For example, in fitting logistic regressions, we found that distance worked well as a linear predictor, whereas arsenic level fit better on the logarithmic scale. A simple utility analysis would suggest that both these factors should come in linearly, and the transformation for arsenic suggests that people are (incorrectly) perceiving the risks on a logarithmic scale—seeing the difference between 4 to 8, say, as no worse than the difference between 1 and 2. (In addition, our residual plot showed the complication that people seem to underestimate risks from arsenic levels very close to 0.5. And behind this is the simplifying assumption that all wells with arsenic levels below 0.5 are "safe.")

We can also use the utility model to interpret the coefficient for education in the model—more educated people are more likely to switch, indicating that their costs of switching are lower, or their perceived benefits from reducing arsenic exposure are higher. Interactions correspond to dependence among the latent utility parameters in the population.

The model could also be elaborated to consider the full range of individual options, which include doing nothing, switching to an existing private well, switching to an existing community well, or digging a new private well. The decision depends on the cost of walking, perception of health risks, financial resources, and future plans.

6.9 Bibliographic note

The concept of the generalized linear model was introduced by Nelder and Wedderburn (1972) and developed further, with many examples, by McCullagh and Nelder (1989). Dobson (2001) is an accessible introductory text. For more on overdispersion, see Anderson (1988) and Liang and McCullagh (1993). Fienberg (1977) and Agresti (2002) are other useful references.

The death penalty example comes from Gelman, Liebman, et al. (2004). Models for traffic accidents are discussed by Chapman (1973) and Hauer, Ng, and Lovell (1988). For more on the New York City police example, see Spitzer (1999) and Gelman, Fagan, and Kiss (2005).

Maddala (1983) presents discrete-data regressions and choice models from an econometric perspective, and McCullagh (1980) considers general forms for latent-parameter models for ordered data. Amemiya (1981) discusses the factor of 1.6 for converting from logit to probit coefficients.

Walker and Duncan (1967) introduce the ordered logistic regression model, and Imai and van Dyk (2003) discuss the models underlying multinomial logit and probit regression. The storable votes example comes from Casella, Gelman, and Palfrey (2006). See Agresti (2002) and Imai and van Dyk (2003) for more on categorical regression models, ordered and unordered.

Robust regression using the t distribution is discussed by Zellner (1976) and

Lange, Little, and Taylor (1989), and the robit model is introduced by Liu (2004). See Stigler (1977) and Mosteller and Tukey (1977) for further discussions of robust inference from an applied perspective. Wiens (1999) and Newton et al. (2001) discuss the gamma and lognormal models for positive continuous data. For generalized additive models and other nonparametric methods, see Hastie and Tibshirani (1990) and Hastie, Tibshirani, and Friedman (2002).

Connections between logit/probit regressions and choice models have been studied in psychology, economics, and political science; some important references are Thurstone (1927a, b), Wallis and Friedman (1942), Mosteller (1951), Bradley and Terry (1952), and McFadden (1973). Tobit models are named after Tobin (1958) and are covered in econometrics texts such as Woolridge (2001).

6.10 Exercises

1. Poisson regression: the folder `risky.behavior` contains data from a randomized trial targeting couples at high risk of HIV infection. The intervention provided counseling sessions regarding practices that could reduce their likelihood of contracting HIV. Couples were randomized either to a control group, a group in which just the woman participated, or a group in which both members of the couple participated. One of the outcomes examined after three months was "number of unprotected sex acts."

   (a) Model this outcome as a function of treatment assignment using a Poisson regression. Does the model fit well? Is there evidence of overdispersion?

   (b) Next extend the model to include pre-treatment measures of the outcome and the additional pre-treatment variables included in the dataset. Does the model fit well? Is there evidence of overdispersion?

   (c) Fit an overdispersed Poisson model. What do you conclude regarding effectiveness of the intervention?

   (d) These data include responses from both men and women from the participating couples. Does this give you any concern with regard to our modeling assumptions?

2. Multinomial logit: using the individual-level survey data from the 2000 National Election Study (data in folder `nes`), predict party identification (which is on a five-point scale) using ideology and demographics with an ordered multinomial logit model.

   (a) Summarize the parameter estimates numerically and also graphically.

   (b) Explain the results from the fitted model.

   (c) Use a binned residual plot to assess the fit of the model.

3. Comparing logit and probit: take one of the data examples from Chapter 5. Fit these data using both logit and probit model. Check that the results are essentially the same (after scaling by factor of 1.6; see Figure 6.2 on page 118).

4. Comparing logit and probit: construct a dataset where the logit and probit models give *different* estimates.

5. Tobit model for mixed discrete/continuous data: experimental data from the National Supported Work example are available in the folder `lalonde`. Use the treatment indicator and pre-treatment variables to predict post-treatment (1978) earnings using a tobit model. Interpret the model coefficients.

6. Robust linear regression using the t model: The folder `congress` has the votes for the Democratic and Republican candidates in each U.S. congressional district in 1988, along with the parties' vote proportions in 1986 and an indicator for whether the incumbent was running for reelection in 1988. For your analysis, just use the elections that were contested by both parties in both years.

   (a) Fit a linear regression (with the usual normal-distribution model for the errors) predicting 1988 Democratic vote share from the other variables and assess model fit.

   (b) Fit a t-regression model predicting 1988 Democratic vote share from the other variables and assess model fit; to fit this model in R you can use the `tlm()` function in the `hett` package. (See the end of Section C.2 for instructions on loading R packages.)

   (c) Which model do you prefer?

7. Robust regression for binary data using the robit model: Use the same data as the previous example with the goal instead of predicting for each district whether it was won by the Democratic or Republican candidate.

   (a) Fit a standard logistic or probit regression and assess model fit.

   (b) Fit a robit regression and assess model fit.

   (c) Which model do you prefer?

8. Logistic regression and choice models: using the individual-level survey data from the election example described in Section 4.7 (data available in the folder `nes`), fit a logistic regression model for the choice of supporting Democrats or Republicans. Then interpret the output from this regression in terms of a utility/choice model.

9. Multinomial logistic regression and choice models: repeat the previous exercise but now with three options: Democrat, no opinion, Republican. That is, fit an ordered logit model and then express it as a utility/choice model.

10. Spatial voting models: suppose that competing political candidates A and B have positions that can be located spatially in a one-dimensional space (that is, on a line). Suppose that voters have "ideal points" with regard to these positions that are normally distributed in this space, defined so that voters will prefer candidates whose positions are closest to their ideal points. Further suppose that voters' ideal points can be modeled as a linear regression given inputs such as party identification, ideology, and demographics.

    (a) Write this model in terms of utilities.

    (b) Express the probability that a voter supports candidate S as a probit regression on the voter-level inputs.

    See Erikson and Romero (1990) and Clinton, Jackman, and Rivers (2004) for more on these models.

11. Multinomial choice models: Pardoe and Simonton (2006) fit a discrete choice model to predict winners of the Academy Awards. Their data are in the folder `academy.awards`.

    (a) Fit your own model to these data.

    (b) Display the fitted model on a plot that also shows the data.

    (c) Make a plot displaying the uncertainty in inferences from the fitted model.

Part 1B: Working with regression inferences

We now discuss how to go beyond simply looking at regression coefficients, first by using simulation to summarize and propagate inferential uncertainty, and then by considering how regression can be used for causal inference.

## CHAPTER 7

Simulation of probability models and statistical inferences

Whenever we represent inferences for a parameter using a point estimate and standard error, we are performing a data reduction. If the estimate is normally distributed, this summary discards no information because the normal distribution is completely defined by its mean and variance. But in other cases it can be useful to represent the uncertainty in the parameter estimation by a set of random simulations that represent possible values of the parameter vector (with more likely values being more likely to appear in the simulation). By *simulation*, then, we mean summarizing inferences by random numbers rather than by point estimates and standard errors.

7.1 Simulation of probability models

In this section we introduce simulation for two simple probability models. The rest of the chapter discusses how to use simulations to summarize and understand regressions and generalize linear models, and the next chapter applies simulation to model checking and validation. Simulation is important in itself and also prepares for multilevel models, which we fit using simulation-based inference, as described in Part 2B.

A simple example of discrete predictive simulation

How many girls in 400 births? The probability that a baby is a girl or boy is 48.8% or 51.2%, respectively. Suppose that 400 babies are born in a hospital in a given year. How many will be girls?

We can simulate the 400 births using the binomial distribution:

```
n.girls <- rbinom (1, 400, .488)                                    R code
print (n.girls)
```

which shows us what could happen in 400 births. To get a sense of the *distribution* of what could happen, we simulate the process 1000 times (after first creating the vector n.girls to store the simulations):

```
n.sims <- 1000                                                      R code
n.girls <- rep (NA, n.sims)
for (s in 1:n.sims){
  n.girls[s] <- rbinom (1, 400, .488)}
hist (n.girls)
```

which yields the histogram in Figure 7.1 representing the probability distribution for the number of girl births. The 1000 simulations capture the uncertainty.[1]

---

[1] In this example, we performed all the simulations in a loop. It would also be possible to simulate 1000 draws from the binomial distribution directly:

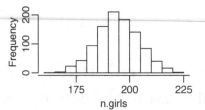

Figure 7.1 *Histogram of 1000 simulated values for the number of girls born in a hospital out of 400 babies, as simulated from the binomial probability distribution with probability 0.488.*

Accounting for twins. We can complicate the model in various ways. For example, there is a 1/125 chance that a birth event results in fraternal twins, of which each has an approximate 49.5% chance of being a girl, and a 1/300 chance of identical twins, which have an approximate 49.5% chance of being girls. We can simulate 400 birth events as follows:

R code
```
birth.type <- sample (c("fraternal twin","identical twin","single birth"),
   size=400, replace=TRUE, prob=c(1/125, 1/300, 1 - 1/125 - 1/300))
girls <- rep (NA, 400)
for (i in 1:400){
  if (birth.type[i]=="single birth"){
    girls[i] <- rbinom (1, 1, .488)}
  else if (birth.type[i]=="identical twin"){
    girls[i] <- 2*rbinom (1, 1, .495)}
  else if (birth.type[i]=="fraternal twin"){
    girls[i] <- rbinom (1, 2, .495)}
}
n.girls <- sum (girls)
```

Here, `girls` is a vector of length 400, of 0's, 1's, and 2's (mostly 0's and 1's) representing the number of girls in each birth event.[2] To approximate the *distribution* of the number of girls in 400 births, we put the simulation in a loop and repeat it 1000 times:

R code
```
n.girls <- rep (NA, n.sims)
for (s in 1:n.sims){
   birth.type <-sample(c("fraternal twin","identical twin","single birth"),
      size=400, replace=TRUE, prob=c(1/125, 1/300, 1 - 1/125 - 1/300))
   girls <- rep (NA, 400)
   for (i in 1:400){
     if (birth.type[i]=="single birth"){
       girls[i] <- rbinom (1, 1, .488)}
     else if (birth.type[i]=="identical twin"){
       girls[i] <- 2*rbinom (1, 1, .495)}
     else if (birth.type[i]=="fraternal twin"){
       girls[i] <- rbinom (1, 2, .495)}
```

```
n.girls <- rbinom (n.sims, 400, .488)
```
In other settings one can write the simulation as a function and perform the looping implicitly using the `replicate()` function in R, as we illustrate on page 139.

[2] Again, this calculation could also be performed without looping using vector operations in R:
```
girls <- ifelse (birth.type=="single birth", rbinom (400, 1, .488), ifelse (
  birth.type="identical twins", 2*rbinom (400, 1, .495), rbinom (400, 2, .495)))
```
We have used looping in the main text to emphasize the parallel calculation for the 400 birth events, but the vectorized computation is faster and can be more convenient when part of a larger computation.

```
   }
   n.girls[s] <- sum (girls)
}
```

This nested looping is characteristic of simulations of complex data structures and can also be implemented using custom R functions and the `replicate()` function, as we discuss shortly.

A simple example of continuous predictive simulation

Similarly, we can program R to simulate continuous random variables. For example, 52% of adults in the United States are women and 48% are men. The heights of the men are approximately normally distributed with mean 69.1 inches and standard deviation 2.9 inches; women with mean 63.7 and standard deviation 2.7.

Suppose we select 10 adults at random. What can we say about their average height?

```
sex <- rbinom (10, 1, .52)                                      R code
height <- ifelse (sex==0, rnorm (10, 69.1, 2.9), rnorm (10, 64.5, 2.7))
avg.height <- mean (height)
print (avg.height)
```

To simulate the distribution of `avg.height`, we loop the simulation 1000 times:

```
n.sims <- 1000                                                  R code
avg.height <- rep (NA, n.sims)
for (s in 1:n.sims){
  sex <- rbinom (10, 1, .52)
  height <- ifelse (sex==0, rnorm (10, 69.1, 2.9), rnorm (10, 64.5, 2.7))
  avg.height[s] <- mean (height)
}
hist (avg.height, main="Average height of 10 adults")
```

What about the maximum height of the 10 people? To determine this, just add the following line within the loop:

```
max.height[s] <- max (height)                                  R code
```

and before the loop, initialize `max.height`:

```
max.height <- rep (NA, n.sims)                                 R code
```

Then, after the loop, make a histogram of `max.height`.

Simulation in R using custom-made functions

The coding for simulations becomes cleaner if we express the steps for a single simulation as a function in R. We illustrate with the simulation of average heights. First, the function:

```
Height.sim <- function (n.adults){                             R code
  sex <- rbinom (n.adults, 1, .52)
  height <- ifelse (sex==0, rnorm (10, 69.5, 2.9), rnorm (10, 64.5, 2.7))
  return (mean(height))
}
```

(For simplicity we have "hard-coded" the proportion of women and the mean and standard deviation of men's and women's heights, but more generally these could be supplied as arguments to the function.)

We then use the `replicate()` function to call `Height.sim()` 1000 times:

R code
```
avg.height <- replicate (1000, Height.sim (n.adults=10))
hist (avg.height)
```

See Section 20.5 for a more elaborate example of the use of functions in R, in a fake-data simulation to perform a power calculation.

7.2 Summarizing linear regressions using simulation: an informal Bayesian approach

In a regression setting, we can use simulation to capture both predictive uncertainty (the error term in the regression model) and inferential uncertainty (the standard errors of the coefficients and uncertainty about the residual error). We first discuss the simplest case of simulating prediction errors, then consider inferential uncertainty and the combination of both sources of variation.

Simulation to represent predictive uncertainty

We illustrate predictive uncertainty with the problem of predicting the earnings of a 68-inch-tall man, using model (4.2) on page 63.

Obtaining the point and interval predictions automatically. The predictive estimate and confidence interval can easily be accessed using the regression software in R:

R code
```
x.new <- data.frame (height=68, male=1)
pred.interval <- predict (earn.logmodel.3, x.new, interval="prediction",
  level=.95)
```

and then exponentiating to get the predictions on the original (unlogged) scale of earnings:

R code
```
exp (pred.interval)
```

Constructing the predictive interval using simulation. We now discuss how to obtain predictive intervals "manually" using simulations derived from the fitted regression model. In this example it would be easier to simply use the predict() function as just shown; however, simulation is a general tool that we will be able to apply in more complicated predictive settings, as we illustrate later in this chapter and the next.

- The point estimate for log earnings is $8.4 + 0.017 \cdot 68 - 0.079 \cdot 1 + 0.007 \cdot 68 \cdot 1 = 9.95$, with a standard deviation of 0.88. To put these on the original (unlogged) scale, we exponentiate to yield a geometric mean of $e^{9.95} = 21000$ and a geometric standard deviation of $e^{0.88} = 2.4$.

  Then, for example, the 68% predictive interval is $[21000/2.4, 21000 \cdot 2.4] = [8800, 50000]$, and the 95% interval is $[21000/2.4^2, 21000 \cdot 2.4^2] = [3600, 121000]$

- The simulation prediction is a set of random numbers whose logarithms have mean 9.95 and standard deviation 0.88. For example, in R, we can summarize the predictive distribution using the following command:

R code
```
pred <- exp (rnorm (1000, 9.95, .88))
```

which tells R to draw 1000 random numbers from a normal distribution with mean 9.95 and variance 0.88, and then exponentiate these values.

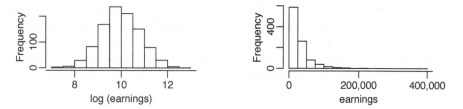

Figure 7.2 *Histogram of 1000 simulated values from the predictive distribution of the earnings of a 68-inch-tall man from a fitted regression model, on the logarithmic and original scales.*

We can display the simulations as a histogram (see Figure 7.2) and also compute various numerical summaries, for example,

- mean: `mean(pred)`
- median: `quantile(pred,.5)`
- 50% interval: `quantile(pred,c(.25,.75))`
- 95% interval: `quantile(pred,c(.025,.975))`

(These calculations ignore uncertainty in the regression parameters and thus are only approximate; we describe a more complete computational procedure later in this section.)

Why do we need simulation for predictive inferences?

For many purposes, point estimates, standard errors, and the intervals obtained from the `predict()` function in R are sufficient because the Central Limit Theorem ensures that for all but the smallest sample sizes and for reasonably well-behaved error distributions, coefficient estimates are approximately normally distributed (see page 14). Accounting for the uncertainty in the standard-error estimates, the t-distribution with $n-k$ degrees of freedom (where k is the number of predictors in the model) is a reliable approximation for the appropriate uncertainty distribution for the coefficients. Analytic procedures can also be used to get uncertainty for linear combinations of parameters and predictions. (An example of a linear combination of predictions is to use one of the models in Chapter 3 to predict the average test score of a group of 100 children whose mothers' educations and IQs are known.)

For more general predictions, however, the easiest and most reliable way to compute uncertainties is by simulation. For example, suppose we have a 68-inch-tall woman and a 68-inch-tall man, and we would like to use model (4.2) to predict the difference of their earnings. As a point estimate, we can use the difference of the point predictions: $\exp(8.4 + 0.017 \cdot 68 - 0.079 \cdot 1 + 0.007 \cdot 68 \cdot 1) - \exp(8.4 + 0.017 \cdot 68 - 0.079 \cdot 0 + 0.007 \cdot 68 \cdot 0) = 6900$. The simplest way to get a standard error or uncertainty interval for this prediction is to use simulation:

```
pred.man <- exp (rnorm (1000, 8.4 + .017*68 - .079*1 + .007*68*1, .88))     R code
pred.woman <- exp (rnorm (1000, 8.4 + .017*68 - .079*0 + .007*68*0, .88))
pred.diff <- pred.man - pred.woman
pred.ratio <- pred.man/pred.woman
```

We can summarize the distribution of this difference using a histogram or numerical summaries such as `mean(pred.diff)`, `quantile(pred.ratio,c(.25,.75))`, and so forth.

More generally, simulation is valuable because it can be used to summarize *any* function of estimated and predicted values. This is important partly for practical purposes in summarizing predictions and also because it allows us to fit complicated models in which the ultimate objects of interest are more complicated than a set of regression coefficients or linear combination of coefficients. Simulation will also be crucial when working with nonlinear models such as logistic regression.

Simulation to represent uncertainty in regression coefficients

The usual summary of a fitted regression gives standard errors along with estimates for each coefficient, and these give a sense of the uncertainty in estimation (see Figure 3.7 on page 40). When going beyond inferences for individual coefficients, however, it is helpful to summarize inferences by simulation, which gives us complete flexibility in propagating uncertainty about combinations of parameters and predictions.

For classical linear regressions and generalized linear models, we implement these simulations using the `sim()` function in R. For example, if we do

R code
```
n.sims <- 1000
fit.1 <- lm (log.earn ~ height + male + height:male)
sim.1 <- sim (fit.1, n.sims)
```

then `sim.1$beta` is a matrix with 1000 rows and 4 columns (representing 1000 independent simulations of the vector $(\beta_0, \beta_1, \beta_2, \beta_3)$), and `sim.1$sigma` is a vector of length 1000 (representing the estimation uncertainty in the residual standard-deviation parameter σ).

We can check that these simulations are equivalent to the regression computations, for example by the following commands in R, which print the mean, standard deviation, and 95% interval for the coefficient for `height` in the fitted model:

R code
```
height.coef <- sim.1$beta[,2]
mean (height.coef)
sd (height.coef)
quantile (height.coef, c(.025,.975))
```

For a more interesting example, consider the question: In this interaction model, what can be said about the coefficient of height among men? We cannot directly answer this question using the regression output: the slope for men is a sum of the `height` and `height:male` coefficients, and there is no simple way to compute its standard error given the information in the regression table. The most direct approach is to compute the 95% interval directly from the inferential simulations:

R code
```
height.for.men.coef <- sim.1$beta[,2] + sim.1$beta[,4]
quantile (height.for.men.coef, c(.025,.975))
```

The result is $[-0.003, 0.049]$, that is, $[-0.3\%, 4.9\%]$. Statistical significance is not the object of the analysis—our conclusions should not be greatly changed if, for example, the 95% interval instead were $[0.1\%, 5.3\%]$—but it is important to have a sense of the uncertainty of estimates, and it is convenient to be able to do this using the inferential simulations. The powers of inferential simulations are demonstrated more effectively when combined with prediction, as illustrated in Section 7.3.

Details of the simulation procedure

To get `n.sims` simulation draws (for example, 1000 is typically more than enough; see Chapter 17), we apply the following procedure based on Bayesian inference (see Chapter 18).

1. Use classical regression of n data points on k predictors to compute the vector $\hat{\beta}$ of estimated parameters, the unscaled estimation covariance matrix V_β, and the residual variance $\hat{\sigma}^2$.

2. Create `n.sims` random simulations of the coefficient vector β and the residual standard deviation σ. For each simulation draw:

   (a) Simulate $\sigma = \hat{\sigma}\sqrt{(n-k)/X}$, where X is a random draw from the χ^2 distribution with $n-k$ degrees of freedom.

   (b) Given the random draw of σ, simulate β from a multivariate normal distribution with mean $\hat{\beta}$ and variance matrix $\sigma^2 V_\beta$.

   These simulations are centered about the estimates $\hat{\beta}$ and $\hat{\sigma}$ with variation representing estimation uncertainty in the parameters. (For example, approximately 68% of the simulations of β_1 will be within ± 1 standard error of $\hat{\beta}_1$, approximately 95% will be within ± 2 standard errors, and so forth.)

These steps are performed automatically by our R function `sim()`, which pulls out $n, k, \hat{\beta}, V_\beta, \hat{\sigma}$ from the fitted linear model and then performs a loop over the n_{sims} simulations:

```
for (s in 1:n.sims){                                          R code
  sigma[s] <- sigma.hat*sqrt((n-k)/rchisq(1,n-k))
  beta[s,] <- mvrnorm (1, beta.hat, V.beta*sigma[s]^2)
}
```

The `sim()` function then returns the vector of simulations of σ and the $n_{\text{sims}} \times k$ matrix of simulations of β:

```
return (list (beta=beta, sigma=sigma))                         R code
```

The list items are given names so they can be accessed using these names from the simulation object. The function works similarly for generalized linear models such as logistic and Poisson regressions, adjusting for any overdispersion by using the standard errors of the coefficient estimates, which are scaled for overdispersion if that is included in the model.

Informal Bayesian inference

Bayesian inference refers to statistical procedures that model unknown parameters (and also missing and latent data) as random variables. As described in more detail in Sections 16.2 and 18.3, Bayesian inference starts with a *prior distribution* on the unknown parameters and updates this with the *likelihood* of the data, yielding a *posterior distribution* which is used for inferences and predictions.

Part 2 of this book discusses how Bayesian inference is appropriate for multilevel modeling—in which it is natural to fit probability distributions to batches of parameters. For the classical models considered in Part 1, Bayesian inference is simpler, typically starting with a "noninformative" or uniform prior distribution on the unknown parameters. We will not explore the technical issues further here except to note that the simulations presented here correspond to these noninformative prior distributions. It can also be helpful to think of these simulations as representing configurations of parameters and predictions that are compatible with the observed data—in the same sense that a classical confidence interval contains a range of parameter values that are not contradicted by the data.

Congressional elections in 1988

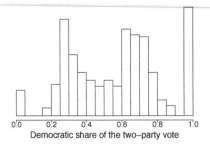

Democratic share of the two–party vote

Figure 7.3 *Histogram of congressional election data from 1988. The spikes at the left and right ends represent uncontested Republicans and Democrats, respectively.*

7.3 Simulation for nonlinear predictions: congressional elections

We illustrate nonlinear predictions in the context of a model of elections for the U.S. Congress. We first construct a model to predict the 1988 election from the 1986 election. Then we apply the model to predict 1990 from 1988. (It is convenient, when learning about a method, to predict outcomes that have already occurred, so the predictions can be compared to reality.)

Background

The United States is divided into 435 congressional districts, and we define the outcome y_i, for $i = 1, \ldots, n = 435$ to be the Democratic Party's share of the two-party vote (that is, excluding the votes for parties other than the Democrats and the Republicans) in district i in 1988. Figure 7.3 shows a histogram of the data y.

How can the variation in the data be understood? What information would be relevant in predicting the outcome of a congressional election? First of all, it is useful to know whether both parties are contesting the election; the spikes at the two ends of the histogram reveal that many of the elections were uncontested. After that, it would seem to make sense to use the outcome of the most recent previous election, which was in 1988. In addition, we use the knowledge of whether the *incumbent*—the current occupant of the congressional seat—is running for reelection.

Our regression model has the following predictors:

- A constant term

- The Democratic share of the two-party vote in district i in the previous election

- Incumbency: an indicator that equals $+1$ if district i is currently (as of 1988) occupied by a Democrat who is running for reelection, -1 if a Republican is running for reelection, and 0 if the election is *open*—that is, if neither of the two candidates is currently occupying the seat.

Because the incumbency predictor is categorical, we can display the data in a single scatterplot using different symbols for Republican incumbents, Democratic incumbents, and open seats; see Figure 7.4a.

We shall fit a linear regression. The data—number of votes for each candidate—are discrete, so it might at first seem appropriate to fit a generalized linear model such as an overdispersed binomial. But the number of votes within each district is large enough that the vote proportions are essentially continuous, so nothing would be gained by attempting to model the discreteness in the data.

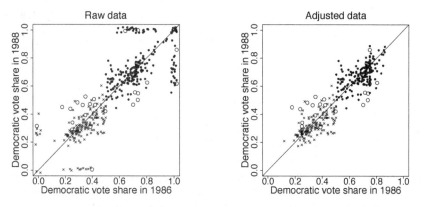

Figure 7.4 *(a) Congressional election data from 1986 and 1988. Crosses correspond to elections with Republican incumbents running in 1988, dots correspond to Democratic incumbents, and open circles correspond to open seats. The "incumbency" predictor in the regression model equals 0 for the circles, +1 for the dots, and −1 for the crosses. Uncontested election outcomes (at 0 and 1) have been jittered slightly. (b) Data for the regression analysis, with uncontested 1988 elections removed and uncontested 1986 election values replaced by 0.25 and 0.75. The y = x line is included as a comparison on both plots.*

Data issues

Many of the elections were uncontested in 1988, so that $y_i = 0$ or 1 exactly; for simplicity, we exclude these from our analysis. Thus, we are predicting the 1988 results given the outcome in the previous election and the knowledge of (a) whether the incumbent is running for reelection, and (b) whether the election will be contested. Primary elections are typically in September, and so it is reasonable to expect to have this information about two months before the November general election. We also exclude any elections that were won by third parties, yielding $n = 343$ congressional elections for our analysis.

In addition, many elections were uncontested in 1986, so the previous election outcome X_{2i} is 0 or 1 exactly. It would be possible to simply include these in the model as is; however, instead we impute the value 0.25 for uncontested Republicans and 0.75 for uncontested Democrats, which are intended to represent approximately the proportion of votes received by the Democratic candidate had the election actually been contested. (More generally, we can impute random values from the distribution of contested election outcomes preceding an uncontested race, but for our purposes here the simple imputation is sufficient.) The adjusted dataset is displayed in Figure 7.4b.

Fitting the model

First we fit the regression model in R: we label the adjusted variables as `vote.88`, `vote.86`, `incumbency.88`, subset them to include only the elections that were contested in 1988, and fit a linear model:

```
fit.88 <- lm (vote.88 ~ vote.86 + incumbency.88)
```
R code

Displaying yields

	coef.est	coef.se
(Intercept)	0.20	0.02
vote.86	0.58	0.04

R output

sim	σ	β_0	β_1	β_2	$\tilde{y}_1$	$\tilde{y}_2$	$\cdots$	$\tilde{y}_{55}$	$\cdots$	$\tilde{y}_{435}$	$\sum_i I(\tilde{y}_i > 0.5)$
1	.065	.19	.62	.067	.69	.57	$\cdots$	NA	$\cdots$	.79	251
2	.069	.25	.50	.097	.75	.63	$\cdots$	NA	$\cdots$	.76	254
$\vdots$	$\vdots$	$\vdots$	$\vdots$	$\vdots$	$\vdots$	$\vdots$	$\vdots$	$\vdots$	$\ddots$	$\vdots$	$\vdots$
1000	.067	.23	.51	.089	.73	.57	$\cdots$	NA	$\cdots$	.69	251
median	.068	.20	.58	.077	.73	.65	$\cdots$	NA	$\cdots$	.72	253
mean	.067	.20	.58	.078	.73	.65	$\cdots$	NA	$\cdots$	.72	252.4
sd	.003	.02	.04	.007	.07	.07	$\cdots$	NA	$\cdots$	.07	3.1

Figure 7.5 *Simulation results for the congressional election forecasting model. The predicted values $\tilde{y}_i$ correspond to the 1990 election. The NAs are for a district that was uncontested in 1990, so it was not predicted by the regression model.*

```
incumbency.88      0.08      0.01
  n = 343, k = 3
  residual sd = 0.067, R-Squared = 0.88
```

This model has serious problems, as can be seen, for example, by careful examination of the plot of residuals or even of the before-after plot in Figure 7.4b (for example, the jump between the average y-values just below and just above $x = 0.5$ is not completely fit by the incumbency.88 predictor). Better models can be fit to these data (see Exercise 9.13), but the simple regression fit here is sufficient to demonstrate the principles of simulation-based predictive inference.

Simulation for inferences and predictions of new data points

The first five columns of Figure 7.5 show a set of simulation results for the parameters in the fitted model. We use these, along with the data from 1988 and incumbency information in 1990, to predict the district-by-district election outcome in 1990. We start by creating a new matrix of predictors, $\tilde{X}$:

R code
```
n.tilde <- length (vote.88)
X.tilde <- cbind (rep(1,n.tilde), vote.88, incumbency.90)
```

We then simulate $n_{\text{sims}} = 1000$ predictive simulations of the vector of $\tilde{n}$ new data points with $\tilde{n} \times k$ matrix of predictors $\tilde{X}$. For each simulation, we compute the predicted value $\tilde{X}\beta$ and add normal errors:

R code
```
n.sims <- 1000
sim.88 <- sim (fit.88, n.sims)
y.tilde <- array (NA, c(n.sims, n.tilde))
for (s in 1:n.sims){
  y.tilde[s,] <- rnorm (n.tilde, X.tilde %*% sim.88$beta[s,],
    sim.88$sigma[s])}
```

This last matrix multiplication works because X.tilde is a $\tilde{n} \times 3$ matrix and sim.88$beta is a $n_{\text{sims}} \times 3$ matrix; thus the selected row, sim.88$beta[s,], is a vector of length 3, and the product X.tilde%*%sim.88$beta[s,] is a vector of length $\tilde{n}$ that represents the vector of predicted values for that particular simulation draw.

Predictive simulation for a nonlinear function of new data

For the congressional elections example, we perform inference on the summary measure $\sum_{i=1}^{\tilde{n}} I(\tilde{y}_i > 0.5)$—the number of elections won by the Democrats in 1990, by summing over the rows in the matrix:[3]

```
dems.tilde <- rowSums (y.tilde > .5)
```
R code

The last column of Figure 7.5 shows the results. Each row shows the outcome of a different random simulation.

The lower lines of the table in Figure 7.5 show the median, mean, and standard deviation of each simulated outcome. The means and medians of the parameters σ and β are nearly identical to the point estimates (the differences are due to variation because there are only 1000 simulation draws). The future election outcome in each district has a predictive uncertainty of about 0.07, which makes sense since the estimated standard deviation from the regression is $\hat{\sigma} = 0.07$. (The predictive uncertainties are slightly higher than $\hat{\sigma}$, but by only a very small amount since the number of data points in the original regression is large, and the x-values for the predictions are all within the range of the original data.)

Finally, the entries in the lower-right corner of Figure 7.5 give a predictive mean of 252.4 and standard error of 3.1 for the number of districts to be won by the Democrats. This estimate and standard error *could not* simply be calculated from the estimates and uncertainties for the individual districts. Simulation is the only practical method of assessing the predictive uncertainty for this nonlinear function of the predicted outcomes.

Incidentally, the actual number of seats won by the Democrats in 1990 was 262. This is more than 3 standard deviations away from the mean, which suggests that the model is not quite applicable to the 1990 election—this makes sense since it does not allow for national partisan swings of the sort that happen from election to election.

Implementation using functions

We could also compute these predictions by writing a custom R function:

```
Pred.88 <- function (X.pred, lm.fit){              R code
  n.pred <- dim(X.pred)[1]
  sim.88 <- sim (lm.fit, 1)
  y.pred <- rnorm (n.pred, X.pred %*% t(sim.88$beta), sim.88$sigma)
  return (y.pred)
}
```

and then creating 1000 simulations using the `replicate()` function in R:

```
y.tilde <- replicate (1000, Pred.88 (X.tilde, fit.88))     R code
```

To predict the total number of seats won by the Democrats, we can add a wrapper:

```
dems.tilde <- replicate (1000, Pred.88 (X.tilde, fit.88) > .5)   R code
```

Computing using `replicate()` (or related functions such as `apply()` and `sapply()`) results in faster and more compact R code which, depending on one's programming experience can appear either simpler or more mysterious than explicit looping. We sometimes find it helpful to perform computations both ways when we are uncertain about the programming.

---

[3] We could also calculate this sum in a loop as
```
dems.tilde <- rep (NA, n.sims)
for (s in 1:n.sims){
  dems.tilde[s] <- sum (y.tilde[s,] > .5)}
```

Combining simulation and analytic calculations

In some settings it is helpful to supplement simulation-based inference with mathematical analysis. For example, in the election prediction model, suppose we want to estimate the probability that the election in a particular district will be tied, or within one vote of being exactly tied. (This calculation is relevant, for example, in estimating the probability that an individual vote will be decisive, and comparing these probabilities can be relevant for parties' decisions for allocating campaign resources.)

Consider a district, i, with n_i voters. For simplicity we suppose n_i is even. This district's election will be tied if the future vote outcome, $\tilde{y}_i$, is exactly 0.5. We have approximated the distribution of $\tilde{y}$ as continuous—which is perfectly reasonable given that the n_i's are in the tens or hundreds of thousands—and so a tie is equivalent to $\tilde{y}_i$ being in the range $[\frac{1}{2} - \frac{1}{2n_i}, \frac{1}{2} + \frac{1}{2n_i}]$.

How can we compute this probability by simulation? The most direct way is to perform many predictive simulations and count the proportion for which $\tilde{y}_i$ falls in the range $0.5 \pm 1/(2n_i)$. Unfortunately, for realistic n_i's, this range is so tiny that thousands or millions of simulations could be required to estimate this probability accurately. (For example, it would not be very helpful to learn that 0 out of 1000 simulations fell within the interval.)

A better approach is to combine simulation and analytical results: first compute 1000 simulations of $\tilde{y}$, as shown, then for each district compute the proportion of simulations that fall between 0.49 and 0.51, say, and divide by $0.02n_i$ (that is, the number of intervals of width $1/n_i$ that fit between 0.49 and 0.51). Or compute the proportion falling between 0.45 and 0.55, and divide by $0.1n_i$. For some districts, the probability will still be estimated at zero after 1000 simulation draws, but in this case the estimated zero is much more precise.

Estimated probabilities for extremely rare events can be computed in this example using the fact that predictive distributions from a linear regression follow the t distribution with $n-k$ degrees of freedom. We can use 1000 simulations to compute a predictive mean and standard deviation for each $\tilde{y}_i$, then use tail probabilities of the t_{340} distribution (in this example, $n = 343$ and $k = 3$) to compute the probability of falling in the range $0.5 \pm 1/(2n_i)$.

7.4 Predictive simulation for generalized linear models

As with linear regression, we simulate inference for generalized linear models in two steps: first using sim() to obtain simulations for the coefficients, then simulating predictions from the appropriate model, given the linear predictor.

Logistic regression

We illustrate for one of the models from Chapter 5 of the probability of switching wells given the distance from the nearest safe well.

Simulating the uncertainty in the estimated coefficients. Figure 7.6a shows the uncertainty in the regression coefficients, computed as follows:

R code
```
sim.1 <- sim (fit.1, n.sims=1000)
plot (sim.1$beta[,1], sim.1$beta[,2], xlab=expression(beta[0]),
  ylab=expression(beta[1]))
```

and Figure 7.6b shows the corresponding uncertainty in the logistic regression curve, displayed as follows:

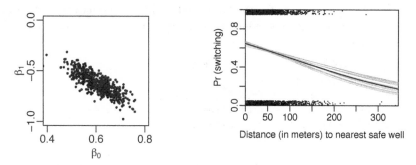

Figure 7.6 *(a) Uncertainty in the estimated coefficients β_0 and β_1 in the logistic regression,* $\Pr(\text{switching wells}) = logit^{-1}(\beta_0 - \beta_0 \cdot \text{dist100})$. *(b) Graphical expression of the best-fit model,* $\Pr(\text{switching wells}) = logit^{-1}(0.61 - 0.62 \cdot \text{dist100})$, *with (jittered) data overlain. Light lines represent estimation uncertainty in the logistic regression coefficients, corresponding to the distribution of β shown to the left. Compare to Figure 5.9 on page 89.*

sim	β_0	β_1	$\tilde{y}_1$	$\tilde{y}_2$	$\cdots$	$\tilde{y}_{10}$
1	0.68	−0.007	1	0	$\cdots$	1
2	0.61	−0.005	0	0	$\cdots$	1
$\vdots$	$\vdots$	$\vdots$	$\vdots$	$\vdots$	$\ddots$	$\vdots$
1000	0.69	−0.006	1	1	$\cdots$	1
mean	0.61	−0.006	0.60	0.59	$\cdots$	0.52

Figure 7.7 *Simulation results for ten hypothetical new households in the well-switching example, predicting based only on distance to the nearest well. The inferences for (β_0, β_1) are displayed as a scatterplot in Figure 7.6a. The bottom row—the mean of the simulated values of $\tilde{y}_i$ for each household i—gives the estimated probabilities of switching.*

```
plot (dist, switch)                                              R code
for (s in 1:20){
  curve (invlogit (sim.1$beta[s,1] + sim.1$beta[s,2]*x), col="gray",
    add=TRUE)}
curve (invlogit (fit.1$coef[1] + fit.1$coef[2]*x), add=TRUE)
```

Predictive simulation using the binomial distribution. Now suppose, for example, that we would like to predict the switching behavior for $\tilde{n}$ new households, given a predictor matrix $\tilde{X}$ (which will have $\tilde{n}$ rows and, in this example, two columns, corresponding to the constant term and the distance to the nearest safe well). As with linear regression, we can use simulation to account for the predictive uncertainty. In this case, we use the binomial distribution to simulate the prediction errors:

```
n.tilde <- nrow (X.tilde)                                        R code
y.tilde <- array (NA, c(n.sims, n.tilde))
for (s in 1:n.sims){
  p.tilde <- invlogit (X.tilde %*% sim.1$beta[s,])
  y.tilde[s,] <- rbinom (n.tilde, 1, p.tilde)
}
```

Figure 7.7 shows an example set of `n.sims` = 1000 simulations corresponding to `n.tilde` = 10 new households.

Predictive simulation using the latent logistic distribution. An alternative way to simulate logistic regression predictions uses the latent-data formulation (see Section

5.3). We obtain simulations for the latent data $\tilde{z}$ by adding independent errors $\tilde{\epsilon}$ to the linear predictor, and then convert to binary data by setting $\tilde{y}_i = 1$ if $\tilde{z}_i > 0$ for each new household i:

R code
```
y.tilde <- array (NA, c(n.sims, n.tilde))
for (s in 1:n.sims){
  epsilon.tilde <- logit (runif (n.tilde, 0, 1))
  z.tilde <- X.tilde %*% t(sim.1$beta) + epsilon.tilde
  y.tilde[s,] <- ifelse (z.tilde>0, 1, 0)
}
```

Other generalized linear models

We can do similar computations with Poisson regression: inference just as before, and predictive simulations using `rpois()`.

For overdispersed Poisson regression, the function `rnegbin()` samples from the negative binomial distribution. Another option is to sample first from the gamma, then the Poisson. For overdispersed binomial, simulations from the beta-binomial distribution can be obtained by drawing first from the beta distribution, then the binomial.

Compound models

Simulation is the easiest way of summarizing inferences from more complex models. For example, as discussed in Section 6.7, we can model earnings from height in two steps:

$$\Pr(\text{earnings} > 0) = \text{logit}^{-1}(-3.76 + 0.08 \cdot \text{height} + 1.70 \cdot \text{male})$$

If earnings > 0, then earnings $= \exp(8.15 + 0.02 \cdot \text{height} + 0.42 \cdot \text{male} + \epsilon)$,

with the error term ϵ having a normal distribution with mean 0 and standard deviation 0.88.

We can simulate the earnings of a randomly chosen 68-inch tall man. We first show, for simplicity, the simulation ignoring uncertainty in the regression coefficients:

R code
```
fit.1a <- glm (earn.pos ~ height + male, family=binomial(link="logit"))
fit.1b <- lm (log.earn ~ height + male, subset = earnings>0)
x.new <- c (1, 68, 1)       # constant term=1, height=68, male=1

n.sims <- 1000
prob.earn.pos <- invlogit (coef(fit.1a) %*% x.new)
earn.pos.sim <- rbinom (n.sims, 1, prob.earn.pos)
earn.sim <- ifelse (earn.pos.sim==0, 0,
  exp (rnorm (n.sims, coef(fit.1b) %*% x.new, sigma.hat(fit.1b))))
```

More generally, we can use the simulated values of the coefficient estimates:

R code
```
sim.1a <- sim (fit.1a, n.sims)
sim.1b <- sim (fit.1b, n.sims)
for (s in 1:n.sims){
  prob.earn.pos <- invlogit (sim.1a$beta %*% x.new)
  earn.pos.sim <- rbinom (n.sims, 1, prob.earn.pos)
  earn.sim[s] <- ifelse (earn.pos.sim==0, 0
    exp (rnorm (n.sims, sim.1b$beta %*% x.new, sim.1b$sigma)))
}
```

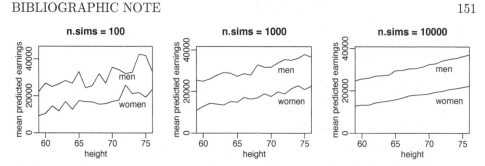

Figure 7.8 *Mean predicted earnings as a function of height and sex for the two-stage model (logistic regression for the probability earnings being positive, followed by linear regression for the logarithms of positive earnings), as computed using 100, 1000, or 10000 simulations.*

Now suppose we want to understand this compound model by plotting the mean predicted earnings as a function of height and sex. We can first put the computations into a function:

```
Mean.earn <- function (height, male, sim.a, sim.b){
  x.new <- c (1, height, male)
  prob.earn.pos <- invlogit (sim.a$beta %*% x.new)
  earn.pos.sim <- rbinom (n.sims, 1, prob.earn.pos)
  earn.sim <- ifelse (earn.pos.sim==0, 0,
    exp (rnorm (n.sims, sim.b$beta %*% x.new, sim.b$sigma)))
  return (mean(earn.sim))
}
```
R code

and then evaluate the function in a loop using the `sapply()` function:[4]

```
heights <- seq (60, 75, 1)
mean.earn.female <- sapply (heights, Mean.earn, male=0, sim.1a, sim.1b)
mean.earn.male <- sapply (heights, Mean.earn, male=1, sim.1a, sim.1b)
```
R code

The plots of `mean.earn.female` and `mean.earn.male` versus `heights` appear in Figure 7.8, for three different values of n_{sims}. The general pattern is clear from 100 simulations, but more simulations are helpful to avoid being distracted by random noise.

7.5 Bibliographic note

Random simulation for performing computations in probability and statistics was one of the first applications of computers, dating back to the 1940s. As computing power became more dispersed since the 1970s, simulation has been used increasingly frequently for summarizing statistical inferences; Rubin (1980) is an early example.

Our simulation-based approach to computation is described in Gelman et al. (2003), and a recent implementation in R appears in Kerman and Gelman (2006). The congressional election analysis in Section 7.3 uses a simplified version of the models of Gelman and King (1990, 1994a).

---

[4] Alternatively, the looping could be programmed explicitly:
```
heights <- seq (60, 75, 1)
k <- length(heights)
mean.earn.female <- rep (NA, k)
mean.earn.male <- rep (NA, k)
for (i in 1:k) {
  mean.earn.female[i] <- Mean.earn (heights[i], 0, sim.1a, sim.1b)
  mean.earn.male[i] <- Mean.earn (heights[i], 1, sim.1a, sim.1b)
}
```

7.6 Exercises

1. Discrete probability simulation: suppose that a basketball player has a 60% chance of making a shot, and he keeps taking shots until he misses two in a row. Also assume his shots are independent (so that each shot has 60% probability of success, no matter what happened before).

   (a) Write an R function to simulate this process.

   (b) Put the R function in a loop to simulate the process 1000 times. Use the simulation to estimate the mean, standard deviation, and distribution of the total number of shots that the player will take.

   (c) Using your simulations, make a scatterplot of the number of shots the player will take and the proportion of shots that are successes.

2. Continuous probability simulation: the logarithms of weights (in pounds) of men in the United States are approximately normally distributed with mean 5.13 and standard deviation 0.17; women with mean 4.96 and standard deviation 0.20. Suppose 10 adults selected at random step on an elevator with a capacity of 1750 pounds. What is the probability that the elevator cable breaks?

3. Propagation of uncertainty: we use a highly idealized setting to illustrate the use of simulations in combining uncertainties. Suppose a company changes its technology for widget production, and a study estimates the cost savings at $5 per unit, but with a standard error of $4. Furthermore, a forecast estimates the size of the market (that is, the number of widgets that will be sold) at 40,000, with a standard error of 10,000. Assuming these two sources of uncertainty are independent, use simulation to estimate the total amount of money saved by the new product (that is, savings per unit, multiplied by size of the market).

4. Predictive simulation for linear regression: take one of the models from Exercise 3.5 or 4.8 that predicts course evaluations from beauty and other input variables. You will do some simulations.

   (a) Instructor A is a 50-year-old woman who is a native English speaker and has a beauty score of −1. Instructor B is a 60-year-old man who is a native English speaker and has a beauty score of −0.5. Simulate 1000 random draws of the course evaluation rating of these two instructors. In your simulation, account for the uncertainty in the regression parameters (that is, use the sim() function) as well as the predictive uncertainty.

   (b) Make a histogram of the difference between the course evaluations for A and B. What is the probability that A will have a higher evaluation?

5. Predictive simulation for linear regression: using data of interest to you, fit a linear regression model. Use the output from this model to simulate a predictive distribution for observations with a particular combination of levels of all the predictors in the regression.

6. Repeat the previous exercise using a logistic regression example.

7. Repeat the previous exercise using a Poisson regression example.

8. Inference for the ratio of parameters: a (hypothetical) study compares the costs and effectiveness of two different medical treatments.

   - In the first part of the study, the difference in costs between treatments A and B is estimated at $600 per patient, with a standard error of $400, based on a regression with 50 degrees of freedom.

- In the second part of the study, the difference in effectiveness is estimated at 3.0 (on some relevant measure), with a standard error of 1.0, based on a regression with 100 degrees of freedom.

- For simplicity, assume that the data from the two parts of the study were collected independently.

Inference is desired for the *incremental cost-effectiveness ratio*: the difference between the average costs of the two treatments, divided by the difference between their average effectiveness. (This problem is discussed further by Heitjan, Moskowitz, and Whang, 1999.)

(a) Create 1000 simulation draws of the cost difference and the effectiveness difference, and make a scatterplot of these draws.

(b) Use simulation to come up with an estimate, 50% interval, and 95% interval for the incremental cost-effectiveness ratio.

(c) Repeat this problem, changing the standard error on the difference in effectiveness to 2.0.

9. Summarizing inferences and predictions using simulation: Exercise 6.5 used a Tobit model to fit a regression with an outcome that had mixed discrete and continuous data. In this exercise you will revisit these data and build a two-step model: (1) logistic regression for zero earnings versus positive earnings, and (2) linear regression for level of earnings given earnings are positive. Compare predictions that result from each of these models with each other.

10. How many simulation draws are needed: take the model from Exercise 3.5 that predicts course evaluations from beauty and other input variables. Use `display()` to summarize the model fit. Focus on the estimate and standard error for the coefficient of beauty.

(a) Use `sim()` with `n.iter = 10000`. Compute the mean and standard deviations of the 1000 simulations of the coefficient of beauty, and check that these are close to the output from `display`.

(b) Repeat with `n.iter = 1000`, `n.iter = 100`, and `n.iter = 10`. Do each of these a few times in order to get a sense of the simulation variability.

(c) How many simulations were needed to give a good approximation to the mean and standard error for the coefficient of beauty?

## CHAPTER 8

Simulation for checking statistical procedures and model fits

This chapter describes a variety of ways in which probabilistic simulation can be used to better understand statistical procedures in general, and the fit of models to data in particular. In Sections 8.1–8.2, we discuss *fake-data simulation*, that is, controlled experiments in which the parameters of a statistical model are set to fixed "true" values, and then simulations are used to study the properties of statistical methods. Sections 8.3–8.4 consider the related but different method of *predictive simulation*, where a model is fit to data, then replicated datasets are simulated from this estimated model, and then the replicated data are compared to the actual data.

The difference between these two general approaches is that, in fake-data simulation, estimated parameters are compared to true parameters, to check that a statistical method performs as advertised. In predictive simulation, replicated datasets are compared to an actual dataset, to check the fit of a particular model.

8.1 Fake-data simulation

Simulation of fake data can be used to validate statistical algorithms and to check the properties of estimation procedures. We illustrate with a simple regression model, where we simulate fake data from the model, $y = \alpha + \beta x + \epsilon$, refit the model to the simulated data, and check the coverage of the 68% and 95% intervals for the coefficent β.

First we set up the true values of the parameters—which we arbitrarily set to $\alpha = 1.4$, $\beta = 2.3$, $\sigma = 0.9$—and set up the predictors, which we arbitrarily set to $(1, 2, 3, 4, 5)$:

```
a <- 1.4
b <- 2.3
sigma <- 0.9
x <- 1:5
n <- length(x)
```
R code

We then simulate a vector y of fake data and fit a regression model to these data. The fitting makes no use of the true values of α, β, and σ.

```
y <- a + b*x + rnorm (n, 0, sigma)
lm.1 <- lm (y ~ x)
display (lm.1)
```
R code

Here is the regression output:

```
lm(formula = y ~ x)
            coef.est coef.se
(Intercept)     0.92    1.09
x               2.62    0.33
  n = 5, k = 2
  residual sd = 1.04, R-Squared = 0.95
```
R output

Comparing the estimated coefficients to the true values 1.4 and 2.3, the fit seems reasonable enough: the estimates are not exact but are within the margin of error. We can perform this comparison more formally by extracting from the regression object the estimate and standard error of β (the second coefficient in the model):

R code
```
b.hat <- coef (lm.1)[2]           # "b" is the 2nd coef in the model
b.se <- se.coef (lm.1)[2]         # "b" is the 2nd coef in the model
```

and then checking whether the true β falls within the estimated 68% and 95% confidence intervals obtained by taking the estimate ± 1 or ± 2 standard errors (recall Figure 3.7 on page 40):

R code
```
cover.68 <- abs (b - b.hat) < b.se      # this will be TRUE or FALSE
cover.95 <- abs (b - b.hat) < 2*b.se    # this will be TRUE or FALSE
cat (paste ("68% coverage: ", cover.68, "\n"))
cat (paste ("95% coverage: ", cover.95, "\n"))
```

So, the confidence intervals worked once, but do they have the correct coverage probabilities? We can check by embedding the data simulation, model fitting, and coverage checking in a loop and running 1000 times:[1]

R code
```
n.fake <- 1000
cover.68 <- rep (NA, n.fake)
cover.95 <- rep (NA, n.fake)
for (s in 1:n.fake){
  y <- a + b*x + rnorm (n, 0, sigma)
  lm.1 <- lm (y ~ x)
  b.hat <- coef (lm.1)[2]
  b.se <- se.coef (lm.1)[2]
  cover.68[s] <- abs (b - b.hat) < b.se
  cover.95[s] <- abs (b - b.hat) < 2*b.se
}
cat (paste ("68% coverage: ", mean(cover.68), "\n"))
cat (paste ("95% coverage: ", mean(cover.95), "\n"))
```

The following appears on the console:

R output
```
68% coverage:  0.61
95% coverage:  0.85
```

That is, mean(cover.68) = 0.61 and mean(cover.95) = 0.85. This does not seem right: only 61% of the 68% intervals and 85% of the 95% intervals covered the true parameter value!

Our problem is that the ± 1 and ± 2 standard-error intervals are appropriate for the normal distribution, but with such a small sample size our inferences should use the t distribution, in this case with 3 degrees of freedom (5 data points, minus 2 coefficients estimated; see Section 3.4). We repeat our simulation but using t_3 confidence intervals:

R code
```
n.fake <- 1000
cover.68 <- rep (NA, n.fake)
cover.95 <- rep (NA, n.fake)
t.68 <- qt (.84, n-2)
t.95 <- qt (.975, n-2)
for (s in 1:n.fake){
  y <- a + b*x + rnorm (n, 0, sigma)
```

---

[1] This and other loops in this chapter could also be performed implicitly using the `replicate()` function in R, as illustrated on pages 139 and 147.

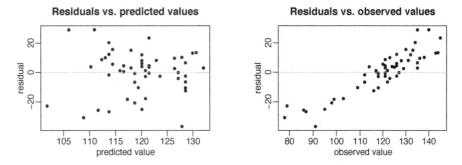

Figure 8.1 *From a model predicting final exam grades from midterms: plots of regression residuals versus predicted and versus observed values. The left plot looks reasonable but the right plot shows strong patterns. How to understand these? An exploration using fake data (see Figure 8.2) shows that, even if the model were correct, we would expect the right plot to show strong patterns. The plot of residuals versus observed thus does not indicate a problem with the model.*

```
    lm.1 <- lm (y ~ x)
    b.hat <- coef (lm.1)[2]
    b.se <- se.coef (lm.1)[2]
    cover.68[s] <- abs (b - b.hat) < t.68*b.se
    cover.95[s] <- abs (b - b.hat) < t.95*b.se
}
cat (paste ("68% coverage ", mean(cover.68), "\n"))
cat (paste ("95% coverage: ", mean(cover.95), "\n"))
```

and now we obtain coverages of 67% and 96%, as predicted (within the expected level of variation based on 1000 simulations; see Exercise 7.10).

8.2 Example: using fake-data simulation to understand residual plots

For another illustration of the power of fake data, we simulate from a regression model to get insight into residual plots, in particular, to understand why we plot residuals versus fitted values rather than versus observed values (see Section 3.6).

We illustrate with a simple model predicting final exam scores from midterms in an introductory statistics class:

```
    lm.1 <- lm (final ~ midterm)
```
R code

yielding

```
              coef.est coef.se
(Intercept)     64.5     17.0
midterm          0.7      0.2
  n = 52, k = 2
  residual sd = 14.8, R-Squared = 0.18
```
R output

We construct fitted values $\hat{y} = X\hat{\beta}$ and residuals $y - X\hat{\beta}$:

```
n <- length (final)
X <- cbind (rep(1,n), midterm)
predicted <- X %*% coef (lm.1)
```
R code

Figure 8.1 shows the residuals from this model, plotted in two different ways: (a) residuals versus fitted values, and (b) residuals versus observed values. The first

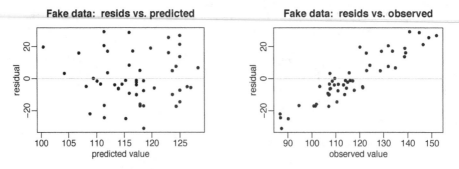

Figure 8.2 *From fake data: plots of regression residuals versus predicted and versus observed values. The data were simulated from the fitted family of regression models, and so we know that the strong pattern in the right panel does not represent any sort of model failure. This is an illustration of the use of fake data to evaluate diagnostic plots. Compare to the corresponding plots of real data in Figure 8.1.*

plot looks reasonable: the residuals are centered around zero for all fitted values. But the second plot looks troubling.

It turns out that the first plot is what we should be looking at, and the second plot is misleading. This can be understood using probability theory (from the regression model, the errors ϵ should be independent of the predictors x, not the data y) but a perhaps more convincing demonstration uses fake data, as we now illustrate.

For this example, we set the regression coefficients and residual standard error to reasonable values given the model estimates, and then simulate fake data:

R code
```
a <- 65
b <- 0.7
sigma <- 15
y.fake <- a + b*midterm + rnorm (n, 0, 15)
```

Next we fit the regression model to the fake data and compute fitted values and residuals:

R code
```
lm.fake <- lm (y.fake ~ midterm)
predicted.fake <- X %*% coef (lm.fake)
resid.fake <- y.fake - predicted.fake
```

(The predicted values could also be obtained in R using `fitted(lm.fake)`; here we explicitly multiply the predictors by the coefficients to emphasize the computations used in creating the fake data.) Figure 8.2 shows the plots of `resid.fake` versus `predicted.fake` and `y.fake`. These are the sorts of residual plots we would see *if the model were correct*. This simulation shows why we prefer, as a diagnostic plot, to view residuals versus predicted rather than observed values.

8.3 Simulating from the fitted model and comparing to actual data

So far we have considered several uses of simulation: exploring the implications of hypothesized probability models (Section 7.1); exploring the implications of statistical models that were fit to data (Sections 7.2–7.4); studying the properties of statistical procedures by comparing to known true values of parameters (Sections 8.1–8.2). Here we introduce yet another twist: simulating replicated data under the fitted model (as with the predictions in Sections 7.2–7.4) and then comparing these to the observed data (rather than comparing estimates to true parameter values as in Section 8.1).

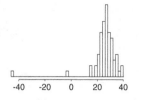

Figure 8.3 *Histogram of Simon Newcomb's measurements for estimating the speed of light, from Stigler (1977). The data represent the amount of time required for light to travel a distance of 7442 meters and are recorded as deviations from 24,800 nanoseconds.*

Example: comparing data to replications from a fitted normal distribution

The most fundamental way to check model fit is to display replicated datasets and compare them to the actual data. Here we illustrate with a simple case, from a famous historical dataset that did not fit the normal distribution. The goal of this example is to demonstrate how the lack of fit can be seen using predictive replications.

Figure 8.3 shows the data, a set of measurements taken by Simon Newcomb in 1882 as part of an experiment to estimate the speed of light. We (inappropriately) fit a normal distribution to these data, which in the regression context can be done by fitting a linear regression with no predictors:

```
light <- lm (y ~ 1)
```
R code

The next step is to simulate 1000 replications from the parameters in the fitted model (in this case, simply the constant term β_0 and the residual standard deviation σ):

```
n.sims <- 1000
sim.light <- sim (light, n.sims)
```
R code

We can then use these simulations to create 1000 fake datasets of 66 observations each:

```
n <- length (y)
y.rep <- array (NA, c(n.sims, n))
for (s in 1:n.sims){
  y.rep[s,] <- rnorm (n, sim.light$beta[s], sim.light$sigma[s])
}
```
R code

Visual comparison of actual and replicated datasets.   Figure 8.4 shows a plot of 20 of the replicated datasets, produced as follows:

```
par (mfrow=c(5,4))
for (s in 1:20){
  hist (y.rep[s,])
}
```
R code

The systematic differences between data and replications are clear. In more complicated problems, more effort may be needed to effectively display the data and replications for useful comparisons, but the same general idea holds.

Checking model fit using a numerical data summary.   Data displays can suggest more focused test statistics with which to check model fit, as we illustrate in Section 24.2. Here we demonstrate a simple example with the speed-of-light measurements. The graphical check in Figures 8.3 and 8.4 shows that the data have some extremely low values that do not appear in the replications. We can formalize this check by defining a *test statistic*, $T(y)$, equal to the minimum value of the data, and then calculating $T(y^{\text{rep}})$ for each of the replicated datasets:

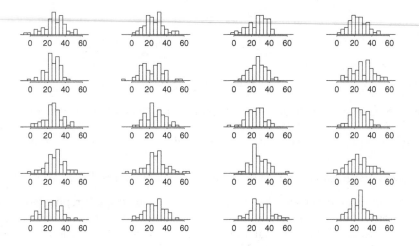

Figure 8.4 *Twenty replications, y^{rep}, of the speed-of-light data from the predictive distribution under the normal model; compare to observed data, y, in Figure 8.3. Each histogram displays the result of drawing 66 independent values y_i^{rep} from a common normal distribution with mean and standard deviation (μ, σ) estimated from the data.*

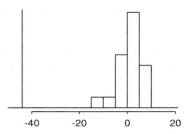

Figure 8.5 *Smallest observation of Newcomb's speed-of-light data (the vertical line at the left of the graph), compared to the smallest observations from each of 20 posterior predictive simulated datasets displayed in Figure 8.4.*

R code
```
Test <- function (y){
  min (y)
}
test.rep <- rep (NA, n.sims)
for (s in 1:n.sims){
  test.rep[s] <- Test (y.rep[s,])
}
```

We then plot a histogram of the minima of the replicated datasets, with a vertical line indicating the minimum of the observed data:

R code
```
hist (test.rep, xlim=range (Test(y), test.rep))
lines (rep (Test(y), 2), c(0,n))
```

Figure 8.5 shows the result: the smallest observations in each of the hypothetical replications are all much larger than Newcomb's smallest observation, which is indicated by a vertical line on the graph. The normal model clearly does not capture the variation that Newcomb observed. A revised model might use an asymmetric contaminated normal distribution or a symmetric long-tailed distribution in place of the normal measurement model.

Example: zeroes in count data

For a more complicated example, we consider a study of the effect of integrated pest management on reducing cockroach levels in urban apartments. In this experiment, the treatment and control were applied to 160 and 104 apartments, respectively, and the outcome measurement y_i in each apartment i was the number of roaches caught in a set of traps. Different apartments had traps for different numbers of days, and we label as u_i the number of trap-days. The natural model for the roach counts is then $y_i \sim \text{Poisson}(u_i \exp(X_i\beta))$, where X represents the regression predictors (in this case, a pre-treatment roach level, a treatment indicator, and an indicator for whether the apartment is in a "senior" building restricted to the elderly, and the constant term). The logarithm of the exposure, $\log(u_i)$, plays the role of the "offset" in the Poisson regression (see model (6.3) on page 111).

We fit the model

```
glm.1 <- glm (y ~ roach1 + treatment + senior, family=poisson,
   offset=log(exposure2))
```
R code

which yields

```
            coef.est coef.se
(Intercept)   -0.46    0.02
roach1         0.24    0.00
treatment     -0.48    0.02
senior        -0.40    0.03
  n = 264, k = 4
  residual deviance = 11753.3, null deviance = 17354 (difference = 5600.7)
```
R output

The treatment appears to be effective in reducing roach counts—we shall return to this issue in a later chapter with a fuller exploration of this study. For now, we are simply interested in evaluating the model as a description of the data, without worrying about causal issues or the interpretation of the coefficients.

Comparing the data, y, to a replicated dataset, y^{rep}. How well does this model fit the data? We explore by simulating a replicated dataset y^{rep} that might be seen if the model were true and the study were performed again:

```
n <- length (y)
X <- cbind (rep(1,n), roach1, treatment, senior)
y.hat <- exposure2 * exp (X %*% coef (glm.1))
y.rep <- rpois (n, y.hat)
```
R code

We can compare the replicated data y^{rep} to the original data y in various ways. We illustrate with a simple test of the number of zeroes in the data:

```
print (mean (y==0))
print (mean (y.rep==0))
```
R code

which reveals that 36% of the observed data points, but none of the replicated data points, equal zero. This suggests a potential problem with the model: in reality, many apartments have zero roaches, but this would not be happening if the model were true, at least to judge from one simulation.

Comparing the data y to 1000 replicated datasets y^{rep}. To perform this model check more formally, we simulate 1000 replicated datasets y^{rep}, which we store in a matrix:

R code
```
n.sims <- 1000
sim.1 <- sim (glm.1, n.sims)
y.rep <- array (NA, c(n.sims, n))
for (s in 1:n.sims){
  y.hat <- exposure2 * exp (X %*% sim.1$beta[s,])
  y.rep[s,] <- rpois (n, y.hat)
}
```

For each of these replications, we then compute a test statistic: the proportion of zeroes in the (hypothetical) dataset:

R code
```
Test <- function (y){
  mean (y==0)
}
test.rep <- rep (NA, n.sims)
for (s in 1:n.sims){
  test.rep[k] <- Test (y.rep[s,])
}
```

The 1000 values of `test.rep` vary from 0 to 0.008—all of which are much lower than the observed test statistic of 0.36. Thus the Poisson regression model does not replicate the frequency of zeroes in the data.

Checking the overdispersed model

We probably should have just started with an overdispersed Poisson regression:

R code
```
glm.2 <- glm (y ~ roach1 + treatment + senior, family=quasipoisson,
    offset=log(exposure2))
```

which yields

R output
```
glm(formula = y ~ roach1 + treatment + senior, family = quasipoisson,
        offset = log(exposure2))
              coef.est coef.se
(Intercept)    -0.46    0.17
roach1          0.24    0.03
treatment      -0.48    0.20
senior         -0.40    0.27
  n = 264, k = 4
  residual deviance = 11753.3, null deviance = 17354 (difference = 5600.7)
  overdispersion parameter = 66.6
```

As discussed in Section 6.2, the coefficient estimates are the same as before but the standard errors are much larger, reflecting the variation that is now being modeled. Again, we can test the model by simulating 1000 replicated datasets:

R code
```
n.sims <- 1000
sim.2 <- sim (glm.2, n.sims)
y.rep <- array (NA, c(n.sims, n))
for (s in 1:n.sims){
  y.hat <- exposure2 * exp (X %*% sim.2$beta[s,])
  a <- y.hat/(overdisp-1)             # Using R's parameterization for
  y.rep[s,] <- rnegbin (n, y.hat, a) # the negative-binomial distribution
}
```

and again computing the test statistic for each replication. This time, the proportion of zeroes in the replicated datasets varies from 18% to 48%, with a 95% interval

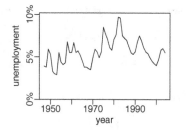

Figure 8.6 *Time series of U.S. unemployment rates from 1947 to 2004. We fit a first-order autoregression to these data and then simulated several datasets, shown in Figure 8.7, from the fitted model.*

of $[0.22, 0.40]$. The observed 36% fits right in, which tells us that this aspect of the data is reasonably fit by the model.

However, other aspects of the data might not be so well fit, as could be discovered by looking at other test statistics. We discuss this in Chapter 24 in the context of a more elaborate example.

8.4 Using predictive simulation to check the fit of a time-series model

Predictive simulation is more complicated in time-series models, which are typically set up so that the distribution for each point depends on the earlier data. We illustrate with a simple autoregressive model.

Fitting a first-order autoregression to the unemployment series

Figure 8.6 shows the time series of annual unemployment rates in the United States from 1947 to 2004. We would like to see how well these data are fit by a first-order autoregression, that is, a regression on last year's unemployment rate. Such a model is easy to set up and fit:[2]

```
n <- length (y)                                                      R code
y.lag <- c (NA, y[1:(n-1)])
lm.lag <- lm (y ~ y.lag)
```

yielding the following fit:

```
            coef.est coef.se                                         R output
(Intercept) 1.43      0.50
y.lag       0.75      0.09
  n = 57, k = 2
  residual sd = 0.99, R-Squared = 0.57
```

This information is potentially informative but does not tell us whether the model is a reasonable fit to the data. To examine fit, we will simulate replicated data from the fitted model.

Simulating replicated datasets

Using a point estimate of the fitted model. We first simulate replicated data in a slightly simplified way, using the following point estimate from the fitted model:

---

[2] Another option is to use some of the special time-series features in R, but it is simpler for us here to just fit as an ordinary regression.

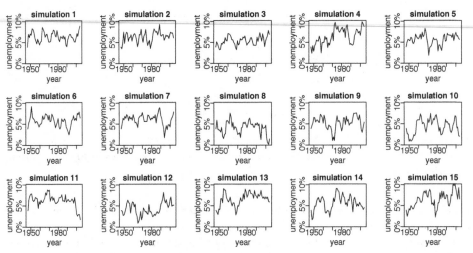

Figure 8.7 *Simulated replications of the unemployment series from the fitted autoregressive model. The replications capture many of the features of the actual data in Figure 8.6 but show slightly more short-term variation.*

R code
```
b.hat <- coef (lm.lag)          # vector of 2 regression coefs
s.hat <- sigma.hat (lm.lag)     # residual standard deviation
```

We start each of the simulated time series at the observed value y_1 (the actual unemployment rate in 1947) and then use the model, step by step, to simulate each year's value from the last:

R output
```
n.sims <- 100
y.rep <- array (NA, c(n.sims, n))
for (s in 1:n.sims){
  y.rep[s,1] <- y[1]
  for (t in 2:n){
    prediction <- c (1, y.rep[s,t-1]) %*% b.hat
    y.rep[s,t] <- rnorm (1, prediction, s.hat)
  }
}
```

Including the uncertainty in the estimated parameters. It is slightly better to propagate the estimation uncertainty by using simulations from the fitted model (as in Section 7.2), and then using these draws of β and σ to simulate replicated datasets:

R code
```
lm.lag.sim <- sim (lm.lag, n.sims)  # simulations of beta and sigma
for (s in 1:n.sims){
  y.rep[s,1] <- y[1]
  for (t in 2:n){
    prediction <- c (1, y.rep[s,t-1]) %*% lm.lag.sim$beta[s,]
    y.rep[s,t] <- rnorm (1, prediction, lm.lag.sim$sigma[s])
  }
}
```

Visual and numerical comparisons of replicated to actual data

Our first step in model checking is to plot some simulated datasets, which we do in Figure 8.7, and compare them visually to the actual data in Figure 8.6. The 15 simulations show different patterns, with many of them capturing the broad features of the data—its range, lack of overall trend, and irregular rises and falls. This autoregressive model clearly can represent many different sorts of time-series patterns.

Looking carefully at Figure 8.7, we see one pattern in all these replicated data that was not in the original data in 8.6, and that is a jaggedness, a level of short-term ups and downs that contrasts to the smoother appearance of the actual time series.

To quantify this discrepancy, we define a test statistic that is the frequency of "switches"—the number of years in which an increase in unemployment is immediately followed by a decrease, or vice versa:

```
Test <- function (y){
  n <- length (y)
  y.lag <- c (NA, y[1:(n-1)])
  y.lag2 <- c (NA, NA, y[1:(n-2)])
  sum (sign(y-y.lag) != sign(y.lag-y.lag2), na.rm=TRUE)
}
```
R code

As with the examples in the previous section, we compute this test for the data and for the replicated datasets:

```
print (Test(y))
test.rep <- rep (NA, n.sims)
for (s in 1:n.sims){
  test.rep[s] <- Test (y.rep[s,])
}
```
R code

The actual unemployment series featured 23 switches. Of the 1000 replications, 97% had more than 23 switches, implying that this aspect of the data was not captured well by the model.

8.5 Bibliographic note

Fake-data simulation is commonly used to validate statistical models and procedures. Two recent papers from a Bayesian perspective are Geweke (2004) and Cook, Gelman, and Rubin (2006). The predictive approach to model checking is described in detail in Gelman et al. (2003, chapter 6) and Gelman, Meng, and Stern (1996), deriving from the ideas of Rubin (1984). Gelman (2004a) connects graphical model checks to exploratory data analysis (Tukey, 1977). Examples of simulation-based model checking appear throughout the statistical literature, especially for highly structured models; see, for example, Bush and Mosteller (1955) and Ripley (1988).

8.6 Exercises

1. Fitting the wrong model: suppose you have 100 data points that arose from the following model: $y = 3 + 0.1 x_1 + 0.5 x_2 + \text{error}$, with errors having a t distribution with mean 0, scale 5, and 4 degrees of freedom. We shall explore the implications of fitting a standard linear regression to these data.

   (a) Simulate data from this model. For simplicity, suppose the values of x_1 are simply the integers from 1 to 100, and that the values of x_2 are random and equally likely to be 0 or 1.[3] Fit a linear regression (with normal errors) to these

---

[3] In R, you can define x.1 <- 1:100, simulate x.2 using rbinom(), then create the linear predictor, and finally simulate the random errors in y using the rt() function.

   data and see if the 68% confidence intervals for the regression coefficients (for each, the estimates ±1 standard error) cover the true values.

   (b) Put the above step in a loop and repeat 1000 times. Calculate the confidence coverage for the 68% intervals for each of the three coefficients in the model.

   (c) Repeat this simulation, but instead fit the model using t errors (see Exercise 6.6).

2. Predictive checks: using data of interest to you, fit a model of interest.

   (a) Simulate replicated datasets and visually compare to the actual data.

   (b) Summarize the data by a numerical test statistic, and compare to the values of the test statistic in the replicated datasets.

3. Using simulation to check the fit of a time-series model: find time-series data and fit a first-order autoregression model to it. Then use predictive simulation to check the fit of this model as in Section 8.4.

4. Model checking for count data: the folder `risky.behavior` contains data from a study of behavior of couples at risk for HIV; see Exercise 6.1.

   (a) Fit a Poisson regression model predicting number of unprotected sex acts from baseline HIV status. Perform predictive simulation to generate 1000 datasets and record both the percent of observations that are equal to 0 and the percent that are greater than 10 (the third quartile in the observed data) for each. Compare these values to the observed value in the original data.

   (b) Repeat (a) using an overdispersed Poisson regression model.

   (c) Repeat (b), also including ethnicity and baseline number of unprotected sex acts as input variables.

## CHAPTER 9

Causal inference using regression on the treatment variable

9.1 Causal inference and predictive comparisons

So far, we have been interpreting regressions *predictively*: given the values of several inputs, the fitted model allows us to predict y, considering the n data points as a simple random sample from a hypothetical infinite "superpopulation" or probability distribution. Then we can make comparisons across different combinations of values for these inputs.

This chapter and the next consider *causal inference*, which concerns what *would happen* to an outcome y as a result of a hypothesized "treatment" or intervention. In a regression framework, the treatment can be written as a variable T:[1]

$$T_i = \begin{cases} 1 & \text{if unit } i \text{ receives the "treatment"} \\ 0 & \text{if unit } i \text{ receives the "control,"} \end{cases}$$

or, for a continuous treatment,

$$T_i = \text{level of the "treatment" assigned to unit } i.$$

In the usual regression context, predictive inference relates to comparisons *between* units, whereas causal inference addresses comparisons of different treatments if applied to the *same* units. More generally, causal inference can be viewed as a special case of prediction in which the goal is to predict what *would have happened* under different treatment options. We shall discuss this theoretical framework more thoroughly in Section 9.2. Causal interpretations of regression coefficients can only be justified by relying on much stricter assumptions than are needed for predictive inference.

To motivate the detailed study of regression models for causal effects, we present two simple examples in which predictive comparisons do not yield appropriate causal inferences.

Hypothetical example of zero causal effect but positive predictive comparison

Consider a hypothetical medical experiment in which 100 patients receive the treatment and 100 receive the control condition. In this scenario, the causal effect represents a comparison between what would have happened to a given patient had he or she received the treatment compared to what would have happened under control. We first suppose that the treatment would have no effect on the health status of any given patient, compared with what would have happened under the control. That is, the *causal effect* of the treatment is zero.

However, let us further suppose that treated and control groups systematically differ, with healthier patients receiving the treatment and sicker patients receiving

---

[1] We use a capital letter for the vector T (violating our usual rule of reserving capitals for matrices) in order to emphasize the treatment as a key variable in causal analyses, and also to avoid potential confusion with t, which we sometimes use for "time."

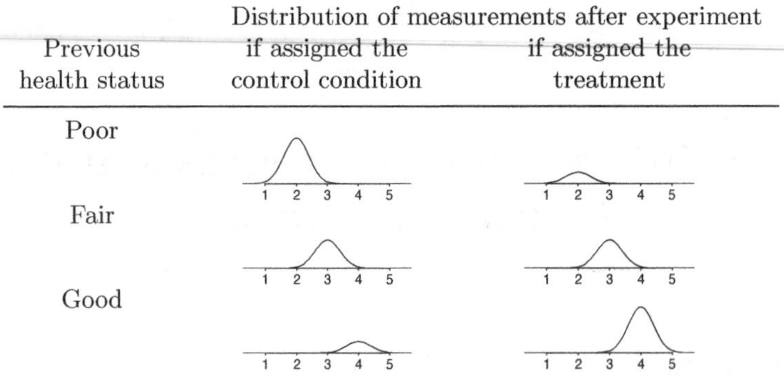

| Previous health status | Distribution of measurements after experiment | |
	if assigned the control condition	if assigned the treatment
Poor		
Fair		
Good		

Figure 9.1 *Hypothetical scenario of* zero causal effect *of treatment: for any value of previous health status, the distributions of potential outcomes are* identical *under control and treatment. However, the* predictive comparison *between treatment and control could be positive, if healthier patients receive the treatment and sicker patients receive the control condition.*

| Previous health status | Distribution of measurements after experiment | |
	if assigned the control condition	if assigned the treatment
Poor		
Fair		
Good		

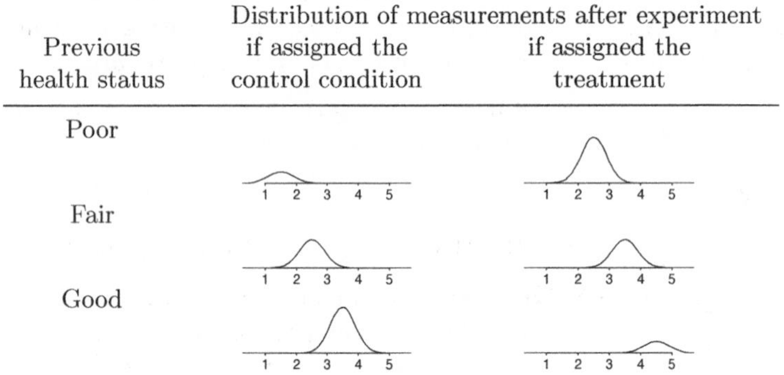

Figure 9.2 *Hypothetical scenario of* positive causal effect *of treatment: for any value of previous health status, the distributions of potential outcomes are* centered at higher values *for treatment than for control. However, the* predictive comparison *between treatment and control could be zero, if sicker patients receive the treatment and healthier patients receive the control condition. Compare to Figure 9.1.*

the control. This scenario is illustrated in Figure 9.1, where the distribution of outcome health status measurements is centered at the same place for the treatment and control conditions within each previous health status category (reflecting the lack of causal effect) but the heights of each distribution reflect the differential proportions of the sample that fell in each condition. This scenario leads to a positive *predictive comparison* between the treatment and control groups, even though the causal effect is zero. This sort of discrepancy between the predictive comparison and the causal effect is sometimes called self-selection bias, or simply selection bias, because participants are selecting themselves into different treatments.

Hypothetical example of positive causal effect but zero positive predictive comparison

Conversely, it is possible for a truly nonzero treatment effect to not show up in the predictive comparison. Figure 9.2 illustrates. In this scenario, the treatment has a positive effect for all patients, whatever their previous health status, as displayed by outcome distributions that for the treatment group are centered one point to the right of the corresponding (same previous health status) distributions in the control group. So, for any given unit, we would expect the outcome to be better under treatment than control. However, suppose that this time, sicker patients are given the treatment and healthier patients are assigned to the control condition, as illustrated by the different heights of these distributions. It is then possible to see equal average outcomes of patients in the two groups, with sick patients who received the treatment canceling out healthy patients who received the control.

Previous health status plays an important role in both these scenarios because it is related both to treatment assignment and future health status. If a causal estimate is desired, simple comparisons of average outcomes across groups that ignore this variable will be misleading because the effect of the treatment will be "confounded" with the effect of previous health status. For this reason, such predictors are sometimes called *confounding covariates*.

Adding regression predictors; "omitted" or "lurking" variables

The preceding theoretical examples illustrate how a simple predictive comparison is not necessarily an appropriate estimate of a causal effect. In these simple examples, however, there is a simple solution, which is to compare treated and control units conditional on previous health status. Intuitively, the simplest way to do this is to compare the averages of the current health status measurements across treatment groups only within each previous health status category; we discuss this kind of subclassification strategy in Section 10.2.

Another way to estimate the causal effect in this scenario is to regress the outcome on two inputs: the treatment indicator and previous health status. If health status is the only confounding covariate—that is, the only variable that predicts both the treatment and the outcome—and if the regression model is properly specified, then the coefficient of the treatment indicator corresponds to the average causal effect in the sample. In this example a simple way to avoid possible misspecification would be to discretize health status using indicator variables rather than including it as a single continuous predictor.

In general, then, causal effects can be estimated using regression if the model includes all confounding covariates (predictors that can affect treatment assignment or the outcome) and if the model is correct. If the confounding covariates are all observed (as in this example), then accurate estimation comes down to proper modeling and the extent to which the model is forced to extrapolate beyond the support of the data. If the confounding covariates are not observed (for example, if we suspect that healthier patients received the treatment, but no accurate measure of previous health status is included in the model), then they are "omitted" or "lurking" variables that complicate the quest to estimate causal effects.

We consider these issues in more detail in the rest of this chapter and the next, but first we will provide some intuition in the form of an algebraic formula.

Formula for omitted variable bias

We can quantify the bias incurred by excluding a confounding covariate in the context where a simple linear regression model is appropriate and there is only one confounding covariate. First define the "correct" specification as

$$y_i = \beta_0 + \beta_1 T_i + \beta_2 x_i + \epsilon_i \tag{9.1}$$

where T_i is the treatment and x_i is the covariate for unit i.

If instead the confounding covariate, x_i, is ignored, one can fit the model

$$y_i = \beta_0^* + \beta_1^* T_i + \epsilon_i^*$$

What is the relation between these models? To understand, it helps to define a third regression,

$$x_i = \gamma_0 + \gamma_1 T_i + \nu_i$$

If we substitute this representation of x into the original, correct, equation, and rearrange terms, we get

$$y_i = \beta_0 + \beta_2 \gamma_0 + (\beta_1 + \beta_2 \gamma_1) T_i + \epsilon_i + \beta_2 \nu_i \tag{9.2}$$

Equating the coefficients of T in (9.1) and (9.2) yields

$$\beta_1^* = \beta_1 + \beta_2^* \gamma_1$$

This correspondence helps demonstrate the definition of a confounding covariate. If there is no association between the treatment and the purported confounder (that is, $\gamma_1 = 0$) or if there is no association between the outcome and the confounder (that is, $\beta_2 = 0$) then the variable is not a confounder because there will be no bias ($\beta_2^* \gamma_1 = 0$).

This formula is commonly presented in regression texts as a way of describing the bias that can be incurred if a model is specified incorrectly. However, this term has little meaning outside of a context in which one is attempting to make causal inferences.

9.2 The fundamental problem of causal inference

We begin by considering the problem of estimating the causal effect of a treatment compared to a control, for example in a medical experiment. Formally, the *causal effect* of a treatment T on an outcome y for an observational or experimental unit i can be defined by comparisons between the outcomes that would have occurred under each of the different treatment possibilities. With a binary treatment T taking on the value 0 (control) or 1 (treatment), we can define *potential outcomes*, y_i^0 and y_i^1 for unit i as the outcomes that would be observed under control and treatment conditions, respectively.[2] (These ideas can also be directly generalized to the case of a treatment variable with multiple levels.)

The problem

For someone assigned to the treatment condition (that is, $T_i = 1$), y_i^1 is observed and y_i^0 is the unobserved *counterfactual* outcome—it represents what *would have* happened to the individual if assigned to control. Conversely, for control units, y_i^0 is observed and y_i^1 is counterfactual. In either case, a simple treatment effect for unit i can be defined as

$$\text{treatment effect for unit } i = y_i^1 - y_i^0$$

Figure 9.3 displays hypothetical data for an experiment with 100 units (and thus 200 potential outcomes). The top panel displays the data we would like to be able to see in order to determine causal effects for each person in the dataset—that is, it includes both potential outcomes for each person.

[2] The word "counterfactual" is sometimes used here, but we follow Rubin (1990) and use the term "potential outcome" because some of these potential data are actually observed.

(Hypothetical) complete data:

Unit, i	Pre-treatment inputs X_i			Treatment indicator T_i	Potential outcomes y_i^0	y_i^1	Treatment effect $y_i^1 - y_i^0$
1	2	1	50	0	**69**	75	6
2	3	1	98	0	**111**	108	−3
3	2	2	80	1	92	**102**	10
4	3	1	98	1	112	**111**	−1
⋮	⋮	⋮	⋮	⋮	⋮	⋮	⋮
100	4	1	104	1	111	**114**	3

Observed data:

Unit, i	Pre-treatment inputs X_i			Treatment indicator T_i	Potential outcomes y_i^0	y_i^1	Treatment effect $y_i^1 - y_i^0$
1	2	1	50	0	69	?	?
2	3	1	98	0	111	?	?
3	2	2	80	1	?	102	?
4	3	1	98	1	?	111	?
⋮	⋮	⋮	⋮	⋮	⋮	⋮	⋮
100	4	1	104	1	?	114	?

Figure 9.3 *Illustration of the fundamental problem of causal inference. For each unit, we have observed some pre-treatment inputs, and then the treatment ($T_i = 1$) or control ($T_i = 0$) is applied. We can then observe only one of the potential outcomes, (y_i^0, y_i^1). As a result, we cannot observe the treatment effect, $y_i^1 - y_i^0$, for any of the units.*
The top table shows what the complete data might look like, if it were possible to observe both potential outcomes on each unit. For each pair, the observed outcome is displayed in boldface. The bottom table shows what would actually be observed.

The so-called *fundamental problem of causal inference* is that at most one of these two potential outcomes, y_i^0 and y_i^1, can be observed for each unit i. The bottom panel of Figure 9.3 displays the data that can actually be observed. The y_i^1 values are "missing" for those in the control group and the y_i^0 values are "missing" for those in the treatment group.

Ways of getting around the problem

We cannot observe *both* what happens to an individual after taking the treatment (at a particular point in time) *and* what happens to that same individual after not taking the treatment (at the same point in time). Thus we can never measure a causal effect directly. In essence, then, we can think of causal inference as a prediction of what would happen to unit i if $T_i = 0$ or $T_i = 1$. It is thus predictive inference in the potential-outcome framework. Viewed this way, estimating causal effects requires one or some combination of the following: close substitutes for the potential outcomes, randomization, or statistical adjustment. We discuss the basic strategies here and go into more detail in the remainder of this chapter and the next.

Close substitutes. One might object to the formulation of the fundamental problem of causal inference by noting situations where it appears one can actually measure both y_i^0 and y_i^1 on the same unit. Consider, for example drinking tea one evening and milk another evening, and then measuring the amount of sleep each time. A careful consideration of this example reveals the implicit assumption that there are no systematic differences between days that could also affect sleep. An additional assumption is that applying the treatment on one day has no effect on the outcome on another day.

More pristine examples can generally be found in the natural and physical sciences. For instance, imagine dividing a piece of plastic into two parts and then exposing each piece to a corrosive chemical. In this case, the hidden assumption is that pieces are identical in how they would respond with and without treatment, that is, $y_1^0 = y_2^0$ and $y_1^1 = y_2^1$.

As a third example, suppose you want to measure the effect of a new diet by comparing your weight before the diet and your weight after. The hidden assumption here is that the pre-treatment measure can act as a substitute for the potential outcome under control, that is, $y_i^0 = x_i$.

It is not unusual to see studies that attempt to make causal inferences by substituting values in this way. It is important to keep in mind the strong assumptions often implicit in such strategies.

Randomization and experimentation. A different approach to causal inference is the "statistical" idea of using the outcomes observed on a sample of units to learn about the distribution of outcomes in the population.

The basic idea is that since we cannot compare treatment and control outcomes for the same units, we try to compare them on similar units. Similarity can be attained by using randomization to decide which units are assigned to the treatment group and which units are assigned to the control group. We will discuss this strategy in depth in the next section.

Statistical adjustment. For a variety of reasons, it is not always possible to achieve close similarity between the treated and control groups in a causal study. In observational studies, units often end up treated or not based on characteristics that are predictive of the outcome of interest (for example, men enter a job training program because they have low earnings and future earnings is the outcome of interest). Randomized experiments, however, can be impractical or unethical, and even in this context imbalance can arise from small-sample variation or from unwillingness or inability of subjects to follow the assigned treatment.

When treatment and control groups are not similar, modeling or other forms of statistical adjustment can be used to fill in the gap. For instance, by fitting a regression (or more complicated model), we may be able to estimate what would have happened to the treated units had they received the control, and vice versa. Alternately, one can attempt to divide the sample into subsets within which the treatment/control allocation mimics an experimental allocation of subjects. We discuss regression approaches in this chapter. We discuss imbalance and related issues more thoroughly in Chapter 10 along with a description of ways to help observational studies mimic randomized experiments.

9.3 Randomized experiments

We begin with the cleanest scenario, an experiment with units randomly assigned to receive treatment and control, and with the units in the study considered as a random sample from a population of interest. The random sampling and random treatment assignment allow us to estimate the average causal effect of the treatment in the population, and regression modeling can be used to refine this estimate.

Average causal effects and randomized experiments

Although we cannot estimate individual-level causal effects (without making strong assumptions, as discussed previously), we can design studies to estimate the population average treatment effect:

$$\text{average treatment effect} = \text{avg}\,(y_i^1 - y_i^0),$$

for the units i in a larger population. The cleanest way to estimate the population average is through a randomized experiment in which each unit has a positive chance of receiving each of the possible treatments.[3] If this is set up correctly, with treatment assignment either entirely random or depending only on recorded data that are appropriately modeled, the coefficient for T in a regression corresponds to the causal effect of the treatment, among the population represented by the n units in the study.

Considered more broadly, we can think of the control group as a group of units that could just as well have ended up in the treatment group, they just happened not to get the treatment. Therefore, on average, their outcomes represent what would have happened to the treated units had they not been treated; similarly, the treatment group outcomes represent what might have happened to the control group had they been treated. Therefore the control group plays an essential role in a causal analysis.

For example, if n_0 units are selected at random from the population and given the control, and n_1 other units are randomly selected and given the treatment, then the observed sample averages of y for the treated and control units can be used to estimate the corresponding population quantities, $\text{avg}(y^0)$ and $\text{avg}(y^1)$, with their difference estimating the average treatment effect (and with standard error $\sqrt{s_0^2/n_0 + s_1^2/n_1}$; see Section 2.3). This works because the y_i^0's for the control group are a random sample of the values of y_i^0 in the entire population. Similarly, the y_i^1's for the treatment group are a random sample of the y_i^1's in the population.

Equivalently, if we select $n_0 + n_1$ units at random from the population, and then randomly assign n_0 of them to the control and n_1 to the treatment, we can think of each of the sample groups as representing the corresponding population of control or treated units. Therefore the control group mean can act as a counterfactual for the treatment group (and vice versa).

What if the $n_0 + n_1$ units are selected nonrandomly from the population but then the treatment is assigned at random within this sample? This is common practice, for example, in experiments involving human subjects. Experiments in medicine, for instance, are conducted on volunteers with specified medical conditions who are willing to participate in such a study, and experiments in psychology are often conducted on university students taking introductory psychology courses. In this case, causal inferences are still justified, but inferences no longer generalize to the entire population. It is usual instead to consider the inference to be appropriate to a hypothetical superpopulation from which the experimental subjects were drawn. Further modeling is needed to generalize to any other population. A study

---

[3] Ideally, each unit should have a nonzero probability of receiving each of the treatments, because otherwise the appropriate counterfactual (potential) outcome cannot be estimated for units in the corresponding subset of the population. In practice, if the probabilities are highly unequal, the estimated population treatment effect will have a high standard error due to the difficulty of reliably estimating such a rare event.

Test scores in control classes   Test scores in treated classes

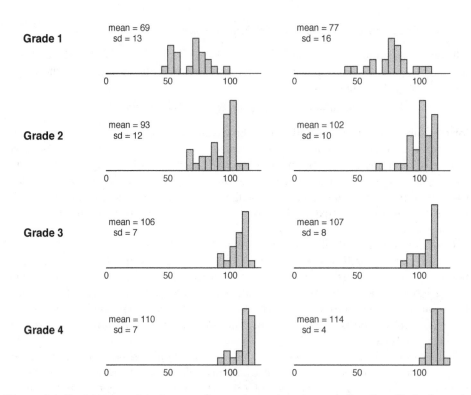

Figure 9.4 *Post-treatment test scores from an experiment measuring the effect of an educational television program, The Electric Company, on children's reading abilities. The experiment was applied on a total of 192 classrooms in four grades. At the end of the experiment, the average reading test score in each classroom was recorded.*

in which causal inferences are merited for a specific sample or population is said to have *internal validity*, and when those inferences can be generalized to a broader population of interest the study is said to have *external validity*.

We illustrate with a simple binary treatment (that is, two treatment levels, or a comparison of treatment to control) in an educational experiment. We then briefly discuss more general categorical, continuous, and multivariate treatments.

Example: showing children an educational television show

Figure 9.4 summarizes data from an educational experiment performed around 1970 on a set of elementary school classes. The treatment in this experiment was exposure to a new educational television show called The Electric Company. In each of four grades, the classes were randomized into treated and control groups. At the end of the school year, students in all the classes were given a reading test, and the average test score within each class was recorded. Unfortunately, we do not have data on individual students, and so our entire analysis will be at the classroom level.

Figure 9.4 displays the distribution of average post-treatment test scores in the control and treatment group for each grade. (The experimental treatment was applied to classes, not to schools, and so we treat the average test score in each class as a single measurement.) We break up the data by grade for convenience and because it is reasonable to suppose that the effects of this show could vary by grade.

Analysis as a completely randomized experiment. The experiment was performed in two cities (Fresno and Youngstown). For each city and grade, the experimenters selected a small number of schools (10–20) and, within each school, they selected the two poorest reading classes of that grade. For each pair, one of these classes was randomly assigned to continue with its regular reading course and the other was assigned to view the TV program.

This is called a *paired comparisons* design (which in turn is a special case of a *randomized block* design, with exactly two units within each block). For simplicity, however, we shall analyze the data here as if the treatment assignment had been completely randomized within each grade. In a *completely randomized experiment* on n units (in this case, classrooms), one can imagine the units mixed together in a bag, completely mixed, and then separated into two groups. For example, the units could be labeled from 1 to n, and then permuted at random, with the first n_1 units receiving the treatment and the others receiving the control. Each unit has the same probability of being in the treatment group and these probabilities are independent of each other.

Again, for the rest of this chapter we pretend that the Electric Company experiment was completely randomized within each grade. In Section 23.1 we return to the example and present an analysis appropriate to the paired design that was actually used.

Basic analysis of a completely randomized experiment

When treatments are assigned completely at random, we can think of the different treatment groups (or the treatment and control groups) as a set of random samples from a common population. The population average under each treatment, $\text{avg}(y^0)$ and $\text{avg}(y^1)$, can then be estimated by the sample average, and the population average difference between treatment and control, $\text{avg}(y^1) - \text{avg}(y^0)$—that is, the average causal effect—can be estimated by the difference in sample averages, $\bar{y}_1 - \bar{y}_0$.

Equivalently, the average causal effect of the treatment corresponds to the coefficient θ in the regression, $y_i = \alpha + \theta T_i + \text{error}_i$. We can easily fit the four regressions (one for each grade) in R:

```
for (k in 1:4) {                                                    R code
    display (lm (post.test ~ treatment, subset=(grade==k)))
}
```

The estimates and uncertainty intervals for the Electric Company experiment are graphed in the left panel of Figure 9.5. The treatment appears to be generally effective, perhaps more so in the low grades, but it is hard to be sure given the large standard errors of estimation.

Controlling for pre-treatment predictors

In this study, a pre-test was given in each class at the beginning of the school year (before the treatment was applied). In this case, the treatment effect can also be estimated using a regression model: $y_i = \alpha + \theta T_i + \beta x_i + \text{error}_i$ on the pre-treatment predictor x.[4] Figure 9.6 illustrates for the Electric Company experiment. For each

---

[4] We avoid the term *confounding covariates* when describing adjustment in the context of a randomized experiment. Predictors are included in this context to increase precision. We expect

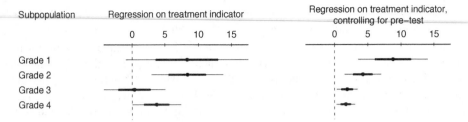

Figure 9.5 *Estimates, 50%, and 95% intervals for the effect of the Electric Company tele-vision show (see data in Figures 9.4 and 9.6) as estimated in two ways: first, from a regression on treatment alone, and second, also controlling for pre-test data. In both cases, the coefficient for treatment is the estimated causal effect. Including pre-test data as a predictor increases the precision of the estimates.*

Displaying these coefficients and intervals as a graph facilitates comparisons across grades and across estimation strategies (controlling for pre-test or not). For instance, the plot highlights how controlling for pre-test scores increases precision and reveals decreasing effects of the program for the higher grades, a pattern that would be more difficult to see in a table of numbers.

Sample sizes are approximately the same in each of the grades. The estimates for higher grades have lower standard errors because the residual standard deviations of the regressions are lower in these grades; see Figure 9.6.

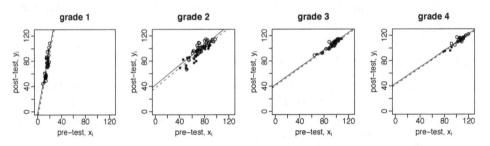

Figure 9.6 *Pre-test/post-test data for the Electric Company experiment. Treated and control classes are indicated by circles and dots, respectively, and the solid and dotted lines represent parallel regression lines fit to the treatment and control groups, respectively. The solid lines are slightly higher than the dotted lines, indicating slightly positive estimated treatment effects. Compare to Figure 9.4, which displays only the post-test data.*

grade, the difference between the regression lines for the two groups represents the treatment effect as a function of pre-test score. Since we have not included any interaction in the model, this treatment effect is assumed constant over all levels of the pre-test score.

For grades 2–4, the pre-test was the same as the post-test, and so it is no surprise that all the classes improved whether treated or not (as can be seen from the plots). For grade 1, the pre-test was a subset of the longer test, which explains why the pre-test scores for grade 1 are so low. We can also see that the distribution of post-test scores for each grade is similar to the next grade's pre-test scores, which makes sense.

In any case, for estimating causal effects (as defined in Section 9.2) we are interested in the difference between treatment and control conditions, not in the simple improvement from pre-test to post-test. The pre-post improvement is not a

them to be related to the outcome but not to the treatment assignment due to the randomization. Therefore they are not confounding covariates.

causal effect (except under the assumption, unreasonable in this case, that under the control there would be no change from pre-post change).

In the regression

$$y_i = \alpha + \theta T_i + \beta x_i + \text{error}_i \qquad (9.3)$$

the coefficient for the treatment indicator still represents the average treatment effect, but controlling for pre-test can improve the efficiency of the estimate. (More generally, the regression can control for multiple pre-treatment predictors, in which case the model has the form $y_i = \alpha + \theta T_i + X_i \beta + \text{error}_i$, or alternatively α can be removed from the equation and considered as a constant term in the linear predictor $X\beta$.)

The estimates for the Electric Company study appear in the right panel of Figure 9.5. It is now clear that the treatment is effective, and it appears to be more effective in the lower grades. A glance at Figure 9.6 suggests that in the higher grades there is less room for improvement; hence this particular test might not be the most effective for measuring the benefits of The Electric Company in grades 3 and 4.

It is only appropriate to control for pre-treatment predictors, or, more generally, predictors that would not be affected by the treatment (such as race or age). This point will be illustrated more concretely in Section 9.7.

Gain scores

An alternative way to specify a model that controls for pre-test measures is to use these measures to transform the response variable. A simple approach is to subtract the pre-test score, x_i, from the outcome score, y_i, thereby creating a "gain score," g_i. Then this score can be regressed on the treatment indicator (and other predictors if desired), $g_i = \alpha + \theta T_i + \text{error}_i$. (In the simple case with no other predictors, the regression estimate is simply $\hat{\theta} = \bar{g}^T - \bar{g}^C$, the average difference of gain scores in the treatment and control groups.)

In some cases the gain score can be more easily interpreted than the original outcome variable y. Using gain scores is most effective if the pre-treatment score is comparable to the post-treatment measure. For instance, in our Electric Company example it would not make sense to create gain scores for the classes in grade 1 since their pre-test measure was based on only a subset of the full test.

One perspective on this model is that it makes an unnecessary assumption, namely, that $\beta = 1$ in model (9.3). On the other hand, if this assumption is close to being true then θ may be estimated more precisely. One way to resolve this concern about misspecification would simply be to include the pre-test score as a predictor as well, $g_i = \alpha + \theta T_i + \gamma x_i + \text{error}_i$. However, in this case, $\hat{\theta}$, the estimate of the coefficient for T, is equivalent to the estimated coefficient from the original model, $y_i = \alpha + \theta T_i + \beta x_i + \text{error}_i$ (see Exercise 9.7).

More than two treatment levels, continuous treatments, and multiple treatment factors

Going beyond a simple treatment-and-control setting, multiple treatment effects can be defined relative to a baseline level. With random assignment, this simply follows general principles of regression modeling.

If treatment levels are numerical, the treatment level can be considered as a continuous input variable. To conceptualize randomization with a continuous treatment variable, think of choosing a random number that falls anywhere in the continuous range. As with regression inputs in general, it can make sense to fit more compli-

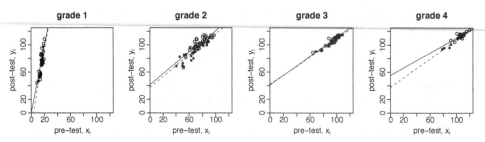

Figure 9.7 *Pre-test/post-test data for the Electric Company experiment. Treated and control classes are indicated by circles and dots, respectively, and the solid and dotted lines represent separate regression lines fit to the treatment and control groups, respectively. For each grade, the difference between the solid and dotted lines represents the estimated treatment effect as a function of pre-test score.*

cated models if suggested by theory or supported by data. A linear model—which estimates the average effect on y for each additional unit of T—is a natural starting point, though it may need to be refined.

With several discrete treatments that are unordered (such as in a comparison of three different sorts of psychotherapy), we can move to multilevel modeling, with the group index indicating the treatment assigned to each unit, and a second-level model on the group coefficients, or treatment effects. We shall illustrate such modeling in Section 13.5 with an experiment from psychology. We shall focus more on multilevel modeling as a tool for fitting data, but since the treatments in that example are randomly assigned, their coefficients can be interpreted as causal effects.

Additionally, different combinations of multiple treatments can be administered randomly. For instance, depressed individuals could be randomly assigned to receive nothing, drugs, counseling sessions, or a combination of drugs and counseling sessions. These combinations could be modeled as two treatments and their interaction or as four distinct treatments.

The assumption of no interference between units

Our discussion so far regarding estimation of causal effects using experiments is contingent upon another, often overlooked, assumption. We must assume also that the treatment assignment for one individual (unit) in the experiment does not affect the outcome for another. This has been incorporated into the "stable unit treatment value assumption" (SUTVA). Otherwise, we would need to define a different potential outcome for the ith unit not just for each treatment received by that unit but for each combination of treatment assignments received by every other unit in the experiment. This would enormously complicate even the definition, let alone the estimation, of individual causal effects. In settings such as agricultural experiments where interference between units is to be expected, it can be modeled directly, typically using spatial interactions.

9.4 Treatment interactions and poststratification

Interactions of treatment effect with pre-treatment inputs

Once we include pre-test in the model, it is natural to allow it to interact with treatment effect. The treatment is then allowed to affect both the intercept and the slope of the pre-test/post-test regression. Figure 9.7 shows the Electric Company data with separate regression lines estimated for the treatment and control groups. As with Figure 9.6, for each grade the difference between the regression lines is the estimated treatment effect as a function of pre-test score.

We illustrate in detail for grade 4. First, we fit the simple model including only the treatment indicator:

```
lm(formula = post.test ~ treatment, subset=(grade==4))
              coef.est coef.se
(Intercept)    110.4     1.3
treatment        3.7     1.8
  n = 42, k = 2
  residual sd = 6.0, R-Squared = 0.09
```
R output

The estimated treatment effect is 3.7 with a standard error of 1.8. We can improve the efficiency of the estimator by controlling for the pre-test score:

```
lm(formula = post.test ~ treatment + pre.test, subset=(grade==4))
              coef.est coef.se
(Intercept)     42.0     4.3
treatment        1.7     0.7
pre.test         0.7     0.0
  n = 42, k = 3
  residual sd = 2.2, R-Squared = 0.88
```
R output

The new estimated treatment effect is 1.7 with a standard error of 0.7. In this case, controlling for the pre-test reduced the estimated effect. Under a clean randomization, controlling for pre-treatment predictors in this way should reduce the standard errors of the estimates.[5] (Figure 9.5 shows the estimates for the Electric Company experiment in all four grades.)

Complicated arise when we include the interaction of treatment with pre-test:

```
lm(formula = post.test ~ treatment + pre.test + treatment:pre.test,
     subset=(grade==4))
                    coef.est coef.se
(Intercept)          37.84    4.90
treatment            17.37    9.60
pre.test              0.70    0.05
treatment:pre.test   -0.15    0.09
  n = 42, k = 4
  residual sd = 2.1, R-Squared = 0.89
```
R output

The estimated treatment effect is now $17 - 0.15x$, which is difficult to interpret without knowing the range of x. From Figure 9.7 we see that pre-test scores range from approximately 80 to 120; in this range, the estimated treatment effect varies from $17 - 0.15 \cdot 80 = 5$ for classes with pre-test scores of 80 to $17 - 0.15 \cdot 120 = -1$ for classes with pre-test scores of 120. This range represents the *variation* in estimated treatment effects as a function of pre-test score, *not* uncertainty in the estimated treatment effect.

To get a sense of the uncertainty, we can plot the estimated treatment effect as a function of x, overlaying random simulation draws to represent uncertainty:

---

[5] Under a clean randomization, controlling for pre-treatment predictors in this way does not change what we are estimating. If the randomization was less than pristine, however, the addition of predictors to the equation may help us control for unbalanced characteristics across groups. Thus, this strategy has the potential to move us from estimating a noncausal estimand (due to lack of randomization) to estimating a causal estimand by in essence "cleaning" the randomization.

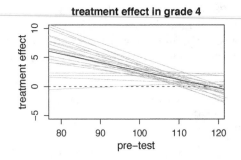

Figure 9.8 *Estimate and uncertainty for the effect of viewing The Electric Company (compared to the control treatment) for fourth-graders. Compare to the data in the rightmost plot in Figure 9.7. The dark line here—the estimated treatment effect as a function of pre-test score—is the difference between the two regression lines in the grade 4 plot in Figure 9.7. The gray lines represent 20 random draws from the uncertainty distribution of the treatment effect.*

R code
```
lm.4 <- lm (post.test ~ treatment + pre.test + treatment:pre.test,
   subset=(grade==4))
lm.4.sim <- sim (lm.4)
plot (0, 0, xlim=range (pre.test[grade==4]), ylim=c(-5,10),
   xlab="pre-test", ylab="treatment effect",
   main="treatment effect in grade 4")
abline (0, 0, lwd=.5, lty=2)
for (i in 1:20){
   curve (lm.4.sim$beta[i,2] + lm.4.sim$beta[i,4]*x, lwd=.5, col="gray",
      add=TRUE)}
curve (coef(lm.4)[2] + coef(lm.4)[4]*x, lwd=.5, add=TRUE)
```

This produces the graph shown in Figure 9.8.

Finally, we can estimate a mean treatment effect by averaging over the values of x in the data. If we write the regression model as $y_i = \alpha + \theta_1 T_i + \beta x_i + \theta_2 T_i x_i + \text{error}_i$, then the treatment effect is $\theta_1 + \theta_2 x$, and the summary treatment effect in the sample is $\frac{1}{n}\sum_{i=1}^n(\theta_1 + \theta_2 x_i)$, averaging over the n fourth-grade classrooms in the data. We can compute the average treatment effect as follows:

R code
```
n.sims <- nrow(lm.4.sim$beta)
effect <- array (NA, c(n.sims, sum(grade==4)))
for (i in 1:n.sims){
   effect[i,] <- lm.4.sim$beta[i,2] + lm.4.sim$beta[i,4]*pre.test[grade==4]
}
avg.effect <- rowMeans (effect)
```

The `rowMeans()` function averages over the grade 4 classrooms, and the result of this computation, `avg.effect`, is a vector of length `n.sims` representing the uncertainty in the average treatment effect. We can summarize with the mean and standard error:

R code
```
print (c (mean(avg.effect), sd(avg.effect)))
```

The result is 1.8 with a standard deviation of 0.7—quite similar to the result from the model controlling for pre-test but with no interactions. In general, for a linear regression model, the estimate obtained by including the interaction, and then averaging over the data, reduces to the estimate with no interaction. The motivation for including the interaction is thus to get a better idea of how the treatment effect varies with pre-treatment predictors, not simply to estimate an average effect.

Poststratification

We have discussed how treatment effects interact with pre-treatment predictors (that is, regression inputs). To estimate an average treatment effect, we can post-stratify—that is, average over the population.[6]

For example, suppose we have treatment variable T and pre-treatment control variables x_1, x_2, and our regression predictors are x_1, x_2, T, and the interactions $x_1 T$ and $x_2 T$, so that the linear model is: $y = \beta_0 + \beta_1 x_1 + \beta_2 x_2 + \beta_3 T + \beta_4 x_1 T + \beta_5 x_2 T +$ error. The estimated treatment effect is then $\beta_3 + \beta_4 x_1 + \beta_5 x_2$, and its average, in a linear regression, is simply $\beta_3 + \beta_4 \mu_1 + \beta_5 \mu_2$, where μ_1 and μ_2 are the averages of x_1 and x_2 in the population. These population averages might be available from another source, or else they can be estimated using the averages of x_1 and x_2 in the data at hand. Standard errors for summaries such as $\beta_3 + \beta_4 \mu_1 + \beta_5 \mu_2$ can be determined analytically, but it is easier to simply compute them using simulations.

Modeling interactions is important when we care about differences in the treatment effect for different groups, and poststratification then arises naturally if a population average estimate is of interest.

9.5 Observational studies

In theory, the simplest solution to the fundamental problem of causal inference is, as we have described, to randomly sample a different set of units for each treatment group assignment from a common population, and then apply the appropriate treatments to each group. An equivalent approach is to randomly assign the treatment conditions among a selected set of units. Either of these approaches ensures that, on average, the different treatment groups are *balanced* or, to put it another way, that the $\bar{y}^0$ and $\bar{y}^1$ from the sample are estimating the average outcomes under control and treatment for the same population.

In practice, however, we often work with *observational data* because, compared to experiments, observational studies can be more practical to conduct and can have more realism with regard to how the program or treatment is likely to be "administered" in practice. As we have discussed, however, in observational studies treatments are observed rather than assigned (for example, comparisons of smokers to nonsmokers), and it is not at all reasonable to consider the observed data under different treatments as random samples from a common population. In an observational study, there can be systematic differences between groups of units that receive different treatments—differences that are outside the control of the experimenter—and they can affect the outcome, y. In this case we need to rely on more data than just treatments and outcomes and implement a more complicated analysis strategy that will rely upon stronger assumptions. The strategy discussed in this chapter, however, is relatively simple and relies on controlling for confounding covariates through linear regression. Some alternative approaches are described in Chapter 10.

---

[6] In survey sampling, *stratification* refers to the procedure of dividing the population into disjoint subsets (strata), sampling separately within each stratum, and then combining the stratum samples to get a population estimate. Poststratification is the analysis of an unstratified sample, breaking the data into strata and reweighting as would have been done had the survey actually been stratified. Stratification can adjust for potential differences between sample and population using the survey design; poststratification makes such adjustments in the data analysis.

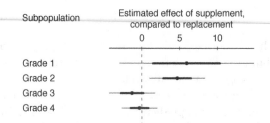

Figure 9.9 *Estimates, 50%, and 95% intervals for the effect of The Electric Company as a supplement rather than a replacement, as estimated by a regression on the supplement/replacement indicator also controlling for pre-test data. For each grade, the regression is performed only on the treated classes; this is an observational study embedded in an experiment.*

Electric Company example

Here we illustrate an observational study for which a simple regression analysis, controlling for pre-treatment information, may yield reasonable causal inferences.

The educational experiment described in Section 9.3 actually had an embedded observational study. Once the treatments had been assigned, the teacher for each class assigned to the Electric Company treatment chose to either *replace* or *supplement* the regular reading program with the Electric Company television show. That is, all the classes in the treatment group watched the show, but some watched it instead of the regular reading program and others got it in addition.[7]

The simplest starting point to analyzing these observational data (now limited to the randomized treatment group) is to consider the choice between the two treatment options—"replace" or "supplement"—to be randomly assigned conditional on pre-test scores. This is a strong assumption but we use it simply as a starting point. We can then estimate the treatment effect by regression, as with an actual experiment. In the R code, we create a variable called supp that equals 0 for the replacement form of the treatment, 1 for the supplement, and NA for the controls. We then estimate the effect of the supplement, as compared to the replacement, for each grade:

R code
```
for (k in 1:4) {
  ok <- (grade==k) & (!is.na(supp))
  lm.supp <- lm (post.test ~ supp + pre.test, subset=ok)
}
```

The estimates are graphed in Figure 9.9. The uncertainties are high enough that the comparison is inconclusive except in grade 2, but on the whole the pattern is consistent with the reasonable hypothesis that supplementing is more effective than replacing in the lower grades.

Assumption of ignorable treatment assignment

As opposed to making the same assumption as the completely randomized experiment, the key assumption underlying the estimate is that, *conditional* on the confounding covariates used in the analysis (here as inputs in the regression analysis), the distribution of units across treatment conditions is, in essence, "random"

---

[7] This procedural detail reveals that the treatment effect for the randomized experiment is actually more complicated than described earlier. As implemented, the experiment estimated the effect of making the program available, either as a supplement or replacement for the current curriculum.

(in this case, pre-test score) with respect to the potential outcomes. To help with the intuition here, one could envision units being randomly assigned to treatment conditions conditional on the confounding covariates; however, of course, no actual randomized assigment need take place.

Ignorability is often formalized by the conditional independence statement,

$$y^0, y^1 \perp T \mid X.$$

This says that the distribution of the potential outcomes, (y^0, y^1), is the same across levels of the treatment variable, T, once we condition on confounding covariates X.

This assumption is referred to as *ignorability* of the treatment assignment in the statistics literature and *selection on observables* in econometrics. Said another way, we would not necessarily expect any two classes to have had the same probability of receiving the supplemental version of the treatment. However, we expect any two classes at the same levels of the confounding covariates (that is, pre-treatment variables; in our example, average pre-test score) to have had the same probability of receiving the supplemental version of the treatment. A third way to think about the ignorability assumption is that it requires that we control for all confounding covariates, the pre-treatment variables that are associated with both the treatment and the outcome.

If ignorability holds, then causal inferences can be made without modeling the treatment assignment process—that is, we can *ignore* this aspect of the model as long as analyses regarding the causal effects condition on the predictors needed to satisfy ignorability. Randomized experiments represent a simple case of ignorability. Completely randomized experiments need not condition on any pre-treatment variables—this is why we can use a simple difference in means to estimate causal effects. Randomized experiments that block or match satisfy ignorability conditional on the design variables used to block or match, and therefore these variables need to be included when estimating causal effects.

In the Electric Company supplement/replacement example, an example of a *non-ignorable assignment mechanism* would be if the teacher of each class chose the treatment that he or she believed would be more effective for that particular class based on unmeasured characteristics of the class that were related to their subsequent test scores. Another nonignorable assignment mechanism would be if, for example, supplementing was more likely to be chosen by more "motivated" teachers, with teacher motivation also associated with the students' future test scores.

For ignorability to hold, it is not necessary that the two treatments be equally likely to be picked, but rather that the probability that a given treatment is picked should be equal, conditional on our confounding covariates.[8] In an experiment, one can control this at the design stage by using a random assignment mechanism. In an observational study, the "treatment assignment" is not under the control of the statistician, but one can aim for ignorability by conditioning in the analysis stage on as much pre-treatment information in the regression model as possible. For example, if teachers' motivation might affect treatment assignment, it would be advisable to have a pre-treatment measure of teacher motivation and include this as an input in the regression model. This would increase the plausibility of the ignorability assumption. Realistically, this may be a difficult characteristic to

---

[8] As further clarification, consider two participants of a study for which ignorability holds. If we define the probability of treatment participation as $\Pr(T = 1|X)$, then this probability must be equal for these two individuals. However, suppose there exists another variable, w, that is associated with treatment participation (conditional on X) but not with the outcome (conditional on X). We do not require that $\Pr(T = 1 \mid X, W)$ be the same for these two participants.

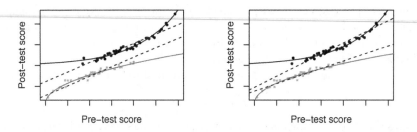

Figure 9.10 *Hypothetical before/after data demonstrating the potential problems in using linear regression for causal inference. The dark dots and line correspond to the children who received the educational supplement; the lighter dots and line correspond to the children who did not receive the supplement. The dashed lines are regression lines fit to the observed data. The model shown in the right panel allows for an interaction between receiving the supplement and pre-test scores.*

measure, but other teacher characteristics such as years of experience and schooling might act as partial proxies.

In general, one can never prove that the treatment assignment process in an observational study is ignorable—it is always possible that the choice of treatment depends on relevant information that has not been recorded. In an educational study this information could be characteristics of the teacher or school that are related both to treatment assignment and to post-treatment test scores. Thus, if we interpret the estimates in Figure 9.9 as causal effects, we do so with the understanding that we would prefer to have further pre-treatment information, especially on the teachers, in order to be more confident in ignorability.

If we believe that treatment assignments depend on information not included in the model, then we should choose a different analysis strategy. We discuss some options at the end of the next chapter.

Judging the reasonableness of regression as a modeling approach, assuming ignorability

Even if the ignorability assumption appears to be justified, this does not mean that simple regression of our outcomes on confounding covariates and a treatment indicator is necessarily the best modeling approach for estimating treatment effects. There are two primary concerns related to the distributions of the confounding covariates across the treatment groups: lack of complete overlap and lack of balance. For instance, consider our initial hypothetical example of a medical treatment that is supposed to affect subsequent health measures. What if there were no treatment observations among the group of people whose pre-treatment health status was highest? Arguably, we could not make any causal inferences about the effect of the treatment on these people because we would have no empirical evidence regarding the counterfactual state. Lack of overlap and balance forces stronger reliance on our modeling than if covariate distributions were the same across treatment groups. We provide a brief illustration in this chapter and discuss in greater depth in Chapter 10.

Suppose we are interested in the effect of a supplementary educational activity (such as viewing The Electric Company) that was not randomly assigned. Suppose, however, that only one predictor, pre-test score, is necessary to satisfy ignorability— that is, there is only one confounding covariate. Suppose further, though, that those individuals who participate in the supplementary activity tend to have higher pre-

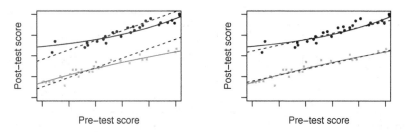

Figure 9.11 *Hypothetical before/after data demonstrating the potential problems in using linear regression for causal inference. The dark dots and line correspond to the children who received the educational supplement; the lighter dots and line correspond to the children who did not receive the supplement. The dashed lines are regression lines fit to the observed data. Plots are restricted to observations in the region where there is overlap in terms of the pre-treatment test score across treatment and control groups. The left panel shows only the portion of the plot in Figure 9.10 where there is overlap. The right panel shows regression lines fit only using observations in this overlapping region.*

test scores, on average, than those who do not participate. One realization of this hypothetical scenario is illustrated in Figure 9.10. The dark line represents the true relation between pre-test scores (x-axis) and post-test scores (y-axis) for those who receive the supplement. The lighter line represents the true relation between pre-test scores and post-test scores for those who do not receive the supplement. Estimated linear regression lines are superimposed for these data. The linear model has problems fitting the true nonlinear regression relation—a problem that is compounded by the lack of overlap of the two groups in the data. Because there are no "control" children with high test scores and virtually no "treatment" children with low test scores, these linear models, to create counterfactual predictions, are forced to extrapolate over portions of the space where there are no data to support them. These two problems combine to create, in this case, a substantial underestimate of the true average treatment effect. Allowing for an interaction, as illustrated in the right panel, does not solve the problem.

In the region of pre-test scores where there are observations from both treatment groups, however, even the incorrectly specified linear regression lines do not provide such a bad fit to the data. And no model extrapolation is required, so diagnosing this lack of fit would be possible. This is demonstrated in the left panel of Figure 9.11 by restricting the plot from the left panel of Figure 9.10 to the area of overlap. Furthermore, if the regression lines are fit only using this restricted sample they fit quite well in this region, as is illustrated in the right panel of Figure 9.11. Some of the strategies discussed in the next chapter use this idea of limiting analyses to observations with the region of complete overlap.

Examining overlap in the Electric Company embedded observational study

For the Electric Company data we can use plots such as in Figure 9.10–9.11 to assess the appropriateness of the modeling assumptions and the extent to which we are relying on unsupported model extrapolations. For the most part, Figure 9.12 reveals a reasonable amount of overlap in pre-test scores across treatment groups within each grade. Grade 3, however, has some classrooms with average pre-test scores that are lower than the bulk of the sample, all of which received the supplement. It might be appropriate to decide that no counterfactual classrooms exist in our data for these classrooms and thus the data cannot support causal inferences for these

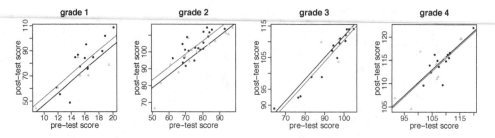

Figure 9.12 *Pre-test/post-test data examining the overlap in pre-test scores across treatment groups as well as the extent to which models are being extrapolated to regions where there is no support in the data. Classrooms that watched The Electric Company as a supplement are represented by the dark points and regression line; classrooms that watched The Electric Company as a replacement are represented by the lighter points and regression line. No interactions were included when estimating the regression lines.*

classrooms. The sample sizes for each grade make it difficult to come to any firm conclusions one way or another, however.

Therefore, we must feel confident in the (probably relatively minor) degree of model extrapolation relied upon by these estimates in order to trust a causal interpretation.

9.6 Understanding causal inference in observational studies

Sometimes the term "observational study" refers to a situation in which a specific intervention was offered nonrandomly to a population or in which a population was exposed nonrandomly to a well-defined treatment. The primary characteristic that distinguishes causal inference in these settings from causal inference in randomized experiments is the inability to identify causal effects without making assumptions such as ignorability. (Other sorts of assumptions will be discussed in the next chapter.)

Often, however, observational studies refer more broadly to survey data settings where no intervention has been performed. In these settings, there are other aspects of the research design that need to be carefully considered as well. The first is the mapping between the "treatment" variable in the data and a policy or intervention. The second considers whether it is possible to separately identify the effects of multiple treatment factors. When attempting causal inference using observational data, it is helpful to formalize exactly what the experiment might have been that would have generated the data, as we discuss next.

Defining a "treatment" variable

A causal effect needs to be defined with respect to a cause, or an intervention, on a particular set of experimental units. We need to be able to conceive of each unit as being able to experience each level of the treatment variable for which causal effects will be defined for that unit. Thus, the "effect" of height on earnings is ill-defined without reference to a treatment that could change one's height. Otherwise what does it mean to define a potential outcome for a person that would occur *if* he or she had been shorter or taller?

More subtly, consider the effect of single-motherhood on children's outcomes. We might be able to envision several different kinds of interventions that could change a mother's marital status either before or after birth: changes in tax laws, participation in a marriage encouragement program for unwed parents, new child support enforcement policies, divorce laws, and so on. These potential "treatments" vary in the timing of marriage relative to birth and even the strength of the marriages that might result, and consequently might be expected to have different effects on the children involved. Therefore, this conceptual mapping to a hypothetical intervention can be important for choice of study design, analysis, and interpretation of results.

Consider, for instance, a study that examines Korean children who were randomly assigned to American families for adoption. This "natural experiment" allows for fair comparisons across conditions such as being raised in one-parent versus two-parent households. However, this is a different kind of treatment altogether than considering whether a couple should get married. There is no attempt to compare *parents* who are similar to each other; instead, it is the *children* who are similar on average at the outset. The treatment in question then has to do with the child's placement in a family. This addresses an interesting although perhaps less policy-relevant question (at least in terms of policies that affect incentives for marriage formation or dissolution).

Multiple treatment factors

It is difficult to directly interpret more than one input variable causally in an observational study. Suppose we have two variables, A and B, whose effects we would like to estimate from a single observational study. To estimate causal effects, we must consider implicit treatments—and to estimate both effects at once, we would have to imagine a treatment that affects A while leaving B unchanged, and a treatment that affects B while leaving A unchanged. In examples we have seen, it is generally difficult to envision both these interventions: if A comes before B in time or logical sequence, then we can estimate the effect of B controlling for A but not the reverse (because of the problem with controlling for post-treatment variables, which we discuss in greater detail in the next section).

More broadly, for many years a common practice when studying a social problem (for example, poverty) was to compare people with different outcomes, throwing many inputs into a regression to see which was the strongest predictor. As opposed to the way we have tried to frame causal questions thus far in this chapter, as the effect of causes, this is a strategy that searches for the causes of an effect. This is an ill-defined notion that we will avoid for exactly the kind of reasons discussed in this chapter.[9]

Thought experiment: what would be an ideal randomized experiment?

If you find yourself confused about what can be estimated and how the various aspects of your study should be defined, a simple strategy is to try to formalize the randomized experiment you would have liked to have done to answer your causal question. A perfect mapping rarely exists between this experimental ideal and your data so often you will be forced instead to figure out, given the data you have, what randomized experiment could be thought to have generated such data.

---

[9] Also, philosophically, looking for the most important cause of an outcome is a confusing framing for a research question because one can always find an earlier cause that affected the "cause" you determine to be the strongest from your data. This phenomenon is sometimes called the "infinite regress of causation."

For instance, if you were interested in the effect of breastfeeding on children's cognitive outcomes, what randomized experiment would you want to perform assuming no practical, legal, or moral barriers existed? We could imagine randomizing mothers to either breastfeed their children exclusively or bottle-feed them formula exclusively. We would have to consider how to handle those who do not adhere to their treatment assignment, such as mothers and children who are not able to breastfeed, and children who are allergic to standard formula. Moreover, what if we want to separately estimate the physiological effects of the breast milk from the potential psychological implications (to both mother and child) of nursing at the breast and the more extended physical contact that is often associated with breastfeeding? In essence, then, we think that perhaps breastfeeding represents several concurrent treatments. Perhaps we would want to create a third treatment group of mothers who feed their babies with bottles of expressed breast milk. This exercise of considering the randomized experiment helps to clarify what the true nature of the intervention is that we are using our treatment variable to represent.

Just as in a randomized experiment, all causal inference requires a comparison of at least two treatments (counting "control" as a treatment). For example, consider a study of the effect on weight loss of a new diet. The treatment (following the diet) may be clear but the control is not. Is it to try a different diet? To continue eating "normally"? To exercise more? Different control conditions imply different counterfactual states and thus induce different causal effects.

Finally, thinking about hypothetical randomized experiments can help with problems of trying to establish a causal link between two variables when neither has temporal priority and when they may have been simultaneously determined. For instance, consider a regression of crime rates in each of 50 states using a cross section of data, where the goal is to determine the "effect" of the number of police officers while controlling for the social, demographic, and economic features of each state as well as characteristics of the state (such as the crime rate) that might affect decisions to increase the size of the police force. The problem is that it may be difficult (if not impossible) to disentangle the "effect" of the size of the police force on crime from the "effect" of the crime rate on the size of the police force.

If one is interested in figuring out policies that can affect crime rates, it might be more helpful to conceptualize both "number of police officers" and "crime rate" as outcome variables. Then one could imagine different treatments (policies) that could affect these outcomes. For example, the number of police officers could be affected by a bond issue to raise money earmarked for hiring new police, or a change in the retirement age, or a reallocation of resources within local and state government law enforcement agencies. These different treatments could have different effects on the crime rate.

9.7 Do not control for post-treatment variables

As illustrated in the examples of this chapter, we recommend controlling for pre-treatment covariates when estimating causal effects in experiments and observational studies. However, it is generally not a good idea to control for variables measured *after* the treatment. In this section and the next we explain why controlling for a post-treatment variable messes up the estimate of total treatment effect, and also the difficulty of using regression on "mediators" or "intermediate outcomes" (variables measured post-treatment but generally prior to the primary outcome of interest) to estimate so-called mediating effects.

Consider a hypothetical study of a treatment that incorporates a variety of social

unit, i	treatment, T_i	observed intermediate outcome, z_i	potential intermediate outcomes,		final outcome, y_i
			z_i^0	z_i^1	
1	0	0.5	**0.5**	*0.7*	y_1
2	1	0.5	*0.3*	**0.5**	y_2
⋮	⋮	⋮	⋮	⋮	⋮

Figure 9.13 *Hypothetical example illustrating the problems with regressions that control on a continuous intermediate outcome. If we control for z when regressing y on T, we will be essentially making comparisons between units such as 1 and 2 above, which differ in T but are identical in z. The trouble is that such units are not, in fact, comparable, as can be seen by looking at the potential outcomes, z^0 and z^1 (which can never both be observed, but which we can imagine for the purposes of understanding this comparison). Unit 1, which received the control, has higher potential outcomes than unit 2, which received the treatment. Matching on the observed z inherently leads to misleading comparisons as measured by the potential outcomes, which are the more fundamental quantity.*
The coefficient θ in regression (9.6) thus in general represents an inappropriate comparison of units that fundamentally differ. See Figure 9.14 for a similar example with a discrete intermediate outcome.

services including high-quality child care and home visits by trained professionals. We label y as the child's IQ score, z as the parenting quality, T as the *randomly assigned* binary treatment, and x as a pre-treatment background variable (which could in general be a vector). The goal here is to measure the effect of T on y, and we shall explain why it is not a good idea to control for the intermediate outcome, z, in making this estimate.

To keep things clean, we shall assume a linear regression for the intermediate outcome:

$$z = 0.3 + 0.2T + \gamma x + \text{error}, \tag{9.4}$$

with independent errors.[10] We further suppose that the pre-treatment variable x has been standardized to have mean 0. Then, on average, we would see parenting quality at 0.3 for the controls and 0.5 for the treated parents. Thus the causal effect of the treatment on parenting quality is 0.2. An interaction of T and x could be easily added and interpreted as well if it is desired to estimate systematic variation of treatment effects.

Similarly, a model for y given T and x—excluding z—is straightforward, with the coefficient of T representing the total effect of the treatment on the child's cognitive outcome:

$$\text{regression estimating the treatment effect:} \quad y = \theta T + \beta x + \epsilon. \tag{9.5}$$

The difficulty comes if z is added to this model. Adding z as a predictor could improve the model fit, explaining much of the variation in y:

$$\text{regression including intermediate outcome:} \quad y = \theta^* T + \beta^* x + \delta^* z + \epsilon^*. \tag{9.6}$$

We add the asterisks here because adding a new predictor changes the interpretation of each of the parameters. Unfortunately, the new coefficient θ^* does *not*, in general, estimate the effect of T.

Figure 9.13 illustrates the problem with controlling for an intermediate outcome.

---

[10] We use the notation γ for the coefficient of x because we are saving β for the regression of y; see model (9.5).

The coefficient of T in regression (9.6) corresponds to a comparison of units that are identical in x and z but differ in T. The trouble is, they will then automatically differ in their *potential outcomes*, z^0 and z^1. For example, consider two families, one with $z = 0.5$ but one with $T = 0$ and one with $T = 1$. Under the (simplifying) assumption that the effect of T is to increase z by exactly 0.2 (recall the assumed model (9.4)), the first family has potential outcomes $z^0 = 0.5, z^1 = 0.7$, and the second family has potential outcomes $z^0 = 0.3, z^1 = 0.5$. Thus, given two families with the same intermediate outcome z, the one that received the treatment has lower underlying parenting skills. Thus, in the regression of y on (x, T, z), the coefficient of T represents a comparison of families that differ in their underlying characteristics. This is an inevitable consequence of controlling for an intermediate outcome.

This reasoning suggests a strategy of estimating treatment effects conditional on the potential outcomes—in this example, including both z^0 and z^1, along with T and x, in the regression. The practical difficulty here (as usual) is that we observe at most one potential outcome for each observation, and thus such a regression would require imputation of z^0 or z^1 for each case (perhaps, informally, by using pre-treatment variables as proxies for z^0 and z^1), and correspondingly strong assumptions.

9.8 Intermediate outcomes and causal paths

Randomized experimentation is often described as a "black box" approach to causal inference. We see what goes into the box (treatments) and we see what comes out (outcomes), and we can make inferences about the relation between these inputs and outputs, without the ability to see what happens *inside* the box. This section discusses what happens when we use standard techniques to try to ascertain the role of post-treatment, or *mediating* variables, in the causal path between treatment and outcomes. We present this material at the end of this chapter because the discussion relies on concepts from the analysis of both randomized experiments and observational studies.

Hypothetical example of a binary intermediate outcome

Continuing the hypothetical experiment on child care, suppose that the randomly assigned treatment increases children's IQ points after three years by an average of 10 points (compared to the outcome under usual care). We would additionally like to know to what extent these positive results were the result of improved parenting practices. This question is sometimes phrased as: "What is the 'direct' effect of the treatment, net the effect of parenting?" Does the experiment allow us to evaluate this question? The short answer is no. At least not without making further assumptions.

Yet it would not be unusual to see such a question addressed by simply running a regression of the outcome on the randomized treatment variable along with a predictor representing (post-treatment) "parenting" added to the equation; recall that this is often called a *mediating* variable or mediator. Implicitly, the coefficient on the treatment variable then creates a comparison between those randomly assigned to treatment and control, within subgroups defined by post-treatment parenting practices. Let us consider what is estimated by such a regression.

For simplicity, assume these parenting practices are measured by a simple categorization as "good" or "poor." The simple comparison of the two groups can mislead, because parents who demonstrate good practices after the treatment is applied are likely to be different, on average, from the parents who would have been classified

Parenting potential	Parenting quality after assigned to control	treat	Child's IQ score after assigned to control	treat	Proportion of sample
Poor parenting either way	Poor	Poor	60	70	0.1
Good parenting if treated	Poor	Good	65	80	0.7
Good parenting either way	Good	Good	90	100	0.2

Figure 9.14 *Hypothetical example illustrating the problems with regressions that control on intermediate outcomes. The table shows, for three categories of parents, their potential parenting behaviors and the potential outcomes for their children under the control and treatment conditions. The proportion of the sample falling into each category is also provided. In actual data, we would not know which category was appropriate for each individual parent—it is the fundamental problem of causal inference that we can observe at most one treatment condition for each person—but this theoretical setup is helpful for understanding the properties of statistical estimates. See Figure 9.13 for a similar example with a continuous intermediate outcome.*

as having good parenting practices even in the absence of the treatment. Therefore such comparisons, in essence, lose the advantages originally imparted by the randomization and it becomes unclear what such estimates represent.

Regression controlling for intermediate outcomes cannot, in general, estimate "mediating" effects

Some researchers who perform these analyses will claim that these models are still useful because, if the estimate of the coefficient on the treatment variable goes to zero after including the mediating variable, then we have learned that the entire effect of the treatment acts through the mediating variable. Similarly, if the treatment effect is cut in half, they might claim that half of the effect of the treatment acts through better parenting practices or, equivalently, that the effect of treatment net the effect of parenting is half the total value. This sort of conclusion is *not* generally appropriate, however, as we illustrate with a hypothetical example.

Hypothetical scenario with direct and indirect effects. Figure 9.14 displays potential outcomes of the children of the three different kinds of parents in our sample: those who will demonstrate poor parenting practices with or without the intervention, those whose parenting will get better if they receive the intervention, and those who will exhibit good parenting practices with or without the intervention. We can think of these categories as reflecting parenting *potential*. For simplicity, we have defined the model deterministically, with no individual variation within the three categories of family.

Here the effect of the intervention is 10 IQ points on children whose parents' parenting practices were unaffected by the treatment. For those parents who would improve their parenting due to the intervention, the children get a 15-point improvement. In some sense, philosophically, it is difficult (some would say impossible) to even define questions such as "what percentage of the treatment effect can be attributed to improved parenting practices" since treatment effects (and fractions attributable to various causes) can differ across people. How can we ever say for those families that have good parenting, if treated, what portion of their treatment effect can be attributed to differences in parenting practices as compared to the effects experienced by the families whose parenting practices would not change based on their treatment assignment? If we assume, however, that the effect on children due to sources other than parenting practices stays constant over different types of people (10 points), then we might say that, at least for those with the potential to have their parenting improved by the intervention, this improved parenting accounts for about $(15 - 10)/15 = 1/3$ of the effect.

A regression controlling for the intermediate outcome does not generally work. However, if one were to try to estimate this effect using a regression of the outcome on the randomized treatment variable and observed parenting behavior, the coefficient on the treatment indicator will be -1.5, falsely implying that the treatment has some sort of negative "direct effect" on IQ scores!

To see what is happening here, recall that this coefficient is based on comparisons of treated and control groups *within* groups defined by *observed* parenting behavior. Consider, for instance, the comparison between treated and control groups within those observed to have poor parenting behavior. The group of parents who did not receive the treatment and are observed to have poor parenting behavior is a mixture of those who would have exhibited poor parenting either way and those who exhibited poor parenting simply because they did not get the treatment. Those in the treatment group who exhibited poor parenting are all those who would have exhibited poor parenting either way. Those whose poor parenting is not changed by the intervention have children with lower test scores on average—under either treatment condition—than those whose parenting would have been affected by the intervention.

The regression controlling for the intermediate outcome thus implicitly compares unlike groups of people and underestimates the treatment effect, because the treatment group in this comparison is made up of lower-performing children, on average. A similar phenomenon occurs when we make comparisons across treatment groups among those who exhibit good parenting. Those in the treatment group who demonstrate good parenting are a mixture of two groups (good parenting if treated and good parenting either way) whereas the control group is simply made up of the parents with the highest-performing children (good parenting either way). This estimate does not reflect the effect of the intervention net the effect of parenting. It does not estimate any causal effect. It is simply a mixture of some nonexperimental comparisons.

This example is an oversimplification, but the basic principles hold in more complicated settings. In short, randomization allows us to calculate causal effects of the variable randomized, but not other variables unless a whole new set of assumptions is made. Moreover, the benefits of the randomization for treatment effect estimation are generally destroyed by including post-treatment variables. These assumptions and the strategies that allow us to estimate the effects conditional on intermediate outcomes in certain situations will be discussed at the end of Chapter 10.

What can be estimated: principal stratification

We noted earlier that questions such as "What proportion of the treatment effect works through variable A?" are in some sense, inherently unanswerable. What can we learn about the role of intermediate outcomes or mediating variables? As we discussed in the context of Figure 9.14, treatment effects can vary depending on the extent to which the mediating variable (in this example, parenting practices) is affected by the treatment. The key theoretical step here is to divide the population into categories based on their potential outcomes for the mediating variable—what would happen under each of the two treatment conditions. In statistical parlance, these categorizations are sometimes called *principal strata*. The problem is that the principal stratum labels are generally unobserved. It is theoretically possible to statistically infer principal-stratum categories based on covariates, especially if the treatment was randomized—because then at least we know that the distribution of principal strata is the same across the randomized groups. In practice, however, this reduces to making the same kinds of assumptions as are made in typical observational studies when ignorability is assumed.

Principal strata are important because they can define, even if only theoretically, the categories of people for whom the treatment effect can be estimated from available data. For example, if treatment effects were nonzero only for the study participants whose parenting practices had been changed, and if we could reasonably exclude other causal pathways, even stronger conclusions could be drawn regarding the role of this mediating variable. We discuss this scenario of *instrumental variables* in greater detail in Section 10.5.

Intermediate outcomes in the context of observational studies

If trying to control directly for mediating variables is problematic in the context of randomized experiments, it should come as no surprise that it generally is also problematic for observational studies. The concern is nonignorability—systematic differences between groups defined conditional on the post-treatment intermediate outcome. In the example above if we could control for the true parenting potential designations, the regression would yield the correct estimate for the treatment effect if we are willing to assume constant effects across groups (or willing to posit a model for how effects change across groups). One conceivably can obtain the same result by controlling sufficiently for covariates that adequately proxy this information.

In observational studies, researchers often already know to control for many predictors. So it is possible that these predictors will mitigate some of the problems we have discussed. On the other hand, studying intermediate outcomes in an observational study involves two ignorability problems to deal with rather than just one, making it all the more challenging to obtain trustworthy results.

Well-switching example.   As an example where the issues discussed in this and the previous section come into play, consider one of the logistic regressions from Chapter 5:

$$\Pr(\text{switch}) = \text{logit}^{-1}(-0.21 - 0.90 \cdot \text{dist100} + 0.47 \cdot \text{arsenic} + 0.17 \cdot \text{educ4}),$$

predicting the probability that a household switches drinking-water wells as a function of distance to the nearest safe well, arsenic level of the current well, and education of head of household.

This model can simply be considered as data description, but it is natural to try to interpret it causally: being further from a safe well makes one less likely to switch, having a higher arsenic level makes switching more likely, and having more education makes one more likely to switch. Each of these coefficients is interpreted with the other two inputs held constant—and this is what we want to do, in isolating the "effects" (as crudely interpreted) of each variable. For example, households that are farther from safe wells turn out to be more likely to have high arsenic levels, and in studying the "effect" of distance, we would indeed like to compare households that are otherwise similar, including in their arsenic level. This fits with a psychological or decision-theoretic model in which these variables affect the perceived costs and benefits of the switching decision (as outlined in Section 6.8).

However, in the well-switching example as in many regression problems, additional assumptions beyond the data are required to justify the convenient interpretation of multiple regression coefficients as causal effects—what would happen to y if a particular input were changed, with all others held constant—and it is rarely appropriate to give more than one coefficient such an interpretation, and then only after careful consideration of ignorability. Similarly, we cannot learn about causal pathways from observational data without strong assumptions.

For example, a careful estimate of the effect of a potential intervention (for example, digging new, safe wells in close proximity to existing high-arsenic households) should include, if not an actual experiment, a model of what would happen in the particular households being affected, which returns us to the principles of observational studies discussed earlier in this chapter.

9.9 Bibliographic note

The fundamental problem of causal inference and the potential outcome notation were introduced by Rubin (1974, 1978). Related earlier work includes Neyman (1923) and Cox (1958). For other approaches to causal inference, see Pearl (2000) along with many of the references in Section 10.8.

The stable unit treatment value assumption was defined by Rubin (1978); see also Sobel (2006) for a more recent discussion in the context of a public policy intervention and evaluation. Ainsley, Dyke, and Jenkyn (1995) and Besag and Higdon (1999) discuss spatial models for interference between units in agricultural experiments. Gelman (2004d) discusses treatment interactions in before/after studies.

Campbell and Stanley (1963) is an early presentation of causal inference in experiments and observational studies from a social science perspective; see also Achen (1986) and Shadish, Cook, and Campbell (2002). Rosenbaum (2002b) and Imbens (2004) present overviews of inference for observational studies. Dawid (2000) offers another perspective on the potential-outcome framework. Leamer (1978, 1983) explores the challenges of relying on regression models for answering causal questions.

Modeling strategies also exist that rely on ignorability but loosen the relatively strict functional form imposed by linear regression. Examples include Hahn (1998), Heckman, Ichimura and Todd (1998), Hirano, Imbens, and Ridder (2003), and Hill and McCulloch (2006).

The example regarding the Korean babies up for adoption was inspired by Sacerdote (2004). The Electric Company experiment is described by Ball and Bogatz (1972) and Ball et al. (1972).

Rosenbaum (1984) provides a good discussion of the dangers outlined in Section 9.8 involved in trying to control for post-treatment outcomes. Raudenbush and Sampson (1999), Rubin (2000), and Rubin (2004) discuss direct and indirect effects for multilevel designs. We do not attempt here to review the vast literature on structural equation modeling; Kenny, Kashy, and Bolger (1998) is a good place to start.

The term "principal stratification" was introduced by Frangakis and Rubin (2002); examples of its application include Frangakis et al. (2003) and Barnard et al. (2003). Similar ideas appear in Robins (1989, 1994).

9.10 Exercises

1. Suppose you are interested in the effect of the presence of vending machines in schools on childhood obesity. What randomized experiment would you want to do (in a perfect world) to evaluate this question?

2. Suppose you are interested in the effect of smoking on lung cancer. What ran-

domized experiment could you plausibly perform (in the real world) to evaluate this effect?

3. Suppose you are a consultant for a researcher who is interested in investigating the effects of teacher quality on student test scores. Use the strategy of mapping this question to a randomized experiment to help define the question more clearly. Write a memo to the researcher asking for needed clarifications to this study proposal.

4. The table below describes a hypothetical experiment on 2400 persons. Each row of the table specifies a category of person, as defined by his or her pre-treatment predictor x, treatment indicator T, and potential outcomes y^0, y^1. (For simplicity, we assume unrealistically that all the people in this experiment fit into these eight categories.)

Category	# persons in category	x	T	y^0	y^1
1	300	0	0	4	6
2	300	1	0	4	6
3	500	0	1	4	6
4	500	1	1	4	6
5	200	0	0	10	12
6	200	1	0	10	12
7	200	0	1	10	12
8	200	1	1	10	12

In making the table we are assuming omniscience, so that we know both y^0 and y^1 for all observations. But the (nonomniscient) investigator would only observe x, T, and y^T for each unit. (For example, a person in category 1 would have $x=0, T=0, y=4$, and a person in category 3 would have $x=0, T=1, y=6$.)

(a) What is the average treatment effect in this population of 2400 persons?

(b) Is it plausible to believe that these data came from a randomized experiment? Defend your answer.

(c) Another population quantity is the mean of y for those who received the treatment minus the mean of y for those who did not. What is the relation between this quantity and the average treatment effect?

(d) For these data, is it plausible to believe that treatment assignment is ignorable given sex? Defend your answer.

5. For the hypothetical study in the previous exercise, figure out the estimate and the standard error of the coefficient of T in a regression of y on T and x.

6. You are consulting for a researcher who has performed a randomized trial where the treatment was a series of 26 weekly therapy sessions, the control was no therapy, and the outcome was self-report of emotional state one year later. However, most people in the treatment group did not attend every therapy session. In fact there was a good deal of variation in the number of therapy sessions actually attended. The researcher is concerned that her results represent "watered down" estimates because of this variation and suggests adding in another predictor to the model: number of therapy sessions attended. What would you advise her?

7. Gain-score models: in the discussion of gain-score models in Section 9.3, we noted that if we include the pre-treatment measure of the outcome in a gain score model, the coefficient on the treatment indicator will be the same as if we had just run a standard regression of the outcome on the treatment indicator and the pre-treatment measure. Show why this is true.

8. Assume that linear regression is appropriate for the regression of an outcome, y, on treatment indicator, T, and a single confounding covariate, x. Sketch hypothetical data (plotting y versus x, with treated and control units indicated by circles and dots, respectively) and regression lines (for treatment and control group) that represent each of the following situations:

   (a) No treatment effect,
   (b) Constant treatment effect,
   (c) Treatment effect increasing with x.

9. Consider a study with an outcome, y, a treatment indicator, T, and a single confounding covariate, x. Draw a scatterplot of treatment and control observations that demonstrates each of the following:

   (a) A scenario where the difference in means estimate would not capture the true treatment effect but a regression of y on x and T would yield the correct estimate.
   (b) A scenario where a linear regression would yield the wrong estimate but a nonlinear regression would yield the correct estimate.

10. The folder sesame contains data from an experiment in which a randomly selected group of children was encouraged to watch the television program Sesame Street and the randomly selected control group was not.

   (a) The goal of the experiment was to estimate the effect on child cognitive development of watching more Sesame Street. In the experiment, encouragement but not actual watching was randomized. Briefly explain why you think this was done. (Hint: think of practical as well as statistical reasons.)
   (b) Suppose that the investigators instead had decided to test the effectiveness of the program simply by examining how test scores changed from before the intervention to after. What assumption would be required for this to be an appropriate causal inference? Use data on just the control group from this study to examine how realistic this assumption would have been.

11. Return to the Sesame Street example from the previous exercise.

   (a) Did encouragement (the variable viewenc in the dataset) lead to an increase in post-test scores for letters (postlet) and numbers (postnumb)? Fit an appropriate model to answer this question.
   (b) We are actually more interested in the effect of watching Sesame Street regularly (regular) than in the effect of being encouraged to watch Sesame Street. Fit an appropriate model to answer this question.
   (c) Comment on which of the two previous estimates can plausibly be interpreted causally.

12. Messy randomization: the folder cows contains data from an agricultural experiment that was conducted on 50 cows to estimate the effect of a feed additive on six outcomes related to the amount of milk fat produced by each cow.

   Four diets (treatments) were considered, corresponding to different levels of the additive, and three variables were recorded before treatment assignment: lactation number (seasons of lactation), age, and initial weight of cow.

   Cows were initially assigned to treatments completely at random, and then the distributions of the three covariates were checked for balance across the treatment groups; several randomizations were tried, and the one that produced the "best" balance with respect to the three covariates was chosen. The treatment assignment is ignorable (because it depends only on fully observed covariates and not on unrecorded variables such as the physical appearances of the cows or the times at which the cows entered the study) but unknown (because the decisions whether to rerandomize are not explained).

We shall consider different estimates of the effect of additive on the mean daily milk fat produced.

(a) Consider the simple regression of mean daily milk fat on the level of additive. Compute the estimated treatment effect and standard error, and explain why this is not a completely appropriate analysis given the randomization used.

(b) Add more predictors to the model. Explain your choice of which variables to include. Compare your estimated treatment effect to the result from (a).

(c) Repeat (b), this time considering additive level as a categorical predictor with four letters. Make a plot showing the estimate (and standard error) of the treatment effect at each level, and also showing the inference the model fit in part (b).

13. The folder `congress` has election outcomes and incumbency for U.S. congressional election races in the 1900s.

(a) Take data from a particular year, t, and estimate the effect of incumbency by fitting a regression of $v_{i,t}$, the Democratic share of the two-party vote in district i, on $v_{i,t-2}$ (the outcome in the previous election, two years earlier), I_{it} (the incumbency status in district i in election t, coded as 1 for Democratic incumbents, 0 for open seats, -1 for Republican incumbents), and P_{it} (the incumbent *party*, coded as 1 if the sitting congressmember is a Democrat and -1 if he or she is a Republican). In your analysis, include only the districts where the congressional election was contested in both years, and do not pick a year ending in "2." (District lines in the United States are redrawn every ten years, and district election outcomes v_{it} and $v_{i,t-2}$ are not comparable across redistrictings, for example, from 1970 to 1972.)

(b) Plot the fitted model and the data, and discuss the political interpretation of the estimated coefficients.

(c) What assumptions are needed for this regression to give a valid estimate of the causal effect of incumbency? In answering this question, define clearly what is meant by incumbency as a "treatment variable."

See Erikson (1971), Gelman and King (1990), Cox and Katz (1996), Levitt and Wolfram (1997), Ansolabehere, Snyder, and Stewart (2000), Ansolabehere and Snyder (2002), and Gelman and Huang (2006) for further work and references on this topic.

14. Causal inference based on data from individual choices: our lives involve trade-offs between monetary cost and physical risk, in decisions ranging from how large a car to drive, to choices of health care, to purchases of safety equipment. Economists have estimated people's implicit balancing of dollars and danger by comparing different jobs that are comparable but with different risks, fitting regression models predicting salary given the probability of death on the job. The idea is that a riskier job should be compensated with a higher salary, with the slope of the regression line corresponding to the "value of a statistical life."

(a) Set up this problem as an individual choice model, as in Section 6.8. What are an individual's options, value function, and parameters?

(b) Discuss the assumptions involved in assigning a causal interpretation to these regression models.

See Dorman and Hagstrom (1998), Costa and Kahn (2002), and Viscusi and Aldy (2002) for different perspectives of economists on assessing the value of a life, and Lin et al. (1999) for a discussion in the context of the risks from radon exposure.

## CHAPTER 10

Causal inference using more advanced models

Chapter 9 discussed situations in which it is dangerous to use a standard linear regression of outcome on predictors and an indicator variable for estimating causal effects: when there is imbalance or lack of complete overlap or when ignorability is in doubt. This chapter discusses these issues in more detail and provides potential solutions for each.

10.1 Imbalance and lack of complete overlap

In a study comparing two treatments (which we typically label "treatment" and "control"), causal inferences are cleanest if the units receiving the treatment are comparable to those receiving the control. Until Section 10.5, we shall restrict ourselves to ignorable models, which means that we only need to consider observed pre-treatment predictors when considering comparability.

For ignorable models, we consider two sorts of departures from comparability—*imbalance* and *lack of complete overlap*. Imbalance occurs if the distributions of relevant pre-treatment variables differ for the treatment and control groups. Lack of complete overlap occurs if there are regions in the space of relevant pre-treatment variables where there are treated units but no controls, or controls but no treated units.

Imbalance and lack of complete overlap are issues for causal inference largely because they force us to rely more heavily on model specification and less on direct support from the data.

When treatment and control groups are *unbalanced*, the simple comparison of group averages, $\bar{y}_1 - \bar{y}_0$, is not, in general, a good estimate of the average treatment effect. Instead, some analysis must be performed to adjust for pre-treatment differences between the groups.

When treatment and control groups do not completely *overlap*, the data are inherently limited in what they can tell us about treatment effects in the regions of nonoverlap. No amount of adjustment can create direct treatment/control comparisons, and one must either restrict inferences to the region of overlap, or rely on a model to extrapolate outside this region.

Thus, lack of complete overlap is a more serious problem than imbalance. But similar statistical methods are used in both scenarios, so we discuss these problems together here.

Imbalance and model sensitivity

When attempting to make causal inferences by comparing two samples that differ in terms of the "treatment" or causing variable of interest (participation in a program, taking a drug, engaging in some activity) but that also differ in terms of confounding covariates (predictors related both to the treatment and outcome), we can be misled

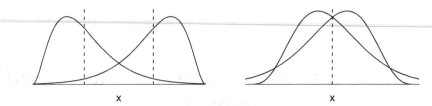

Figure 10.1 *Imbalance in distributions across treatment and control groups. (a) In the left panel, the groups differ in their averages (dotted vertical lines) but cover the same range of x. (b) The right panel shows a more subtle form of imbalance, in which the groups have the same average but differ in their distributions.*

if we do not appropriately control for those confounders. The examples regarding the effect of a treatment on health outcomes in Section 9.1 illustrated this point in a simple setting.

Even when all the confounding covariates are measured (hence ignorability is satisfied), however, it can be difficult to properly control for them if the distributions of the predictors are not similar across groups. Broadly speaking, any differences across groups can be referred to as lack of *balance* across groups. The terms "imbalance" and "lack of balance" are commonly used as a shorthand for differences in averages, but more broadly they can refer to more general differences in distributions across groups. Figure 10.1 provides two examples of imbalance. In the first case the groups have different means (dotted vertical lines) and different skews. In the second case groups have the same mean but different skews. In both examples the standard deviations are the same across groups though differences in standard deviation might be another manifestation of imbalance.

Imbalance creates problems primarily because it forces us to rely more on the correctness of our model than we would have to if the samples were balanced. To see this, consider what happens when we try to make inferences about the effect of a treatment variable, for instance a new reading program, on test score, y, while controlling for a crucial confounding covariate, pre-test score, x. Suppose that the true treatment effect is θ and the relations between the response variable, y, and the sole confounding covariate, x, is quadratic, as indicated by the following regressions, written out separately for the members of each treatment group:

$$\text{treated: } y_i = \beta_0 + \beta_1 x_i + \beta_2 x_i^2 + \theta + \text{error}_i$$
$$\text{controls: } y_i = \beta_0 + \beta_1 x_i + \beta_2 x_i^2 + \text{error}_i$$

Averaging over each treatment group separately, solving the second equation for β_0, plugging back into the first, and solving for θ yields the estimate

$$\hat{\theta} = \bar{y}_1 - \bar{y}_0 - \beta_1(\bar{x}_1 - \bar{x}_0) - \beta_2(\overline{x_1^2} - \overline{x_0^2}), \tag{10.1}$$

where $\bar{y}_1$ and $\bar{y}_0$ denote the average of the outcome test scores in the treatment and control groups respectively, $\bar{x}_1$ and $\bar{x}_0$ represent average pre-test scores for treatment and control groups respectively, and $\overline{x_1^2}$ and $\overline{x_0^2}$ represent these averages for squared pre-test scores. Ignoring x (that is, simply using the raw treatment/control comparison $\bar{y}_1 - \bar{y}_0$) is a poor estimate of the treatment effect: it will be off by the amount $\beta_1(\bar{x}_1 - \bar{x}_0) + \beta_2(\overline{x_1^2} - \overline{x_0^2})$, which corresponds to systematic pre-treatment differences between groups 0 and 1. The magnitude of this bias depends on how different the distribution of x is across treatment and control groups (specifically with regard to variance in this case) and how large β_1 and β_2 are. The closer the

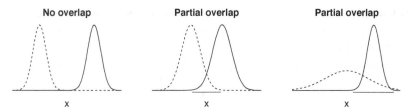

Figure 10.2 *Lack of complete overlap in distributions across treatment and control groups. Dashed lines indicate distributions for the control group; solid lines indicate distributions for the treatment group. (a) Two distributions with no overlap; (b) two distributions with partial overlap; (c) a scenario in which the* range *of one distribution is a subset of the* range *of the other.*

distributions of pre-test scores across treatment and control groups, the smaller this bias will be.

Moreover, a linear model regression using x as a predictor would also yield the wrong answer; it will be off by the amount $\beta_2(\overline{x_1^2} - \overline{x_0^2})$. The closer the distributions of pre-test scores across treatment and control groups, however, the smaller $(\overline{x_1^2} - \overline{x_0^2})$ will be, and the less worried we need to be about correctly specifying this model as quadratic rather than linear.

Lack of complete overlap and model extrapolation

Overlap describes the extent to which the range of the data is the same across treatment groups. There is *complete overlap* if this range is the same in the two groups. Figure 10.1 illustrated treatment and control confounder distributions with complete overlap.

As discussed briefly in the previous chapter, lack of complete overlap creates problems because it means that there are treatment observations for which we have no counterfactuals (that is, control observations with the same covariate distribution) and vice versa. A model fitted to data such as these is forced to extrapolate beyond the support of the data. The illustrations in Figure 10.2 display several scenarios that exhibit lack of complete overlap.

If these are distributions for an important confounding covariate, then areas where there is no overlap represent observations about which we may not want to make causal inferences. Observations in these areas have no empirical counterfactuals. Thus, any inferences regarding these observations would have to rely on modeling assumptions in place of direct support from the data. Adhering to this structure would imply that in the setting of Figure 10.2a, it would be impossible to make data-based causal inferences about any of the observations. Figure 10.2b shows a scenario in which data-based inferences are only possible for the region of overlap, which is underscored on the plot. In Figure 10.2c, causal inferences are possible for the full treatment group but only for a subset of the control group (again indicated by the underscored region).

Example: evaluating the effectiveness of high-quality child care

We illustrate with data collected regarding the development of nearly 4500 children born in the 1980s. A subset of 290 of these children who were premature and with low birth weight (between 1500 and 2500 grams) received special services in the first few years of life, including high-quality child care (five full days a week) in the

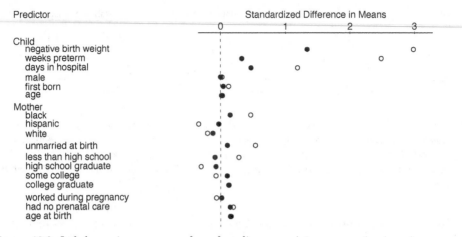

Figure 10.3 *Imbalance in averages of confounding covariates across treatment groups. Open circles represent differences in averages for the unmatched groups standardized by the pooled within-group standard deviations for unmatched groups. Solid circles represent differences in averages for matched groups standardized by the pooled within-group standard deviation for unmatched groups to facilitate comparisons. Negative birth weight is defined as 2500 grams minus the child's weight at birth.*

second and third years of life as part of a formal intervention (the Infant Health and Development Program). We want to evaluate the impact of this intervention on the children's subsequent cognitive outcomes by comparing the outcomes for children in the intervention group to the outcomes in a comparison group of 4091 children who did not participate in the program. The outcome of interest is test score at age 3; this test is similar to an IQ measure so we simplistically refer to these scores as IQ scores from now on.

Missing data. Incomplete data arise in virtually all observational studies. For this sample dataset, we imputed missing data once, using a model-based random imputation (see Chapter 25 for a general discussion of this approach). We excluded the most severely low-birth-weight children (those at or below 1500 grams) from the sample because they are so different from the comparison sample. For these reasons, results presented here do not exactly match the complete published analysis, which multiply imputed the missing values.

Examining imbalance for several covariates

To illustrate the ways in which the treated and comparison groups differ, the open circles in Figure 10.3 display the standardized differences in mean values (differences in averages divided by the pooled within-group standard deviations for the treatment and control groups) for a set of confounding covariates that we think predict both program participation and subsequent test scores. Many of these differences are large given that they are shown in standard-deviation units.

Setting up the plot to reveal systematic patterns of imbalance. In Figure 10.3, the characteristics of this sample are organized by whether they pertain to the child or to the mother. Additionally, continuous and binary predictors have been coded when possible such that the larger values are typically associated with lower test scores for children. For instance, "negative birth weight" is defined as the child's birth weight subtracted from 2500 grams, the cutoff for the official designation of

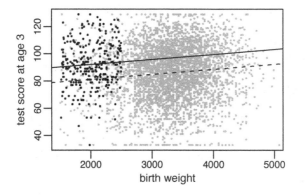

Figure 10.4 *Data from an intervention targeting low birth weight, premature children (black dots), and data from a comparison group of children (gray dots). Test scores at age 3 are plotted against birth weight. The solid line and dotted lines are regressions fit to the black and gray points, respectively.*

low birth weight. Therefore, high values of this predictor reflect children whom we would expect to have lower test scores than children with lower values for negative birth weight. Categorical variables have been broken out into indicators for each category and organized so that the category associated with lowest test scores comes first.

Displaying the confounders in this way and plotting standardized averages—rather than displaying a table of numbers—facilitate comparisons across predictors and methods (the dark points, to be described later, correspond to results obtained from another strategy) and allow us to more clearly identify trends when they exist. For instance, compared to the control group, the at-risk treatment group generally has characteristics associated with lower test scores—such as low birth weight for the child (coded as high "negative birth weight"), mother unmarried at birth, and mother not a high school graduate.

Figure 10.4, which shows a scatterplot and regression lines of test scores on birth weight, illustrates that, not only do the average birth weights differ in the two groups (lack of balance), but there are many control observations (gray dots) who have birth weights far out of the range of birth weights experienced in the treatment population (black dots). This is an example of *lack of complete overlap* in this predictor across groups. If birth weight is a confounding covariate that we need to control for to achieve ignorability, Figure 10.4 demonstrates that if we want to make inferences about the effect of the program on children with birth weights above 2500 grams, we will have to rely on model extrapolations that may be inappropriate.

Imbalance is not the same as lack of overlap

Figure 10.5 illustrates the distinction between balance and overlap. Imbalance does not necessarily imply lack of complete overlap; conversely, lack of complete overlap does not necessarily necessarily result in imbalance in the sense of different average values in the two groups. Ultimately, lack of overlap is a more serious problem, corresponding to a lack of data that limits the causal conclusions that can be made without uncheckable modeling assumptions.

Figure 10.5a demonstrates complete overlap across groups in terms of mother's education. Each category includes observations in each treatment group. However,

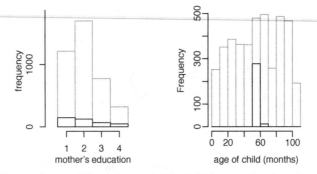

Figure 10.5 *Comparisons of the treatment (black histogram bars) and control (gray histogram bars) groups for the child-intervention study, with respect to two of the pretreatment variables. There is lack of complete overlap for child age, but the averages are similar across groups. In contrast, mother's education shows complete overlap, but imbalance exists in that the distributions differ for the two groups.*

the percentages falling in each category (and the overall average, were we to code these categories as 1–4) differ when comparing treatment and control groups—thus there is clearly imbalance.

Figure 10.5b shows balance in mean values but without complete overlap. As the histograms show, the averages of children's ages differ little across treatment groups, but the vast majority of control children have ages that are not represented in the treatment group. Thus there is a lack of complete overlap across groups for this variable. More specifically, there is complete overlap in terms of the treatment observations, but not in terms of the control observations. If we believe age to be a crucial confounding covariate, we probably would not want to make inferences about the full set of controls in this sample.

10.2 Subclassification: effects and estimates for different subpopulations

Assuming we are willing to trust the ignorability assumption, how can we assess whether we are relying too strongly on modeling assumptions? And if we are uncertain of our assumptions, how can we proceed cautiously? Section 9.5 illustrated a check for overlap in one continuous predictor across treatment groups. In this section we demonstrate a check that accommodates many predictors and discuss options for more flexible modeling.

Subclassification

We saw in Chapter 3 that mother's educational attainment is an important predictor of her child's test scores. Education level also traditionally is associated with participation in interventions such as this program for children with low birth weights. Let us make the (unreasonable) assumption for the moment that this is the only confounding covariate (that is, the only predictor associated with both participation in this program and test scores). How would we want to estimate causal effects? In this case a simple solution would be to estimate the difference in mean test scores within each subclass defined by mother's education. These averages as well as the associated standard error and sample size in each subclass are displayed in Figure 10.6. These point to positive effects for all participants, though not all

Mother's education	Treatment effect estimate ± s.e.	Sample size treated	controls
Not a high school grad	9.3 ± 1.3	126	1358
High school graduate	4.0 ± 1.8	82	1820
Some college	7.9 ± 2.3	48	837
College graduate	4.6 ± 2.1	34	366

Figure 10.6 *Estimates ± standard errors of the effect on children's test scores of a child care intervention, for each of four subclasses formed by mother's educational attainment. The study was of premature infants with low birth weight, most of whom were born to mothers with low levels of education.*

effects are statistically significant, with by far the largest effects for the children whose mothers had not graduated from high school.

Recall that there is overlap on this variable across the treatment and control groups as is evidenced by the sample sizes for treated and control observations within each subclass in Figure 10.6. If there were a subclass with observations only from one group, we would not be able to make inferences for this type of person. Also, if there were a subclass with only a small number of observations in either the treatment group or the control group, we would probably be wary of making inferences for these children as well.

To get an estimate of the overall effect for those who participated in the program, the subclass-specific estimates could be combined using a weighted average where the weights are defined by the number of children in each subclass who participated in the program:

$$\text{Est. effect on the treated} = \frac{9.3 \cdot 126 + 4.0 \cdot 82 + 7.9 \cdot 48 + 4.6 \cdot 34}{126 + 82 + 48 + 34} = 7.0, \quad (10.2)$$

with a standard error of $\sqrt{\frac{1.3^2 \cdot 126^2 + 1.8^2 \cdot 82^2 + 2.3^2 \cdot 5.3^2 + 2.1^2 \cdot 34^2}{(126+82+48+34)^2}} = 0.9$.

This analysis is similar to a regression with interactions between the treatment and mother's educational attainment. To calculate the average treatment effect for program participants, we would have to poststratify—that is, estimate the treatment effect separately for each category of mother's education, and then average these effects based on the distribution of mother's education in the population.

This strategy has the advantage of imposing overlap and, moreover, forcing the control sample to have roughly the same covariate distribution as the treated sample. This reduces reliance on the type of model extrapolations discussed previously. Moreover, one can choose to avoid modeling altogether after subclassifying, and simply can take a difference in averages across treatment and control groups to perform inferences, therefore completely avoiding making assumptions about the parametric relation between the response and the confounding covariates.

One drawback of subclassifying, however, is that when controlling for a continuous variable, some information may be lost when discretizing the variable. A more substantial drawback is that it is difficult to control for many variables at once.

Average treatment effects: whom do we average over?

Figure 10.6 demonstrated how treatment effects can vary over different subpopulations. Why did we weight these subclass-specific estimates by the number of treated children in each subclass rather than the total number of children in each subclass?

For this application, we are interested in the effect of the intervention *for the sort of children who would have participated in it.* Weighting using the number of treatment children in each subclass forces the estimate implicitly to be representative of the treatment children we observe. The effect we are trying to estimate is sometimes called the *effect of the treatment on the treated.*

If we had weighted instead by the number of control children in each subclass, we could estimate the effect of the treatment on the controls. However, this particular intervention was designed for the special needs of low-birth-weight, premature children—not for typical children—and there is little interest in its effect on comparison children who would not have participated.

The effect of the intervention might vary, for instance, for children with different initial birth weights, and since we know that the mix of children's birth weights differs in treatment and comparison groups, the average effects across these groups could also differ. Moreover, we saw in Figure 10.4 that there are so many control observations with no counterfactual observations in the treatment group with regard to birth weight that these data are likely inappropriate for drawing inferences about the control group either directly (the effect of the treatment on the controls) or as part of an average effect across the entire sample.

Again, this is related to poststratification. We can think of the estimate of the effect of the treatment on the treated as a poststratified version of the estimate of the average causal effect. As the methods we discuss in this section rely on more and more covariates, however, it can be more attractive to apply methods that more directly estimate the effect of the treatment on the treated, as we discuss next.

10.3 Matching: subsetting the data to get overlapping and balanced treatment and control groups

Matching refers to a variety of procedures that restrict and reorganize the original sample in preparation for a statistical analysis. In the simplest form of matching, one-to-one matching, the data points are divided into pairs—each containing one treated and one control unit—with the two units matched into a pair being as similar as possible on relevant pre-treatment variables. The number of units in the two groups will not in general be equal—typically there are more controls than treated units, as in Figure 10.5, for example—and so there will be some leftover units unmatched. In settings with poor overlap, there can be unmatched units from both groups, so that the matched pairs represent the region of data space where the treatment and control groups overlap.

Once the matched units have been selected out of the larger dataset, they can be analyzed by estimating a simple difference in average outcomes across treatment groups or by using regression methods to estimate the effect of the treatment in the area of overlap.

Matching and subclassification

Matching on one variable is similar to subclassification except that it handles continuous variables more precisely. For instance, a treatment observation might be matched to control observations that had the closest age to their own as opposed to being grouped into subclasses based on broader age categories. Thus, matching has the same advantages of stratification in terms of creating balance and forcing overlap, and may even be able to create slightly better balance. However, many matching methods discard observations even when they are within the range of overlap, which is likely inefficient.

Matching has some advantages over subclassification when controlling for many variables at once. Exact matching is difficult with many confounders, but "nearest-neighbor" matching is often still possible. This strategy matches treatment units to control units that are "similar" in terms of their confounders where the metric for similarity can be defined in any variety of ways, one of the most popular being the *Mahalanobis distance*, which is defined in matrix notation as $d(x^{(1)}, x^{(2)}) = (x^{(1)} - x^{(2)})^t \Sigma^{-1} (x^{(1)} - x^{(2)})$, where $x^{(1)}$ and $x^{(2)}$ represent the vectors of predictors for points 1 and 2, and Σ is the covariance of the predictors in the dataset. Recently, other algorithms have been introduced to accomplish this same task—finding similar treatment and control observations—that rely on algorithms originally created for genetic or data mining applications. Another matching approach, which we describe next, compares the input variables for treatment and control cases in order to find an effective scale on which to match.

Propensity score matching

One way to simplify the issue of matching or subclassifying on many confounding covariates at once is to create a one-number summary of all the covariates and then use this to match or subclassify. We illustrate using a popular summary, the propensity score, with our example of the intervention for children with low birth weights. It seems implausible that mother's education, for example, is the only predictor we need to satisfy the ignorability assumption in our example. We would like to control for as many predictors as possible to allow for the possibility that any of them is a confounding covariate. We also want to maintain the beneficial properties of matching. How can we match on many predictors at once?

Propensity score matching provides a solution to this problem. The *propensity score* for the ith individual is defined as the probability that he or she receives the treatment given everything we observe before the treatment (that is, all the confounding covariates for which we want to control). Propensity scores can be estimated using standard models such as logistic regression, where the outcome is the treatment indicator and the predictors are all the confounding covariates. Then matches are found by choosing for each treatment observation the control observation with the closest propensity score.

In our example we randomly ordered the treatment observations, and then each time a control observation was chosen as a match for a given treatment observation it could not be used again. More generally, methods have been developed for matching multiple control units to a single treated unit, and vice versa; these ideas can be effective, especially when there is overlap but poor balance (so that, for example, some regions of predictor space contain many controls and few treated units, or the reverse). From this perspective, matching can be thought of as a way of discarding observations so that the remaining data show good balance and overlap.

The goal of propensity score matching is not to ensure that each *pair of matched observations* is similar in terms of all their covariate values, but rather that the matched groups are similar *on average* across all their covariate values. Thus, the adequacy of the model used to estimate the propensity score can be evaluated by examining the balance that results on average across the matched groups.

Computation of propensity score matches

The first step in creating matches is to fit a model to predict who got the intervention based on the set of predictors we think are necessary to achieve ignorability (confounding covariates). A natural starting point would be a logistic regression, something like

R code
```
ps.fit.1 <- glm (treat ~ as.factor(educ) + as.factor(ethnic) + b.marr  +
    work.dur + prenatal + mom.age + sex + first + preterm + age +
    dayskidh + bw + unemp.rt, data=cc2, family=binomial(link="logit"))
```

In our example, we evaluated several different model fits before settling on one that provided balance that seemed adequate. In each case we evaluated the adequacy of the model by evaluating the balance that resulted from matching on the estimated propensity scores from that model. Model variations tried excluding variables and including interactions and quadratic terms. We finally settled on

R code
```
ps.fit.2 <- glm (treat ~ bwg + as.factor(educ) + bwg:as.factor(educ) +
    as.factor(ethnic) + b.marr + as.factor(ethnic):b.marr +
    work.dur + prenatal + preterm + age + mom.age + sex + first,
    data=cc2, family=binomial(link="logit"))
```

We then create predicted values:[1]

R code
```
pscores <- predict (ps.fit.2, type="link")
```

The regression model is messy, but we are not concerned with all its coefficients; we are only using it as a tool to construct a balanced comparison between treatment and control groups. We used the estimated propensity scores to create matches, using a little R function called **matching** that finds for each treatment unit in turn the control unit (not previously chosen) with the closest propensity score:[2]

R code
```
matches <- matching (z=cc2$treat, score=pscores)
matched <- cc2[matches$matched,]
```

Then the full dataset was reduced to only the treated observations and only those control observations that were chosen as matches.

The differences between treated and control averages, for the matched subset, are displayed by the solid dots in Figure 10.3. The imbalance has decreased noticeably compared to the unmatched sample. Certain variables (birth weight and the number of days the children were in the hospital after being born) still show imbalance, but none of our models succeeded in balancing those variables. We hope the other variables are more important in predicting future test scores (which appears to be reasonable from the previous literature on this topic).

The process of fitting, assessing, and selecting a model for the propensity scores has completely ignored the outcome variable. We have judged the model solely by the balance that results from subsequent matches on the associated propensity scores. This helps the researcher to be "honest" when fitting the propensity score model because a treatment effect estimate is not automatically produced each time a new model is fit.

[1] We use the `type="link"` option to get predictions on the scale of the linear predictor, that is, $\tilde{X}\beta$. If we wanted predictions on the probability scale, we would set `type="response"`. In this example, similar results would arise from using either approach.

[2] Here we have performed the matching mostly "manually" in the sense of setting up a regression on the treatment variable and then using the predicted probabilities to select a subset of matched units for the analysis. Various more automatic methods for propensity score estimation, matching, and balancing have be implemented in R and other software packages; see the end of this chapter for references.

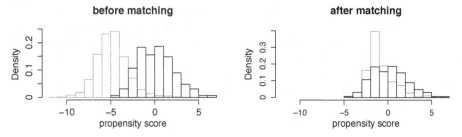

Figure 10.7 *(a) Distribution of logit propensity scores for treated (dark lines) and control groups (gray lines) before matching. (b) Distributions of logit propensity scores for treated (dark lines) and control groups (gray lines) after matching.*

Having created and checked appropriateness of the matches by examining balance, we fit a regression model just on the matched data including all the predictors considered so far, along with an indicator to estimate the treatment effect:

```
reg.ps <- lm (ppvtr.36 ~ treat + hispanic + black + b.marr + lths +
    hs + ltcoll + work.dur + prenatal + mom.age + sex + first +
    preterm + age + dayskidh + bw, data=matched)
```
R code

Given the balance and overlap that the matching procedure has achieved, we are less concerned than in the standard regression context about issues such as deviations from linearity and model extrapolation. Our estimated treatment effect from the matched dataset is 10.2 (with a standard error of 1.6), which can be compared to the standard regression estimate of 11.7 (with standard error of 1.3) based on the full dataset.

If we fully believed in the linear model and were confident that it could be extrapolated to the areas of poor overlap, we would use the regression based on all the data. Realistically, however, we prefer to construct comparable groups and restrict our attention to the range of overlap.

Insufficient overlap? What happens if there are observations about which we want to make inferences but there are no observations with similar propensity scores in the other group? For instance, suppose we are interested in the effect of the treatment on the treated but there are some treated observations with propensity scores far from the propensity scores of all the control observations. One option is to accept some lack of comparability (and corresponding level of imbalance in covariates). Another option is to eliminate the problematic treated observations. If the latter choice is made it is important to be clear about the change in the population about whom inferences will now generalize. It is also helpful to try "profile" the observations that are omitted from the analysis.

Matched pairs? Although matching often results in pairs of treated and control units, we typically ignore the pairing in the analysis of the matched data. Propensity score matching works well (in appropriate settings) to create matched groups, but it does not necessarily created closely matched *pairs*. It is not generally appropriate to add the complication of including the pairing in the model, because the pairing in the matching is performed in the analysis, not the data collection. However, pairing in this way does affect variance calculations, as we shall discuss.

The propensity score as a one-number summary used to assess balance and overlap

A quick way of assessing whether matching has achieved increased balance and overlap is to plot histograms of propensity scores across treated and control groups.

Figure 10.7 displays these histograms for unmatched and matched samples. (We plot the propensity scores on the logit scale to better display their variation at the extremes, which correspond to probabilities near 0 and 1.) The decreased imbalance and increased overlap illustrated in the histograms for the matched groups do not ensure that all predictors included in the model will be similarly matched, but they provide some indication that these distribution will have closer balance in general than before matching.

Geographic information

We have excluded some important information from these analyses. We have access to indicators reflecting the state in which each child resides. Given the tremendous variation in test scores and child care quality[3] across states, it seems prudent to control for this variable as well. If we redo the propensity score matching by including state indicators in both the propensity score model and final regression model, we get an estimate of 8.8 (with standard error of 2.1), which is even lower than our original estimate of 10.2. Extending the regression analysis on the full dataset to include state indicators changes the estimate only from 11.7 to 11.6.

We include results from this analyses using classical regression to adjust for states because it would be a standard approach given these data. A better approach would be to include states in a multilevel model, as we discuss in Chapter 23.

Experimental benchmark by which to evaluate our estimates

It turns out that the researchers evaluating this intervention did not need to rely on a comparison group strategy to assess its impact on test scores. The intervention was evaluated using a randomized experiment. In the preceding example, we simply replaced the true experimental control group with a comparison group pulled from the National Longitudinal Survey of Youth. The advantage of this setup as an illustration of propensity score matching is that we can compare the estimates obtained from the observational study that we have "constructed" to the estimates found using the original randomized experiment. For this sample, the experimental estimate is 7.4. Thus, both propensity score estimates are much closer to the best estimate of the true effect than the standard regression estimates.

Subclassification on mother's education alone yields an estimated treatment effect of 7.0, which happens to be close to the experimental benchmark. However, this does not imply that subclassifying on one variable is generally the best strategy overall. In this example, failure to control for all confounding covariates leads to many biases (some negative and some positive—the geographic variables complicate this picture), and unadjusted differences in average outcomes yield estimates that are lower than the experimental benchmark. Controlling for one variable appears to work well for this example because the biases caused by the imbalances in the other variables just happen to cancel. We would not expect this to happen in general.

Other matching methods, matching on all covariates, and subclassification

The method we have illustrated is called *matching without replacement* because any given control observation cannot be used as a match for more than one treatment

---

[3] Variation in quality of child care is important because it reflects one of the most important alternatives that can be chosen by the parents in the control group.

observation. This can work well in situations when there is a large enough control group to provide adequate overlap. It has the advantage of using each control observation only once, which maximizes our sample size (assuming a constraint of one match per treatment unit) and makes variance calculations a bit easier; see the discussion of standard errors at the end of this section.

However, situations arise when there are not enough controls in the overlapping region to fully provide one match per treated unit. In this case it can help to use some control observations as matches for more than one treated unit. This approach is often called *matching with replacement*, a term which commonly refers to with one-to-one matching but could generalize to multiple control matches for each control. Such strategies can create better balance, which should yield estimates that are closer to the truth on average. Once such data are incorporated into a regression, however, the multiple matches reduce to single data points, which suggests that matching with replacement has limitations as a general strategy.

A limitation of one-to-one matching is that it may end up "throwing away" many informative units if the control group is substantially bigger than the treatment group. One way to make better use of the full sample is simply to subclassify based on values of the propensity score—perhaps discarding some noncomparable units in the tails of the propensity score distribution. Then separate analyses can be performed within each subclass (for example, difference in outcome averages across treatment groups or linear regressions of the outcome on an indicator variable for treatment and other covariates). The estimated treatment effects from each of the subclasses then can either be reported separately or combined in a weighted average with different weights used for different estimands. For instance, when estimating the effect of the treatment on the treated, the number of treated observations in each subclass would be used as the weight, just as we did for the simple subclassification of mother's education in model (10.2) on page 205.

A special case of subclassification called *full matching* can be conceptualized as a fine stratification of the units where each statum has either (1) one treated unit and one control unit, (2) one treated unit and multiple control units, or (3) multiple treated units and one control unit. "Optimal" versions of this matching algorithm have the property of minimizing the average distance between treatment and control units. Strategies with nonoverlapping strata such as subclassification and full matching have the advantage of being more easily incorporated into larger models. This enables strata to be modeled as groups in any number of ways.

Other uses for propensity scores

Some researchers use the propensity score in other ways. For instance, the inverse of estimated propensity scores can be used to create a weight for each point in the data, with the goal that weighted averages of the data should look, in effect, like what would be obtained from a randomized experiment. For instance, to obtain an estimate of an average treatment effect, one would use weights of $1/p_i$ and $1/(1 - p_i)$ for treated and control observations i, respectively, where the p_i's are the estimated propensity scores. To obtain an estimate of the effect of the treatment on the treated, one would use weights of 1 for the treated and $p_i/(1 - p_i)$ for the controls. These weights can be used to calculate simple means or can be included within a regression framework. In our example, this method yielded a treatment effect estimate of 7.8 (when including state information), which is close to the experimental benchmark.

These strategies have the advantage (in terms of precision) of retaining the full sample. However, the weights may have wide variability and may be sensitive to model specification, which could lead to instability. Therefore, these strategies work best when care is taken to create stable weights and to use robust or nonparametric models to estimate the weights. Such methods are beyond the scope of this book.

More simply, propensity scores can be used in a regression of the outcome on the treatment and the scores rather than the full set of covariates. However, if observations that lie in areas where there is no overlap across treatment groups are not removed, the same problems regarding model extrapolation will persist. Also, this method once again places a great deal of faith in precise and correct estimation of the propensity score.

Finally, generalizations of the binary treatment setup have been formalized to accommodate multiple-category or continuous treatment variables.

Standard errors

The standard errors presented for the analyses fitted to matched samples are not technically correct. First, matching induces correlation among the matched observations. The regression model, however, if correctly specified, should account for this by including the variables used to match. Second, our uncertainty about the true propensity score is not reflected in our calculations. This issue has no perfect solution to date and is currently under investigation by researchers in this field. Moreover, more complicated matching methods (for example, matching with replacement and many-to-one matching methods) generally require more sophisticated approaches to variance estimation. Ultimately, one good solution may be a multilevel model that includes treatment interactions so that inferences explicitly recognize the decreased precision that can be obtained outside the region of overlap.

10.4 Lack of overlap when the assignment mechanism is known: regression discontinuity

Simple regression works to estimate treatment effects under the assumption of ignorable treatment assignment if the model is correct, or if the confounding covariates are well balanced with respect to the treatment variable, so that regression serves as a fine-tuning compared to a simple difference of averages. But if the treated and control groups are very different from each other, it can be more appropriate to identify the subset of the population with overlapping values of the predictor variables for both treatment and control conditions, and to estimate the causal effect (and the regression model) in this region only. Propensity score matching is one approach to lack of overlap.

If the treatment and control groups do not overlap at all in key confounding covariates, it can be prudent to abandon causal inferences altogether. However, sometimes a clean lack of overlap arises from a covariate that itself was used to assign units to treatment conditions. *Regression discontinuity analysis* is an approach for dealing with this extreme case of lack of overlap in which the assignment mechanism is clearly defined.

Regression discontinuity and ignorability

A particularly clear case of imbalance sometimes arises in which there is some pretreatment variable x, with a cutoff value C so that one of the treatments applies for all units i for which $x_i < C$, and the other treatment applies for all units for

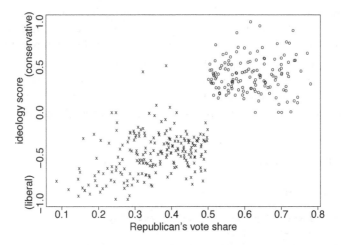

Figure 10.8 *Example of a regression discontinuity analysis: political ideology of members of the 1993–1994 House of Representatives versus Republican share of the two-party vote in the district's congressional election in 1992. Democrats and Republicans are indicated by crosses and circles, respectively. For the purpose of estimating the effect of electing a Democrat or Republican, there is no overlap between the "treatment" (the congressmember's party) and the pre-treatment control variable on the x-axis.*

which $x_i > C$. This could occur, for example, in a medical experiment in which a risky new treatment is only given to patients who are judged to be in particularly bad condition. But the usual setting is in observational studies, where a particular event or "treatment" only occurs under certain specified conditions. For example, in a two-candidate election, a candidate wins if and only if he or she receives more than half the vote.

In a setting where one treatment occurs only for $x < C$ and the other only for $x > C$, it is still possible to estimate the treatment effect for units with x in the neighborhood of C, *if* we assume that the regression function—the average value of the outcome y, given x and the treatment—is a continuous function of x near the cutoff value C.

In this scenario, the mechanism that assigns observations to treatment or control is known, and so we need not struggle to set up a model in which the ignorability assumption is reasonable. All we need to do is control for the input(s) used to determine treatment assignment—these are our confounding covariates. The disadvantage is that, by design, there is no overlap on this covariate across treatment groups. Therefore, to "control for" this variable we must make stronger modeling assumptions because we will be forced to extrapolate our model out of the range of our data. To mitigate such extrapolations, one can limit analyses to observations that fall just above and below the threshold for assignment.

Example: political ideology of congressmembers

Figure 10.8 shows an example, where the goal is to estimate one aspect of the effect of electing a Republican, as compared to a Democrat, in the U.S. House of Representatives. The graph displays political ideologies (as computed using a separate statistical analysis of congressional roll-call votes) for Republican and Democratic congressmembers, plotted versus the vote received by the Republican candidate in the previous election. There is no overlap because the winner in each district necessarily received at least 50% of the vote. (For simplicity, we are only considering districts where an incumbent was running for reelection, so that different districts with the same congressional vote share can be considered as comparable.)

Regression discontinuity analysis.   If we wish to consider the effect of the winning party on the political ideology of the district's congressmember, then a simple regression discontinuity analysis would consider a narrow range—for example, among all the districts where x lies between 0.45 and 0.55, and then fit a model of the form

$$y_i = \beta_0 + \theta T_i + \beta_1 x_i + \text{error}_i$$

where T_i is the "treatment," which we can set to 1 for Republicans and 0 for Democrats.

Here is the result of the regression:

R output
```
lm(formula = score1 ~ party + x, subset=overlap)
            coef.est coef.se
(Intercept)   -1.21    0.62
party          0.73    0.07
x              1.65    1.31
  n = 68, k = 3
  residual sd = 0.15, R-Squared = 0.88
```

The effect of electing a Republican (compared to a Democrat) is 0.73 (on a scale in which the most extreme congressmembers are at ± 1; see Figure 10.8) after controlling for the party strength in the district. The coefficient of x is estimated to be positive—congressmembers in districts with higher Republican votes tend to be more conservative, after controlling for party—but this coefficient is not statistically significant. The large uncertainty in the coefficient for x is no surprise, given that we have restricted our analysis to the subset of data for which x lies in the narrow range from 0.45 to 0.55.

Regression fit to all the data.   Alternatively, we could fit the model to the whole dataset:

R output
```
lm(formula = score1 ~ party + x)
            coef.est coef.se
(Intercept)   -0.68    0.05
party          0.69    0.04
x              0.64    0.13
  n = 357, k = 3
  residual sd = 0.21, R-Squared = 0.8
```

The coefficient on x is estimated much more precisely, which makes sense given that we have more leverage on x (see Figure 10.8).

Regression with interactions.   However, a closer look at the figure suggests different slopes for the two parties, and so we can fit a model interacting x with party:

R output
```
lm(formula = score1 ~ party + x + party:x)
            coef.est coef.se
(Intercept)   -0.76    0.06
party          1.13    0.16
x              0.87    0.15
party:x       -0.81    0.29
  n = 357, k = 4
  residual sd = 0.21, R-Squared = 0.81
```

Everything is statistically significant, but it is difficult to interpret these coefficients. We shall reparameterize and define

```
z <- x - 0.5                                                    R code
```

so that when $z = 0$, we are at the point of discontinuity. We then reparameterize the interaction slope as separate slopes for the Democrats (`party==0`) and Republicans (`party==1`):

```
lm(formula = score1 ~ party + I(z*(party==0)) + I(z*(party==1)))      R output
                   coef.est coef.se
(Intercept)          -0.33    0.03
party                 0.73    0.04
I(z * (party == 0))   0.87    0.15
I(z * (party == 1))   0.06    0.24
  n = 357, k = 4
  residual sd = 0.21, R-Squared = 0.81
```

We see a strong positive slope of z among Democrats but not Republicans, and an estimate of 0.73 for the effect of party at the discontinuity point.

Comparison of regression discontinuity analysis to the model with interactions using all the data. In this example, the analysis fit to the entire dataset gives similar results (but with a much lower standard error) as the regression discontinuity analysis that focused on the region of near overlap. In general, however, the model fit just to the area of overlap may be considered more trustworthy.

Partial overlap

What happens when the discontinuity is not so starkly defined? This is sometimes called a "fuzzy" discontinuity, as opposed to the "sharp" discontinuity discussed thus far. Consider, for instance, a situation where the decision whether to promote children to the next grade is made based upon results from a standardized test (or set of standardized tests). Theoretically this should create a situation with no overlap in these test scores across those children forced to repeat their grade and those promoted to the next grade (the treatment and control groups). In reality, however, there is some "slippage" in the assignment mechanism. Some children may be granted waivers from the official policy based on any of several reasons, including parental pressure on school administrators, a teacher who advocates for the child, and designation of the child as learning-disabled.

This situation creates partial overlap between the treatment and control groups in terms of the supposed sole confounding covariate, promotion test scores. Unfortunately, this overlap arises from deviations from the stated assignment mechanism. If the reasons for these deviations are well defined (and measurable), then ignorability can be maintained by controlling for the appropriate child, parent, or school characteristics. Similarly, if the reasons for these deviations are independent of the potential outcomes of interest, there is no need for concern. If not, inferences could be compromised by failure to control for important omitted confounders.

10.5 Estimating causal effects indirectly using instrumental variables

There are situations when the ignorability assumption seems inadequate because the dataset does not appear to capture all inputs that predict both the treatment and the outcomes. In this case, controlling for observed confounding covariates through regression, subclassification, or matching will not be sufficient for calculating valid causal estimates because unobserved variables could be driving differences in outcomes across groups.

When ignorability is in doubt, the method of *instrumental variables* (IV) can sometimes help. This method requires a special variable, the *instrument*, which is predictive of the treatment and brings with it a new set of assumptions.

Example: a randomized-encouragement design

Suppose we want to estimate the effect of watching an educational television program (this time the program is Sesame Street) on letter recognition. We might consider implementing a randomized experiment where the participants are preschool children, the treatment of interest is watching Sesame Street, the control condition is not watching,[4] and the outcome is the score on a test of letter recognition. It is not possible here for the experimenter to force children to watch a TV show or to refrain from watching (the experiment took place while Sesame Street was on the air). Thus *watching* cannot be randomized. Instead, when this study was actually performed, what was randomized was *encouragement* to watch the show—this is called a randomized encouragement design.

A simple comparison of randomized groups in this study will yield an estimate of the effect of *encouraging* these children to watch the show, not an estimate of the effect of actually viewing the show. In this setting the simple randomized comparison is an estimate of a quantity called the *intent-to-treat* (ITT) effect. However, we may be able to take advantage of the randomization to estimate a causal effect for at least some of the people in the study by using the randomized encouragement as an "instrument." An instrument is a variable thought to randomly induce variation in the treatment variable of interest.

Assumptions for instrumental variables estimation

Instrumental variables analyses rely on several key assumptions, one combination of which we will discuss in this section in the context of a simple example with binary treatment and instrument:

- Ignorability of the instrument,
- Nonzero association between instrument and treatment variable,
- Monotonicity,
- Exclusion restriction.

In addition, the model assumes no interference between units (the stable unit treatment value assumption) as with most other causal analyses, an issue we have already discussed at the end of Section 9.3.

Ignorability of the instrument

The first assumption in the list above is *ignorability of the instrument* with respect to the potential outcomes (both for the primary outcome of interest and the treatment variable). This is trivially satisfied in a randomized experiment (assuming the randomization was pristine). In the absence of a randomized experiment (or natural experiment) this property may be more difficult to satisfy and often requires conditioning on other predictors.

---

[4] Actually the researchers in this study recorded four viewing categories: (1) rarely watched, (2) watched once or twice a week, (3) watched 3-5 times a week, and (4) watched more than 5 times a week on average. Since there is no a category for "never watched," for the purposes of this illustration we treat the lowest viewing category ("rarely watched") as if it were equivalent to "never watched."

Nonzero association between instrument and treatment variable

To demonstrate how we can use the instrument to obtain a causal estimate of the treatment effect in our example, first consider that about 90% of those encouraged watched the show regularly; by comparison, only 55% of those not encouraged watched the show regularly. Therefore, if we are interested in the effect of actually viewing the show, we should focus on the 35% of the treatment population who decided to watch the show because they were encouraged but who otherwise would not have watched the show. If the instrument (encouragement) did not affect regular watching, then we could not proceed. Although a nonzero association between the instrument and the treatment is an assumption of the model, fortunately this assumption is empirically verifiable.

Monotonicity and the exclusion restrictions

Those children whose viewing patterns could be altered by encouragement are the only participants in the study for whom we can conceptualize counterfactuals with regard to viewing behavior—under different experimental conditions they might have been observed either viewing or not viewing, so a comparison of these potential outcomes (defined in relation to randomized encouragement) makes sense. We shall label these children "induced watchers"; these are the only children for whom we will make inferences about the effect of watching Sesame Street.

For the children who were encouraged to watch but did not, we might plausibly assume that they also would not have watched if not encouraged—we shall label this type of child a "never-watcher." We cannot directly estimate the effect of viewing for these children since in this context they would never be observed watching the show. Similarly, for the children who watched Sesame Street even though not encouraged, we might plausibly assume that if they had been encouraged they would have watched as well, again precluding an estimate of the effect of viewing for these children. We shall label these children "always-watchers."

Monotonicity.   In defining never-watchers and always-watchers, we assumed that there were no children who would watch if they were not encouraged but who would *not* watch if they *were* encouraged. Formally this is called the *monotonicity assumption*, and it need not hold in practice, though there are many situations in which it is defensible.

Exclusion restriction.   To estimate the effect of viewing for those children whose viewing behavior would have been affected by the encouragement (the induced watchers), we must make another important assumption, called the *exclusion restriction*. This assumption says for those children whose behavior would not have been changed by the encouragement (never-watchers and always-watchers) there is no effect of encouragement on outcomes. So for the never-watchers (children who would not have watched either way), for instance, we assume encouragement to watch did not affect their outcomes. And for the always-watchers (children who would have watched either way), we assume encouragement to watch did not affect their outcomes.[5]

It is not difficult to tell a story that violates the exclusion restriction. Consider, for instance, the conscientious parents who do not let their children watch television

---

[5] Technically, the assumptions regarding always-watchers and never-watchers represent distinct exclusion restrictions. In this simple framework, however, the analysis suffers if either assumption is violated. Using more complicated estimation strategies, it can be helpful to consider these assumptions separately as it may be possible to weaken one or the other or both.

Unit i	T_i^0	T_i^1	Potential viewing outcomes	Encouragement indicator z_i	Potential test outcomes y_i^0	y_i^1	Encouragement effect $y_i^1 - y_i^0$
1	**0**	1	(induced watcher)	0	**67**	76	9
2	**0**	1	(induced watcher)	0	**72**	80	8
3	**0**	1	(induced watcher)	0	**74**	81	7
4	**0**	1	(induced watcher)	0	**68**	78	10
5	**0**	0	(never-watcher)	0	**68**	68	0
6	**0**	0	(never-watcher)	0	**70**	70	0
7	**1**	1	(always-watcher)	0	**76**	76	0
8	**1**	1	(always-watcher)	0	**74**	74	0
9	**1**	1	(always-watcher)	0	**80**	80	0
10	**1**	1	(always-watcher)	0	**82**	82	0
11	0	**1**	(induced watcher)	1	67	**76**	9
12	0	**1**	(induced watcher)	1	72	**80**	8
13	0	**1**	(induced watcher)	1	74	**81**	7
14	0	**1**	(induced watcher)	1	68	**78**	10
15	0	**0**	(never-watcher)	1	68	**68**	0
16	0	**0**	(never-watcher)	1	70	**70**	0
17	1	**1**	(always-watcher)	1	76	**76**	0
18	1	**1**	(always-watcher)	1	74	**74**	0
19	1	**1**	(always-watcher)	1	80	**80**	0
20	1	**1**	(always-watcher)	1	82	**82**	0

Figure 10.9 *Hypothetical complete data in a randomized encouragement design. Units have been ordered for convenience. For each unit, the students are encouraged to watch Sesame Street ($z_i = 1$) or not ($z_i = 0$). This reveals which of the potential viewing outcomes (T_i^0, T_i^1) and which of the potential test outcomes (y_i^0, y_i^1) we get to observe. The observed outcomes are displayed in boldface. Here, potential outcomes are what we would observe under either encouragement option. The exclusion restriction forces the potential outcomes to be the same for those whose viewing would not be affected by the encouragement. The effect of watching for the "induced watchers" is equivalent to the intent-to-treat effect (encouragement effect over the whole sample) divided by the proportion induced to view; thus, 3.4/0.4 = 8.5.*

and are concerned with providing their children with a good start educationally. The materials used to encourage them to have their children watch Sesame Street for its educational benefits might instead have motivated them to purchase other types of educational materials for their children or to read to them more often.

Derivation of instrumental variables estimation with complete data (including unobserved potential outcomes)

To illustrate the instrumental variables approach, however, let us proceed as if the exclusion restriction were true (or at least approximately true). In this case, if we think about individual-level causal effects, the answer becomes relatively straightforward.

Figure 10.9 illustrates with hypothetical data based on the concepts in this real-life example by displaying for each study participant not only the observed data (encouragement and viewing status as well as observed outcome test score) but also the unobserved categorization, c_i, into always-watcher, never-watcher, or induced watcher based on potential watching behavior as well as the counterfactual test outcomes (the potential outcome corresponding to the treatment not received). Here, potential outcomes are the outcomes we would have observed under either *encouragement* option. Because of the exclusion restriction, for the always-watchers and the never-watchers the potential outcomes are the same no matter the encouragement (really they need not be *exactly* the same, just distributionally the same, but this simplifies the exposition).

The true intent-to-treat effect for these 20 observations is then an average of the effects for the 8 induced watchers, along with 12 zeroes corresponding to the encouragement effects for the always-watchers and never-watchers:

$$
\begin{aligned}
\text{ITT} &= \frac{9 + 8 + 7 + 10 + 9 + 8 + 7 + 10 + 0 + \cdots + 0}{20} \\
&= 8.5 \cdot \frac{8}{20} + 0 \cdot \frac{12}{20} \\
&= 8.5 \cdot 0.4.
\end{aligned}
\tag{10.3}
$$

The effect of watching Sesame Street for the induced watchers is 8.5 points on the letter recognition test. This is algebraically equivalent to the intent-to-treat effect (3.4) divided by the proportion of induced watchers ($8/20 = 0.40$).

Instrumental variables estimate

We can calculate an estimate of the effect of watching Sesame Street for the induced watchers with the actual data using the same principles.

We first estimate the percentage of children actually induced to watch Sesame Street by the intervention, which is the coefficient of the treatment (`encouraged`), in the following regression:

```
fit.1a <- lm (watched ~ encouraged)
```
R code

The estimated coefficient of `encouraged` here is 0.36 (which, in this regression with a single binary predictor, is simply the proportion of induced watchers in the data).

We then compute the intent-to-treat estimate, obtained in this case using the regression of outcome on treatment:

```
fit.1b <- lm (y ~ encouraged)
```
R code

The estimated coefficient of `encouraged` in this regression is 2.9, which we then "inflate" by dividing by the percentage of children affected by the intervention:

```
iv.est <- coef(fit.1a)[,"encouraged"]/coef(fit.1b)[,"encouraged"]
```
R code

The estimated effect of regularly viewing Sesame Street is thus $2.9/0.36 = 7.9$ points on the letter recognition test. This ratio is sometimes called the *Wald estimate*.

Local average treatment effects

The instrumental variables strategy here does not estimate an overall causal effect of watching Sesame Street across everyone in the study. The exclusion restriction implies that there is no effect of the instrument (encouragement) on the outcomes for always-watchers and for never-watchers. Given that the children in these groups cannot be induced to change their watching behavior by the instrument, we cannot estimate the causal effect of watching Sesame Street for these children. Therefore the causal estimates apply only to the "induced watchers."

We are estimating (a special case of) what has been called a *local average treatment effect* (LATE). Some researchers argue that intent-to-treat effects are more interesting from a policy perspective because they accurately reflect that not all targeted individuals will participate in the intended program. However, the intent-to-treat effect only parallels a true policy effect if in the subsequent policy implementation the compliance rate remains unchanged. We recommend estimating both the intent-to-treat effect and the local average treatment effect to maximize what we can learn about the intervention.

10.6 Instrumental variables in a regression framework

Instrumental variables models and estimators can also be derived using regression, allowing us to more easily extend the basic concepts discussed in the previous section. A general instrumental variables model with continuous instrument, z, and treatment, d, can be written as

$$
\begin{aligned}
y &= \beta_0 + \beta_1 T + \epsilon_i \\
T &= \gamma_0 + \gamma_1 z + \nu_i
\end{aligned}
\tag{10.4}
$$

The assumptions can now be expressed in a slightly different way. The first assumption is that z_i is uncorrelated with both ϵ_i and ν_i, which translates informally into the ignorability assumption and exclusion restriction (here often expressed informally as "the instrument only affects the outcome *through* its effect on the treatment"). Also the correlation between z_i and t_i must be nonzero (parallel to the monotonicity assumption from the previous section). We next address how this framework identifies the causal effect of T on y.

Identifiability with instrumental variables

Generally speaking, *identifiability* refers to whether the data contain sufficient information for unique estimation of a given parameter or set of parameters in a particular model. For example, in our formulation of the instrumental variables model, the causal parameter is not identified without assuming the exclusion restriction (although more generally the exclusion restriction is not the only assumption that could be used to achieve identifiability).

What if we did not impose the exclusion restriction for our basic model? The model (ignoring covariate information, and switching to mathematical notation for simplicity and generalizability) can be written as

$$
\begin{aligned}
y &= \beta_0 + \beta_1 T + \beta_2 z + \text{error} \\
T &= \gamma_0 + \gamma_1 z + \text{error},
\end{aligned}
\tag{10.5}
$$

where y is the response variable, z is the instrument, and T is the treatment of interest. Our goal is to estimate β_1, the treatment effect. The difficulty is that T has not been randomly assigned; it is observational and, in general, can be correlated with the error in the first equation; thus we cannot simply estimate β_1 by fitting a regression of y on T and z.

However, as described in the previous section, we can estimate β_1 using instrumental variables. We derive the estimate here algebraically, in order to highlight the assumptions needed for identifiability.

Substituting the equation for T into the equation for y yields

$$
\begin{aligned}
y &= \beta_0 + \beta_1 T + \beta_2 z + \text{error} \\
&= \beta_0 + \beta_1(\gamma_0 + \gamma_1 z) + \beta_2 z + \text{error} \\
&= (\beta_0 + \beta_1\gamma_0) + (\beta_1\gamma_1 + \beta_2)z + \text{error}. \tag{10.6}
\end{aligned}
$$

We now show how to estimate β_1, the causal effect of interest, using the slope of this regression, along with the regressions (10.5) and the exclusion restriction.

The first step is to express (10.6) in the form

$$y = \delta_0 + \delta_1 z + \text{error}.$$

From this equation we need δ_1, which can be estimated from a simple regression of y on z. We can now solve for β_1 in the following equation:

$$\delta_1 = \beta_1\gamma_1 + \beta_2,$$

which we can rearrange to get

$$\beta_1 = (\delta_1 - \beta_2)/\gamma_2. \tag{10.7}$$

We can directly estimate the denominator of this expression, γ_2, from the regression of T on z in (10.5)—this is not a problem since we are assuming that the instrument, z, is randomized.

The only challenge that remains in estimating β_1 from (10.7) is to estimate β_2, which in general cannot simply be estimated from the top equation of (10.5) since, as already noted, the error in that equation can be correlated with T. However, under the exclusion restriction, we know that β_2 is zero, and so $\beta_1 = \delta_1/\gamma_1$, leaving us with the standard instrumental variables estimate.

Other models.   There are other ways to achieve identifiability in this two-equation setting. Approaches such as selection correction models rely on functional form specifications to identify the causal effects even in the absence of an instrument. For example, a probit specification could be used for the regression of T on z. The resulting estimates of treatment effects are often unstable if a true instrument is not included as well.

Two-stage least squares

The Wald estimate discussed in the previous section can be used with this formulation of the model as well. We now describe a more general estimation strategy, *two-stage least squares*.

To illustrate we return to our Sesame Street example. The first step is to regress the "treatment" variable—an indicator for regular watching (watched)—on the randomized instrument, encouragement to watch (encouraged). Then we plug predicted values of encouraged into the equation predicting the letter recognition outcome, y:

```
fit.2a <- lm (watched ~ encouraged)
watched.hat <- fit.2a$fitted
fit.2b <- lm (y ~ watched.hat)
```
R code

The result is

```
              coef.est coef.se
(Intercept)      20.6     3.9
watched.hat       7.9     4.9
  n = 240, k = 2
  residual sd = 13.3, R-Squared = 0.01
```
R output

where now the coefficient on `watched.hat` is the estimate of the causal effect of watching Sesame Street on letter recognition for those induced to watch by the experiment. This two-stage estimation strategy is especially useful for more complicated versions of the model, for instance, when multiple instruments are included.

This second-stage regression does not give the correct standard error, however, as we discuss at the bottom of this page.

Adjusting for covariates in an instrumental variables framework

It turns out that the randomization for this experiment took place within sites and settings; it is therefore appropriate to control for these covariates in estimating the treatment effect. Additionally, pre-test scores are available that are highly predictive of post-test scores. Our preferred model would control for all of these predictors. We can calculate the same ratio (intent-to-treat effect divided by effect of encouragement on viewing) as before using models that include these additional predictors but pulling out only the coefficients on `encouraged` for the ratio.

Here we equivalently perform this analysis using two-stage least squares:

R code
```
fit.3a <- lm (watched ~ encouraged + pretest + as.factor(site) + setting)
watched.hat <- fit.3a$fitted
fit.3b <- lm (y ~ watched.hat + pretest + as.factor(site) + setting)
display (fit.3b)
```

yielding

R output
```
                  coef.est coef.se
(Intercept)            1.2     4.8
watched.hat           14.0     4.0
pretest                0.7     0.1
as.factor(site)2       8.4     1.8
as.factor(site)3      -3.9     1.8
as.factor(site)4       0.9     2.5
as.factor(site)5       2.8     2.9
setting                1.6     1.5
  n = 240, k = 8
  residual sd = 9.7, R-Squared = 0.49
```

The estimated effect of watching Sesame Street on the induced watchers is about 14 points on the letter recognition test. Again, we do not trust this standard error and will discuss later how to appropriately adjust it for the two stages of estimation.

Since the randomization took place within each combination of site (five categories) and setting (two categories), it would be appropriate to interact these variables in our equations. Moreover, it would probably be interesting to estimate variation of effects across sites and settings. However, for simplicity of illustration (and also due to the complication that one site × setting combination has no observations) we only include main effects for this discussion. We return to this example using multilevel models in Chapter 23. It turns out that the estimated average treatment effect changes only slightly (from 14.0 to 14.1) with the model that includes site × setting interactions.

Standard errors for instrumental variables estimates

The second step of two-stage regression yields the instrumental variables estimate, but the standard-error calculation is complicated because we cannot simply look at the second regression in isolation. We show here how to adjust the standard error to account for the uncertainty in both stages of the model. We illustrate with the model we have just fitted.

The regression of compliance on treatment and other covariates (model fit.3a) is unchanged. We then regress the outcome on predicted compliance and covariance, this time saving the predictor matrix, X, from this second-stage regression (which we do using the x=TRUE option in the lm call):

```
fit.3b <- lm (y ~ watched.hat+pretest+as.factor(site)+setting, x=TRUE)
```
R code

We next compute the standard deviation of the adjusted residuals, $r_i^{\mathrm{adj}} = y_i - X_i^{\mathrm{adj}}\hat{\beta}$, where X^{adj} is the predictor matrix from fit.3b but with the column of predicted treatment values replaced by observed treatment values:

```
X.adj <- fit.2$x
X.adj[,"watched.hat"] <- watched
residual.sd.adj <- sd (y - X.adj %*% coef(fit.3b))
```
R code

Finally, we compute the adjusted standard error for the two-stage regression esti-mate by taking the standard error from fit.3b and scaling by the adjusted residual standard deviation, divided by the residual standard deviation from fit.3b itself:

```
se.adj <-se.coef(fit.3b)["watched.hat"]*residual.sd.adj/sigma.hat(fit.3b)
```
R code

So the adjusted standard errors are calculated as the square roots of the diag-onal elements of $(X^t X)^{-1}\hat{\sigma}_{\mathrm{TSLS}}^2$ rather than $(X^t X)^{-1}\hat{\sigma}^2$, where $\hat{\sigma}$ is the residual standard deviation from fit.3b and $\hat{\sigma}_{\mathrm{TSLS}}$ is calculated using the residuals from an equation predicting the outcome from watched (not watched.hat) using the two-stage least squares estimate of the coefficient, not the coefficient that would have been obtained in a least squares regression of the outcome on watched).

The resulting standard-error estimate for our example is 3.9, which is actually a bit smaller than the unadjusted estimate (which is not unusual for these correc-tions).

Performing two-stage least squares automatically using the tsls function

We have illustrated the key concepts in our instrumental variables discussion using basic R commands with which you were already familiar so that the steps were transparent. There does exist, however, a package available in R called sem that has a function, tsls(), that automates this process, including calculating appropriate standard errors.

To calculate the effect of regularly watching Sesame Street on post-treatment letter recognition scores using encouragement as an instrument, we specify both equations:

```
iv1 <- tsls (postlet ~ regular, ~ encour, data=sesame)
display (iv1)
```
R code

where in the second equation it is assumed that the "treatment" (in econometric parlance, the *endogenous* variable) for which encour is an instrument is whatever predictor from the first equation that is not specified as a predictor in the second. Fitting and displaying the two-stage least squares model yields

```
            Estimate  Std. Error
(Intercept)   20.6       3.7
watched        7.9       4.6
```
R output

To incorporate other pre-treatment variables as controls, we must include them in both equations; for example,

R code

```
iv2 <- tsls (postlet ~ watched + prelet + as.factor(site) + setting,
    ~ encour + prelet + as.factor(site) + setting, data=sesame)
display(iv2)
```

yielding

R output

	Estimate	Std. Error
(Intercept)	1.2	4.6
watched	14.0	3.9
prelet	0.7	0.1
as.factor(site)2	8.4	1.8
as.factor(site)3	-3.9	1.7
as.factor(site)4	0.9	2.4
as.factor(site)5	2.8	2.8
setting	1.6	1.4

The point estimate of the treatment calculated this way is the same as with the preceding step-by-step procedure, but now we automatically get correct standard errors.

More than one treatment variable; more than one instrument

In the experiment discussed in Section 10.3, the children randomly assigned to the intervention group received several services ("treatments") that the children in the control group did not receive, most notably, access to high-quality child care and home visits from trained professionals. Children assigned to the intervention group did not make full use of these services. Simply conceptualized, some children participated in the child care while some did not, and some children received home visits while others did not. Can we use the randomization to treatment or control groups as an instrument for these two treatments? The answer is no.

Similar arguments as those used in Section 10.6 can be given to demonstrate that a single instrument cannot be used to identify more than one treatment variable. In fact, as a general rule, we need to use at least as many instruments as treatment variables in order for all the causal estimates to be identifiable.

Continuous treatment variables or instruments

When using two-stage least squares, the models we have discussed can easily be extended to accommodate continuous treatment variables and instruments, although at the cost of complicating the interpretation of the causal effects.

Researchers must be careful, however, in the context of binary instruments and continuous treatment variables. A binary instrument cannot in general identify a continuous treatment or "dosage" effect (without further assumptions). If we map this back to a randomized experiment, the randomization assigns someone only to be encouraged or not. This encouragement may lead to different dosage levels, but for those in the intervention group these levels will be chosen by the subject (or subject's parents in this case). In essence this is equivalent to a setting with many different treatments (one at each dosage level) but only one instrument—therefore causal effects for all these treatments are not identifiable (without further assumptions). To identify such dosage effects, one would need to randomly assign encouragement levels that lead to the different dosages or levels of participation.

Have we really avoided the ignorability assumption? Natural experiments and instrumental variables

We have motivated instrumental variables using the cleanest setting, within a controlled, randomized experiment. The drawback of illustrating instrumental variables using this example is that it de-emphasizes one of the most important assumptions of the instrumental variables model, *ignorability of the instrument*. In the context of a randomized experiment, this assumption should be trivially satisfied (assuming the randomization was pristine). However, in practice an instrumental variables strategy potentially is more useful in the context of a *natural experiment*, that is, an observational study context in which a "randomized" variable (instrument) appears to have occurred naturally. Examples of this include:

- The draft lottery in the Vietnam War as an instrument for estimating the effect of military service on civilian health and earnings,

- The weather in New York as an instrument for estimating the effect of supply of fish on their price,

- The sex of a second child (in an analysis of people who have at least two children) as an instrument when estimating the effect of number of children on labor supply.

In these examples we have simply traded one ignorability assumption (ignorability of the treatment variable) for another (ignorability of the instrument) that we believe to be more plausible. Additionally, we must assume monotonicity and the exclusion restriction.

Assessing the plausibility of the instrumental variables assumptions

How can we assess the plausibility of the assumptions required for causal inference from instrumental variables? As a first step, the "first stage" model (the model that predicts the treatment using the instrument) should be examined closely to ensure both that the instrument is strong enough and that the sign of the coefficient makes sense. This is the only assumption that can be directly tested. If the association between the instrument and the treatment is weak, instrumental variables can yield incorrect estimates of the treatment effect even if all the other assumptions are satisfied. If the association is not in the expected direction, then closer examination is required because this might be the result of a mixture of two different mechanisms, the expected process and one operating in the opposite direction, which could in turn imply a violation of the monotonicity assumption.

Another consequence of a weak instrument is that it exacerbates the bias that can result from failure to satisfy the monotonicity and exclusion restrictions. For instance, for a binary treatment and instrument, when the exclusion restriction is not satisfied, our estimates will be off by a quantity that is equal to the effect of encouragement on the outcomes of noncompliers (in our example, never-watchers and always-watchers) multiplied by the ratio of noncompliers to compliers (in our example, induced watchers). The bias when monotonicity is not satisfied is slightly more complicated but also increases as the percentage of compliers decreases.

The two primary assumptions of instrumental variables (ignorability, exclusion) are not directly verifiable, but in some examples we can work to make them more plausible. For instance, if unconditional ignorability of the instrument is being assumed, yet there are differences in important pre-treatment characteristics across groups defined by the instrument, then these characteristics should be included in the model. This will not ensure that ignorability is satisfied, but it removes the *observed* problem with the ignorability assumption.

Example: Vietnam War draft lottery study.   One strategy to assess the plausibility of the exclusion restriction is to calculate an estimate within a sample that would not be expected to be affected by the instrument. For instance, researchers estimated the effect of military service on earnings (and other outcomes) using, as an instrument, the draft lottery number for young men eligible for the draft during the Vietnam War. This number was assigned randomly and strongly affected the probability of military service. It was hoped that the lottery number would only have an effect on earnings for those who served in the military only because they were drafted (as determined by a low enough lottery number). Satisfaction of the exclusion restriction is not certain, however, because, for instance, men with low lottery numbers may have altered their educational plans so as to avoid or postpone military service. So the researchers also ran their instrumental variables model for a sample of men who were assigned numbers so late that the war ended before they ever had to serve. This showed no significant relation between lottery number and earnings, which provides some support for the exclusion restriction.

Structural equation models

A goal in many areas of social science is to infer causal relations among many variables, a generally difficult problem (as discussed in Section 9.8). *Structural equation modeling* is a family of methods of multivariate data analysis that are sometimes used for causal inference.[6] In that setting, structural equation modeling relies on conditional independence assumptions in order to identify causal effects, and the resulting inferences can be sensitive to strong parametric assumptions (for instance, linear relationships and multivariate normal errors). Instrumental variables can be considered to be a special case of a structural equation model. As we have just discussed, even in a relatively simple instrumental variables model, the assumptions needed to identify causal effects are difficult to satisfy and largely untestable. A structural equation model that tries to estimate many causal effects at once multiplies the number of assumptions required with each desired effect so that it quickly becomes difficult to justify all of them. Therefore we do not discuss the use of structural equation models for causal inference in any greater detail here. We certainly have no objection to complicated models, as will become clear in the rest of this book; however, we are cautious about attempting to estimate complex causal structures from observational data.

10.7 Identification strategies that make use of variation within or between groups

Comparisons within groups—so-called fixed effects models

What happens when you want to make a causal inference but no valid instrument exists and ignorability does not seem plausible? Do alternative strategies exist? Sometimes repeated observations within groups or within individuals over time can provide a means for controlling for unobserved characteristics of these groups or individuals. If comparisons are made across the observations within a group or

---

[6] Structural equation modeling is also used to estimate latent factors in noncausal regression settings with many inputs, and sometimes many outcome variables, which can be better understood by reducing to a smaller number of linear combinations.

persons, implicitly such comparisons "hold constant" all characteristics intrinsic to the group or individual that do not vary across observations (across members of the group or across measures over time for the same person).

For example, suppose you want to examine the effect of low birth weight on children's mortality and other health outcomes. One difficulty in establishing a causal effect here is that children with low birth weight are also typically disadvantaged in genetic endowments and socioeconomic characteristics of the family, some of which may not be easy or possible to measure. Rather than trying to directly control for all of these characteristics, however, one could implicitly control for them by comparing outcomes across twins. Twins share many of the same genetic endowments (all if identical) and, in most cases, live in exactly the same household. However, there are physiological reasons (based, for instance, on position in the uterus) why one child in the pair may be born with a markedly different birth weight than the sibling. So we may be able to consider birth weight to be randomly assigned (ignorable) *within* twin pairs. Theoretically, if there is enough variation in birth weight, within sets of twins, we can estimate the effect of birth weight on subsequent outcomes. In essence each twin acts as a counterfactual for his or her sibling.

A regression model that is sometimes used to approximate this conceptual comparison simply adds an indicator variable for each of the groups to the standard regression model that might otherwise have been fit. So, for instance, in our twins example one might regress outcomes on birth weight (the "treatment" variable) and one indicator variable for each pair of twins (keeping one pair as a baseline category to avoid collinearity). More generally, we could control for the groups using a multilevel model, as we discuss in Part 2. In any case, the researcher might want to control for other covariates to improve the plausibility of the ignorability assumption (to control for the fact that the treatment may not be strictly randomly assigned even within each group—here, the pair of twins). In this particular example, however, it is difficult to find child-specific predictors that vary across children within a pair but can still be considered "pre-treatment."

In examples where the treatment is dichotomous, a substantial portion of the data may not exhibit any variation at all in "treatment assignment" within groups. For instance, if this strategy is used to estimate the effect of maternal employment on child outcomes by including indicators for each family (set of siblings) in the dataset, then in some families the mother may not have varied her employment status across children. Therefore, no inferences about the effect of maternal employment status can be made for these families. We can only estimate effects for the type of family where the mother varied her employment choice across the children (for example, working after her first child was born but staying home from work after the second).

Conditioning on post-treatment outcomes. Still more care must be taken when considering variation over time. Consider examining the effect of marriage on men's earnings by looking at data that follows men over time and tracks marital status, earnings, and predictors of each (confounding covariates such as race, education, and occupation). Problems can easily arise in a model that includes an indicator for each person and also controls for covariates at each time point (to help satisfy ignorability). In this case the analysis would be implicitly conditioning on post-treatment variables, which, as we know from Section 9.8, can lead to bias.

Better suited for a multilevel model framework? This model with indicators for each group is often (particularly in the economics literature) called a "fixed effects" model. We dislike this terminology because it is interpreted differently in different settings, as discussed in Section 11.4. Further, this model is hierarchically structured, so from our perspective it is best analyzed using a multilevel model. This is not completely straightforward, however, because one of the key assumptions of a simple multilevel model is that the individual-level effects are independent of the other predictors in the model—a condition that is particularly problematic in this setting where we are expecting that unobserved characteristics of the individuals may be associated with observed characteristics of the individuals. In Chapter 23 we discuss how to appropriately extend this model to the multilevel framework while relaxing this assumption.

Comparisons within and between groups: difference-in-differences estimation

Almost all causal strategies make use of comparisons across groups: one or more that were exposed to a treatment, and one or more that were not. *Difference-in-difference* strategies additionally make use of another source of variation in outcomes, typically time, to help control for potential (observed and unobserved) differences across these groups. For example, consider estimating the effect of a newly introduced school busing program on housing prices in a school district where some neighborhoods were affected by the program and others were not. A simple comparison of housing prices across affected and unaffected areas sometime after the busing program went into effect might not be appropriate because these neighborhoods might be different in other ways that might be related to housing prices. A simple before-after comparison of housing prices may also be inappropriate if other changes that occurred during this time period (for example, a recession) might also be influencing housing prices. A difference-in-differences approach would instead calculate the difference in the before-after *change* in housing prices in exposed and unexposed neighborhoods. An important advantage of this strategy is that the units of observation (in this case, houses) need not be the same across the two time periods.

The assumption needed with this strategy is a weaker than the (unconditional) ignorability assumption because rather than assuming that potential outcomes are the same across treatment groups, one only has to assume that the potential *gains* in potential outcomes over time are the same across groups (for example, exposed and unexposed neighborhoods). Therefore we need only believe that the difference in housing prices over time would be the same across the two types of neighborhoods, not that the average post-program potential housing prices if exposed or unexposed would be the same.

Panel data. A special case of difference-in-differences estimation occurs when the same set of units are observed at both time points. This is also a special case of the so-called fixed effects model that includes indicators for treatment groups and for time periods. A simple way to fit this model is with a regression of the outcome on an indicator for the groups, an indicator for the time period, and the interaction between the two. The coefficient on the interaction is the estimated treatment effect.

In this setting, however, the advantages of the difference-in-differences strategy are less apparent because an alternative model would be to include an indicator for treatment exposure but then simply regress on the pre-treatment version of the outcome variable. In this framework it is unclear if the assumption of randomly assigned *changes* in potential outcome is truly weaker than the assumption of randomly assigned potential outcomes for those with the same value of the pre-treatment variable.[7]

---

[7] Strictly speaking, we need not assume actual random manipulation of treatment assignment for either assumption to hold, only results that would be consistent with such manipulation.

Do not condition on post-treatment outcomes.   Once again, to make the (new) ignorability assumption more plausible it may be desirable to condition on additional predictor variables. For models where the variation takes place over time—for instance, the differences-in-differences estimate that includes both pre-treatment and post-treatment observations on the same units—a standard approach is to include changes in characteristics for each observation over time. Implicitly, however, this conditions on post-treatment variables. If these predictors can be reasonably assumed to be unchanged by the treatment, then this is reasonable. However, as discussed in Section 9.8, it is otherwise inappropriate to control for post-treatment variables. A better strategy would be to control for pre-treatment variables only.

10.8 Bibliographic note

We have more references here than for any of the other chapters in this book because causal inference is a particularly contentious and active research area, with methods and applications being pursued in many fields, including statistics, economics, public policy, and medicine.

Imbalance and lack of complete overlap have been discussed in many places; see, for example, Cochran and Rubin (1973), and King and Zeng (2006). The intervention for low-birth-weight children is described by Brooks-Gunn, Liaw, and Klebanov (1992) and Hill, Brooks-Gunn, and Waldfogel (2003). Imbalance plots such as Figure 10.3 are commonly used; see Hansen (2004), for example.

Subclassification and its connection to regression are discussed by Cochran (1968). Imbens and Angrist (1994) introduce the local average treatment effect. Cochran and Rubin (1973), Rubin (1973), Rubin (1979), Rubin and Thomas (2000), and Rubin (2006) discuss the use of matching, followed by regression, for causal inference. Dehejia (2003) discusses an example of the interpretation of a treatment effect with interactions.

Propensity scores were introduced by Rosenbaum and Rubin (1983a, 1984, 1985). A discussion of common current usage is provided by D'Agostino (1998). Examples across several fields include Lavori, Keller, and Endicott (1995), Lechner (1999), Hill, Waldfogel, and Brooks-Gunn (2002), Vikram et al. (2003), and O'Keefe (2004). Rosenbaum (1989) and Hansen (2004) discuss full matching. Diamond and Sekhon (2005) present a genetic matching algorithm. Drake (1993) discusses robustness of treatment effect estimates to misspecification of the propensity score model. Joffe and Rosenbaum (1999), Imbens (2000), and Imai and van Dyk (2004) generalize the propensity score beyond binary treatments. Rubin and Stuart (2005) extend to matching with multiple control groups. Imbens (2004) provides a recent review of methods for estimating causal effects assuming ignorability using matching and other approaches.

Use of propensity scores as weights is discussed by Rosenbaum (1987), Ichimura and Linton (2001), Hirano, Imbens, and Ridder (2003), and Frolich (2004) among others. This work has been extended to a "doubly-robust" framework by Robins and Rotnitzky (1995), Robins, Rotnitzsky, and Zhao (1995), and Robins and Ritov (1997).

As far as we are aware, LaLonde (1986) was the first use of so-called constructed observational studies as a testing ground for nonexperimental methods. Other examples include Friedlander and Robins (1995), Heckman, Ichimura, and Todd (1997), Dehejia and Wahba (1999), Michalopoulos, Bloom, and Hill (2004), and Agodini and Dynarski (2004). Dehejia (2005a, b), in response to Smith and Todd (2005), provides useful guidance regarding appropriate uses of propensity scores (the need to think hard about ignorability and to specify propensity score models that are specific to any given dataset). The constructed observational analysis presented in this chapter is based on a more complete analysis presented in Hill, Reiter, and Zanutto (2004).

Interval estimation for treatment effect estimates obtained via propensity score matching is discussed in Hill and Reiter (2006). Du (1998) and Tu and Zhou (2003) discuss intervals for estimates obtained via propensity score subclassification. Hill and McCulloch (2006) present a Bayesian nonparametric method for matching.

Several packages exist that automate different combinations of the propensity score steps described here and are available as supplements to R and other statistical software. We mention some of these here without intending to provide a comprehensive list. There is a program available for R called MatchIt that is available at gking.harvard.edu/matchit/ that implements several different matching methods including full matching (using software called OptMatch; Hansen, 2006). Three packages available for Stata are psmatch2, pscore, and nnmatch; any of these can be installed easily using the "net search" (or comparable) feature in Stata. Additionally, nnmatch produces valid standard errors for matching. Code is also available in SAS for propensity score matching or subclassification; see, for example, www.rx.uga.edu/main/home/cas/faculty/propensity.pdf.

Regression discontinuity analysis is described by Thistlethwaite and Campbell (1960). Recent work in econometrics includes Hahn, Todd, and van der Klaauw (2001) and Linden (2006). The political ideology example in Section 10.4 is derived from Poole and Rosenthal (1997) and Gelman and Katz (2005); see also Lee, Moretti, and Butler (2004) for related work. The example regarding children's promotion in school was drawn from work by Jacob and Lefgren (2004).

Instrumental variables formulations date back to work in the economics literature by Tinbergen (1930) and Haavelmo (1943). Angrist and Krueger (2001) present an upbeat applied review of instrumental variables. Imbens (2004) provides a review of statistical methods for causal inference that is a little less enthusiastic about instrumental variables. Woolridge (2001, chapter 5) provides a crisp overview of instrumental variables from a classical econometric perspective; Lancaster (2004, chapter 8) uses a Bayesian framework. The "always-watcher," "induced watcher," and "never-watcher" categorizations here are alterations of the "never-taker," "complier," and "always-taker" terminology first used by Angrist, Imbens, and Rubin (1996), who reframe the classic econometric presentation of instrumental variables in statistical language and clarify the assumptions and the implications when the assumptions are not satisfied. For a discussion of all of the methods discussed in this chapter from an econometric standpoint, see Angrist and Krueger (1999).

The Vietnam draft lottery example comes from several papers including Angrist (1990). The weather and fish price example comes from Angrist, Graddy, and Imbens (2000). The sex of child example comes from Angrist and Evans (1998).

For models that link instrumental variables with the potential-outcomes framework described in Chapter 9, see Angrist, Imbens, and Rubin (1996). Glickman and Normand (2000) derive an instrumental variables estimate using a latent-data model; see also Carroll et al. (2004).

Imbens and Rubin (1997) discuss a Bayesian approach to instrumental variables in the context of a randomized experiment with noncompliance. Hirano et al. (2000) extend this framework to include covariates. Barnard et al. (2003) describe further extensions that additionally accommodate missing outcome and covariate data. For discussions of prior distributions for instrumental variables models, see Dreze

(1976), Maddala (1976), Kleibergen and Zivot (2003), and Hoogerheide, Kleibergen and van Dijk (2006).

For a discussion of use of instrumental variables models to estimate bounds for the average treatment effect (as opposed to the local average treatment effect), see Robins (1989), Manski (1990), and Balke and Pearl (1997). Robins (1994) discusses estimation issues.

For more on the Sesame Street encouragement study, see Bogatz and Ball (1971) and Murphy (1991).

Wainer, Palmer, and Bradlow (1998) provide a friendly introduction to selection bias. Heckman (1979) and Diggle and Kenward (1994) are influential works on selection models in econometrics and biostatistics, respectively. Rosenbaum and Rubin (1983b), Rosenbaum (2002a), and Greenland (2005) consider sensitivity of inferences to ignorability assumptions.

Sobel (1990, 1998) discusses the assumptions needed for structural equation modeling more generally.

Ashenfelter, Zimmerman, and Levine (2003) discuss "fixed effects" and difference-in-differences methods for causal inference. The twins and birth weight example was based on a paper by Almond, Chay, and Lee (2005). Another interesting twins example examining the returns from education on earnings can be found in Ashenfelter and Krueger (1994). Aaronson (1998) and Chay and Greenstone (2003) provide further examples of the application of these approaches. The busing and housing prices example is from Bogart and Cromwell (2000). Card and Krueger (1994) discuss a classic example of a difference-in-differences model that uses panel data.

10.9 Exercises

1. Constructed observational studies: the folder `lalonde` contains data from an observational study constructed by LaLonde (1986) based on a randomized experiment that evaluated the effect on earnings of a job training program called National Supported Work. The constructed observational study was formed by replacing the randomized control group with a comparison group formed using data from two national public-use surveys: the Current Population Survey (CPS) and the Panel Study in Income Dynamics.

   Dehejia and Wahba (1999) used a subsample of these data to evaluate the potential efficacy of propensity score matching. The subsample they chose removes men for whom only one pre-treatment measure of earnings is observed. (There is substantial evidence in the economics literature that controlling for earnings from only one pre-treatment period is insufficient to satisfy ignorability.) This exercise replicates some of Dehejia and Wahba's findings based on the CPS comparison group.

   (a) Estimate the treatment effect from the experimental data in two ways: (i) a simple difference in means between treated and control units, and (ii) a regression-adjusted estimate (that is, a regression of outcomes on the treatment indicator as well as predictors corresponding to the pre-treatment characteristics measured in the study).

   (b) Now use a regression analysis to estimate the causal effect from Dehejia and Wahba's subset of the constructed observational study. Examine the sensitivity of the model to model specification (for instance, by excluding the employed indicator variables or by including interactions). How close are these estimates to the experimental benchmark?

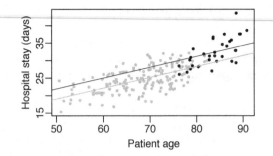

Figure 10.10 *Hypothetical data of length of hospital stay and age of patients, with separate points and regression lines plotted for each treatment condition: the new procedure in gray and the old procedure in black.*

(c) Now estimate the causal effect from the Dehejia and Wahba subset using propensity score matching. Do this by first trying several different specifications for the propensity score model and choosing the one that you judge to yield the best balance on the most important covariates.

Perform this propensity score modeling *without* looking at the estimated treatment effect that would arise from each of the resulting matching procedures.

For the matched dataset you construct using your preferred model, report the estimated treatment effects using the difference-in-means and regression-adjusted methods described in part (a) of this exercise. How close are these estimates to the experimental benchmark (about $1800)?

(d) Assuming that the estimates from (b) and (c) can be interpreted causally, what causal effect does each estimate? (Hint: what populations are we making inferences about for each of these estimates?)

(e) Redo both the regression and the matching exercises, excluding the variable for earnings in 1974 (two time periods before the start of this study). How important does the earnings-in-1974 variable appear to be in terms of satisfying the ignorability assumption?

2. Regression discontinuity analysis: suppose you are trying to evaluate the effect of a new procedure for coronary bypass surgery that is supposed to help with the postoperative healing process. The new procedure is risky, however, and is rarely performed in patients who are over 80 years old. Data from this (hypothetical) example are displayed in Figure 10.10.

(a) Does this seem like an appropriate setting in which to implement a regression discontinuity analysis?

(b) The folder **bypass** contains data for this example: **stay** is the length of hospital stay after surgery, **age** is the age of the patient, and **new** is the indicator variable indicating that the new surgical procedure was used. Preoperative disease severity (**severity**) was unobserved by the researchers, but we have access to it for illustrative purposes. Can you find any evidence using these data that the regression discontinuity design is inappropriate?

(c) Estimate the treatment effect using a regression discontinuity estimate (ignoring) severity. Estimate the treatment effect in any way you like, taking advantage of the information in severity. Explain the discrepancy between these estimates.

3. Instrumental variables: come up with a hypothetical example in which it would be appropriate to estimate treatment effects using an instrumental variables strategy. For simplicity, stick to an example with a binary instrument and binary treatment variable.

  (a) Simulate data for this imaginary example if all the assumptions are met. Estimate the local average treatment effect for the data by dividing the intent-to-treat effect by the percentage of compliers. Show that two-stage least squares yields the same point estimate.

  (b) Now simulate data in which the exclusion restriction is not met (so, for instance, those whose treatment level is left unaffected by the instrument have a treatment effect of half the magnitude of the compliers) but the instrument is strong (say, 80% of the population are compliers), and see how far off your estimate is.

  (c) Finally, simulate data in which the exclusion restriction is violated in the same way, but where the instrument is weak (only 20% of the population are compliers), and see how far off your estimate is.

4. In Exercise 9.13, you estimated the effect of incumbency on votes for Congress. Now consider an additional variable: money raised by the congressional candidates. Assume this variable has been coded in some reasonable way to be positive in districts where the Democrat has raised more money and negative in districts where the Republican has raised more.

  (a) Explain why it is inappropriate to include money as an additional input variable to "improve" the estimate of incumbency advantage in the regression in Exercise 9.13.

  (b) Suppose you are interested in estimating the effect of money on the election outcome. Set this up as a causal inference problem (that is, define the treatments and potential outcomes).

  (c) Explain why it is inappropriate to simply estimate the effect of money using instrumental variables, with incumbency as the instrument. Which of the instrumental variables assumptions would be reasonable in this example and which would be implausible?

  (d) How could you estimate the effect of money on congressional election outcomes?

  See Campbell (2002) and Gerber (2004) for more on this topic.

Part 2A: Multilevel regression

We now introduce multilevel linear and generalized linear models, including issues such as varying intercepts and slopes and non-nested models. We view multilevel models either as regressions with potentially large numbers of coefficients that are themselves modeled, or as regressions with coefficients that can vary by group.

## CHAPTER 11

Multilevel structures

As we illustrate in detail in subsequent chapters, multilevel models are extensions of regression in which data are structured in groups and coefficients can vary by group. In this chapter, we illustrate basic multilevel models and present several examples of data that are collected and summarized at different levels. We start with simple grouped data—persons within cities—where some information is available on persons and some information is at the city level. We then consider examples of repeated measurements, time-series cross sections, and non-nested structures. The chapter concludes with an outline of the costs and benefits of multilevel modeling compared to classical regression.

11.1 Varying-intercept and varying-slope models

With grouped data, a regression that includes indicators for groups is called a *varying-intercept model* because it can be interpreted as a model with a different intercept within each group. Figure 11.1a illustrates with a model with one continuous predictor x and indicators for $J = 5$ groups. The model can be written as a regression with 6 predictors or, equivalently, as a regression with two predictors (x and the constant term), with the intercept varying by group:

$$\text{varying-intercept model: } y_i = \alpha_{j[i]} + \beta x_i + \epsilon_i.$$

Another option, shown in Figure 11.1b, is to let the slope vary with constant intercept:

$$\text{varying-slope model: } y_i = \alpha + \beta_{j[i]} x_i + \epsilon_i.$$

Finally, Figure 11.1c shows a model in which both the intercept and the slope vary by group:

$$\text{varying-intercept, varying-slope model: } y_i = \alpha_{j[i]} + \beta_{j[i]} x_i + \epsilon_i.$$

The varying slopes are interactions between the continuous predictor x and the group indicators.

As we discuss shortly, it can be challenging to estimate all these α_j's and β_j's, especially when inputs are available at the group level. The first step of multilevel modeling is to set up a regression with varying coefficients; the second step is to set up a regression model for the coefficients themselves.

11.2 Clustered data: child support enforcement in cities

With multilevel modeling we need to go beyond the classical setup of a data vector y and a matrix of predictors X (as shown in Figure 3.6 on page 38). Each level of the model can have its own matrix of predictors.

We illustrate multilevel data structures with an observational study of the effect of city-level policies on enforcing child support payments from unmarried fathers. The treatment is at the group (city) level, but the outcome is measured on individual families.

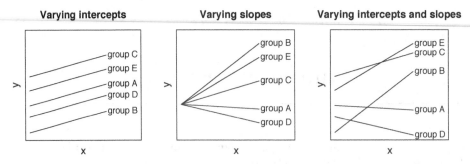

Figure 11.1 *Linear regression models with (a) varying intercepts ($y = \alpha_j + \beta x$), (b) varying slopes ($y = \alpha + \beta_j x$), and (c) both ($y = \alpha_j + \beta_j x$). The varying intercepts correspond to group indicators as regression predictors, and the varying slopes represent interactions between x and the group indicators.*

ID	dad age	mom race	informal support	city ID	city name	enforce intensity	benefit level	city indicators 1	2	$\cdots$	20
1	19	hisp	1	1	Oakland	0.52	1.01	1	0	$\cdots$	0
2	27	black	0	1	Oakland	0.52	1.01	1	0	$\cdots$	0
3	26	black	1	1	Oakland	0.52	1.01	1	0	$\cdots$	0
$\vdots$	$\vdots$	$\vdots$	$\vdots$	$\vdots$	$\vdots$	$\vdots$	$\vdots$	$\vdots$	$\vdots$		$\vdots$
248	19	white	1	3	Baltimore	0.05	1.10	0	0	$\cdots$	0
249	26	black	1	3	Baltimore	0.05	1.10	0	0	$\cdots$	0
$\vdots$	$\vdots$	$\vdots$	$\vdots$	$\vdots$	$\vdots$	$\vdots$	$\vdots$	$\vdots$	$\vdots$		$\vdots$
1366	21	black	1	20	Norfolk	-0.11	1.08	0	0	$\cdots$	1
1367	28	hisp	0	20	Norfolk	-0.11	1.08	0	0	$\cdots$	1

Figure 11.2 *Some of the data from the child support study, structured as a single matrix with one row for each person. These indicators would be used in classical regression to allow for variation among cities. In a multilevel model they are not necessary, as we code cities using their index variable ("city ID") instead. We prefer separating the data into individual-level and city-level datasets, as in Figure 11.3.*

Studying the effectiveness of child support enforcement

Cities and states in the United States have tried a variety of strategies to encourage or force fathers to give support payments for children with parents who live apart. In order to study the effectiveness of these policies for a particular subset of high-risk children, an analysis was done using a sample of 1367 noncohabiting parents from the Fragile Families study, a survey of unmarried mothers of newborns in 20 cities. The survey was conducted by sampling from hospitals which themselves were sampled from the chosen cities, but here we ignore the complexities of the data collection and consider the mothers to have been sampled at random (from their demographic category) in each city.

To estimate the effect of child support enforcement policies, the key "treatment" predictor is a measure of enforcement policies, which is available at the city level. The researchers estimated the probability that the mother received informal support, given the city-level enforcement measure and other city- and individual-level predictors.

ID	dad age	mom race	informal support	city ID
1	19	hisp	1	1
2	27	black	0	1
3	26	black	1	1
⋮	⋮	⋮	⋮	⋮
248	19	white	1	3
249	26	black	1	3
⋮	⋮	⋮	⋮	⋮
1366	21	black	1	20
1367	28	hisp	0	20

city ID	city name	enforce- ment	benefit level
1	Oakland	0.52	1.01
2	Austin	0.00	0.75
3	Baltimore	−0.05	1.10
⋮	⋮	⋮	⋮
20	Norfolk	−0.11	1.08

Figure 11.3 *Data from the child support study, structured as two matrices, one for persons and one for cities. The inputs at the different levels are now clear. Compare to Figure 11.2.*

A data matrix for each level of the model

Figure 11.2 shows the data for the analysis as it might be stored in a computer package, with information on each of the 1367 mothers surveyed. To make use of the multilevel structure of the data, however, we need to construct *two* data matrices, one for each level of the model, as Figure 11.3 illustrates. At the left is the person-level data matrix, with one row for each survey respondent, and their cities are indicated by an index variable; at the right is the city data matrix, giving the name and other information available for each city.

At a practical level, the two-matrix format of Figure 11.3 has the advantage that it contains each piece of information exactly once. In contrast, the single large matrix in Figure 11.2 has each city's data repeated several times. Computer memory is cheap so this would not seem to be a problem; however, if city-level information needs to be added or changed, the single-matrix format invites errors.

Conceptually, the two-matrix, or multilevel, data structure has the advantage of clearly showing which information is available on individuals and which on cities. It also gives more flexibility in fitting models, allowing us to move beyond the classical regression framework.

Individual- and group-level models

We briefly outline several possible ways of analyzing these data, as a motivation and lead-in to multilevel modeling.

Individual-level regression. In the most basic analysis, informal support (as reported by mothers in the survey) is the binary outcome, and there are several individual- and city-level predictors. Enforcement is considered as the treatment, and a logistic regression is used, also controlling for other inputs. This is the starting point of the observational study.

Using classical regression notation, the model is $\Pr(y_i = 1) = \text{logit}^{-1}(X_i\beta)$, where X includes the constant term, the treatment (enforcement intensity), and the other predictors (father's age and indicators for mother's race at the individual level; and benefit level at the city level). X is thus constructed from the data matrix of Figure 11.2. This individual-level regression has the problem that it ignores city-level variation beyond that explained by enforcement intensity and benefit level, which are the city-level predictors in the model.

city ID	city name	enforcement	benefit level	# in sample	avg. age	prop. black	proportion with informal support
1	Oakland	0.52	1.01	78	25.9	0.67	0.55
2	Austin	0.00	0.75	91	25.8	0.42	0.54
3	Baltimore	−0.05	1.10	101	27.0	0.86	0.67
⋮	⋮	⋮	⋮	⋮	⋮	⋮	⋮
20	Norfolk	−0.11	1.08	31	27.4	0.84	0.65

Figure 11.4 *City-level data from child support study (as in the right panel of Figure 11.3), also including sample sizes and sample averages from the individual responses.*

Group-level regression on city averages. Another approach is to perform a city-level analysis, with individual-level predictors included using their group-level averages. Figure 11.4 illustrates: here, the outcome, y_j, would be the average total support among the respondents in city j, the enforcement indicator would be the treatment, and the other variables would also be included as predictors. Such a regression—in this case, with 20 data points—has the advantage that its errors are automatically at the city level. However, by aggregating, it removes the ability of individual predictors to predict individual outcomes. For example, it is possible that older fathers give more informal support—but this would not necessarily translate into average father's age being predictive of more informal support at the city level.

Individual-level regression with city indicators, followed by group-level regression of the estimated city effects. A slightly more elaborate analysis proceeds in two steps, first fitting a logistic regression to the individual data y given individual predictors (in this example, father's age and indicators for mother's race) along with indicators for the 20 cities. This first-stage regression then has 22 predictors. (The constant term is *not* included since we wish to include indicators for all the cities; see the discussion at the end of Section 4.5.)

The next step in this two-step analysis is to perform a *linear* regression at the city level, considering the estimated coefficients of the city indicators (in the individual model that was just fit) as the "data" y_j. This city-level regression has 20 data points and uses, as predictors, the city-level data (in this case, enforcement intensity and benefit level). Each of the predictors in the model is thus included in one of the two regressions.

The two-step analysis is reasonable in this example but can run into problems when sample sizes are small in particular groups, or when there are interactions between individual- and group-level predictors. Multilevel modeling is a more general approach that can include predictors at both levels at once.

Multilevel models

The multilevel model looks something like the two-step model we have described, except that both steps are fitted at once. In this example, a simple multilevel model would have two components: a logistic regression with 1369 data points predicting the binary outcome given individual-level predictors and with an intercept that can vary by city, and a linear regression with 20 data points predicting the city intercepts from city-level predictors. In the multilevel framework, the key link between the individual and city levels is the city indicator—the "city ID" variable in Figure 11.3, which takes on values between 1 and 20.

For this example, we would have a logistic regression at the data level:

$$\Pr(y_i = 1) = \text{logit}^{-1}(X_i\beta + \alpha_{j[i]}), \text{ for } i = 1, \ldots, n, \tag{11.1}$$

where X is the matrix of individual-level predictors and $j[i]$ indexes the city where person i resides. The second part of the model—what makes it "multilevel"—is the regression of the city coefficients:

$$\alpha_j \sim \text{N}(U_j\gamma, \sigma_\alpha^2), \text{ for } j = 1, \ldots, 20, \tag{11.2}$$

where U is the matrix of city-level predictors, γ is the vector of coefficients for the city-level regression, and σ_α is the standard deviation of the unexplained group-level errors.

The model for the α's in (11.2) allows us to include all 20 of them in model (11.1) without having to worry about collinearity. The key is the group-level variation parameter σ_α, which is estimated from the data (along with α, β, and a) in the fitting of the model. We return to this point in the next chapter.

Directions for the observational study

The "treatment" variable in this example is not randomly applied; hence it is quite possible that cities that differ in enforcement intensities could differ in other important ways in the political, economic, or cultural dimensions. Suppose the goal were to estimate the effects of potential interventions (such as increased enforcement), rather than simply performing a comparative analysis. Then it would make sense to set this up as an observational study, gather relevant pre-treatment information to capture variation among the cities, and perhaps use a matching approach to estimate effects. In addition, good pre-treatment measures on individuals should improve predictive power, thus allowing treatment effects to be estimated more accurately. The researchers studying these child support data are also looking at other outcomes, including measures of the amity between the parents as well as financial and other support.

Along with the special concerns of causal inference, the usual recommendations of regression analysis apply. For example, it might make sense to consider interactions in the model (to see if enforcement is more effective for older fathers, for example).

11.3 Repeated measurements, time-series cross sections, and other non-nested structures

Repeated measurements

Another kind of multilevel data structure involves repeated measurements on persons (or other units)—thus, measurements are clustered within persons, and predictors can be available at the measurement or person level. We illustrate with a model fitted to a longitudinal dataset of about 2000 Australian adolescents whose smoking patterns were recorded every six months (via questionnaire) for a period of three years. Interest lay in the extent to which smoking behavior can be predicted based on parental smoking and other background variables, and the extent to which boys and girls pick up the habit of smoking during their teenage years. Figure 11.5 illustrates the overall rate of smoking among survey participants.

A multilevel logistic regression was fit, in which the probability of smoking depends on sex, parental smoking, the wave of the study, and an individual parameter

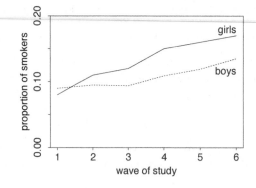

Figure 11.5 *Prevalence of regular (daily) smoking among participants responding at each wave in the study of Australian adolescents (who were on average 15 years old at wave 1).*

| person | | parents smoke? | | wave 1 | | wave 2 | | |
ID	sex	mom	dad	age	smokes?	age	smokes?	$\cdots$
1	f	Y	Y	15:0	N	15:6	N	$\cdots$
2	f	N	N	14:7	N	15:1	N	$\cdots$
3	m	Y	N	15:1	N	15:7	Y	$\cdots$
4	f	N	N	15:3	N	15:9	N	$\cdots$
$\vdots$	$\vdots$	$\vdots$	$\vdots$	$\vdots$	$\vdots$	$\vdots$	$\vdots$	$\ddots$

Figure 11.6 *Data from the smoking study as they might be stored in a single computer file and read into R as a matrix, data. (Ages are in years:months.) These data have a multilevel structure, with observations nested within persons.*

for the person. For person j at wave t, the modeled probability of smoking is

$$\Pr(y_{jt} = 1) = \text{logit}^{-1}(\beta_0 + \beta_1 \text{psmoke}_j + \beta_2 \text{female}_j + \\ + \beta_3(1 - \text{female}_j) \cdot t + \beta_4 \text{female}_j \cdot t + \alpha_j), \qquad (11.3)$$

where `psmoke` is the number of the person's parents who smoke and `female` is an indicator for females, so that β_3 and β_4 represent the time trends for boys and girls, respectively.[1]

Figures 11.6 and 11.7 show two ways of storing the smoking data, either of which would be acceptable for a multilevel analysis. Figure 11.6 shows a single data matrix, with one row for each person in the study. We could then pull out the smoking outcome $y = (y_{jt})$ in R, as follows:

R code
```
y <- data[,seq(6,16,2)]
female <- ifelse (data[,2]=="f", 1, 0)
mom.smoke <- ifelse (data[,3]=="Y", 1, 0)
dad.smoke <- ifelse (data[,4]=="Y", 1, 0)
psmoke <- mom.smoke + dad.smoke
```

and from there fit the model (11.3).

Figure 11.7 shows an alternative approach using two data matrices, one with a

---

[1] Alternatively, we could include a main effect for time and an interaction between time and sex, $\Pr(y_{jt} = 1) = \text{logit}^{-1}(\beta_0 + \beta_1 \cdot \text{psmoke}_j + \beta_2 \cdot \text{female}_j + \beta_3 \cdot t + \beta_4 \cdot \text{female}_j \cdot t + \alpha_j)$, so that the time trends for boys and girls are β_3 and $\beta_3 + \beta_4$, respectively. This parameterization is appropriate to the extent that the comparison between the sexes is of interest; in this case we used (11.3) so that we could easily interpret β_3 and β_4 symmetrically.

age	smokes?	person ID	wave
15:0	N	1	1
14.7	N	2	1
15:1	N	3	1
15:3	N	4	1
⋮	⋮	⋮	⋮
15:6	N	1	2
15:1	N	2	2
15:7	Y	3	2
15:9	N	4	2
⋮	⋮	⋮	⋮

person ID	sex	parents smoke? mom	dad
1	f	Y	Y
2	f	N	N
3	m	Y	N
4	f	N	N
⋮	⋮	⋮	⋮

Figure 11.7 *Data from the smoking study, with observational data written as a single long matrix, `obs.data`, with person indicators, followed by a shorter matrix, `person.data`, of person-level information. Compare to Figure 11.6.*

row for each observation and one with a row for each person. To model these data, one could use R code such as

```
y <- obs.data[,2]
person <- obs.data[,3]
wave <- obs.data[,4]
female <- ifelse (person.data[,2]=="f", 1, 0)
mom.smoke <- ifelse (person.data[,3]=="Y", 1, 0)
dad.smoke <- ifelse (person.data[,4]=="Y", 1, 0)
psmoke <- mom.smoke + dad.smoke
```
R code

and then parameterize the model using the index i to represent individual observations, with $j[i]$ and $t[i]$ indicating the person and wave associated with observation i:

$$\Pr(y_i = 1) = \text{logit}^{-1}(\beta_0 + \beta_1 \text{psmoke}_{j[i]} + \beta_2 \text{female}_{j[i]} + \\ + \beta_3(1 - \text{female}_{j[i]}) \cdot t[i] + \beta_4 \text{female}_{j[i]} \cdot t[i] + \alpha_{j[i]}). \quad (11.4)$$

Models (11.3) and (11.4) are equivalent, and both can be fit in Bugs (as we describe in Part 2B). Choosing between them is a matter of convenience. For data in a simple two-way structure (each adolescent is measured at six regular times), it can make sense to work with the double-indexed outcome variable, (y_{jt}). For a less rectangular data structure (for example, different adolescents measured at irregular intervals) it can be easier to string together a long data vector (y_i), with person and time recorded for each measurement, and with a separate matrix of person-level information (as in Figure 11.7).

Time-series cross-sectional data

In settings where overall time trends are important, repeated measurement data are sometimes called *time-series cross-sectional*. For example, Section 6.3 introduced a study of the proportion of death penalty verdicts that were overturned, in each of 34 states in the 23 years, 1973–1995. The data come at the state × year levels but we are also interested in studying variation among states and over time.

Time-series cross-sectional data are typically (although not necessarily) "rectangular" in structure, with observations at regular time intervals. In contrast, general repeated measurements could easily have irregular patterns (for example, in the smoking study, some children could be measured only once, others could be measured monthly and others yearly). In addition, time-series cross-sectional data commonly have overall time patterns, for example, the steady expansion of the death penalty from the 1970s through the early 1990s. In this context one must consider the state-year data as clustered within states and also within years, with the potential for predictors at all three levels. We discuss such non-nested models in Section 13.5.

Other non-nested structures

Non-nested data also arise when individuals are characterized by overlapping categories of attributes. For example, consider a study of earnings given occupation and state of residence. A survey could include, say, 1500 persons in 40 job categories in 50 states, and a regression model could predict log earnings given individual demographic predictors X, 40 indicators for job categories, and 50 state indicators. We can write the model generalizing the notation of (11.1)–(11.2):

$$y_i = X_i\beta + \alpha_{j[i]} + \gamma_{k[i]} + \epsilon_i, \text{ for } i = 1,\ldots,n, \qquad (11.5)$$

where $j[i]$ and $k[i]$ represent the job category and state, respectively, for person i. The model becomes multilevel with regressions for the job and state coefficients. For example,

$$\alpha_j \sim N(U_j a, \sigma_\alpha^2), \text{ for } j = 1,\ldots,40, \qquad (11.6)$$

where U is a matrix of occupation-level predictors (for example, a measure of social status and an indicator for whether it is supervisory), a is a vector of coefficients for the job model, and σ_α is the standard deviation of the model errors at the level of job category. Similarly, for the state coefficients:

$$\gamma_k \sim N(V_k g, \sigma_\gamma^2) \text{ for } k = 1,\ldots,50. \qquad (11.7)$$

The model defined by regressions (11.5)–(11.7) is non-nested because neither the job categories $j[i]$ nor the states $k[i]$ are subsets of the other.

As this example illustrates, regression notation can become awkward with multilevel models because of the need for new symbols (U, V, a, g, and so forth) to denote data matrices, coefficients, and errors at each level.

11.4 Indicator variables and fixed or random effects

Classical regression: including a baseline and $J - 1$ indicator variables

As discussed at the end of Section 4.5, when including an input variable with J categories into a classical regression, standard practice is to choose one of the categories as a baseline and include indicators for the other $J - 1$ categories. For example, if controlling for the $J = 20$ cities in the child support study in Figure 11.2 on page 238, one could set city 1 (Oakland) as the baseline and include indicators for the other 19. The coefficient for each city then represents its comparison to Oakland.

Multilevel regression: including all J indicators

In a multilevel model it is unnecessary to do this arbitrary step of picking one of the levels as a baseline. For example, in the child support study, one would include indicators for all 20 cities as in model (11.1). In a classical regression these could not all be included because they would be collinear with the constant term, but in a multilevel model this is not a problem because they are themselves modeled by a group-level distribution (which itself can be a regression, as in (11.2)). We discuss on page 393 how the added information removes the collinearity that is present in the simple least squares estimate.

Fixed and random effects

The varying coefficients (α_j's or β_j's) in a multilevel model are sometimes called *random effects*, a term that refers to the randomness in the probability model for the group-level coefficients (as, for example, in (11.2) on page 241).

The term *fixed effects* is used in contrast to random effects—but not in a consistent way! Fixed effects are usually defined as varying coefficients that are not themselves modeled. For example, a classical regression including $J - 1 = 19$ city indicators as regression predictors is sometimes called a "fixed-effects model" or a model with "fixed effects for cities." Confusingly, however, "fixed-effects models" sometimes refer to regressions in which coefficients do *not* vary by group (so that they are fixed, not random).[2]

A question that commonly arises is when to use fixed effects (in the sense of varying coefficients that are unmodeled) and when to use random effects. The statistical literature is full of confusing and contradictory advice. Some say that fixed effects are appropriate if group-level coefficients are of interest, and random effects are appropriate if interest lies in the underlying population. Others recommend fixed

---

[2] Here we outline five definitions that we have seen of fixed and random effects:

1. Fixed effects are constant across individuals, and random effects vary. For example, in a growth study, a model with random intercepts α_i and fixed slope β corresponds to parallel lines for different individuals i, or the model $y_{it} = \alpha_i + \beta t$. Kreft and De Leeuw (1998, p. 12) thus distinguish between fixed and random coefficients.

2. Effects are fixed if they are interesting in themselves or random if there is interest in the underlying population. Searle, Casella, and McCulloch (1992, section 1.4) explore this distinction in depth.

3. "When a sample exhausts the population, the corresponding variable is *fixed*; when the sample is a small (i.e., negligible) part of the population the corresponding variable is *random*" (Green and Tukey, 1960).

4. "If an effect is assumed to be a realized value of a random variable, it is called a random effect" (LaMotte, 1983).

5. Fixed effects are estimated using least squares (or, more generally, maximum likelihood) and random effects are estimated with shrinkage ("linear unbiased prediction" in the terminology of Robinson, 1991). This definition is standard in the multilevel modeling literature (see, for example, Snijders and Bosker, 1999, section 4.2) and in econometrics.

In a multilevel model, this definition implies that fixed effects β_j are estimated conditional on a group-level variance $\sigma_\beta = \infty$ and random effects β_j are estimated conditional on σ_β estimated from data.

Of these definitions, the first clearly stands apart, but the other four definitions differ also. Under the second definition, an effect can change from fixed to random with a change in the goals of inference, even if the data and design are unchanged. The third definition differs from the others in defining a finite population (while leaving open the question of what to do with a large but not exhaustive sample), while the fourth definition makes no reference to an actual (rather than mathematical) population at all. The second definition allows fixed effects to come from a distribution, as long as that distribution is not of interest, whereas the fourth and fifth do not use any distribution for inference about fixed effects. The fifth definition has the virtue of mathematical precision but leaves unclear when a given set of effects should be considered fixed or random. In summary, it is easily possible for a factor to be "fixed" according to some definitions above and "random" for others. Because of these conflicting definitions, it is no surprise that "clear answers to the question 'fixed or random?' are not necessarily the norm" (Searle, Casella, and McCulloch, 1992, p. 15).

effects when the groups in the data represent all possible groups, and random effects when the population includes groups not in the data. These two recommendations (and others) can be unhelpful. For example, in the child support example, we are interested in these particular cities and also the country as a whole. The cities are only a sample of cities in the United States—but if we were suddenly given data from all the other cities, we would not want then to change our model.

Our advice (elaborated upon in the rest of this book) is to *always* use multilevel modeling ("random effects"). Because of the conflicting definitions and advice, we avoid the terms "fixed" and "random" entirely, and focus on the description of the model itself (for example, varying intercepts and constant slopes), with the understanding that batches of coefficients (for example, $\alpha_1, \ldots, \alpha_J$) will themselves be modeled.

11.5 Costs and benefits of multilevel modeling

Quick overview of classical regression

Before we go to the effort of learning multilevel modeling, it is helpful to briefly review what can be done with classical regression:

- Prediction for continuous or discrete outcomes,
- Fitting of nonlinear relations using transformations,
- Inclusion of categorical predictors using indicator variables,
- Modeling of interactions between inputs,
- Causal inference (under appropriate conditions).

Motivations for multilevel modeling

There are various reasons why it might be worth moving to a multilevel model, whether for purposes of causal inference, the study of variation, or prediction of future outcomes:

- Accounting for individual- and group-level variation in estimating *group-level* regression coefficients. For example, in the child support study in Section 11.2, interest lies in a city-level predictor (child support enforcement), and in classical regression it is not possible to include city indicators along with city-level predictors.
- Modeling variation among *individual-level* regression coefficients. In classical regression, one can do this using indicator variables, but multilevel modeling is convenient when we want to model the variation of these coefficients across groups, make predictions for new groups, or account for group-level variation in the uncertainty for individual-level coefficients.
- Estimating regression coefficients for *particular* groups. For example, in the next chapter, we discuss the problem of estimating radon levels from measurements in several counties in Minnesota. With a multilevel model, we can get reasonable estimates even for counties with small sample sizes, which would be difficult using classical regression.

One or more of these reasons might apply in any particular study.

Complexity of multilevel models

A potential drawback to multilevel modeling is the additional complexity of coefficients varying by group. We do not mind this complexity—in fact, we embrace it in its realism—however, it does create new difficulties in understanding and summarizing the model, issues we explore in Part 3 of this book.

Additional modeling assumptions

As we discuss in the next few chapters, a multilevel model requires additional assumptions beyond those of classical regression—basically, each level of the model corresponds to its own regression with its own set of assumptions such as additivity, linearity, independence, equal variance, and normality.

We usually don't mind. First, it can be possible to check these assumptions. Perhaps more important, classical regressions can typically be identified with particular special cases of multilevel models with hierarchical variance parameters set to zero or infinity—these are the *complete pooling* and *no pooling* models discussed in Sections 12.2 and 12.3. Our ultimate justification, which can be seen through examples, is that the assumptions pay off in practice in allowing more realistic models and inferences.

When does multilevel modeling make a difference?

The usual alternative to multilevel modeling is classical regression—either ignoring group-level variation, or with varying coefficients that are estimated classically (and not themselves modeled)—or combinations of classical regressions such as the individual and group-level models described on page 239.

In various limiting cases, the classical and multilevel approaches coincide. When there is very little group-level variation, the multilevel model reduces to classical regression with no group indicators; conversely, when group-level coefficients vary greatly (compared to their standard errors of estimation), multilevel modeling reduces to classical regression with group indicators.

When the number of groups is small (less than five, say), there is typically not enough information to accurately estimate group-level variation. As a result, multilevel models in this setting typically gain little beyond classical varying-coefficient models.

These limits give us a sense of where we can gain the most from multilevel modeling—where it is worth the effort of expanding a classical regression in this way. However, there is little risk from applying a multilevel model, assuming we are willing to put in the effort to set up the model and interpret the resulting inferences.

11.6 Bibliographic note

Several introductory books on multilevel models have been written in the past decade in conjunction with specialized computer programs (see Section 1.5), including Raudenbush and Bryk (2002), Goldstein (1995), and Snijders and Bosker (1999). Kreft and De Leeuw (1998) provide an accessible introduction and a good place to start (although we do not agree with all of their recommendations). These books have a social science focus, perhaps because it is harder to justify the use of linear models in laboratory sciences where it is easier to isolate the effects of individual factors and so the functional form of responses is better understood. Giltinan and Davidian (1995) and Verbeke and Molenberghs (2000) are books on nonlinear multilevel models focusing on biostatistical applications.

Another approach to regression with multilevel data structures is to use classical estimates and then correct the standard errors to deal with the dependence in the data. We briefly discuss the connection between multilevel models and correlated-error models in Section 12.5 but do not consider these other inferential methods, which include *generalized estimating equations* (see Carlin et al., 2001, for a comparison to multilevel models) and *panel-corrected standard errors* (see Beck and Katz, 1995, 1996).

The articles in the special issue of *Political Analysis* devoted to multilevel modeling (Kedar and Shively, 2005) illustrate several different forms of analysis of multilevel data, including two-level classical regression and multilevel modeling.

Gelman (2005) discusses difficulties with the terms "fixed" and "random" effects. See also Kreft and De Leeuw (1998, section 1.3.3), for a discussion of the multiplicity of definitions of fixed and random effects and coefficients, and Robinson (1998) for a historical overview.

The child support example comes from Nepomnyaschy and Garfinkel (2005). The teenage smoking example comes from Carlin et al. (2001), who consider several different models, including a multilevel logistic regression.

11.7 Exercises

1. The file `apt.dat` in the folder `rodents` contains data on rodent infestation in a sample of New York City apartments (see codebook `rodents.doc`). The file `dist.dat` contains data on the 55 "community districts" (neighborhoods) in the city.

   (a) Write the notation for a varying-intercept multilevel logistic regression (with community districts as the groups) for the probability of rodent infestation using the individual-level predictors but no group-level predictors.

   (b) Expand the model in (a) by including the variables in `dist.dat` as group-level predictors.

2. Time-series cross-sectional data: download data with an outcome y and predictors X in each of J countries for a series of K consecutive years. The outcome should be some measure of educational achievement of children and the predictors should be a per capita income measure, a measure of income inequality, and a variable summarizing how democratic the country is. For these countries, also create country-level predictors that are indicators for the countries' geographic regions.

   (a) Set up the data as a wide matrix of countries × measurements (as in Figure 11.6).

   (b) Set up the data as two matrices as in Figure 11.7: a long matrix with JK rows with all the measurements, and a matrix with J rows, with information on each country.

   (c) Write a multilevel regression as in (11.5)–(11.7). Explain the meaning of all the variables in the model.

3. The folder `olympics` has seven judges' ratings of seven figure skaters (on two criteria: "technical merit" and "artistic impression") from the 1932 Winter Olympics.

   (a) Construct a $7 \times 7 \times 2$ array of the data (ordered by skater, judge, and judging criterion).

   (b) Reformulate the data as a 98×4 array (similar to the top table in Figure 11.7), where the first two columns are the technical merit and artistic impression scores, the third column is a skater ID, and the fourth column is a judge ID.

(c) Add another column to this matrix representing an indicator variable that equals 1 if the skater and judge are from the same country, or 0 otherwise.

4. The folder cd4 has CD4 percentages for a set of young children with HIV who were measured several times over a period of two years. The dataset also includes the ages of the children at each measurement.

(a) Graph the outcome (the CD4 percentage, on the square root scale) for each child as a function of time.

(b) Each child's data has a time course that can be summarized by a linear fit. Estimate these lines and plot them for all the children.

(c) Set up a model for the children's slopes and intercepts as a function of the treatment and age at baseline. Estimate this model using the two-step procedure–first estimate the intercept and slope separately for each child, then fit the between-child models using the point estimates from the first step.

## CHAPTER 12

Multilevel linear models: the basics

Multilevel modeling can be thought of in two equivalent ways:

- We can think of a generalization of linear regression, where intercepts, and possibly slopes, are allowed to vary by group. For example, starting with a regression model with one predictor, $y_i = \alpha + \beta x_i + \epsilon_i$, we can generalize to the varying-intercept model, $y_i = \alpha_{j[i]} + \beta x_i + \epsilon_i$, and the varying-intercept, varying-slope model, $y_i = \alpha_{j[i]} + \beta_{j[i]} x_i + \epsilon_i$ (see Figure 11.1 on page 238).

- Equivalently, we can think of multilevel modeling as a regression that includes a categorical input variable representing group membership. From this perspective, the group index is a factor with J levels, corresponding to J predictors in the regression model (or $2J$ if they are interacted with a predictor x in a varying-intercept, varying-slope model; or $3J$ if they are interacted with two predictors $X_{(1)}, X_{(2)}$; and so forth).

In either case, $J-1$ linear predictors are added to the model (or, to put it another way, the constant term in the regression is replaced by J separate intercept terms). The crucial multilevel modeling step is that these J coefficients are then themselves given a model (most simply, a common distribution for the J parameters α_j or, more generally, a regression model for the α_j's given group-level predictors). The group-level model is estimated simultaneously with the data-level regression of y.

This chapter introduces multilevel linear regression step by step. We begin in Section 12.2 by characterizing multilevel modeling as a compromise between two extremes: *complete pooling*, in which the group indicators are not included in the model, and *no pooling*, in which separate models are fit within each group. After laying out some notational difficulties in Section 12.5, we discuss in Section 12.6 the different roles of the individual- and group-level regressions. Chapter 13 continues with more complex multilevel structures.

12.1 Notation

We briefly review the notation for classical regression and then outline how it can be generalized for multilevel models. As we illustrate in the examples, however, no single notation is appropriate for all problems. We use the following notation for classical regression:

- Units $i = 1, \ldots, n$. By *units*, we mean the smallest items of measurement.

- Outcome measurements $y = (y_1, \ldots, y_n)$. These are the unit-level data being modeled.

- Regression predictors are represented by an $n \times k$ matrix X, so that the vector of predicted values is $\hat{y} = X\beta$, where $\hat{y}$ and β are column vectors of length n and k, respectively. We include in X the constant term (unless it is explicitly excluded from the model), so that the first column of X is all 1's. We usually label the coefficients as $\beta_0, \ldots, \beta_{k-1}$, but sometimes we index from 1 to k.

- For each individual unit i, we denote its row vector of predictors as X_i. Thus, $\hat{y}_i = X_i\beta$ is the prediction for unit i.

- For each predictor κ, we label the $(\kappa+1)^{st}$ column of X as $X_{(\kappa)}$ (assuming that $X_{(0)}$ is a column of 1's).

- Any information contained in the unit labels i should be coded in the regression inputs. For example, if $i = 1, \ldots, n$ represents the order in which persons i enrolled in a study, we should create a time variable t_i and, for example, include it in the matrix X of regression predictors. Or, more generally, consider transformations and interactions of this new input variable.

For multilevel models, we label:

- Groups $j = 1, \ldots, J$. This works for a single level of grouping (for example, students within schools, or persons within states).

- We occasionally use $k = 1, \ldots, K$ for a second level of grouping (for example, students within schools within districts; or, for a non-nested example, test responses that can be characterized by person or by item). In any particular example, we have to distinguish this k from the number of predictors in X. For more complicated examples we develop idiosyncratic notation as appropriate.

- Index variables $j[i]$ code group membership. For example, if $j[35] = 4$, then the 35th unit in the data ($i = 35$) belongs to group 4.

- Coefficients are sometimes written as a vector β, sometimes as α, β (as in Figure 11.1 on page 238), with group-level regression coefficients typically called γ.

- We make our R and Bugs code more readable by typing α, β, γ as `a,b,g`.

- We write the varying-intercept model with one additional predictor as $y_i = \alpha_{j[i]} + \beta x_i + \epsilon_i$ or $y_i \sim N(\alpha_{j[i]} + \beta x_i, \sigma_y^2)$. Similarly, the varying-intercept, varying-slope model is $y_i = \alpha_{j[i]} + \beta_{j[i]} x_i + \epsilon_i$ or $y_i \sim N(\alpha_{j[i]} + \beta_{j[i]} x_i, \sigma_y^2)$.

- With multiple predictors, we write $y_i = X_i B + \epsilon_i$, or $y_i \sim N(X_i B, \sigma_y^2)$. B is a matrix of coefficients that can be modeled using a general varying-intercept, varying-slope model (as discussed in the next chapter).

- Standard deviation is σ_y for data-level errors and $\sigma_\alpha, \sigma_\beta$, and so forth, for group-level errors.

- Group-level predictors are represented by a matrix U with J rows, for example, in the group-level model, $\alpha_j \sim N(U_j \gamma, \sigma_\alpha^2)$. When there is a single group-level predictor, we label it as lowercase u.

12.2 Partial pooling with no predictors

As noted in Section 1.3, multilevel regression can be thought of as a method for compromising between the two extremes of excluding a categorical predictor from a model (*complete pooling*), or estimating separate models within each level of the categorical predictor (*no pooling*).

Complete-pooling and no-pooling estimates of county radon levels

We illustrate with the home radon example, which we introduced in Section 1.2 and shall use throughout this chapter. Consider the goal of estimating the distribution of radon levels of the houses within each of 85 counties in Minnesota.[1] This seems

---

[1] Radon levels are always positive, and it is reasonable to suppose that effects will be multiplicative; hence it is appropriate to model the data on the logarithmic scale (see Section 4.4). For some purposes, though, such as estimating total cancer risk, it makes sense to estimate averages on the original, unlogged scale; we can obtain these inferences using simulation, as discussed at the end of Section 12.8.

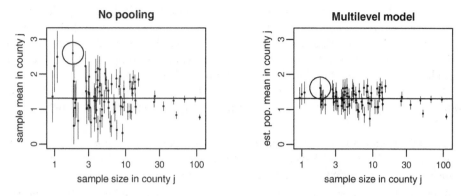

Figure 12.1 *Estimates ± standard errors for the average log radon levels in Minnesota counties plotted versus the (jittered) number of observations in the county: (a) no-pooling analysis, (b) multilevel (partial pooling) analysis, in both cases with no house-level or county-level predictors. The counties with fewer measurements have more variable estimates and larger higher standard errors. The horizontal line in each plot represents an estimate of the average radon level across all counties. The left plot illustrates a problem with the no-pooling analysis: it systematically causes us to think that certain counties are more extreme, just because they have smaller sample sizes.*

simple enough. One estimate would be the average that completely pools data across all counties. This ignores variation among counties in radon levels, however, so perhaps a better option would be simply to use the average log radon level in each county. Figure 12.1a plots these averages against the number of observations in each county.

Whereas complete pooling ignores variation between counties, the no-pooling analysis overstates it. To put it another way, the no-pooling analysis overfits the data within each county. To see this, consider Lac Qui Parle County (circled in the plot), which has the highest average radon level of all 85 counties in the data. This average, however, is estimated using only two data points. Lac Qui Parle may very well be a high-radon county, but do we really believe it is *that* high? Maybe, but probably not: given the variability in the data we would not have much trust in an estimate based on only two measurements.

To put it another way, looking at all the counties together: the estimates from the no-pooling model overstate the variation among counties and tend to make the individual counties look more different than they actually are.

Partial-pooling estimates from a multilevel model

The multilevel estimates of these averages, displayed in Figure 12.1b, represent a compromise between these two extremes. The goal of estimation is the average log radon level α_j among all the houses in county j, for which all we have available are a random sample of size n_j. For this simple scenario with no predictors, the multilevel estimate for a given county j can be approximated as a weighted average of the mean of the observations in the county (the unpooled estimate, $\bar{y}_j$) and the mean over all counties (the completely pooled estimate, $\bar{y}_{\text{all}}$):

$$\hat{\alpha}_j^{\text{multilevel}} \approx \frac{\frac{n_j}{\sigma_y^2}\bar{y}_j + \frac{1}{\sigma_\alpha^2}\bar{y}_{\text{all}}}{\frac{n_j}{\sigma_y^2} + \frac{1}{\sigma_\alpha^2}}, \tag{12.1}$$

where n_j is the number of measured houses in county j, σ_y^2 is the within-county variance in log radon measurements, and σ_α^2 is the variance among the average log radon levels of the different counties. We could also allow the within-county variance to vary by county (in which case σ_y would be replaced by σ_{yj} in the preceding formula) but for simplicity we assume it is constant.

The weighted average (12.1) reflects the relative amount of information available about the individual county, on one hand, and the average of all the counties, on the other:

- Averages from counties with smaller sample sizes carry less information, and the weighting pulls the multilevel estimates closer to the overall state average. In the limit, if $n_j = 0$, the multilevel estimate is simply the overall average, $\bar{y}_{\text{all}}$.

- Averages from counties with larger sample sizes carry more information, and the corresponding multilevel estimates are close to the county averages. In the limit as $n_j \to \infty$, the multilevel estimate is simply the county average, $\bar{y}_j$.

- In intermediate cases, the multilevel estimate lies between the two extremes.

To actually apply (12.1), we need estimates of the variation within and between counties. In practice, we estimate these variance parameters together with the α_j's, either with an approximate program such as lmer() (see Section 12.4) or using fully Bayesian inference, as implemented in Bugs and described in Part 2B of this book. For now, we present inferences (as in Figure 12.1) without dwelling on the details of estimation.

12.3 Partial pooling with predictors

The same principle of finding a compromise between the extremes of complete pooling and no pooling applies for more general models. This section considers partial pooling for a model with unit-level predictors. In this scenario, no pooling might refer to fitting a separate regression model within each group. However, a less extreme and more common option that we also sometimes refer to as "no pooling" is a model that includes group indicators and estimates the model classically.[2]

As we move on to more complicated models, we present estimates graphically but do not continue with formulas of the form (12.1). However, the general principle remains that multilevel models compromise between pooled and unpooled estimates, with the relative weights determined by the sample size in the group and the variation within and between groups.

Complete-pooling and no-pooling analyses for the radon data, with predictors

Continuing with the radon data, Figure 12.2 shows the logarithm of the home radon measurement versus floor of measurement[3] for houses sampled from eight of 85 counties in Minnesota. (We fit our model to the data from all 85 counties, including a total of 919 measurements, but to save space we display the data and estimates for a selection of eight counties, chosen to capture a range of the sample sizes in the survey.)

In each graph of Figure 12.2, the dashed line shows the linear regression of log

[2] This version of "no pooling" does not pool the estimates for the intercepts—the parameters we focus on in the current discussion—but it does completely pool estimates for any slope coefficients (they are forced to have the same value across all groups) and also assumes the residual variance is the same within each group.

[3] Measurements were taken in the lowest living area of each house, with basement coded as 0 and first floor coded as 1.

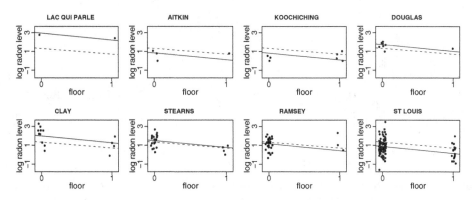

Figure 12.2 *Complete-pooling (dashed lines, $y = \alpha + \beta x$) and no-pooling (solid lines, $y = \alpha_j + \beta x$) regressions fit to radon data from 85 counties in Minnesota, and displayed for eight of the counties. The estimated slopes β differ slightly for the two models, but here our focus is on the intercepts.*

radon, given the floor of measurement, using a model that pools all counties together (so the same line appears in all eight plots), and the solid line shows the no-pooling regressions, obtained by including county indicators in the regression (with the constant term removed to avoid collinearity; we also could have kept the constant term and included indicators for all but one of the counties). We can write the complete-pooling regression as $y_i = \alpha + \beta x_i + \epsilon_i$ and the no-pooling regression as $y_i = \alpha_{j[i]} + \beta x_i + \epsilon_i$, where $j[i]$ is the county corresponding to house i. The solid lines then plot $y = \hat{\alpha} + \hat{\beta} x$ from the complete-pooling model, and the dashed lines show $y = \hat{\alpha}_j + \hat{\beta} x$, for $j = 1, \ldots, 8$, from the no-pooling model.

Here is the complete-pooling regression for the radon data:

```
lm(formula = y ~ x)                                    R output
            coef.est coef.se
(Intercept)   1.33     0.03
x            -0.61     0.07
  n = 919, k = 2
  residual sd = 0.82
```

To fit the no-pooling model in R, we include the county index (a variable named county that takes on values between 1 and 85) as a factor in the regression—thus, predictors for the 85 different counties. We add "-1" to the regression formula to remove the constant term, so that all 85 counties are included. Otherwise, R would use county 1 as a baseline.

```
lm(formula = y ~ x + factor(county) - 1)               R output
                   coef.est coef.sd
x                   -0.72     0.07
factor(county)1      0.84     0.38
factor(county)2      0.87     0.10
  .  .  .
factor(county)85     1.19     0.53
  n = 919, k = 86
  residual sd = 0.76
```

The estimated slopes β differ slightly for the two regressions. The no-pooling model includes county indicators, which can change the estimated coefficient for x, if the proportion of houses with basements varies among counties. This is just

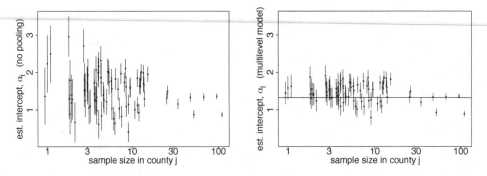

Figure 12.3 *(a) Estimates $\pm$ standard errors for the county intercepts α_j in the model*
$y_i = \alpha_{j[i]} + \beta x_i + error_i$, *for the no-pooling analysis of the radon data, plotted versus num-*
ber of observations from the county. The counties with fewer measurements have more
variable estimates with higher standard errors. This graph illustrates a problem with clas-
sical regression: it systematically causes us to think that certain counties are more extreme,
just because they have smaller sample sizes.
(b) Multilevel (partial pooling) estimates $\pm$ standard errors for the county intercepts α_j
for the radon data, plotted versus number of observations from the county. The horizontal
line shows the complete pooling estimate. Comparing to the left plot (no pooling), which is
on the same scale, we see that the multilevel estimate is typically closer to the complete-
pooling estimate for counties with few observations, and closer to the no-pooling estimates
for counties with many observations.
These plots differ only slightly from the no-pooling and multilevel estimates without the
house-level predictor, as displayed in Figure 12.1.

a special case of the rule that adding new predictors in a regression can change
the estimated coefficient of x, if these new predictors are correlated with x. In
the particular example shown in Figure 12.2, the complete-pooling and no-pooling
estimates of β differ only slightly; in the graphs, the difference can be seen most
clearly in Stearns and Ramsey counties.

Problems with the no-pooling and complete-pooling analyses

Both the analyses shown in Figure 12.2 have problems. The complete-pooling analysis ignores any variation in average radon levels between counties. This is unde-
sirable, particularly since the goal of our analysis was to identify counties with
high-radon homes. We do not want to pool away the main subject of our study!

The no-pooling analysis has problems too, however, which we can again see in
Lac Qui Parle County. Even after controlling for the floors of measurement, this
county has the highest fitted line (that is, the highest estimate $\hat{\alpha}_j$), but again we
do not have much trust in an estimate based on only two observations.

More generally, we would expect the counties with the least data to get more
extreme estimates $\hat{\alpha}_j$ in the no-pooling analyses. Figure 12.3a illustrates with the
estimates $\pm$ standard errors for the county intercepts α_j, plotted versus the sample
size in each county j.

Multilevel analysis

The simplest multilevel model for the radon data with the floor predictor can be
written as

$$y_i \sim N(\alpha_{j[i]} + \beta x_i, \sigma_y^2), \quad \text{for } i = 1, \ldots, n, \tag{12.2}$$

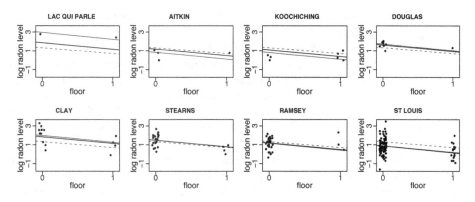

Figure 12.4 *Multilevel (partial pooling) regression lines $y = \alpha_j + \beta x$ fit to radon data from Minnesota, displayed for eight counties. Light-colored dashed and solid lines show the complete-pooling and no-pooling estimates, respectively, from Figure 12.3a.*

which looks like the no-pooling model but with one key difference. In the no-pooling model, the α_j's are set to the classical least squares estimates, which correspond to the fitted intercepts in a model run separately in each county (with the constraint that the slope coefficient equals β in all models). Model (12.2) also looks a little like the complete-pooling model except that, with complete pooling, the α_j's are given a "hard constraint"—they are all fixed at a common α.

In the multilevel model, a "soft constraint" is applied to the α_j's: they are assigned a probability distribution,

$$\alpha_j \sim \mathrm{N}(\mu_\alpha, \sigma_\alpha^2), \quad \text{for } j = 1, \dots, J, \qquad (12.3)$$

with their mean μ_α and standard deviation σ_α estimated from the data. The distribution (12.3) has the effect of pulling the estimates of α_j toward the mean level μ_α, but not all the way—thus, in each county, a *partial-pooling* compromise between the two estimates shown in Figure 12.2. In the limit of $\sigma_\alpha \to \infty$, the soft constraints do nothing, and there is no pooling; as $\sigma_\alpha \to 0$, they pull the estimates all the way to zero, yielding the complete-pooling estimate.

Figure 12.4 shows, for the radon example, the estimated line from the multilevel model (12.2), which in each county lies between the complete-pooling and no-pooling regression lines. There is strong pooling (solid line closer to complete-pooling line) in counties with small sample sizes, and only weak pooling (solid line closer to no-pooling line) in counties containing many measurements.

Going back to Figure 12.3, the right panel shows the estimates and standard errors for the county intercepts α_j from the multilevel model, plotted versus county sample size. Comparing to the left panel, we see more pooling for the counties with fewer observations. We also see a trend that counties with larger sample sizes have lower radon levels, indicating that "county sample size" is correlated with some relevant county-level predictor.

Average regression line and individual- and group-level variances

Multilevel models typically have so many parameters that it is not feasible to closely examine all their numerical estimates. Instead we plot the estimated group-level models (as in Figure 12.4) and varying parameters (as in Figure 12.3b) to look for patterns and facilitate comparisons across counties. It can be helpful, however, to look at numerical summaries for the *hyperparameters*—those model parameters without group-level subscripts.

For example, in the radon model, the hyperparameters are estimated as $\hat{\mu}_\alpha = 1.46$, $\hat{\beta} = -0.69$, $\hat{\sigma}_y = 0.76$, and $\hat{\sigma}_\alpha = 0.33$. (We show the estimates in Section 12.4.) That is, the estimated average regression line for all the counties is $y = 1.46 - 0.69x$, with error standard deviations of 0.76 at the individual level and 0.33 at the county level. For this dataset, variation within counties (after controlling for the floor of measurement) is comparable to the average difference between measurements in houses with and without basements.

One way to interpret the variation between counties, σ_α, is to consider the variance ratio, $\sigma_\alpha^2/\sigma_y^2$, which in this example is estimated at $0.33^2/0.76^2 = 0.19$, or about one-fifth. Thus, the standard deviation of average radon levels between counties is the same as the standard deviation of the average of 5 measurements within a county (that is, $0.76/\sqrt{5} = 0.33$). The relative values of individual- and group-level variances are also sometimes expressed using the *intraclass correlation*, $\sigma_\alpha^2/(\sigma_\alpha^2 + \sigma_y^2)$, which ranges from 0 if the grouping conveys no information to 1 if all members of a group are identical.

In our example, the group-level model tells us that the county intercepts, α_j, have an estimated mean of 1.46 and standard deviation of 0.33. (What is relevant to our discussion here is the standard deviation, not the mean.) The amount of information in this distribution is the same as that in 5 measurements within a county. To put it another way, for a county with a sample size less than 5, there is more information in the group-level model than in the county's data; for a county with more than 5 observations, the within-county measurements are more informative (in the sense of providing a lower-variance estimate of the county's average radon level). As a result, the multilevel regression line in a county is closer to the complete-pooling estimate when sample size is less than 5, and closer to the no-pooling estimate when sample size exceeds 5. We can see this in Figure 12.4: as sample size increases, the multilevel estimates move closer and closer to the no-pooling lines.

Partial pooling (shrinkage) of group coefficients α_j

Multilevel modeling partially pools the group-level parameters α_j toward their mean level, μ_α. There is more pooling when the group-level standard deviation σ_α is small, and more smoothing for groups with fewer observations. Generalizing (12.1), the multilevel-modeling estimate of α_j can be expressed as a weighted average of the no-pooling estimate for its group ($\bar{y}_j - \beta\bar{x}_j$) and the mean, μ_α:

$$\text{estimate of } \alpha_j \approx \frac{\frac{n_j}{\sigma_y^2}}{\frac{n_j}{\sigma_y^2} + \frac{1}{\sigma_\alpha^2}}(\bar{y}_j - \beta\bar{x}_j) + \frac{\frac{1}{\sigma_\alpha^2}}{\frac{n_j}{\sigma_y^2} + \frac{1}{\sigma_\alpha^2}}\mu_\alpha. \tag{12.4}$$

When actually fitting multilevel models, we do not actually use this formula; rather, we fit models using `lmer()` or Bugs, which automatically perform the calculations, using formulas such as (12.4) internally. Chapter 19 provides more detail on the algorithms used to fit these models.

Classical regression as a special case

Classical regression models can be viewed as special cases of multilevel models. The limit of $\sigma_\alpha \to 0$ yields the complete-pooling model, and $\sigma_\alpha \to \infty$ reduces to the no-pooling model. Given multilevel data, we can estimate σ_α. Therefore we see no reason (except for convenience) to accept estimates that arbitrarily set this parameter to one of these two extreme values.

12.4 Quickly fitting multilevel models in R

We fit most of the multilevel models in this part of the book using the `lmer()` function, which fits linear and generalized linear models with varying coefficients.[4] Part 2B of the book considers computation in more detail, including a discussion of why it can be helpful to make the extra effort and program models using Bugs (typically using a simpler `lmer()` fit as a starting point). The `lmer()` function is currently part of the R package `Matrix`; see Appendix C for details. Here we introduce `lmer()` in the context of simple varying-intercept models.

The lmer function

Varying-intercept model with no predictors. The varying intercept model with no predictors (discussed in Section 12.2) can be fit and displayed using `lmer()` as follows:

```
M0 <- lmer (y ~ 1 + (1 | county))
display (M0)
```
R code

This model simply includes a constant term (the predictor "1") and allows it to vary by county. We next move to a more interesting model including the floor of measurement as an individual-level predictor.

Varying-intercept model with an individual-level predictor. We shall introduce multilevel fitting with model (12.2)–(12.3), the varying-intercept regression with a single predictor. We start with the call to `lmer()`:

```
M1 <- lmer (y ~ x + (1 | county))
```
R code

This expression starts with the no-pooling model, "y ~ x," and then adds "(1 | county)," which allows the intercept (the coefficient of the predictor "1," which is the column of ones—the constant term in the regression) to vary by county.

We can then display a quick summary of the fit:

```
display (M1)
```
R code

which yields

```
lmer(formula = y ~ x + (1 | county))
            coef.est coef.se
(Intercept)  1.46     0.05
x           -0.69     0.07
Error terms:
 Groups    Name        Std.Dev.
 county    (Intercept) 0.33
 Residual              0.76
# of obs: 919, groups: county, 85
deviance = 2163.7
```
R output

---

[4] The name `lmer` stands for "linear mixed effects in R," but the function actually works for generalized linear models as well. The term "mixed effects" refers to random effects (coefficients that vary by group) and fixed effects (coefficients that do not vary). We avoid the terms "fixed" and "random" (see page 245) and instead refer to coefficients as "modeled" (that is, grouped) or "unmodeled."

The top part of this display shows the inference about the intercept and slope for the model, averaging over the counties. The bottom part gives the estimated variation: $\hat{\sigma}_\alpha = 0.33$ and $\hat{\sigma}_y = 0.76$. We also see that the model was fit to 919 houses within 85 counties. We shall ignore the deviance for now.

Estimated regression coefficients

To see the estimated model within each county. We type

R code   `coef (M1)`

which yields

R output
```
$county
      (Intercept)    x
1            1.19 -0.69
2            0.93 -0.69
3            1.48 -0.69
   . . .
85           1.39 -0.69
```

Thus, the estimated regression line is $y = 1.19 - 0.69x$ in county 1, $y = 0.93 + 0.69x$ in county 2, and so forth. The slopes are all identical because they were specified thus in the model. (The specification (1|county) tells the model to allow only the intercept to vary. As we shall discuss in the next chapter, we can allow the slope to vary by specifying (1+x|county) in the regression model.)

Fixed and random effects. Alternatively, we can separately look at the estimated model averaging over the counties—the "fixed effects"—and the county-level errors— the "random effects." Typing

R code   `fixef (M1)`

yields

R output
```
(Intercept)            x
       1.46        -0.69
```

The estimated regression line in an average county is thus $y = 1.46 - 0.69x$. We can then look at the county-level errors:

R code   `ranef (M1)`

which yields

R output
```
      (Intercept)
1           -0.27
2           -0.53
3            0.02
   . . .
85          -0.08
```

These tell us how much the intercept is shifted up or down in particular counties. Thus, for example, in county 1, the estimated intercept is 0.27 lower than average, so that the regression line is $(1.46 - 0.27) - 0.69x = 1.19 - 0.69x$, which is what we saw earlier from the call to coef(). For some applications, it is best to see the estimated model within each group; for others, it is helpful to see the estimated average model and group-level errors.

Uncertainties in the estimated coefficients

We wrote little functions `se.fixef()` and `se.ranef()` for quickly pulling out these standard errors from the model fitted by `lmer()`. In this example,

> `se.fixef (M1)`   R code

yields

```
(Intercept)            x
      0.05         0.07
```
R output

and

> `se.ranef (M1)`   R code

yields,

```
$county
   (Intercept)
1       0.25
2       0.10
3       0.26
 . . .
85      0.28
```
R output

As discussed in Section 12.3, the standard errors differ according to the sample size within each county; for example, counties 1, 2, and 85 have 4, 52, and 2 houses, respectively, in the sample. For the within-county regressions, standard errors are only given for the intercepts, since this model has a common slope for all counties.

Summarizing and displaying the fitted model

We can access the components of the estimates and standard errors using list notation in R. For example, to get a 95% confidence interval for the slope (which, in this model, does not vary by county):

> `fixef(M1)["x"] + c(-2,2)*se.fixef(M1)["x"]`   R code

or, equivalently, since the slope is the second coefficient in the regression,

> `fixef(M1)[2] + c(-2,2)*se.fixef(M1)[2]`   R code

The term "fixed effects" is used for the regression coefficients that do not vary by group (such as the coefficient for x in this example) or for group-level coefficients or group averages (such as the average intercept, μ_α in (12.3)).

Identifying the batches of coefficients. In pulling out elements of the coefficients from `coef()` or `ranef()`, we must first identify the grouping (`county`, in this case). The need for this labeling will become clear in the next chapter in the context of non-nested models, where there are different levels of grouping and thus different structures of varying coefficients.

For example, here is a 95% confidence interval for the intercept in county 26:

> `coef(M1)$county[26,1] + c(-2,2)*se.ranef(M1)$county[26]`   R code

and here is a 95% confidence interval for the error in the intercept in that county (that is, the deviation from the average):

> `as.matrix(ranef(M1)$county)[26] + c(-2,2)*se.ranef(M1)$county[26]`   R code

For a more elaborate example, we make Figure 12.4 using the following commands:

R code
```
a.hat.M1 <- coef(M1)$county[,1]    # 1st column is the intercept
b.hat.M1 <- coef(M1)$county[,2]    # 2nd element is the slope
x.jitter <- x + runif(n,-.05,.05)   # jittered data for plotting
par (mfrow=c(2,4))                  # make a 2x4 grid of plots
for (j in display8){
  plot (x.jitter[county==j], y[county==j], xlim=c(-.05,1.05),
    ylim=y.range, xlab="floor", ylab="log radon level", main=uniq.name[j])
## [uniq.name is a vector of county names that was created earlier]
  curve (coef(lm.pooled)[1] + coef(lm.pooled)[2]*x, lty=2, col="gray10",
    add=TRUE)
  curve (coef(lm.unpooled)[j+1] + coef(lm.unpooled)[1]*x, col="gray10",
    add=TRUE)
  curve (a.hat.M1[j] + b.hat.M1[j]*x, lwd=1, col="black", add=TRUE)
}
```

Here, `lm.pooled` and `lm.unpooled` are the classical regressions that we have already fit.

More complicated models

The `lmer()` function can also handle many of the multilevel regressions discussed in this part of the book, including group-level predictors, varying intercepts and slopes, nested and non-nested structures, and multilevel generalized linear models. Approximate routines such as `lmer()` tend to work well when the sample size and number of groups is moderate to large, as in the radon models. When the number of groups is small, or the model becomes more complicated, it can be useful to switch to Bayesian inference, using the Bugs program, to better account for uncertainty in model fitting. We return to this point in Section 16.1.

12.5 Five ways to write the same model

We begin our treatment of multilevel models with the simplest structures—*nested* models, in which we have observations $i = 1, \ldots, n$ clustered in groups $j = 1, \ldots, J$, and we wish to model variation among groups. Often, predictors are available at the individual and group levels. We shall use as a running example the home radon analysis described above, using as predictors the house-level x_i and a measure of the logarithm of soil uranium as a county-level predictor, u_j. For some versions of the model, we include these both as individual-level predictors and label them as X_{i1} and X_{i2}.

There are several different ways of writing a multilevel model. Rather than introducing a restrictive uniform notation, we describe these different formulations and explain how they are connected. It is useful to be able to express a model in different ways, partly so that we can recognize the similarities between models that only appear to be different, and partly for computational reasons.

Allowing regression coefficients to vary across groups

Perhaps the simplest way to express a multilevel model generally is by starting with the classical regression model fit to all the data, $y_i = \beta_0 + \beta_1 X_{i1} + \beta_2 X_{i2} + \cdots + \epsilon_i$, and then generalizing to allow the coefficients β to vary across groups; thus,

$$y_i = \beta_{0\,j[i]} + \beta_{1\,j[i]} X_{i1} + \beta_{2\,j[i]} X_{i2} + \cdots + \epsilon_i.$$

The "multilevel" part of the model involves assigning a multivariate distribution to the vector of β's within each group, as we discuss in Section 13.1.

For now we will focus on *varying-intercept models*, in which the only coefficient that varies across groups is the constant term β_0 (which, to minimize subscripting, we label α). For the radon data that include the floor and a county-level uranium predictor, the model then becomes

$$y_i = \alpha_{j[i]} + \beta_1 X_{i1} + \beta_2 X_{i2} + \epsilon_i$$

where X_{i1} is the ith element of the vector $X_{(1)}$ representing the first-floor indicators and X_{i2} is the ith element of the vector $X_{(2)}$ representing the uranium measurement in the county containing house i. We can also write this in matrix notation as

$$y_i = \alpha_{j[i]} + X_i \beta + \epsilon_i$$

with the understanding that X includes the first-floor indicator and the county uranium measurement but not the constant term. This is the way that models are built using `lmer()`, including all predictors at the individual level, as we discuss in Section 12.6.

The second level of the model is simply

$$\alpha_j \sim \text{N}(\mu_\alpha, \sigma_\alpha^2). \tag{12.5}$$

Group-level errors. The model (12.5) can also be written as

$$\alpha_j = \mu_\alpha + \eta_j, \quad \text{with } \eta_j \sim \text{N}(0, \sigma_\alpha^2). \tag{12.6}$$

The group-level errors η_j can be helpful in understanding the model; however, we often use the more compact notation (12.5) to reduce the profusion of notation. (We have also toyed with notation such as $\alpha_j = \mu^\alpha + \epsilon_j^\alpha$ in which ϵ is consistently used for regression errors—but the superscripts seem too confusing. As illustrated in Part 2B of this book, we sometimes use such notation when programming models in Bugs.)

Combining separate local regressions

An alternative way to write the multilevel model is as a linking of local regressions in each group. Within each group j, a regression is performed on the local predictors (in this case, simply the first-floor indicator, x_i), with a constant term α that is indexed by group:

$$\text{within county } j: \ y_i \sim \text{N}(\alpha_j + \beta x_i, \sigma_y^2), \quad \text{for } i = 1, \ldots, n_j. \tag{12.7}$$

The county uranium measurement has not yet entered the model since we are imagining separate regressions fit to each county—there would be no way to estimate the coefficient for a county-level predictor from any of these within-county regressions.

Instead, the county-level uranium level, u_j, is included as a predictor in the second level of the model:

$$\alpha_j \sim \text{N}(\gamma_0 + \gamma_1 u_j, \sigma_\alpha^2). \tag{12.8}$$

We can also write the distribution in (12.8) as $\text{N}(U_j \gamma, \sigma_\alpha^2)$, where U has two columns: a constant term, $U_{(0)}$, and the county-level uranium measurement, $U_{(1)}$. The errors in this model (with mean 0 and standard deviation σ_α) represent variation *among counties* that is not explained by the local and county-level predictors.

The multilevel model combines the J local regression models (12.7) in two ways: first, the local regression coefficients β are the same in all J models (an assumption we will relax in Section 13.1). Second, the different intercepts α_j are connected through the group-level model (12.8), with consequences to the coefficient estimates that we discuss in Section 12.6.

Group-level errors. We can write (12.8) as

$$\alpha_j = \gamma_0 + \gamma_1 u_j + \eta_j, \quad \text{with } \eta_j \sim N(0, \sigma_\alpha^2), \tag{12.9}$$

explicitly showing the errors in the county-level regression.

Modeling the coefficients of a large regression model

The identical model can be written as a single regression, in which the local and group-level predictors are combined into a single matrix X:

$$y_i \sim N(X_i \beta, \sigma_y^2), \tag{12.10}$$

where, for our example, X includes vectors corresponding to:

- A constant term, $X_{(0)}$;
- The floor where the measurement was taken, $X_{(1)}$;
- The county-level uranium measure, $X_{(2)}$;
- J (not $J-1$) county indicators, $X_{(3)}, \ldots, X_{(J+2)}$.

At the upper level of the model, the J county indicators (which in this case are $\beta_3, \ldots, \beta_{J+2}$) follow a normal distribution:

$$\beta_j \sim N(0, \sigma_\alpha^2), \quad \text{for } j = 3, \ldots, J+2. \tag{12.11}$$

In this case, we have centered the β_j distribution at 0 rather than at an estimated μ_β because any such μ_β would be statistically indistinguishable from the constant term in the regression. We return to this point shortly.

The parameters in the model (12.10)–(12.11) can be identified exactly with those in the separate local regressions above:

- The local predictor x in model (12.7) is the same as $X_{(1)}$ (the floor) here.
- The local errors ϵ_i are the same in the two models.
- The matrix of group-level predictors U in (12.8) is just $X_{(0)}$ here (the constant term) joined with $X_{(2)}$ (the uranium measure).
- The group-level errors $\eta_1, \ldots, \eta_J$ in (12.9) are identical to $\beta_3, \ldots, \beta_{J+2}$ here.
- The standard-deviation parameters σ_y and σ_α keep the same meanings in the two models.

Moving the constant term around. The multilevel model can be written in yet another equivalent way by moving the constant term:

$$\begin{aligned} y_i &= N(X_i \beta, \sigma_y^2), \quad \text{for } i = 1, \ldots, n \\ \beta_j &\sim N(\mu_\alpha, \sigma_\alpha^2), \quad \text{for } j = 3, \ldots, J+2. \end{aligned} \tag{12.12}$$

In this version, we have removed the constant term from X (so that it now has only $J+2$ columns) and replaced it by the equivalent term μ_α in the group-level model. The coefficients $\beta_3, \ldots, \beta_{J+2}$ for the group indicators are now centered around μ_α rather than 0, and are equivalent to $\alpha_1, \ldots, \alpha_J$ as defined earlier, for example, in model (12.9).

Regression with multiple error terms

Another option is to re-express model (12.10), treating the group-indicator coefficients as error terms rather than regression coefficients, in what is often called a "mixed effects" model popular in the social sciences:

$$y_i \sim N(X_i\beta + \eta_{j[i]}, \sigma_y^2), \text{ for } i = 1, \ldots, n$$
$$\eta_j \sim N(0, \sigma_\alpha^2), \tag{12.13}$$

where $j[i]$ represents the county that contains house i, and X now contains only three columns:

- A constant term, $X_{(0)}$;

- The floor, $X_{(1)}$;

- The county-level uranium measure, $X_{(2)}$.

This is the same as model (12.10)–(12.11), simply renaming some of the β_j's as η_j's. All our tools for multilevel modeling will automatically work for models with multiple error terms.

Large regression with correlated errors

Finally, we can express a multilevel model as a classical regression with correlated errors:

$$y_i = X_i\beta + \epsilon_i^{\text{all}}, \quad \epsilon^{\text{all}} \sim N(0, \Sigma), \tag{12.14}$$

where X is now the matrix with three predictors (the constant term, first-floor indicator, and county-level uranium measure) as in (12.13), but now the errors ϵ_i^{all} have an $n \times n$ covariance matrix Σ. The error ϵ_i^{all} in (12.14) is equivalent to the sum of the two errors, $\eta_{j[i]} + \epsilon_i$, in (12.13). The term $\eta_{j[i]}$, which is the same for all units i in group j, induces correlation in ϵ^{all}.

In multilevel models, Σ is parameterized in some way, and these parameters are estimated from the data. For the nested multilevel model we have been considering here, the variances and covariances of the n elements of ϵ^{all} can be derived in terms of the parameters σ_y and σ_α:

$$\text{For any unit } i: \quad \Sigma_{ii} = \text{var}(\epsilon_i^{\text{all}}) = \sigma_y^2 + \sigma_\alpha^2$$
$$\text{For any units } i, k \text{ within the same group } j: \quad \Sigma_{ik} = \text{cov}(\epsilon_i^{\text{all}}, \epsilon_k^{\text{all}}) = \sigma_\alpha^2$$
$$\text{For any units } i, k \text{ in different groups:} \quad \Sigma_{ik} = \text{cov}(\epsilon_i^{\text{all}}, \epsilon_k^{\text{all}}) = 0.$$

It can also be helpful to express Σ in terms of standard errors and correlations:

$$\text{sd}(\epsilon_i) = \sqrt{\Sigma_{ii}} = \sqrt{\sigma_y^2 + \sigma_\alpha^2}$$
$$\text{corr}(\epsilon_i, \epsilon_k) = \frac{\Sigma_{ik}}{\sqrt{\Sigma_{ii}\Sigma_{kk}}} = \begin{cases} \frac{\sigma_\alpha^2}{\sigma_y^2 + \sigma_\alpha^2} & \text{if } j[i] = j[k] \\ 0 & \text{if } j[i] \neq j[k]. \end{cases}$$

We generally prefer modeling the multilevel effects explicitly rather than burying them as correlations, but once again it is useful to see how the same model can be written in different ways.

12.6 Group-level predictors

Adding a group-level predictor to improve inference for group coefficients α_j

We continue with the radon example from Sections 12.2–12.3 to illustrate how a multilevel model handles predictors at the group as well as the individual levels.

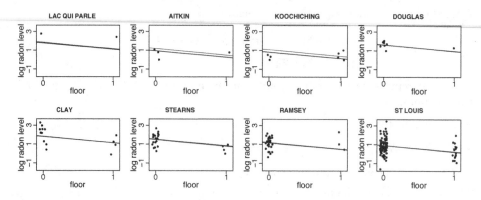

Figure 12.5 *Multilevel (partial pooling) regression lines $y = \alpha_j + \beta x$ fit to radon data, displayed for eight counties, including uranium as a county-level predictor. Light-colored lines show the multilevel estimates, without uranium as a predictor, from Figure 12.4.*

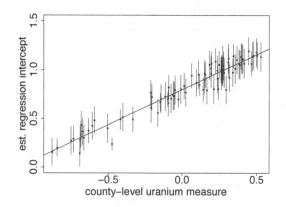

Figure 12.6 *Estimated county coefficients α_j (± 1 standard error) plotted versus county-level uranium measurement u_j, along with the estimated multilevel regression line $\alpha_j = \gamma_0 + \gamma_1 u_j$. The county coefficients roughly follow the line but not exactly; the deviation of the coefficients from the line is captured in σ_α, the standard deviation of the errors in the county-level regression.*

We use the formulation

$$y_i \sim \mathrm{N}(\alpha_{j[i]} + \beta x_i, \sigma_y^2), \text{ for } i = 1, \ldots, n$$
$$\alpha_j \sim \mathrm{N}(\gamma_0 + \gamma_1 u_j, \sigma_\alpha^2), \text{ for } j = 1, \ldots, J, \qquad (12.15)$$

where x_i is the house-level first-floor indicator and u_j is the county-level uranium measure.

R code
```
u.full <- u[county]
M2 <- lmer (y ~ x + u.full + (1 | county))
display (M2)
```

This model includes floor, uranium, and intercepts that vary by county. The `lmer()` function only accepts predictors at the individual level, so we have converted u_j to $u_i^{\text{full}} = u_{j[i]}$ (with the variable `county` playing the role of the indexing $j[i]$), to pull out the uranium level of the county where house i is located.

The display of the `lmer()` fit shows coefficients and standard errors, along with estimated residual variation at the county and individual ("residual") level:

```
lmer(formula = y ~ x + u.full + (1 | county))          R output
            coef.est coef.se
(Intercept)  1.47     0.04
x           -0.67     0.07
u.full       0.72     0.09
Error terms:
 Groups    Name        Std.Dev.
 county    (Intercept) 0.16
 Residual              0.76
# of obs: 919, groups: county, 85
deviance = 2122.9
```

As in our earlier example on page 261, we use `coef()` to pull out the estimated coefficients,

```
coef (M2)                                              R code
```

yielding

```
$county                                                R output
    (Intercept)      x u.full
1          1.45  -0.67   0.72
2          1.48  -0.67   0.72
 . . .
85         1.42  -0.67   0.72
```

Only the intercept varies, so the coefficients for x and u.full are the same for all 85 counties. (Actually, u.full is constant within counties so it cannot have a varying coefficient here.) On page 280 we shall see a similar display for a model in which the coefficient for x varies by county.

As before, we can also examine the estimated model averaging over the counties:

```
fixef (M2)                                             R code
```

yielding

```
(Intercept)             x      u.full                  R output
      1.47         -0.67        0.72
```

and the county-level errors:

```
ranef (M2)                                             R code
```

yielding

```
    (Intercept)                                        R output
1         -0.02
2          0.01
 . . .
85        -0.04
```

The results of `fixef()` and `ranef()` add up to the coefficients in `coef()`: for county 1, $1.47 - 0.02 = 1.45$, for county 2, $1.47 + 0.01 = 1.48$, ..., and for county 85, $1.47 - 0.04 = 1.42$ (up to rounding error).

Interpreting the coefficients within counties

We can add the unmodeled coefficients (the "fixed effects") to the county-level errors to get an intercept and slope for each county. We start with the model that averages over all counties, $y_i = 1.47 - 0.67x_i + 0.72u_{j[i]}$ (as obtained from `display(M2)` or `fixef(M2)`.

Now consider a particular county, for example county 85. We can determine its fitted regression line in two ways from the `lmer()` output, in each case using the log uranium level in county 85, $u_{85} = 0.36$.

First, using the the last line of the display of `coef(M2)`, the fitted model for county 85 is $y_i = 1.42 - 0.67x_i + 0.72u_{85} = (1.42 + 0.72 \cdot 0.36) - 0.67x_i = 1.68 - 0.67x_i$, that is, 1.68 for a house with a basement and 1.01 for a house with no basement. Exponentiating gives estimated geometric mean predictions of 5.4 pCi/L and 2.7 pCi/L for houses in county 85 with and without basements.

Alternatively, we can construct the fitted line for county 85 by starting with the results from `fixef(M2)`—that is, $y_i = 1.47 - 0.67x_i + 0.72u_{j[i]}$, setting $u_{j[i]} = u_{85} = 0.36$—and adding the group-level error from `ranef(M2)`, which for county 85 is -0.04. The resulting model is $y_i = 1.47 - 0.67x_i + 0.72 \cdot 0.36 - 0.04 = 1.68 - 0.67x_i$, the same as in the other calculation (up to rounding error in the last digit of the intercept).

Figure 12.5 shows the fitted line for each of a selection of counties, and Figure 12.6 shows the county-level regression, plotting the estimated coefficients α_j versus the county-level predictor u_j. These two figures represent the two levels of the multilevel model.

The group-level predictor has increased the precision of our estimates of the county intercepts α_j: the ± 1 standard-error bounds are narrower in Figure 12.6 than in Figure 12.3b, which showed α_j's estimated without the uranium predictor (note the different scales on the y-axes of the two plots and the different county variables plotted on the x-axes).

The estimated individual- and county-level standard deviations in this model are $\hat{\sigma}_y = 0.76$ and $\hat{\sigma}_\alpha = 0.16$. In comparison, these residual standard deviations were 0.76 and 0.33 without the uranium predictor. This predictor has left the within-county variation unchanged—which makes sense, since it is a county-level predictor which has no hope of explaining variation within any county—but has drastically reduced the unexplained variation between counties. In fact, the variance ratio is now only $\sigma_\alpha^2/\sigma_y^2 = 0.16^2/0.76^2 = 0.044$, so that the county-level model is as good as $1/0.044 = 23$ observations within any county. The multilevel estimates under this new model will be close to the complete-pooling estimates (with county-level uranium included as a predictor) for many of the smaller counties in the dataset because a county would have to have more than 23 observations to be pulled closer to the no-pooling estimate than the complete-pooling estimate.

Interpreting the coefficient of the group-level predictor

The line in Figure 12.6 shows the prediction of average log radon in a county (for homes with basements—that is, $x_i = 0$—since these are the intercepts α_j), as a function of the log uranium level in the county. This estimated group-level regression line has an estimated slope of about 0.7. Coefficients between 0 and 1 are typical in a log-log regression: in this case, each increase of 1% in uranium level corresponds to a 0.7% predicted increase in radon.

It makes sense that counties higher in uranium have higher radon levels, and it also makes sense that the slope is less than 1. Radon is affected by factors other than soil uranium, and the "uranium" variable in the dataset is itself an imprecise measure of actual soil uranium in the county, and so we would expect a 1% increase in the uranium variable to match to something less than a 1% increase in radon. Compared to classical regression, the estimation of this coefficient is trickier (since the α_j's—the "data" for the county-level regression—are not themselves observed) but the principles of interpretation do not change.

A multilevel model can include county indicators along with a county-level predictor

Users of multilevel models are often confused by the idea of including county indicators along with a county-level predictor. Is this possible? With 85 counties in the dataset, how can a regression fit 85 coefficients for counties, plus a coefficient for county-level uranium? This would seem to induce perfect collinearity into the regression or, to put it more bluntly, to attempt to learn more than the data can tell us. Is it really possible to estimate 86 coefficients from 85 data points?

The short answer is that we really have more than 85 data points. There are hundreds of houses with which to estimate the 85 county-level intercepts, and 85 counties with which to estimate the coefficient of county-level uranium. In a classical regression, however, the 85 county indicators and the county-level predictor would indeed be collinear. This problem is avoided in a multilevel model because of the partial pooling of the α_j's toward the group-level linear model. This is illustrated in Figure 12.6, which shows the estimates of all these 86 parameters—the 85 separate points and the slope of the line. In this model that includes a group-level predictor, the estimated intercepts are pulled toward this group-level regression line (rather than toward a constant, as in Figure 12.3b). The county-level uranium predictor u_j thus helps us estimate the county intercepts α_j but without overwhelming the information in individual counties.

Partial pooling of group coefficients α_j in the presence of group-level predictors

Equation (12.4) on page 258 gives the formula for partial pooling in the simple model with no group-level predictors. Once we add a group-level regression, $\alpha_j \sim N(U_j\gamma, \sigma_\alpha^2)$, the parameters α_j are shrunk toward their regression estimates $\hat{\alpha}_j = U_j\gamma$. Equivalently, we can say that the group-level errors η_j (in the model $\alpha_j = U_j\gamma + \eta_j$) are shrunk toward 0. As always, there is more pooling when the group-level standard deviation σ_α is small, and more smoothing for groups with fewer observations. The multilevel estimate of α_j is a weighted average of the no-pooling estimate for its group ($\bar{y}_j - \overline{X}_j\beta$) and the regression prediction $\hat{\alpha}_j$:

$$\text{estimate of } \alpha_j \quad \approx \quad \frac{\frac{n_j}{\sigma_y^2}}{\frac{n_j}{\sigma_y^2} + \frac{1}{\sigma_\alpha^2}} \cdot (\text{estimate from group } j) +$$

$$+ \frac{\frac{1}{\sigma_\alpha^2}}{\frac{n_j}{\sigma_y^2} + \frac{1}{\sigma_\alpha^2}} \cdot (\text{estimate from regression}). \qquad (12.16)$$

Equivalently, the group-level errors η_j are partially pooled toward zero:

$$\text{estimate of } \eta_j \approx \frac{\frac{n_j}{\sigma_y^2}}{\frac{n_j}{\sigma_y^2} + \frac{1}{\sigma_\alpha^2}}(\bar{y}_j - \overline{X}_j\beta - U_j\gamma) + \frac{\frac{1}{\sigma_\alpha^2}}{\frac{n_j}{\sigma_y^2} + \frac{1}{\sigma_\alpha^2}} \cdot 0.$$

12.7 Model building and statistical significance

From classical to multilevel regression

When confronted with a multilevel data structure, such as the radon measurements considered here or the examples in the previous chapter, we typically start by fitting some simple classical regressions and then work our way up to a full multilevel model. The four natural starting points are:

- Complete-pooling model: a single classical regression completely ignoring the group information—that is, a single model fit to all the data, perhaps including group-level predictors but with no coefficients for group indicators.

- No-pooling model: a single classical regression that includes group indicators (but no group-level predictors) but with no model for the group coefficients.

- Separate models: a separate classical regression in each group. This approach is not always possible if there are groups with small sample sizes. (For example, in Figure 12.4 on page 257, Aitkin County has three measurements in homes with basements and one in a home with no basement. If the sample from Aitkin County had happened to contain only houses with basements, then it would be impossible to estimate the slope β from this county alone.)

- Two-step analysis: starting with either the no-pooling or separate models, then fitting a classical group-level regression using, as "data," the estimated coefficients for each group.

Each of these simpler models can be informative in its own right, and they also set us up for understanding the partial pooling in a multilevel model, as in Figure 12.4.

For large datasets, fitting a model separately in each group can be computationally efficient as well. One might imagine an iterative procedure that starts by fitting separate models, continues with the two-step analysis, and then returns to fitting separate models, but using the resulting group-level regression to guide the estimates of the varying coefficients. Such a procedure, if formalized appropriately, is in fact the usual algorithm used to fit multilevel models, as we discuss in Chapter 17.

When is multilevel modeling most effective?

Multilevel model is most important when it is close to complete pooling, at least for some of the groups (as for Lac Qui Parle County in Figure 12.4 on page 257). In this setting we can allow estimates to vary by group while still estimating them precisely. As can be seen from formula (12.16), estimates are more pooled when the group-level standard deviation σ_α is small, that is, when the groups are similar to each other. In contrast, when σ_α is large, so that groups vary greatly, multilevel modeling is not much better than simple no-pooling estimation.

At this point, it might seem that we are contradicting ourselves. Earlier we motivated multilevel modeling as a compromise between no pooling and complete pooling, but now we are saying that multilevel modeling is effective when it is close to complete pooling, and ineffective when it is close to no pooling. If this is so, why not just always use the complete-pooling estimate?

We answer this question in two ways. First, when the multilevel estimate is close to complete pooling, it still allows variation between groups, which can be important, in fact can be one of the goals of the study. Second, as in the radon example, the multilevel estimate can be close to complete pooling for groups with small sample size and close to no pooling for groups with large sample size, automatically performing well for both sorts of group.

Using group-level predictors to make partial pooling more effective

In addition to being themselves of interest, group-level predictors play a special role in multilevel modeling by reducing the unexplained group-level variation and thus reducing the group-level standard deviation σ_α. This in turn increases the amount of pooling done by the multilevel estimate (see formula (12.16)), giving more precise estimates of the α_j's, especially for groups for which the sample size n_j is small. Following the template of classical regression, multilevel modeling typically proceeds by adding predictors at the individual and group levels and reducing the unexplained variance at each level. (However, as discussed in Section 21.7, adding a group-level predictor can actually increase the unexplained variance in some situations.)

Statistical significance

It is *not* appropriate to use statistical significance as a criterion for including particular group indicators in a multilevel model. For example, consider the simple varying-intercept radon model with no group-level predictor, in which the average intercept μ_α is estimated at 1.46, and the within-group intercepts α_j are estimated at $1.46 - 0.27 \pm 0.25$ for county 1, $1.46 - 0.53 \pm 0.10$ for county 2, $1.46 + 0.02 \pm 0.28$ for county 3, and so forth (see page 261).

County 1 is thus approximately 1 standard error away from the average intercept of 1.46, county 2 is more than 4 standard errors away, ... and county 85 is less than 1 standard error away. Of these three counties, only county 2 would be considered "statistically significantly" different from the average.

However, we should include all 85 counties in the model, and nothing is lost by doing so. The purpose of the multilevel model is not to see whether the radon levels in county 1 are statistically significantly different from those in county 2, or from the Minnesota average. Rather, we seek the best possible estimate in each county, with appropriate accounting for uncertainty. Rather than make some significance threshold, we allow all the intercepts to vary and recognize that we may not have much precision in many of the individual groups. We illustrate this point in another example in Section 21.8.

The same principle holds for the models discussed in the following chapters, which include varying slopes, non-nested levels, discrete data, and other complexities. Once we have included a source of variation, we do not use statistical significance to pick and choose indicators to include or exclude from the model.

In practice, our biggest constraints—the main reasons we do not use extremely elaborate models in which all coefficients can vary with respect to all grouping factors—are fitting and understanding complex models. The lmer() function works well when it works, but it can break down for models with many grouping factors. Bugs is more general (see Part 2B of this book) but can be slow with large datasets or complex models. In the meantime we need to start simple and build up gradually, a process during which we can also build understanding of the models being fit.

12.8 Predictions for new observations and new groups

Predictions for multilevel models can be more complicated than for classical regression because we can apply the model to existing groups or new groups. After a brief review of classical regression prediction, we explain in the context of the radon model.

Review of prediction for classical regression

In classical regression, prediction is simple: specify the predictor matrix $\tilde{X}$ for a set of new observations[5] and then compute the linear predictor $\tilde{X}\beta$, then simulate the predictive data:

- For linear regression, simulate independent normal errors $\tilde{\epsilon}_i$ with mean 0 and standard deviation σ, and compute $\tilde{y} = \tilde{X}\beta + \tilde{\epsilon}$; see Section 7.2.

- For logistic regression, simulate the predictive binary data: $\Pr(\tilde{y}_i) = \text{logit}^{-1}(\tilde{X}_i\beta)$ for each new data point i; see Section 7.4.

- With binomial logistic regression, specify the number of tries $\tilde{n}_i$ for each new unit i, and simulate $\tilde{y}_i$ from the binomial distribution with parameters $\tilde{n}_i$ and $\text{logit}^{-1}(\tilde{X}_i\beta)$; see Section 7.4.

- With Poisson regression, specify the exposures $\tilde{u}_i$ for the new units, and simulate $\tilde{y}_i \sim \text{Poisson}(\tilde{u}_i e^{\tilde{X}_i\beta})$ for each new i; see Section 7.4.

As discussed in Section 7.2, the estimation for a regression in R gives a set of n_{sims} simulation draws. Each of these is used to simulate the predictive data vector $\tilde{y}$, yielding a set of n_{sims} simulated predictions. For example, in the election forecasting example of Figure 7.5 on page 146:

R code
```
model.1 <- lm (vote.88 ~ vote.86 + party.88 + inc.88)
display (model.1)
n.sims <- 1000
sim.1 <- sim (model.1, n.sims)
beta.sim <- sim.1$beta
sigma.sim <- sim.1$sigma
n.tilde <- length (vote.88)
X.tilde <- cbind (rep(1,n.tilde), vote.88, party.90, inc.90)
y.tilde <- array (NA, c(n.sims, n.tilde))
for (s in 1:n.sims) {
  y.tilde[s,] <- rnorm (n.tilde, X.tilde%*%beta.sim[s,], sigma.sim[s])
}
```

This matrix of simulations can be used to get point predictions (for example, `median(y.tilde[,3])` gives the median estimate for $\tilde{y}_3$) or predictive intervals (for example, `quantile(y.tilde[,3],c(.025,.975))`) for individual data points or for more elaborate derived quantities, such as the predicted number of seats won by the Democrats in 1990 (see the end of Section 7.3). For many applications, the `predict()` function in R is a good way to quickly get point predictions and intervals (see page 48); here we emphasize the more elaborate simulation approach which allows inferences for arbitrary quantities.

---

[5] Predictions are more complicated for time-series models: even when parameters are fit by classical regression, predictions must be made sequentially. See Sections 8.4 and 24.2 for examples.

Prediction for a new observation in an existing group

We can make two sorts of predictions for the radon example: predicting the radon level for a new house within one of the counties in the dataset, and for a new house in a new county. We shall work with model (12.15) on page 266, with floor as an individual-level predictor and uranium as a group-level predictor

For example, suppose we wish to predict $\tilde{y}$, the log radon level for a house with no basement (thus, with radon measured on the first floor, so that $\tilde{x} = 1$) in Hennepin County ($j = 26$ of our Minnesota dataset). Conditional on the model parameters, the predicted value has a mean of $\alpha_{26} + \beta$ and a standard deviation of σ_y. That is,

$$\tilde{y}|\theta \sim N(\alpha_{26} + \beta\tilde{x}, \sigma_y^2),$$

where we are using θ to represent the entire vector of model parameters.

Given estimates of α, β, and σ_y, we can create a predictive simulation for $\tilde{y}$ using R code such as

```
x.tilde <- 1
sigma.y.hat <- sigma.hat(M2)$sigma$data
coef.hat <- as.matrix(coef(M2)$county)[26,]
y.tilde <- rnorm (1, coef.hat %*% c(1, x.tilde, u[26]), sigma.y.hat)
```
R code

More generally, we can create a vector of `n.sims` simulations to represent the predictive uncertainty in $\tilde{y}$:

```
n.sims <- 1000
coef.hat <- as.matrix(coef(M2)$county)[26,]
y.tilde <- rnorm (1000, coef.hat %*% c(1, x.tilde, u[26]), sigma.y.hat)
```
R code

Still more generally, we can add in the inferential uncertainty in the estimated parameters, α, β, and σ. For our purposes here, however, we shall ignore inferential uncertainty and just treat the parameters $\alpha, \beta, \sigma_y, \sigma_\alpha$ as if they were estimated perfectly from the data.[6] In that case, the computation gives us 1000 simulation draws of $\tilde{y}$, which we can summarize in various ways. For example,

```
quantile (y.tilde, c(.25,.5,.75))
```
R code

gives us a predictive median of 0.76 and a 50% predictive interval of $[0.26, 1.27]$. Exponentiating gives us a prediction on the original (unlogged) scale of $\exp(0.76) = 2.1$, with a 50% interval of $[1.3, 3.6]$.

For some applications we want the average, rather than the median, of the predictive distribution. For example, the expected risk from radon exposure is proportional to the predictive average or mean, which we can compute directly from the simulations:

```
unlogged <- exp(y.tilde)
mean (unlogged)
```
R code

In this example, the predictive mean is 2.9, which is a bit higher than the median of 2.1. This makes sense: on the unlogged scale, this predictive distribution is skewed to the right.

---

[6] One reason we picked Hennepin County ($j = 26$) for this example is that, with a sample size of 105, its average radon level is accurately estimated from the available data.

Prediction for a new observation in a new group

Now suppose we want to predict the radon level for a house, once again with no basement, but this time in a county not included in our analysis. We then must generate a new county-level error term, $\tilde{\alpha}$, which we sample from its $N(\gamma_0 + \gamma_1 \tilde{u}_j, \sigma_\alpha^2)$ distribution. We shall assume the new county has a uranium level equal to the average of the uranium levels in the observed counties:

R code
```
u.tilde <- mean (u)
```

grab the estimated $\gamma_0, \gamma_1, \sigma_\alpha$ from the fitted model:

R code
```
g.0.hat <- fixef(M2)["(Intercept)"]
g.1.hat <- fixef(M2)["u.full"]
sigma.a.hat <- sigma.hat(M2)$sigma$county
```

and simulate possible intercepts for the new county:

R code
```
a.tilde <- rnorm (n.sims, g.0.hat + g.1.hat*u.tilde, sigma.a.hat)
```

We can then simulate possible values of the radon level for the new house in this county:

R code
```
y.tilde <- rnorm (n.sims, a.tilde + b.hat*x.tilde, sigma.y.hat)
```

Each simulation draw of $\tilde{y}$ uses a different simulation of $\tilde{\alpha}$, thus propagating the uncertainty about the new county into the uncertainty about the new house in this county.

Comparison of within-group and between-group predictions. The resulting prediction will be more uncertain than for a house in a known county, since we have no information about $\tilde{\alpha}$. Indeed, the predictive 50% interval of this new $\tilde{y}$ is $[0.28, 1.34]$, which is slightly wider than the predictive interval of $[0.26, 1.27]$ for the new house in county 26. The interval is only slightly wider because the within-county variation in this particular example is much higher than the between-county variation.

More specifically, from the fitted model on page 266, the within-county (residual) standard deviation σ_y is estimated at 0.76, and the between-county standard deviation σ_α is estimated at 0.16. The log radon level for a new house in an already-measured county can then be measured to an accuracy of about ± 0.76. The log radon level for a new house in a new county can be predicted to an accuracy of about $\pm\sqrt{0.76^2 + 0.16^2} = \pm 0.78$. The ratio $0.78/0.76$ is 1.03, so we would expect the predictive interval for a new house in a new county to be about 3% wider than for a new house in an already-measured county. The change in interval width is small here because the unexplained between-county variance is so small in this dataset.

For another example, the 50% interval for the log radon level of a house with no basement in county 2 is $[0.28, 1.30]$, which is centered in a different place but also is narrower than the predictive interval for a new county.

Nonlinear predictions

Section 7.3 illustrated the use of simulation for nonlinear predictions from classical regression. We can perform similar calculations in multilevel models. For example, suppose we are interested in the average radon level among all the houses in Hennepin County ($j = 26$). We can perform this inference using poststratification, first estimating the average radon level of the houses with and without basements in the county, then weighting these by the proportion of houses in the county that have basements. We can look up this proportion from other data sources on homes, or we can estimate it from the available sample data.

For our purposes here, we shall assume that 90% of all the houses in Hennepin County have basements. The average radon level of all the houses in the county is then 0.1 times the average for the houses in Hennepin County without basements, plus 0.9 times the average for those with basements. To simulate in R:

```
y.tilde.basement <- rnorm (n.sims, a.hat[26], sigma.y.hat)
y.tilde.nobasement <- rnorm (n.sims, a.hat[26] + b.hat, sigma.y.hat)
```
R code

We then compute the estimated mean for 1000 houses of each type in the county (first exponentiating since our model was on the log scale):

```
mean.radon.basement <- mean (exp (y.tilde.basement))
mean.radon.nobasement <- mean (exp (y.tilde.nobasement))
```
R code

and finally poststratify given the proportion of houses of each type in the county:

```
mean.radon <- .9*mean.radon.basement + .1*mean.radon.basement
```
R code

In Section 16.6 we return to the topic of predictions, using simulations from Bugs to capture the uncertainty in parameter estimates and then propagating inferential uncertainty into the predictions, rather than simply using point estimates a.hat, b.hat, and so forth.

12.9 How many groups and how many observations per group are needed to fit a multilevel model?

Advice is sometimes given that multilevel models can only be used if the number of groups is higher than some threshold, or if there is some minimum number of observations per groups. Such advice is misguided. Multilevel modeling includes classical regression as a limiting case (complete pooling when group-level variances are zero, no pooling when group-level variances are large). When sample sizes are small, the key concern with multilevel modeling is the estimation of variance parameters, but it should still work at least as well as classical regression.

How many groups?

When J, the number of groups, is small, it is difficult to estimate the between-group variation and, as a result, multilevel modeling often adds little in such situations, beyond classical no-pooling models. The difficulty of estimating variance parameters is a technical issue to which we return in Section 19.6; to simplify, when σ_α cannot be estimated well, it tends to be overestimated, and so the partially pooled estimates are close to no pooling (this is what happens when σ_α has a high value in (12.16) on page 269).

At the same time, multilevel modeling should not do any worse than no-pooling regression and sometimes can be easier to interpret, for example because one can include indicators for all J groups rather than have to select one group as a baseline category.

One or two groups

With only one or two groups, however, multilevel modeling reduces to classical regression (unless "prior information" is explicitly included in the model; see Section 18.3). Here we usually express the model in classical form (for example, including a single predictor for `female`, rather than a multilevel model for the two levels of the `sex` factor).

Even with only one or two groups in the data, however, multilevel models can be useful for making predictions about new groups. See also Sections 21.2–22.5 for further connections between classical and multilevel models, and Section 22.6 for hierarchical models for improving estimates of variance parameters in settings with many grouping factors but few levels per factor.

How many observations per group?

Even two observations per group is enough to fit a multilevel model. It is even acceptable to have one observation in many of the groups. When groups have few observations, their α_j's won't be estimated precisely, but they can still provide partial information that allows estimation of the coefficients and variance parameters of the individual- and group-level regressions.

Larger datasets and more complex models

As more data arise, it makes sense to add parameters to a model. For example, consider a simple medical study, then separate estimates for men and women, other demographic breakdowns, different regions of the country, states, smaller geographic areas, interactions between demographic and geographic categories, and so forth. As more data become available it makes sense to estimate more. These complexities are latent everywhere, but in small datasets it is not possible to learn so much, and it is not necessarily worth the effort to fit a complex model when the resulting uncertainties will be so large.

12.10 Bibliographic note

Multilevel models have been used for decades in agriculture (Henderson, 1950, 1984, Henderson et al., 1959, Robinson, 1991) and educational statistics (Novick et al., 1972, 1973, Bock, 1989), where it is natural to model animals in groups and students in classrooms. More recently, multilevel models have become popular in many social sciences and have been reviewed in books by Longford (1993), Goldstein (1995), Kreft and De Leeuw (1998), Snijders and Bosker (1999), Verbeke and Molenberghs (2000), Leyland and Goldstein (2001), Hox (2002), and Raudenbush and Bryk (2002). We do not attempt to trace here the many applications of multilevel models in various scientific fields.

It might also be useful to read up on Bayesian inference to understand the theoretical background behind multilevel models.[7] Box and Tiao (1973) is a classic reference that focuses on linear models. It predates modern computational methods but might be useful for understanding the fundamentals. Gelman et al. (2003) and Carlin and Louis (2000) cover applied Bayesian inference including the basics of multilevel modeling, with detailed discussions of computational algorithms. Berger

---

[7] As we discuss in Section 18.3, multilevel inferences can be formulated non-Bayesianly; however, understanding the Bayesian derivations should help with the other approaches too. All multilevel models are Bayesian in the sense of assigning probability distributions to the varying regression coefficients. The distinction between Bayesian and non-Bayesian multilevel models arises only for the question of modeling the other parameters—the nonvarying coefficients and the variance parameters—and this is typically a less important issue, especially when the number of groups is large.

(1985) and Bernardo and Smith (1994) cover Bayesian inference from two different theoretical perspectives.

The R function lmer() is described by Bates (2005a, b) and was developed from the linear and nonlinear mixed effects software described in Pinheiro and Bates (2000).

Multilevel modeling used to be controversial in statistics; see, for example, the discussions of the papers by Lindley and Smith (1972) and Rubin (1980) for some sense of the controversy.

The Minnesota radon data were analyzed by Price, Nero, and Gelman (1996); see also Price and Gelman (2004) for more on home radon modeling.

Statistical researchers have studied partial pooling in many ways; see James and Stein (1960), Efron and Morris (1979), DuMouchel and Harris (1983), Morris (1983), and Stigler (1983). Louis (1984), Shen and Louis (1998), Louis and Shen (1999), and Gelman and Price (1999) discuss some difficulties in the interpretation of partially pooled estimates. Zaslavsky (1993) discusses adjustments for undercount in the U.S. Census from a partial-pooling perspective. Normand, Glickman, and Gatsonis (1997) discuss the use of multilevel models for evaluating health-care providers.

12.11 Exercises

1. Using data of your own that are appropriate for a multilevel model, write the model in the five ways discussed in Section 12.5.

2. Continuing with the analysis of the CD4 data from Exercise 11.4:

   (a) Write a model predicting CD4 percentage as a function of time with varying intercepts across children. Fit using lmer() and interpret the coefficient for time.

   (b) Extend the model in (a) to include child-level predictors (that is, group-level predictors) for treatment and age at baseline. Fit using lmer() and interpret the coefficients on time, treatment, and age at baseline.

   (c) Investigate the change in partial pooling from (a) to (b) both graphically and numerically.

   (d) Compare results in (b) to those obtained in part (c).

3. Predictions for new observations and new groups:

   (a) Use the model fit from Exercise 12.2(b) to generate simulation of predicted CD4 percentages for each child in the dataset at a hypothetical next time point.

   (b) Use the same model fit to generate simulations of CD4 percentages at each of the time periods for a new child who was 4 years old at baseline.

4. Posterior predictive checking: continuing the previous exercise, use the fitted model from Exercise 12.2(b) to simulate a new dataset of CD4 percentages (with the same sample size and ages of the original dataset) for the final time point of the study, and record the average CD4 percentage in this sample. Repeat this process 1000 times and compare the simulated distribution to the observed CD4 percentage at the final time point for the actual data.

5. Using the radon data, include county sample size as a group-level predictor and write the varying-intercept model. Fit this model using lmer().

6. Return to the beauty and teaching evaluations introduced in Exercise 3.5 and 4.8.

(a) Write a varying-intercept model for these data with no group-level predictors. Fit this model using `lmer()` and interpret the results.

(b) Write a varying-intercept model that you would like to fit including three group-level predictors. Fit this model using `lmer()` and interpret the results.

(c) How does the variation in average ratings across instructors compare to the variation in ratings across evaluators for the same instructor?

7. This exercise will use the data you found for Exercise 4.7. This time, rather than repeating the same analysis across each year, or country (or whatever group the data varies across), fit a multilevel model using `lmer()` instead. Compare the results to those obtained in your earlier analysis.

8. Simulate data (outcome, individual-level predictor, group indicator, and group-level predictor) that would be appropriate for a multilevel model. See how partial pooling changes as you vary the sample size in each group and the number of groups.

9. Number of observations and number of groups:

(a) Take a simple random sample of one-fifth of the radon data. (You can create this subset using the `sample()` function in R.) Fit the varying-intercept model with floor as an individual-level predictor and log uranium as a county-level predictor, and compare your inferences to what was obtained by fitting the model to the entire dataset. (Compare inferences for the individual- and group-level standard deviations, the slopes for floor and log uranium, the average intercept, and the county-level intercepts.)

(b) Repeat step (a) a few times, with a different random sample each time, and summarize how the estimates vary.

(c) Repeat step (a), but this time taking a cluster sample: a random sample of one-fifth of the counties, but then all the houses within each sampled county.

## CHAPTER 13

Multilevel linear models: varying slopes, non-nested models, and other complexities

This chapter considers some generalizations of the basic multilevel regression. Models in which slopes and intercepts can vary by group (for example, $y_i = \alpha_{j[i]} + \beta_{j[i]} x_i + \cdots$, where α and β both vary by group j; see Figure 11.1c on page 238) can also be interpreted as interactions of the group index with individual-level predictors.

Another direction is non-nested models, in which a given dataset can be structured into groups in more than one way. For example, persons in a national survey can be divided by demographics or by states. Responses in a psychological experiment might be classified by person (experimental subject), experimental condition, and time.

The chapter concludes with some examples of models with nonexchangeable multivariate structures. We continue with generalized linear models in Chapters 14–15 and discuss how to fit all these models in Chapters 16–19.

13.1 Varying intercepts and slopes

The next step in multilevel modeling is to allow more than one regression coefficient to vary by group. We shall illustrate with the radon model from the previous chapter, which is relatively simple because it only has a single individual-level predictor, x (the indicator for whether the measurement was taken on the first floor).

We begin with a varying-intercept, varying-slope model including x but without the county-level uranium predictor; thus,

$$y_i \sim \mathrm{N}(\alpha_{j[i]} + \beta_{j[i]} x_i, \sigma_y^2), \text{ for } i = 1, \ldots, n$$

$$\begin{pmatrix} \alpha_j \\ \beta_j \end{pmatrix} \sim \mathrm{N}\left(\begin{pmatrix} \mu_\alpha \\ \mu_\beta \end{pmatrix}, \begin{pmatrix} \sigma_\alpha^2 & \rho\sigma_\alpha\sigma_\beta \\ \rho\sigma_\alpha\sigma_\beta & \sigma_\beta^2 \end{pmatrix} \right), \text{ for } j = 1, \ldots, J, \quad (13.1)$$

with variation in the α_j's and the β_j's and also a between-group correlation parameter ρ. In R:

```
M3 <- lmer (y ~ x + (1 + x | county))
display (M3)
```
R code

which yields

```
lmer(formula = y ~ x + (1 + x | county))
            coef.est coef.se
(Intercept)  1.46    0.05
x           -0.68    0.09
Error terms:
 Groups    Name        Std.Dev. Corr
 county    (Intercept) 0.35
           x           0.34     -0.34
```
R output

```
Residual                    0.75
# of obs: 919, groups: county, 85
deviance = 2161.1
```

In this model, the unexplained within-county variation has an estimated standard deviation of $\hat{\sigma}_y = 0.75$; the estimated standard deviation of the county intercepts is $\hat{\sigma}_\alpha = 0.35$; the estimated standard deviation of the county slopes is $\hat{\sigma}_\beta = 0.34$; and the estimated correlation between intercepts and slopes is -0.34.

We then can type

R code   `coef (M3)`

to yield

R output
```
$county
        (Intercept)      x
1            1.14 -0.54
2            0.93 -0.77
3            1.47 -0.67
    . . .
85           1.38 -0.65
```

Or we can separately look at the estimated population mean coefficients μ_α, μ_β and then the estimated errors for each county. First, we type

R code   `fixef (M3)`

to see the estimated average coefficients ("fixed effects"):

R output
```
(Intercept)            x
       1.46        -0.68
```

Then, we type

R code   `ranef (M3)`

to see the estimated group-level errors ("random effects"):

R output
```
        (Intercept)        x
1            -0.32   0.14
2            -0.53  -0.09
3             0.01   0.01
    . . .
85           -0.08   0.03
```

We can regain the estimated intercept and slope α_j, β_j for each county by simply adding the errors to μ_α and μ_β; thus, the estimated regression line for county 1 is $(1.46 - 0.32) + (-0.68 + 0.14)x = 1.14 - 0.54x$, and so forth.

The group-level model for the parameters (α_j, β_j) allows for partial pooling in the estimated intercepts and slopes. Figure 13.1 shows the results—the estimated lines $y = \alpha_j + \beta_j x$—for the radon data in eight different counties.

Including group-level predictors

We can expand the model of (α, β) in (13.1) by including a group-level predictor (in this case, soil uranium):

$$\begin{pmatrix} \alpha_j \\ \beta_j \end{pmatrix} \sim \text{N}\left(\begin{pmatrix} \gamma_0^\alpha + \gamma_1^\alpha u_j \\ \gamma_0^\beta + \gamma_1^\beta u_j \end{pmatrix}, \begin{pmatrix} \sigma_\alpha^2 & \rho\sigma_\alpha\sigma_\beta \\ \rho\sigma_\alpha\sigma_\beta & \sigma_\beta^2 \end{pmatrix} \right), \text{ for } j = 1, \ldots, J. \quad (13.2)$$

The resulting estimates for the α_j's and β_j's are changed slightly from what is displayed in Figure 13.1, but more interesting are the second-level models themselves, whose estimates are shown in Figure 13.2. Here is the result of fitting the model in R:

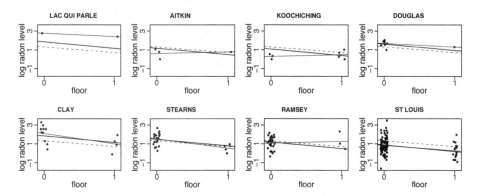

Figure 13.1 *Multilevel (partial pooling) regression lines* $y = \alpha_j + \beta_j x$, *displayed for eight counties* j. *In this model, both the intercept and the slope vary by county. The light solid and dashed lines show the no-pooling and complete pooling regression lines. Compare to Figure 12.4, in which only the intercept varies.*

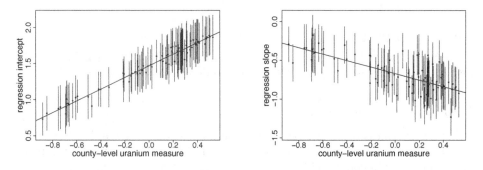

Figure 13.2 *(a) Estimates* $\pm$ *standard errors for the county intercepts* α_j, *plotted versus county-level uranium measurement* u_j, *along with the estimated multilevel regression line,* $\alpha = \gamma_0^\alpha + \gamma_1^\alpha u$. *(b) Estimates* $\pm$ *standard errors for the county slopes* β_j, *plotted versus county-level uranium measurement* u_j, *along with the estimated multilevel regression line,* $\beta = \gamma_0^\beta + \gamma_1^\beta u$. *Estimates and standard errors are the posterior medians and standard deviations, respectively. For each graph, the county coefficients roughly follow the line but not exactly; the discrepancies of the coefficients from the line are summarized by the county-level standard-deviation parameters* $\sigma_\alpha, \sigma_\beta$.

```
lmer(formula = y ~ x + u.full + x:u.full + (1 + x | county))          R output
             coef.est coef.se
(Intercept)   1.47     0.04
x            -0.67     0.08
u.full        0.81     0.09
x:u.full     -0.42     0.23
Error terms:
 Groups    Name        Std.Dev. Corr
 county    (Intercept) 0.12
           x           0.31     0.41
 Residual              0.75
# of obs: 919, groups: county, 85
deviance = 2114.3
```

The parameters $\gamma_0^\alpha, \gamma_0^\beta, \gamma_1^\alpha, \gamma_1^\beta$ in model (13.2) are the coefficients for the intercept, `x`, `u.full`, and `x:u.full`, respectively, in the regression. In particular, the interaction corresponds to allowing uranium to be a predictor in the regression for the slopes.

The estimated coefficients in each group (from `coef(M4)`) are:

R output
```
$county
     (Intercept)     x u.full x:u.full
1           1.46 -0.65   0.81    -0.42
2           1.50 -0.89   0.81    -0.42
. . .
85          1.44 -0.70   0.81    -0.42
```

Or we can display the average coefficients (using `fixef(M4)`):

R output
```
(Intercept)            x     u.full     x:u.full
       1.47        -0.67       0.81        -0.42
```

and the group-level errors for the intercepts and slopes (using `ranef(M4)`):

R output
```
     (Intercept)     x
1          -0.01  0.02
2           0.03 -0.21
. . .
85         -0.02 -0.03
```

The coefficients for the intercept and x vary, as specified in the model. This can be compared to the model on page 267 in which only the intercept varies.

Going from lmer output to intercepts and slopes

As before, we can combine the average coefficients with the group-level errors to compute the intercepts α_j and slopes β_j of model (13.2). For example, the fitted regression model in county 85 is $y_i = 1.47 - 0.67x_i + 0.81u_{85} - 0.42x_iu_{85} - 0.02 - 0.03x_i$. The log uranium level in county 85, u_{85}, is 0.36, and so the fitted regression line in county 85 is $y_i = 1.73 - 0.85x_i$. More generally, we can compute a vector of county intercepts α and slopes β:

R code
```
a.hat.M4 <- coef(M4)[,1] + coef(M4)[,3]*u
b.hat.M4 <- coef(M4)[,2] + coef(M4)[,4]*u
```

Here it is actually useful to have the variable u defined at the county level (as compared to `u.full = u[county]` which was used in the `lmer()` call). We next consider these linear transformations algebraically.

Varying slopes as interactions

Section 12.5 gave multiple ways of writing the basic multilevel model. These same ideas apply to models with varying slopes, which can be considered as interactions between group indicators and an individual-level predictor. For example, consider the model with an individual-level predictor x_i and a group-level predictor u_j,

$$
\begin{aligned}
y_i &= \alpha_{j[i]} + \beta_{j[i]}x_i + \epsilon_i \\
\alpha_j &= \gamma_0^\alpha + \gamma_1^\alpha u_j + \eta_j^\alpha \\
\beta_j &= \gamma_0^\beta + \gamma_1^\beta u_j + \eta_j^\beta .
\end{aligned}
$$

We can re-express this as a single model by substituting the formulas for α_j and β_j into the equation for y_i:

$$
y_i = \left[\gamma_0^\alpha + \gamma_1^\alpha u_{j[i]} + \eta_{j[i]}^\alpha \right] + \left[\gamma_0^\beta + \gamma_1^\beta u_{j[i]} + \eta_{j[i]}^\beta \right] x_i + \epsilon_i . \tag{13.3}
$$

This expression looks messy but it is really just a regression including various interactions. If we define a new individual-level predictor $v_i = u_{j[i]}$ (in the radon example, this is the uranium level in the county where your house is located), we can re-express (13.3) term by term as

$$y_i = a + bv_i + c_{j[i]} + dx_i + ev_ix_i + f_{j[i]}x_i + \epsilon_i.$$

This can be thought of in several ways:

- A varying-intercept, varying-slope model with four individual-level predictors (the constant term, v_i, x_i, and the interaction v_ix_i) and varying intercepts and slopes that are centered at zero.

- A regression model with $4 + 2J$ predictors: the constant term, v_i, x_i, v_ix_i, indicators for the J groups, and interactions between x and the J group indicators.

- A regression model with four predictors and three error terms.

- Or, to go back to the original formulation, a varying-intercept, varying-slope model with one group-level predictor.

Which of these expressions is most useful depends on the context. In the radon analysis, where the goal is to predict radon levels in individual counties, the varying-intercept, varying-slope formulation, as pictured in Figure 13.2, seems most appropriate. But in a problem where interest lies in the regression coefficients for x_i, u_j, and their interaction, it can be more helpful to focus on these predictors and consider the unexplained variation in intercepts and slopes merely as error terms.

13.2 Varying slopes without varying intercepts

Figure 11.1 on page 238 displays a varying-intercept model, a varying-slope model, and a varying-intercept, varying-slope model. Almost always, when a slope is allowed to vary, it makes sense for the intercept to vary also. That is, the graph in the center of Figure 11.1b usually does not make sense. For example, if the coefficient of floor varies with county, then it makes sense to allow the intercept of the regression to vary also. It would be an implausible scenario in which the counties were all identical in radon levels for houses without basements, but differed in their coefficients for x.

A situation in which a constant-intercept, varying-slope model is appropriate

Occasionally it is reasonable to allow the slope but not the intercept to vary by group. For example, consider a study in which J separate experiments are performed on samples from a common population, with each experiment randomly assigning a control condition to half its subjects and a treatment to the other half. Further suppose that the "control" conditions are the same for each experiment but the "treatments" vary. In that case, it would make sense to fix the intercept and allow the slope to vary—thus, a basic model of:

$$
\begin{aligned}
y_i &\sim \ \mathrm{N}(\alpha + \theta_{j[i]}T_i,\ \sigma_y^2) \\
\theta_j &\sim \ \mathrm{N}(\mu_\theta, \sigma_\theta^2),
\end{aligned}
\tag{13.4}
$$

where $T_i = 1$ for treated units and 0 for controls. Individual-level predictors could be added to the regression for y, and any interactions with treatment could also have varying slopes; for example,

$$y_i \sim N\left(\alpha + \beta x_i + \theta_{1,j[i]}T_i + \beta_{2,j[i]}x_iT_i, \ \sigma_y^2\right)$$

$$\begin{pmatrix} \theta_{1,j} \\ \theta_{2,j} \end{pmatrix} \sim N\left(\begin{pmatrix} \mu_1 \\ \mu_2 \end{pmatrix}, \begin{pmatrix} \sigma_1^2 & \rho\sigma_1\sigma_2 \\ \rho\sigma_1\sigma_2 & \sigma_2^2 \end{pmatrix}\right), \text{ for } j = 1, \ldots, J, \quad (13.5)$$

The multilevel model could be further extended with group-level predictors characterizing the treatments.

Fitting in R

To fit such a model in `lmer()`, we must explicitly remove the intercept from the group of coefficients that vary by group; for example, here is model (13.4) including the treatment indicator T as a predictor:

R code   `lmer (y ~ T + (T - 1 | group))`

The varying slope allows a different treatment effect for each group.

And here is model (13.5) with an individual-level predictor `x`:

R code   `lmer (y ~ x + T + (T + x:T - 1 | group))`

Here, the treatment effect and its interaction with x vary by group.

13.3 Modeling multiple varying coefficients using the scaled inverse-Wishart distribution

When more than two coefficients vary (for example, $y_i \sim N(\beta_0 + \beta_1 X_{i1} + \beta_2 X_{i2}, \sigma^2)$, with β_0, β_1, and β_2 varying by group), it is helpful to move to matrix notation in modeling the coefficients and their group-level regression model and covariance matrix.

Simple model with two varying coefficients and no group-level predictors

Starting with the model that begins this chapter, we can rewrite the basic varying-intercept, varying-slope model (13.1) in matrix notation as

$$\begin{aligned} y_i &\sim N(X_i B_{j[i]}, \sigma_y^2), \text{ for } i = 1, \ldots, n \\ B_j &\sim N(M_B, \Sigma_B), \text{ for } j = 1, \ldots, J, \end{aligned} \quad (13.6)$$

where

- X is the $n \times 2$ matrix of predictors: the first column of X is a column of 1's (that is, the constant term in the regression), and the second column is the predictor x. X_i is then the vector of length 2 representing the ith row of X, and $X_i B_{j[i]}$ is simply $\alpha_{j[i]} + \beta_{j[i]}x_i$ from the top line of (13.1).

- $B = (\alpha, \beta)$ is the $J \times 2$ matrix of individual-level regression coefficients. For any group j, B_j is a vector of length 2 corresponding to the jth row of B (although for convenience we consider B_j as a column vector in the product $X_i B_{j[i]}$ in model (13.6)). The two elements of B_j correspond to the intercept and slope, respectively, for the regression model in group j. $B_{j[i]}$ in the first line of (13.6) is the $j[i]^{th}$ row of B, that is, the vector representing the intercept and slope for the group that includes unit i.

- $M_B = (\mu_\alpha, \mu_\beta)$ is a vector of length 2, representing the mean of the distribution of the intercepts and the mean of the distribution of the slopes.

- Σ_B is the 2×2 covariance matrix representing the variation of the intercepts and slopes in the population of groups, as in the second line of (13.1).

We are following our general notation in which uppercase letters represent matrices: thus, the vectors α and β are combined into the matrix B.

In the fitted radon model on page 279, the parameters of the group-level model are estimated at $\widehat{M_B} = (1.46, -0.68)$ and $\widehat{\Sigma}_B = \begin{pmatrix} \hat{\sigma}_a^2 & \hat{\rho}\hat{\sigma}_a\hat{\sigma}_b \\ \hat{\rho}\hat{\sigma}_a\hat{\sigma}_b & \hat{\sigma}_b^2 \end{pmatrix}$, where $\hat{\sigma}_a = 0.35$, $\hat{\sigma}_b = 0.34$, and $\hat{\rho} = -0.34$. The estimated coefficient matrix $\widehat{B}$ is given by the 85×2 array at the end of the display of `coef(M3)` on page 280.

More than two varying coefficients

The same expression as above holds, except that the 2's are replaced by K's, where K is the number of individual-level predictors (including the intercept) that vary by group. As we discuss shortly in the context of the inverse-Wishart model, estimation becomes more difficult when $K > 2$ because of constraints among the correlation parameters of the covariance matrix Σ_B.

Including group-level predictors

More generally, we can have J groups, K individual-level predictors, and L predictors in the group-level regression (including the constant term as a predictor in both cases). For example, $K = L = 2$ in the radon model that has floor as an individual predictor and uranium as a county-level predictor.

We can extend model (13.6) to include group-level predictors:

$$\begin{aligned} y_i &\sim \text{N}(X_i B_{j[i]}, \sigma_y^2), \text{ for } i = 1, \ldots, n \\ B_j &\sim \text{N}(U_j G, \Sigma_B), \text{ for } j = 1, \ldots, J, \end{aligned} \qquad (13.7)$$

where B is the $J \times K$ matrix of individual-level coefficients, U is the $J \times L$ matrix of group-level predictors (including the constant term), and G is the $L \times K$ matrix of coefficients for the group-level regression. U_j is the jth row of U, the vector of predictors for group j, and so $U_j G$ is a vector of length K.

Model (13.1) is a special case with $K = L = 2$, and the coefficients in G are then $\gamma_0^\alpha, \gamma_0^\beta, \gamma_1^\alpha, \gamma_1^\beta$. For the fitted radon model on page 279, the γ's are the four unmodeled coefficients (for the intercept, `x`, `u.full`, and `x:u.full`, respectively), and the two columns of the estimated coefficient matrix $\widehat{B}$ are estimated by `a.hat` and `b.hat`, as defined by the R code on page 282.

Including individual-level predictors whose coefficients do not vary by group

The model can be further expanded by adding unmodeled individual-level coefficients, so that the top line of (13.7) becomes

$$y_i \sim \text{N}(X_i^0 \beta^0 + X_i B_{j[i]}, \sigma_y^2), \text{ for } i = 1, \ldots, n, \qquad (13.8)$$

where X^0 is a matrix of these additional predictors and β^0 is the vector of their regression coefficients (which, by assumption, are common to all the groups).

Model (13.8) is sometimes called a *mixed-effects* regression, where the β^0's and the B's are the *fixed* and *random* effects, respectively. As noted on pages 2 and 245, we avoid these terms because of their ambiguity in the statistical literature. For example, sometimes unvarying coefficients such as the β^0's in model (13.8) are called "fixed," but sometimes the term "fixed effects" refers to intercepts that vary by groups but are not given a multilevel model (this is what we call the "no-pooling model," as pictured, for example, by the solid lines in Figure 12.2 on page 255).

Equivalently, model (13.8) can be written by folding X^0 and X into a common predictor matrix X, folding β^0 and B into a common coefficient matrix B, and using model (13.1), with the appropriate elements in Σ_B set to zero, implying no variation among groups for certain coefficients.

Modeling the group-level covariance matrix using the scaled inverse-Wishart distribution

When the number K of varying coefficients per group is more than two, modeling the correlation parameters ρ is a challenge. In addition to each of the correlations being restricted to fall between -1 and 1, the correlations are jointly constrained in a complicated way—technically, the covariance matrix Σ_β must be positive definite. (An example of the constraint is: if $\rho_{12} = 0.9$ and $\rho_{13} = 0.9$, then ρ_{23} must be at least 0.62.)

Modeling and estimation are more complicated in this jointly constrained space. We first introduce the inverse-Wishart model, then generalize to the scaled inverse-Wishart, which is what we recommend for modeling the covariance matrix of the distribution of varying coefficients.

Inverse-Wishart model. One model that has been proposed for the covariance matrix Σ_β is the *inverse-Wishart* distribution, which has the advantage of being computationally convenient (especially when using Bugs, as we illustrate in Section 17.1) but the disadvantage of being difficult to interpret.

In the model $\Sigma_B \sim \text{Inv-Wishart}_{K+1}(I)$, the two parameters of the inverse-Wishart distribution are the *degrees of freedom* (here set to $K+1$, where K is the dimension of B, that is, the number of coefficients in the model that vary by group) and the *scale* (here set to the $K \times K$ identity matrix).

To understand this model, we consider its implications for the standard deviation and correlations. Recall that if there are K varying coefficients, then Σ_B is a $K \times K$ matrix, with diagonal elements $\Sigma_{kk} = \sigma_k^2$ and off-diagonal-elements $\Sigma_{kl} = \rho_{kl}\sigma_k\sigma_l$ (generalizing models (13.1) and (13.2) to $K > 2$).

Setting the degrees-of-freedom parameter to $K+1$ has the effect of setting a uniform distribution on the individual correlation parameters (that is, they are assumed equally likely to take on any value between -1 and 1).

Scaled inverse-Wishart model. When the degrees of freedom parameter of the inverse-Wishart distribution is set to $K+1$, the resulting model is reasonable for the correlations but is quite constraining on the scale parameters σ_k. This is a problem because we would like to estimate σ_k from the data. Changing the degrees of freedom allows the σ_k's to be estimated more freely, but at the cost of constraining the correlation parameters.

We get around this problem by expanding the inverse-Wishart model with a new vector of scale parameters ξ_k:

$$\Sigma_B = \text{Diag}(\xi)Q\text{Diag}(\xi),$$

with the *unscaled covariance matrix* Q being given the inverse-Wishart model:

$$Q \sim \text{Inv-Wishart}_{K+1}(I).$$

The variances then correspond to the diagonal elements of the unscaled covariance

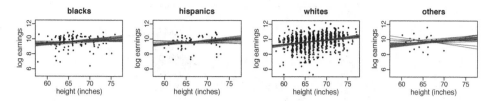

Figure 13.3 *Multilevel regression lines $y = \alpha_j + \beta_j x$ for log earnings on height (among those with positive earnings), in four ethnic categories j. The gray lines indicate uncertainty in the fitted regressions.*

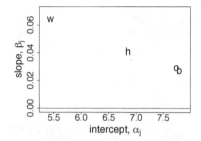

Figure 13.4 *Scatterplot of estimated intercepts and slopes (for whites, hispanics, blacks, and others), (α_j, β_j), for the earnings-height regressions shown in Figure 13.3. The extreme negative correlation arises because the center of the range of height is far from zero. Compare to the coefficients in the rescaled model, as displayed in Figure 13.7.*

matrix Q, multiplied by the appropriate scaling factors ξ:

$$\sigma_k^2 = \Sigma_{kk} = \xi_k^2 Q_{kk}, \text{ for } k = 1, \ldots, K,$$

and the covariances are

$$\Sigma_{kl} = \xi_k \xi_l Q_{kl}, \text{ for } k, l = 1, \ldots, K,$$

We prefer to express in terms of the standard deviations,

$$\sigma_k = |\xi_k| \sqrt{Q_{kk}},$$

and correlations

$$\rho_{kl} = \Sigma_{kl}/(\sigma_k \sigma_l).$$

The parameters in ξ and Q cannot be interpreted separately: they are a convenient way to set up the model, but it is the standard deviations σ_k and the correlations ρ_{kl} that are of interest (and which are relevant for producing partially pooled estimates for the coefficients in B).

As with the unscaled Wishart, the model implies a uniform distribution on the correlation parameters. As we discuss next, it can make sense to transform the data to remove any large correlations that could be expected simply from the structure of the data.

13.4 Understanding correlations between group-level intercepts and slopes

Recall that varying slopes can be interpreted as interactions between an individual-level predictor and group indicators. As with classical regression models with interactions, the intercepts can often be more clearly interpreted if the continuous

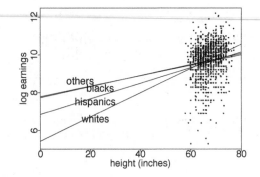

Figure 13.5 *Sketch illustrating the difficulty of simultaneously estimating α and β. The lines show the regressions for the four ethnic groups as displayed in Figure 13.3: the center of the range of x values is far from zero, and so small changes in the slope induce large changes in the intercept.*

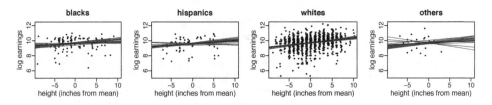

Figure 13.6 *Multilevel regression lines $y = \alpha_j + \beta_j z$, for log earnings given mean-adjusted height ($z_i = x_i - \bar{x}$), in four ethnic groups j. The gray lines indicate uncertainty in the fitted regressions.*

predictor is appropriately centered. We illustrate with the height and earnings example from Chapter 4.

We begin by fitting a multilevel model of log earnings given height, allowing the coefficients to vary by ethnicity. The data and fitted model are displayed in Figure 13.3. (Little is gained by fitting a multilevel model here—with only four groups, a classical no-pooling model would work nearly as well, as discussed in Section 12.9—but this is a convenient example to illustrate a general point.)

Figure 13.4 displays the estimates of (α_j, β_j) for the four ethnic groups, and they have a strong negative correlation: the groups with high values of α have relatively low values of β, and vice versa. This correlation occurs because the center of the x-values of the data is far from zero. The regression lines have to go roughly through the center of the data, and then changes in the slope induce opposite changes in the intercept, as illustrated in Figure 13.5.

There is nothing wrong with a high correlation between the α's and β's, but it makes the estimated intercepts more difficult to interpret. As with interaction models in classical regression, it can be helpful to subtract the average value of the continuous x before including it in the regression; thus, $y_i \sim N(\alpha_{j[i]} + \beta_{j[i]} z_i, \sigma_y^2)$, where $z_i = x_i - \bar{x}$. Figures 13.6 and 13.7 show the results for the earnings regression: the correlation has pretty much disappeared. Centering the predictor x will not necessarily remove correlations between intercepts and slopes—but any correlation that remains can then be more easily interpreted. In addition, centering can speed convergence of the Gibbs sampling algorithm used by Bugs and other software.

We fit this model, and the subsequent models in this chapter, in Bugs (see Chap-

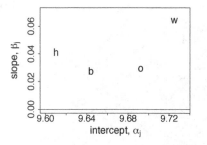

Figure 13.7 *Scatterplot of estimated intercepts and slopes, (α_j, β_j), for the regression of earnings on mean-adjusted height z, for the four groups j displayed in Figure 13.6. The coefficients are no longer strongly correlated (compare to Figure 13.4).*

ter 17 for examples of code) because, as discussed in Section 12.4, the current version of `lmer()` does not work so well when the number of groups is small—and, conversely, with these small datasets, Bugs is not too slow.

13.5 Non-nested models

So far we have considered the simplest hierarchical structure of individuals i in groups j. We now discuss models for more complicated grouping structures such as introduced in Section 11.3.

Example: a psychological experiment with two potentially interacting factors

Figure 13.8 displays data from a psychological experiment of pilots on flight simulators, with $n = 40$ data points corresponding to $J = 5$ treatment conditions and $K = 8$ different airports. The responses can be fit to a *non-nested* multilevel model of the form

$$
\begin{aligned}
y_i &\sim \ \mathrm{N}(\mu + \gamma_{j[i]} + \delta_{k[i]}, \sigma_y^2), \text{ for } i = 1, \ldots, n \\
\gamma_j &\sim \ \mathrm{N}(0, \sigma_\gamma^2), \text{ for } j = 1, \ldots, J \\
\delta_k &\sim \ \mathrm{N}(0, \sigma_\delta^2), \text{ for } k = 1, \ldots, K.
\end{aligned}
\tag{13.9}
$$

The parameters γ_j and δ_k represent treatment effects and airport effects. Their distributions are centered at zero (rather than given mean levels μ_γ, μ_δ) because the regression model for y already has an intercept, μ, and any nonzero mean for the γ and δ distributions could be folded into μ. As we shall see in Section 19.4, it can sometimes be effective for computational purposes to add extra mean-level parameters into the model, but the coefficients in this expanded model must be interpreted with care.

We can perform a quick fit as follows:

```
lmer (y ~ 1 + (1 | group.id) + (1 | scenario.id))
```
R code

where `group.id` and `scenario.id` are the index variables for the five treatment conditions and eight airports, respectively.

When fit to the data in Figure 13.8, the estimated residual standard deviations at the individual, treatment, and airport levels are $\hat{\sigma}_y = 0.23$, $\hat{\sigma}_\gamma = 0.04$, and $\hat{\sigma}_\delta = 0.32$. Thus, the variation among airports is huge—even larger than that among individual measurements—but the treatments vary almost not at all. This general pattern can be seen in Figure 13.8.

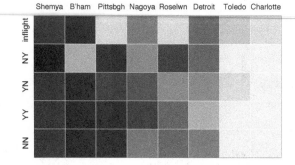

Figure 13.8 *Success rates of pilots training on a flight simulator with five different treatments and eight different airports. Shadings in the 40 cells i represent different success rates y_i, with black and white corresponding to 0 and 100%, respectively. For convenience in reading the display, the treatments and airports have each been sorted in increasing order of average success. These 40 data points have two groupings—treatments and airports—which are not nested.*

	Data in matrix form						Data in vector form		
							y	j	k
airport	treatment conditions								
							0.38	1	1
1	0.38	0.25	0.50	0.14	0.43		0.00	1	2
2	0.00	0.00	0.67	0.00	0.00		0.38	1	3
3	0.38	0.50	0.33	0.71	0.29		0.00	1	4
4	0.00	0.12	0.00	0.00	0.86		0.33	1	5
5	0.33	0.50	0.14	0.29	0.86		1.00	1	6
6	1.00	1.00	1.00	1.00	0.86		0.12	1	7
7	0.12	0.12	0.00	0.14	0.14		1.00	1	8
8	1.00	0.86	1.00	1.00	0.75		0.25	2	1
							...	...	...

Figure 13.9 *Data from Figure 13.8 displayed as an array (y_{jk}) and in our preferred notation as a vector (y_i) with group indicators $j[i]$ and $k[i]$.*

Model (13.9) can also be written more cleanly as $y_{jk} \sim \mathrm{N}(\mu + \gamma_j + \delta_k, \sigma_y^2)$, but we actually prefer the more awkward notation using $j[i]$ and $k[i]$ because it emphasizes the multilevel structure of the model and is not restricted to balanced designs. When modeling a data array of the form (y_{jk}), we usually convert it into a vector with index variables for the rows and columns, as illustrated in Figure 13.9 for the flight simulator data.

Example: regression of earnings on ethnicity categories, age categories, and height

All the ideas of the earlier part of this chapter, introduced in the context of a simple structure of individuals within groups, apply to non-nested models as well. For example, Figure 13.10 displays the estimated regression of log earnings, y_i, on height, z_i (mean-adjusted, for reasons discussed in the context of Figures 13.3–13.6), applied to the $J = 4$ ethnic groups and $K = 3$ age categories. In essence, there is a separate regression model for each age group and ethnicity combination. The multilevel model can be written, somewhat awkwardly, as a data-level model,

$$y_i \sim \mathrm{N}(\alpha_{j[i],k[i]} + \beta_{j[i],k[i]} z_i, \sigma_y^2), \text{ for } i = 1, \ldots, n,$$

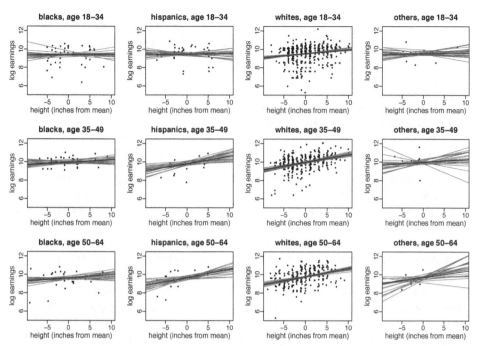

Figure 13.10 *Multilevel regression lines* $y = \beta_{j,k}^0 + \beta_{j,k}^1 z$, *for log earnings* y *given mean-adjusted height* z, *for four ethnic groups* j *and three age categories* k. *The gray lines indicate uncertainty in the fitted regressions.*

a decomposition of the intercepts and slopes into terms for ethnicity, age, and ethnicity × age,

$$\begin{pmatrix} \alpha_{j,k} \\ \beta_{j,k} \end{pmatrix} = \begin{pmatrix} \mu_0 \\ \mu_1 \end{pmatrix} + \begin{pmatrix} \gamma_{0j}^{\mathrm{eth}} \\ \gamma_{1j}^{\mathrm{eth}} \end{pmatrix} + \begin{pmatrix} \gamma_{0k}^{\mathrm{age}} \\ \gamma_{1k}^{\mathrm{age}} \end{pmatrix} + \begin{pmatrix} \gamma_{0jk}^{\mathrm{eth}\times\mathrm{age}} \\ \gamma_{1jk}^{\mathrm{eth}\times\mathrm{age}} \end{pmatrix},$$

and models for variation,

$$\begin{pmatrix} \gamma_{0j}^{\mathrm{eth}} \\ \gamma_{1j}^{\mathrm{eth}} \end{pmatrix} \sim \mathrm{N}\left(\begin{pmatrix} 0 \\ 0 \end{pmatrix}, \Sigma^{\mathrm{eth}} \right), \text{ for } j = 1, \ldots, J$$

$$\begin{pmatrix} \gamma_{0k}^{\mathrm{age}} \\ \gamma_{1k}^{\mathrm{age}} \end{pmatrix} \sim \mathrm{N}\left(\begin{pmatrix} 0 \\ 0 \end{pmatrix}, \Sigma^{\mathrm{age}} \right), \text{ for } k = 1, \ldots, K$$

$$\begin{pmatrix} \gamma_{0jk}^{\mathrm{eth}\times\mathrm{age}} \\ \gamma_{1jk}^{\mathrm{eth}\times\mathrm{age}} \end{pmatrix} \sim \mathrm{N}\left(\begin{pmatrix} 0 \\ 0 \end{pmatrix}, \Sigma^{\mathrm{eth}\times\mathrm{age}} \right), \text{ for } j = 1, \ldots, J; \ k = 1, \ldots, K.$$

Because we have included means μ_0, μ_1 in the decomposition above, we can center each batch of coefficients at 0.

Interpretation of data-level variance. The data-level errors have estimated residual standard deviation $\hat{\sigma}_y = 0.87$. That is, given ethnicity, age group, and height, log earnings can be predicted to within approximately ± 0.87, and so earnings themselves can be predicted to within a multiplicative factor of $e^{0.87} = 2.4$. So earnings cannot be predicted well at all by these factors, which is also apparent from the scatter in Figure 13.10.

Interpretation of group-level variances. The group-level errors can be separated into intercept and slope coefficients. The intercepts have estimated residual stan-

B: 257	E: 230	A: 279	C: 287	D: 202
D: 245	A: 283	E: 245	B: 280	C: 260
E: 182	B: 252	C: 280	D: 246	A: 250
A: 203	C: 204	D: 227	E: 193	B: 259
C: 231	D: 271	B: 266	A: 334	E: 338

Figure 13.11 *Data from a* 5×5 *latin square experiment studying the effects of five ordered treatments on the yields of millet crops, from Snedecor and Cochran (1989). Each cell shows the randomly assigned treatment and the observed yield for the plot.*

dard deviations of $(\widehat{\Sigma}_{00}^{\text{eth}})^{1/2} = 0.08$ at the ethnicity level, $(\widehat{\Sigma}_{00}^{\text{age}})^{1/2} = 0.25$ at the age level, and $(\widehat{\Sigma}_{00}^{\text{eth} \times \text{age}})^{1/2} = 0.11$ at the ethnicity $\times$ age level. Because we have rescaled height to have a mean of zero (see Figure 13.10), we can interpret these standard deviations as the relative importance of each factor (ethnicity, age group, and their interaction) on log earnings at the average height in the population.

This model fits earnings on the log scale and so these standard deviations can be interpreted accordingly. For example, the residual standard deviation of 0.08 for the ethnicity coefficients implies that the predictive effects of ethnic groups in the model are on the order of ± 0.08, which correspond to multiplicative factors from about $e^{-0.08} = 0.92$ to $e^{0.08} = 1.08$.

The slopes have estimated residual standard deviations of $(\widehat{\Sigma}_{11}^{\text{eth}})^{1/2} = 0.03$ at the ethnicity level, $(\widehat{\Sigma}_{11}^{\text{age}})^{1/2} = 0.02$ at the age level, and $(\widehat{\Sigma}_{11}^{\text{eth} \times \text{age}})^{1/2} = 0.02$ at the ethnicity $\times$ age level. These slopes are per inch of height, so, for example, the predictive effects of ethnic groups in the model are in the range of $\pm 3\%$ in income per inch of height. One can also look at the estimated correlation between intercepts and slopes for each factor.

Example: a latin square design with grouping factors and group-level predictors

Non-nested models can also include group-level predictors. We illustrate with data from a 5×5 latin square experiment, a design in which 25 units arranged in a square grid are assigned five different treatments, with each treatment being assigned to one unit in each row and each column. Figure 13.11 shows the treatment assignments and data from a small agricultural experiment. There are three non-nested levels of grouping—rows, columns, and treatments—and each has a natural group-level predictor corresponding to a linear trend. (The five treatments are ordered.)

The corresponding multilevel model can be written as

$$
\begin{aligned}
y_i &\sim \text{N}(\mu + \beta_{j[i]}^{\text{row}} + \beta_{k[i]}^{\text{column}} + \beta_{l[i]}^{\text{treat}}, \sigma_y^2), \text{ for } i = 1, \ldots, 25 \\
\beta_j^{\text{row}} &\sim \text{N}(\gamma^{\text{row}} \cdot (j - 3), \sigma_{\beta\,\text{row}}^2), \text{ for } j = 1, \ldots, 5 \\
\beta_k^{\text{column}} &\sim \text{N}(\gamma^{\text{column}} \cdot (k - 3), \sigma_{\beta\,\text{column}}^2), \text{ for } k = 1, \ldots, 5 \\
\beta_l^{\text{treat}} &\sim \text{N}(\gamma^{\text{treat}} \cdot (l - 3), \sigma_{\beta\,\text{treat}}^2), \text{ for } l = 1, \ldots, 5.
\end{aligned}
\tag{13.10}
$$

Thus j, k, and l serve simultaneously as values of the row, column, and treatment predictors.

By subtracting 3, we have centered the row, column, and treatment predictors at zero; the parameter μ has a clear interpretation as the grand mean of the data, with the different β's supplying deviations for rows, columns, and treatments. As with group-level models in general, the linear trends at each level potentially allow more precise estimates of the group effects, to the extent that these trends are supported by the data. An advantage of multilevel modeling here is that it doesn't force a

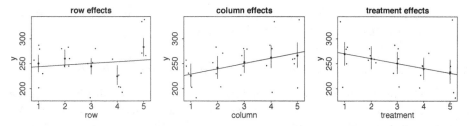

Figure 13.12 *Estimates* ±1 *standard error for the row, column, and treatment effects for the latin square data in Figure 13.11. The five levels of each factor are ordered, and the lines display the estimated group-level regressions,* $y = \mu + \gamma^{\text{row}} \cdot (x-3)$, $y = \mu + \gamma^{\text{column}} \cdot (x-3)$, *and* $y = \mu + \gamma^{\text{treat}} \cdot (x-3)$.

choice between a linear fit and separate estimates for each level of a predictor. (This is an issue we discussed more generally in Chapter 11 in the context of including group indicators as well as group-level predictors.)

Figure 13.12 shows the estimated row, column, and treatment effects on graphs, along with the estimated linear trends. The grand mean μ has been added back to each of these observations so that the plots are on the scale of the original data. This sort of data structure is commonly studied using the analysis of variance, whose connections with multilevel models we discuss fully in Chapter 22, including a discussion of this latin square example in Section 22.5.

13.6 Selecting, transforming, and combining regression inputs

As with classical regression (see Section 4.5), choices must be made in multilevel models about which input variables to include, and how best to transform and combine them. We discuss here how some of these decisions can be expressed as particular choices of parameters in a multilevel model. The topic of formalizing modeling choices is currently an active area of research—key concerns include using information in potential input variables without being overwhelmed by the complexity of the relating model, and including model choice in uncertainty estimates. As discussed in Section 9.5, the assumption of ignorability in observational studies is more plausible when controlling for more pre-treatment inputs, which gives us a motivation to include more regression predictors.

Classical models for regression coefficients

Multilevel modeling includes classical least squares regression as a special case. In a multilevel model, each coefficient is part of a model with some mean and standard deviation. (These mean values can themselves be determined by group-level predictors in a group-level model.) In classical regression, every predictor is either in or out of the model, and each of these options corresponds to a special case of the multilevel model.

- If a predictor is "in," this corresponds to a coefficient model with standard deviation of ∞: no group-level information is used to estimate this parameter, so it is estimated directly using least squares. It turns out that in this case the group-level mean is irrelevant (see formula (12.16) on page 269 for the case $\sigma_\alpha = \infty$); for convenience we often set it to 0.

- If a predictor is "out," this corresponds to a group-level model with group-level mean 0 and standard deviation 0: the coefficient estimate is then fixed at zero (see (12.16) for the case $\sigma_\alpha = 0$) with no uncertainty.

Multilevel modeling as an alternative to selecting regression predictors

Multilevel models can be used to combine inputs into more effective regression predictors, generalizing some of the transformation ideas discussed in Section 4.6. When many potential regression inputs are available, the fundamental approach is to include as many of these inputs as possible, but not necessarily as independent least squares predictors.

For example, Witte et al. (1994) describe a logistic regression in a case-control study of 362 persons, predicting cancer incidence given information on consumption of 87 different foods (and also controlling for five background variables which we do not discuss further here). Each of the foods can potentially increase or decrease the probability of cancer, but it would be hard to trust the result of a regression with 87 predictors fit to only 362 data points, and classical tools for selecting regression predictors do not seem so helpful here. In our general notation, the challenge is to estimate the logistic regression of cancer status y on the 362×87 matrix X of food consumption (and the 362×6 matrix X^0 containing the constant term and the 5 background variables).

More information is available, however, because each of the 87 foods can be characterized by its level of each of 35 nutrients, information that can be expressed as an 87×36 matrix of predictors Z indicating how much of each nutrient is in each food. Witte et al. fit the following multilevel model:

$$
\begin{aligned}
\Pr(y_i = 1) &= \operatorname{logit}^{-1}(X_i^0 \beta^0 + X_i B_{j[i]}), \text{ for } i = 1, \ldots, 362 \\
B_j &\sim \mathrm{N}(Z_j \gamma, \sigma_\beta^2), \text{ for } j = 1, \ldots, 87.
\end{aligned}
\tag{13.11}
$$

The food-nutrient information in Z allows the multilevel model to estimate separate predictive effects for foods, after controlling for systematic patterns associated with nutrients. In the extreme case that $\sigma_\beta = 0$, all the variation associated with the foods is explained by the nutrients. At the other extreme, $\sigma_\beta = \infty$ would imply that the nutrient information is not helping at all.

Model (13.11) is helpful in reducing the number of food predictors from 87 to 35. At this point, Witte et al. used substantive understanding of diet and cancer to understand the result. Ultimately, we would like to have a model that structures the 35 predictors even more, perhaps by categorizing them into batches or combining them in some way. The next example sketches how this might be done; it is currently an active research topic to generally structure large numbers of regression predictors.

Linear transformation and combination of inputs in a multilevel model

For another example, we consider the problem of forecasting presidential elections by state (see Section 1.2). A forecasting model based on 11 recent national elections has more than 500 "data points"—state-level elections—and can then potentially include many state-level predictors measuring factors such as economic performance, incumbency, and popularity. However, at the national level there are really only 11 observations and so one must be parsimonious with national-level predictors. In practice, this means performing some preliminary data analysis to pick a single economic predictor, a single popularity predictor, and maybe one or two other predictors based on incumbency and political ideology.

Setting up a model to allow partial pooling of a set of regression predictors

A more general approach to including national predictors is possible using multilevel modeling. For example, suppose we wish to include five measures of the national economy (for example, change in GDP per capita, change in unemployment, and so forth). The usual approach (which we have followed in the past in this problem) is to choose one of these as the economic predictor, x, thus writing the model as

$$y_i = \alpha + \beta x_i + \cdots, \tag{13.12}$$

where the dots indicate all the rest of the model, including other state-level and national predictors, as well as error terms at the state, regional, and national levels. Here we focus on the economic inputs, for simplicity setting aside the rest of the model.

Instead of choosing just one of the five economic inputs, it would perhaps be better first to standardize each of them (see Section 4.2), orient them so they are in the same direction, label these standardized variables as $X_{(j)}$, for $j = 1, \ldots, 5$, and then average them into a single predictor, defined for each data point as

$$x_i^{\text{avg}} = \frac{1}{5} \sum_{j=1}^{5} \sum X_{ij}, \text{ for } i = 1, \ldots n. \tag{13.13}$$

This new x^{avg} can be included in place of x as the regression predictor in (13.12), or, equivalently,

$$
\begin{aligned}
y_i &= \alpha + \beta x_i^{\text{avg}} + \cdots \\
&= \alpha + \frac{1}{5}\beta X_{i1} + \cdots + \frac{1}{5}\beta X_{i5} + \cdots.
\end{aligned}
$$

The resulting model will represent an improvement to the extent that the average of the five standardized economy measures is a better predictor than the single measure chosen before.

However, model (13.13) is limited in that it restricts the coefficients of the five separate x^j's to be equal. More generally, we can replace (13.13) by a weighted average:

$$x_i^{\text{w.avg}} = \frac{1}{5} \sum_{j=1}^{5} \gamma_j X_{ij}, \text{ for } i = 1, \ldots, n, \tag{13.14}$$

so that the data model becomes

$$
\begin{aligned}
y_i &= \alpha + \beta x_i^{\text{w.avg}} + \cdots \\
&= \alpha + \frac{1}{5}\gamma_1 \beta X_{i1} + \cdots + \frac{1}{5}\gamma_5 \beta X_{i5} + \cdots.
\end{aligned}
\tag{13.15}
$$

We would like to estimate the relative coefficients γ_j from the data, but we cannot simply use classical regression, since this would then be equivalent to estimating a separate coefficient for each of the five predictors, and we have already established that not enough data are available to do a good job of this.

Instead, one can set up a model for the γ_j's:

$$\gamma_j \sim N(1, \sigma_\gamma^2), \text{ for } j = 1, \ldots, 5, \tag{13.16}$$

so that, in the model (13.15), the common coefficient β can be estimated classically, but the relative coefficients γ_j are part of a multilevel model. The hyperparameter σ_γ can be interpreted as follows:

- If $\sigma_\gamma = 0$, the model reduces to the simple averaging (13.14): *complete pooling* of the γ_j's to the common value of 1, so that the combined predictor $x^{\text{w.avg}}$ is simply x^{avg}, the average of the five individual $X_{(j)}$'s.

- If $\sigma_\gamma = \infty$, there is *no pooling*, with the individual coefficients $\frac{1}{5}\gamma_j\beta$ estimated separately using least squares.

- When σ_γ is positive but finite, the γ_j's are *partially pooled*, so that the five predictors x_j have coefficients that are near each other but not identical.

Depending on the amount of data available, σ_γ can be estimated as part of the model or set to a value such as 0.3 that constrains the γ_j's to be fairly close to 1 and thus constrains the coefficients of the individual x^j's toward each other in the data model (13.15).

Connection to factor analysis

A model can include multiplicative parameters for both modeling and computational purposes. For example, we could predict the election outcome in year t in state s within region $r[s]$ as

$$y_{st} = \beta^{(0)}X_{st}^{(0)} + \alpha_1 \sum_{j=1}^{5} \beta_j^{(1)} X_{jt}^{(1)} + \alpha_2\gamma_t + \alpha_3\delta_{r[s],t} + \epsilon_{st},$$

where $X^{(0)}$ is the matrix of state $\times$ year-level predictors, $X^{(1)}$ is the matrix of year-level predictors, and γ, δ, and ϵ are national, regional, and statewide error terms. In this model, the auxiliary parameters α_2 and α_3 exist for purely computational reasons, and they can be estimated, with the understanding that we are interested only in the products $\alpha_2\gamma_t$ and $\alpha_3\delta_{r,t}$. More interestingly, α_1 serves both a computational and modeling role—the $\beta_j^{(1)}$ parameters have a common $N(\frac{1}{5}, \sigma_m^2)$ model, and α_1 has the interpretation as the overall coefficient for the economic predictors.

More generally, we can imagine K batches of predictors, with the data-level regression model using a weighted average from each batch:

$$y = X^{(0)}\beta^{(0)} + \beta_1 x^{\text{w.avg}, 1} + \cdots + \beta_k x^{\text{w.avg}, K} + \cdots,$$

where each predictor $x_k^{\text{w.avg}}$ is a combination of J_k individual predictors x^{jk}:

$$\text{for each } k: \quad x_i^{\text{w.avg}, k} = \frac{1}{J_k} \sum_{j=1}^{J_k} \gamma_{jk} x_i^{jk}, \text{ for } i = 1, \ldots, n.$$

This is equivalent to a regression model on the complete set of available predictors, $x^{11}, \ldots, x^{J_1 1}; x^{12}, \ldots, x^{J_2 2}; \ldots; x^{1K}, \ldots, x^{J_K K}$, where the predictor x^{jk} gets the coefficient $\frac{1}{J_k}\gamma_{jk}\beta_k$. Each batch of relative weights γ is then modeled hierarchically:

$$\text{for each } k: \quad \gamma_{jk} \sim N(1, \sigma_{\gamma k}^2), \text{ for } j = 1, \ldots, J_k,$$

with the hyperparameters $\sigma_{\gamma k}$ estimated from the data or set to low values such as 0.3.

In this model, each combined predictor $x^{\text{w.avg}, k}$ represents a "factor" formed by a linear combination of the J_k individual predictors, β_k represents the importance of that factor, and the γ_{jk}'s give the relative importance of the different components.

As noted at the beginning of this section, these models are currently the subject of active research, and we suggest that they can serve as a motivation to specially tailored models for individual problems rather than as off-the-shelf solutions to generic multilevel problems with many predictors.

13.7 More complex multilevel models

The models we have considered so far can be generalized in a variety of ways. Chapters 14 and 15 discuss multilevel logistic and generalized linear models. Other extensions within multilevel linear and generalized linear models include the following:

- Variances can vary, as parametric functions of input variables, and in a multilevel way by allowing different variances for groups. For example, the model $y_i \sim \mathrm{N}(X_i\beta, \sigma_i^2)$, with $\sigma_i = \exp(X_i\gamma)$, allows the variance to depend on the predictors in a way that can be estimated from the data, and similarly, in a multilevel context, a model such as $\sigma_i = \exp(a_{j[i]} + bx_i)$ allows variances to vary by group. (It is natural to model the parameters σ on the log scale because they are restricted to be positive.)

- Models with several factors can have many potential interactions, which themselves can be modeled in a structured way, for example with larger variances for coefficients of interactions whose main effects are large. This is a model-based, multilevel version of general advice for classical regression modeling.

- Regression models can be set up for multivariate outcomes, so that vectors of coefficients become matrices, with a data-level covariance matrix. These models become correspondingly more complex when multilevel factors are added.

- Time series can be modeled in many ways going beyond simple autoregressions, and these parameters can vary by group with time-series cross-sectional data. This can be seen as a special case of non-nested groupings (for example, country × year), with calendar time being a group-level predictor.

- One way to go beyond linearity is with nonparametric regression, with the simplest version being $y_i = g(X_i, \theta) + \epsilon_i$, and the function g being allowed to have some general form (for example, cubic splines, which are piecewise-continuous third-degree polynomials). Versions of such models can also be estimated using locally weighted regression, and again can be expanded to multilevel structures as appropriate.

- More complicated models are appropriate to data with spatial or network structure. These can be thought of as generalizations of multilevel models in which groups (for example, social networks) are not necessarily disjoint, and in which group membership can be continuous (some connections are stronger than others) rather than simply "in" or "out."

We do not discuss any of these models further here, but we wanted to bring them up to be clear that the particular models presented in this book are just the starting point to our general modeling approach.

13.8 Bibliographic note

The textbooks by Kreft and De Leeuw (1998), Raudenbush and Bryk (2002), and others discuss multilevel models with varying intercepts and slopes. For an early example, see Dempster, Rubin, and Tsutakawa (1981). Non-nested models are discussed by Rasbash and Browne (2003). The flight simulator example comes from Gawron et al. (2003), and the latin square example comes from Snedecor and Cochran (1989).

Models for covariance matrices have been presented by Barnard, McCulloch, and Meng (1996), Pinheiro and Bates (1996), Daniels and Kass (1999, 2001), Daniels and Pourahmadi (2002). Boscardin and Gelman (1996) discuss parametric models for unequal variances in multilevel linear regression. The scaled inverse-Wishart model we recommend comes from O'Malley and Zaslavsky (2005).

The models for combining regression predictors discussed in Section 13.6 appear in Witte et al. (1994), Greenland (2000), Gelman (2004b), and Gustafson and Greenland (2005). See also Hodges et al. (2005) and West (2003) on methods of including many predictors and interactions in a regression. Other work on selecting and combining regression predictors in multilevel models includes Madigan and Raftery (1994), Hoeting et al. (1999), Chipman, George, and McCulloch (2001), and Dunson (2006). The election forecasting example is discussed in Gelman and King (1993) and Gelman et al. (2003, section 15.2); see Fair (1978), Rosenstone (1983), Campbell (1992), and Wlezien and Erikson (2004, 2005) for influential work in this area.

Some references for hierarchical spatial and space-time models include Besag, York, and Mollie (1991), Waller et al. (1997), Besag and Higdon (1999), Wikle et al. (2001), and Bannerjee, Gelfand, and Carlin (2003). Jackson, Best, and Richardson (2006) discuss hierarchical models combining aggregate and survey data in public health. Datta et al. (1999) compare hierarchical time series models; see also Fay and Herriot (1979). Girosi and King (2005) present a multilevel model for estimating trends within demographic subgroups.

For information on nonparametric methods such as lowess, splines, wavelets, hazard regression, generalized additive models, and regression trees, see Hastie, Tibshirani, and Friedman (2002), and, for examples in R, see Venables and Ripley (2002). Crainiceanu, Ruppert, and Wand (2005) fit spline models using Bugs. MacLehose et al. (2006) combine ideas of nonparametric and multilevel models.

13.9 Exercises

1. Fit a multilevel model to predict course evaluations from beauty and other predictors in the beauty dataset (see Exercises 3.5, 4.8, and 12.6) allowing the intercept and coefficient for beauty to vary by course category:

   (a) Write the model in statistical notation.

   (b) Fit the model using lmer() and discuss the results: the coefficient estimates and the estimated standard deviation and correlation parameters. Identify each of the estimated parameters with the notation in your model from (a).

   (c) Display the estimated model graphically in plots that also include the data.

2. Models for adjusting individual ratings: a committee of 10 persons is evaluating 100 job applications. Each person on the committee reads 30 applications (structured so that each application is read by three people) and gives each a numerical rating between 1 and 10.

   (a) It would be natural to rate the applications based on their combined scores; however, there is a worry that different raters use different standards, and we would like to correct for this. Set up a model for the ratings (with parameters for the applicants and the raters).

   (b) It is possible that some persons on the committee show more variation than others in their ratings. Expand your model to allow for this.

3. Non-nested model: continuing the Olympic ratings example from Exercise 11.3:

   (a) Write the notation for a non-nested multilevel model (varying across skaters and judges) for the technical merit ratings and fit using lmer().

(b) Fit the model in (a) using the artistic impression ratings.

(c) Display your results for both outcomes graphically.

(d) Use posterior predictive checks to investigate model fit in (a) and (b).

4. Models with unequal variances: the folder `age.guessing` contains a dataset from Gelman and Nolan (2002) from a classroom demonstration in which 10 groups of students guess the ages of 10 different persons based on photographs. The dataset also includes the true ages of the people in the photographs.

Set up a non-nested model to these data, including a coefficient for each of the persons in the photos (indicating their apparent age), a coefficient for each of the 10 groups (indicating potential systematic patterns of groups guessing high or low), and a separate error variance for each group (so that some groups are more consistent than others).

5. Return to the CD4 data introduced from Exercise 11.4.

(a) Extend the model in Exercise 12.2 to allow for varying slopes for the time predictor.

(b) Next fit a model that does not allow for varying slopes but does allow for different coefficients for each time point (rather than fitting the linear trend).

(c) Compare the results of these models both numerically and graphically.

6. Using the time-series cross-sectional dataset you worked with in Exercise 11.2, fit the model you formulated in part (c) of that exercise.

## CHAPTER 14

Multilevel logistic regression

Multilevel modeling is applied to logistic regression and other generalized linear models in the same way as with linear regression: the coefficients are grouped into batches and a probability distribution is assigned to each batch. Or, equivalently (as discussed in Section 12.5), error terms are added to the model corresponding to different sources of variation in the data. We shall discuss logistic regression in this chapter and other generalized linear models in the next.

14.1 State-level opinions from national polls

Dozens of national opinion polls are conducted by media organizations before every election, and it is desirable to estimate opinions at the levels of individual states as well as for the entire country. These polls are generally based on national random-digit dialing with corrections for nonresponse based on demographic factors such as sex, ethnicity, age, and education.

Here we describe a model developed for estimating state-level opinions from national polls, while simultaneously correcting for nonresponse, for any survey response of interest. The procedure has two steps: first fitting the model and then applying the model to estimate opinions by state:

1. We fit a regression model for the individual response y given demographics and state. This model thus estimates an average response θ_l for each cross-classification l of demographics and state. In our example, we have sex (male or female), ethnicity (African American or other), age (4 categories), education (4 categories), and 51 states (including the District of Columbia); thus $l = 1, \ldots, L = 3264$ categories.

2. From the U.S. Census, we look up the adult population N_l for each category l. The estimated population average of the response y in any state j is then

$$\theta_j = \sum_{l \in j} N_l \theta_l / \sum_{l \in j} N_l, \qquad (14.1)$$

with each summation over the 64 demographic categories l in the state. This weighting by population totals is called *poststratification* (see the footnote on page 181). In the actual analysis we also considered poststratification over the population of eligible voters but we do not discuss this further complication here.

We need many categories because (a) we are interested in estimates for individual states, and (b) nonresponse adjustments force us to include the demographics. As a result, any given survey will have few or no data in many categories. This is not a problem, however, if a multilevel model is fitted. Each factor or set of interactions in the model is automatically given a variance component. This inferential procedure works well and outperforms standard survey estimates when estimating state-level outcomes.

In this demonstration, we choose a single outcome—the probability that a respondent prefers the Republican candidate for president—as estimated by a logistic regression model from a set of seven CBS News polls conducted during the week before the 1988 presidential election.

A simple model with some demographic and geographic variation

We label the survey responses y_i as 1 for supporters of the Republican candidate and 0 for supporters of the Democrat (with undecideds excluded) and model them as independent, with $\Pr(y_i = 1) = \text{logit}^{-1}(X_i\beta)$. Potential input variables include the state index $j[i]$ and the demographics used by CBS in the survey weighting: categorical variables for sex, ethnicity, age, and education.

We introduce multilevel logistic regression with a simple example including two individual predictors—female and black—and the 51 states:

$$\Pr(y_i = 1) = \text{logit}^{-1}\left(\alpha_{j[i]} + \beta^{\text{female}} \cdot \text{female}_i + \beta^{\text{black}} \cdot \text{black}_i\right), \text{ for } i = 1, \ldots, n$$
$$\alpha_j \sim N\left(\mu_\alpha, \sigma^2_{\text{state}}\right), \text{ for } j = 1, \ldots, 51.$$

We can quickly fit the model in R,

R code
```
M1 <- lmer(y ~ black + female + (1|state), family=binomial(link="logit"))
display (M1)
```

and get the following:

R code
```
              coef.est coef.se
(Intercept)   0.4      0.1
black        -1.7      0.2
female       -0.1      0.1
Error terms:
 Groups        Name        Std.Dev.
 state         (Intercept) 0.4
 No residual sd
# of obs: 2015, groups: state, 49
deviance = 2658.7
  overdispersion parameter = 1.0
```

The top part of this display gives the estimate of the average intercept, the co-efficients for black and female, and their standard errors. Reading down, we see that σ_{state} is estimated at 0.4. There is no "residual standard deviation" because the logistic regression model does not have such a parameter (or, equivalently, it is fixed to the value 1.6, as discussed near the end of Section 5.3). The deviance (see page 100) is printed as a convenience but we usually do not look at it. Finally, the model has an overdispersion of 1.0—that is, no overdispersion—because logistic regression with binary data (as compared to count data; see Section 6.3) cannot be overdispersed.

We can also type coef(M1) to examine the estimates and standard errors of the state intercepts α_j, but rather than doing this we shall move to a larger model including additional predictors at the individual and state level. Recall that our ultimate goal here is not to estimate the α's, β's, and σ's, but to estimate the average value of y within each of the poststratification categories, and then to average over the population using the census numbers using equation (14.1).

A fuller model including non-nested factors

We expand the model to use all the demographic predictors used in the CBS weighting, including sex × ethnicity and age × education. We model age and education (with four categories each) with varying intercepts, and also model the 16 levels of the age × education interaction.

At the state level, we include indicators for the 5 regions of the country (Northeast, Midwest, South, West, and D.C., considered as a separate region because of its distinctive voting patterns), along with v.prev, a measure of previous Republican vote in the state (more precisely, the average Republican vote share in the three previous elections, adjusted for home-state and home-region effects in the previous elections).

We shall write the model using indexes j, k, l, m for state, age category, education category, and region:

$$\Pr(y_i = 1) = \text{logit}^{-1}\left(\beta^0 + \beta^{\text{female}} \cdot \text{female}_i + \beta^{\text{black}} \cdot \text{black}_i + \right.$$
$$\left. + \beta^{\text{female.black}} \cdot \text{female}_i \cdot \text{black}_i + \alpha^{\text{age}}_{k[i]} + \alpha^{\text{edu}}_{l[i]} + \alpha^{\text{age.edu}}_{k[i],l[i]} + \alpha^{\text{state}}_{j[i]}\right)$$
$$\alpha^{\text{state}}_j \sim \text{N}\left(\alpha^{\text{region}}_{m[j]} + \beta^{\text{v.prev}} \cdot \text{v.prev}_j, \sigma^2_{\text{state}}\right). \tag{14.2}$$

We also model the remaining multilevel coefficients:

$$\alpha^{\text{age}}_k \sim \text{N}(0, \sigma^2_{\text{age}}), \text{ for } k = 1, \ldots, 4 \tag{14.3}$$
$$\alpha^{\text{edu}}_l \sim \text{N}(0, \sigma^2_{\text{edu}}), \text{ for } l = 1, \ldots, 4$$
$$\alpha^{\text{age.edu}}_{k,l} \sim \text{N}(0, \sigma^2_{\text{age.edu}}), \text{ for } k = 1, \ldots, 4, \ l = 1, \ldots, 4$$
$$\alpha^{\text{region}}_m \sim \text{N}(0, \sigma^2_{\text{region}}), \text{ for } m = 1, \ldots, 5. \tag{14.4}$$

As with the non-nested linear models in Section 13.5, this model can be expressed in equivalent ways by moving the constant term β_0 around. Here we have included β^0 in the data-level regression and included no intercepts in the group-level models for the different batches of α's.

Another approach is to include constant terms in several places in the model, centering the distributions in (14.4) at $\mu_{\text{age}}, \mu_{\text{edu}}, \mu_{\text{age}}, \mu_{\text{age.edu}}$, and μ_{region}. This makes the model nonidentifiable, but it can then be reparameterized in terms of identifiable combinations of parameters. Such a *redundant parameterization* speeds computation and offers some conceptual advantages, and we shall return to it in Section 19.4.

We can quickly fit model (14.2) in R: we first construct the index variable for the age × education interaction and expand the state-level predictors to the data level:

```
age.edu <- n.edu*(age-1) + edu                                    R code
region.full <- region[state]
v.prev.full <- v.prev[state]
```

We then fit and display the full multilevel model, to get:

```
lmer(formula = y ~ black + female + black:female + v.prev.full +   R output
    (1 | age) + (1 | edu) + (1 | age.edu) + (1 | state) +
    (1 | region.full), family = binomial(link = "logit"))
              coef.est coef.se
(Intercept)   -3.5      1.0
black         -1.6      0.3
female        -0.1      0.1
v.prev.full    7.0      1.7
black:female  -0.2      0.4
Error terms:
  Groups         Name         Std.Dev.
```

```
state            (Intercept) 0.2
age.edu          (Intercept) 0.2
region.full      (Intercept) 0.2
edu              (Intercept) 0.1
age              (Intercept) 0.0
No residual sd
# of obs: 2015, groups: state,49; age.edu,16; region.full,5; edu,4; age,4
deviance = 2629.5
  overdispersion parameter = 1.0
```

Quickly reading this regression output:

- The intercept is not easily interpretable since it corresponds to a case in which `black`, `female`, and `v.prev` are all 0—but `v.prev` typically takes on values near 0.5 and is never 0.

- The coefficient for `black` is −1.6. Dividing by 4 (see page 82) yields a rough estimate that African-American men were 40% less likely than other men to support Bush, after controlling for age, education, and state.

- The coefficient for `female` is −0.1. Dividing by 4 yields a rough estimate that non-African-American women were very slightly less likely than non-African-American men to support Bush, after controlling for age, education, and state. However, the standard error on this coefficient is as large as the estimate itself, indicating that our sample size is too small for us to be certain of this pattern in the population.

- The coefficient for `v.prev.full` is 7.0, which, when divided by 4, is 1.7, suggesting that a 1% difference in a state's support for Republican candidates in previous elections mapped to a predicted 1.7% difference in support for Bush in 1988.

- The large standard error on the coefficient for `black:female` indicates that the sample size is too small to estimate this interaction precisely.

- The state-level errors have estimated standard deviation 0.2 on the logit scale. Dividing by 4 tells us that the states differed by approximately ±5% on the probability scale (over and above the differences explained by demographic factors).

- The differences among age-education groups and regions are also approximately ±5% on the probability scale.

- Very little variation is found among age groups or education groups after controlling for the other predictors in the model.

To make more precise inferences and predictions, we shall fit the model using Bugs (as described in Section 17.4), because with so many factors—including some with only 4 or 5 levels—the approximate inference provided by `lmer()` (which does not fully account for uncertainty in the estimated variance parameters) is not so reliable. It is still useful as a starting point, however, and we recommend performing the quick fit if possible before getting to more elaborate inference. In some other settings, it will be difficult to get Bugs to run successfully and we simply use the inferences from `lmer()`.

Graphing the estimated model

We would like to construct summary plots as we did with the multilevel models of Chapters 12 and 13. We alter the plotting strategy in two ways. First, the outcome is binary and so we plot $\Pr(y=1) = \mathrm{E}(y)$ as a function of the predictors; thus the graphs are curved, as are the classical generalized linear models in Chapter 6.

Our second modification of the plots is needed to deal with the many different predictors in our model: instead of plotting $E(y)$ as a function of each of the demographic inputs, we combine them into a linear predictor for demographics, which we shall call linpred_i:

$$\text{linpred}_i = \beta^0 + \beta^{\text{female}} \cdot \text{female}_i + \beta^{\text{black}} \cdot \text{black}_i +$$
$$+ \beta^{\text{female.black}} \cdot \text{female}_i \cdot \text{black}_i + \alpha^{\text{age}}_{k[i]} + \alpha^{\text{edu}}_{l[i]} + \alpha^{\text{age.edu}}_{k[i],l[i]}. \quad (14.5)$$

The estimates, 50% intervals, and 95% intervals for the demographic coefficients are displayed in Figure 14.1. Because all categories of each predictor variable have been included, these estimates can be interpreted directly as the contribution each makes to the sum, $X_i\beta$. So, for instance, if we were to predict the response for someone who is female, age 20, and with no high school diploma, we could simply take the constant term, plus the estimates for the corresponding three main effects plus the interaction between "18–29" and "no high school," plus the corresponding state coefficient, and then take the inverse-logit to obtain the probability of a Republican vote. As can be seen from the graph, the demographic factors other than ethnicity are estimated to have little predictive power. (Recall from Section 5.1 that we can quickly interpret logistic regression coefficients on the probability scale by dividing them by 4.)

For any survey respondent i, the regression prediction can then be written as

$$\Pr(y_i = 1) = \text{logit}^{-1}(\text{linpred}_i + \alpha^{\text{state}}_{j[i]}),$$

where linpred_i is the combined demographic predictor (14.5), and we can plot this for each state. We can do this in R—after first fitting the model in Bugs (as called from R) and attaching the resulting object, which puts arrays into the R workspace representing simulations for all the parameters from the model fit.

We summarize the linear predictor linpred_i from (14.5) by its average over the simulations. Recall that we are using simulations from the fitted model (see Section 17.4), which we shall call M3.bugs. As discussed in Chapter 16, the first step after fitting the model is to attach the Bugs object so that the vectors and arrays of parameter simulations can be accessed within the R workspace. Here is the code to compute the vector linpred:

```
attach.bugs (M3.bugs)                                              R code
linpred <- rep (NA, n)
for (i in 1:n){
  linpred[i] <- mean (b.0 + b.female*female[i] + b.black*black[i] +
    b.female.black*female[i]*black[i] + a.age[age[i]] + a.edu[edu[i]] +
    a.age.edu[age[i],edu[i]])
}
```

We can then make Figure 14.2 given the simulations from the fitted Bugs model:

```
par (mfrow=c(2,4))                                                 R code
for (j in displayed.states){
  plot (0, 0, xlim=range(linpred), ylim=c(0,1), yaxs="i",
        xlab="linear predictor", ylab="Pr (support Bush)",
        main=state.name[j], type="n")
  for (s in 1:20){
    curve (invlogit (a.state[s,j] + x), lwd=.5, add=TRUE, col="gray")}
  curve (invlogit (median (a.state[,j]) + x), lwd=2, add=TRUE)
  if (sum(state==j)>0) points (linpred[state==j], y.jitter[state==j])
}
```

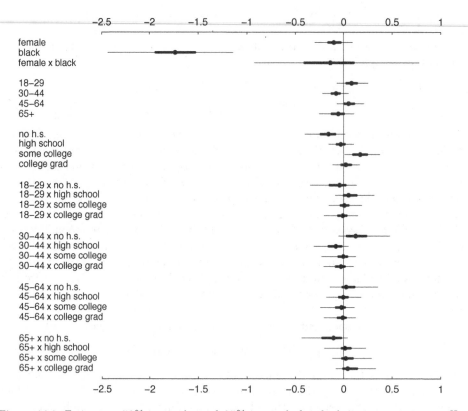

Figure 14.1 *Estimates, 50% intervals, and 95% intervals for the logistic regression coefficients for the demographic predictors in the model predicting the probability of supporting George Bush in polls before the 1988 presidential election. Recall that a change of x on the logistic scale corresponds to a change of at most x/4 on the probability scale. Thus, demographic factors other than ethnicity have small estimated predictive effects on vote preference.*

Figure 14.2 shows the result for a selection of eight states and illustrates a number of points about multilevel models. The solid lines display the estimated logistic regressions: thus, in any state, the probability of supporting Bush ranges from about 10% to 70% depending on the demographic variables—most importantly, ethnicity. Roughly speaking, there is about a 10% probability of supporting Bush for African Americans and about 60% for others, with other demographic variables slightly affecting the predicted probability. The variation among states is fairly small—you have to look at the different plots carefully to see it—but is important in allowing us to estimate average opinion by state, as we shall discuss. Changes of only a few percent in preferences can have large political impact.

The gray lines on the graphs represent uncertainty in the state-level coefficients, α_j^{state}. Alaska has no data at all, but the inference there is still reasonably precise—its α_j^{state} is estimated from its previous election outcome, its regional predictor (Alaska is categorized as a Western state), and from the distribution of the errors from the state-level regression. In general, the larger states such as California have more precise estimates than the smaller states such as Delaware—with more data in a state j, it is possible to estimate α_j^{state} more accurately.

The logistic regression curve is estimated for all states, even those such as Arizona with little range of x in the data (the survey included no black respondents from

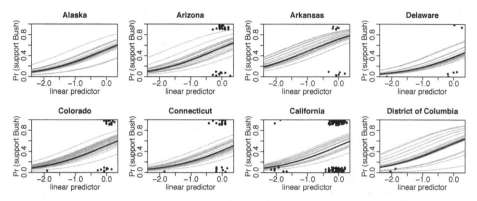

Figure 14.2 *Estimated probability of a survey respondent supporting Bush for president, as a function of the linear predictor for demographics, in each state (displaying only a selection of eight states, ordered by decreasing support for Bush, to save space). Dots show the data (y-jittered for visibility), and the heavy and light lines show the median estimate and 20 random simulation draws from the estimated model.*

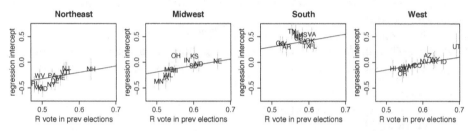

Figure 14.3 *Estimates and 50% intervals for the state coefficients α_j^{state}, plotted versus previous state vote v.prev$_j$, in each of the four regions of the United States. The estimated group-level regression line, $\alpha_j^{\text{state}} = \alpha_{m[j]}^{\text{region}} + \beta_j^{\text{v.prev}} \cdot$ v.prev$_j$, is overlain on each plot (corresponding to regions $m = 1, 2, 3, 4$).*

Arizona). The model is set up so the demographic coefficients are the same for all states, so the estimate of the logistic curve is pooled for all the data. If the model included an interaction between demographics and state, then we would see differing slopes, and more uncertainty about the slope in states such as Arizona that have less variation in their data.

Figure 14.3 displays the estimated logistic regression coefficients for the 50 states, grouping them by region and, within each region, showing the state-level regression on **v.prev**, the measure of Republican vote in the state in previous presidential elections. Region and previous vote give good but not perfect predictions of the state-level coefficients in the public opinion model.

Using the model inferences to estimate average opinion for each state

The logistic regression model gives the probability that any adult will prefer Bush, given the person's sex, ethnicity, age, education level, and state. We can now compute weighted averages of these probabilities to represent the proportion of Bush supporters in any specified subset of the population.

We first extract from the U.S. Census the counts N_l in each of the 3264 cross-classification cells and create a 3264 × 6 data frame, **census**, indicating the sex,

	female	black	age	edu	state	N
1	0	0	1	1	1	66177
2	0	1	1	1	1	32465
3	1	0	1	1	1	59778
4	1	1	1	1	1	27416
5	0	0	2	1	1	83032
. . .						
3262	0	1	4	4	51	5
3263	1	0	4	4	51	2610
3264	1	1	4	4	51	5

Figure 14.4 *The data frame* census *in R used for poststratification in the election polling example. The categories are ordered by ethnicity, sex, age category, education category, and state. The states are in alphabetical order; thus there were, according to the U.S. Census, 66177 non-African-American men between 18 and 29 with less than a high school education in Alabama, ..., and 5 African American women over 65 with a college education in Wyoming.*

ethnicity, age, education, state, and number of people corresponding to each cell, as shown in Figure 14.4.

We then compute the expected response y^{pred}—the probability of supporting Bush for each cell. Assuming we have n.sims simulation draws after fitting the model in Bugs (see Chapter 16), we construct the following n.sims × 3264 matrix:

R code
```
L <- ncol (census)
y.pred <- array (NA, c(n.sims, L))
for (l in 1:L){
  y.pred[,l] <- invlogit(b.0 + b.female*census$female[l] +
    b.black*census$black[l] +
    b.female.black*census$female[l]*census$black[l] +
    a.age[,census$age[l]] + a.edu[,census$edu[l]] +
    a.age.edu[,census$age[l],census$edu[l]] + a.state[,census$state[l]])
}
```

For each state j, we are estimating the average response in the state,

$$y^{\text{pred}}_{\text{state } j} = \frac{\sum_{l \in j} N_l \theta_l}{\sum_{l \in j} N_l},$$

summing over the 64 demographic categories within the state. Here, we are using l as a general stratum indicator (not the same l used to index education categories in model (14.2); we are simply running out of "index"-type letters from the middle of the alphabet). The notation "$l \in j$" is shorthand for "category l represents a subset of state j." In R:

R code
```
y.pred.state <- array (NA, c(n.sims, n.state))
for (s in 1:n.sims){
  for (j in 1:n.state){
    ok <- census$state==j
    y.pred.state[s,j] <- sum(census$N[ok]*y.pred[s,ok])/sum(census$N[ok])
  }
}
```

We can then summarize these n.sims simulations to get a point prediction and a 50% interval for the proportion of adults in each state who supported Bush:

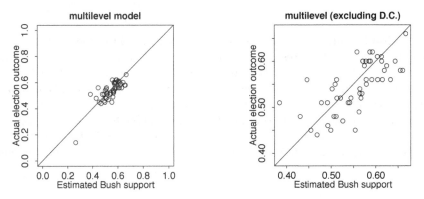

Figure 14.5 *For each state, the proportion of the two-party vote received by George Bush in 1988, plotted versus the support for Bush in the state, as estimated from a multilevel model applied to pre-election polls. The second plot excludes the District of Columbia in order to more clearly show the 50 states.*

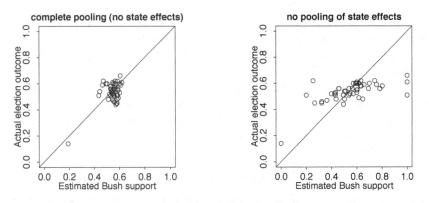

Figure 14.6 *For each state, Bush's vote in 1988 plotted versus his support in the polls, as estimated from (a) the complete-pooling model (using demographics alone with no state predictors), and (b) the no-pooling models (estimating each state separately). The two models correspond to $\sigma_{\text{state}} = \sigma_{\text{region}} = 0$ and ∞, respectively. Compare to Figure 14.5a, which shows results from the multilevel model (with σ_{state} and σ_{region} estimated from data).*

```
state.pred <- array (NA, c(3,n.state))          R code
for (j in 1:n.state){
  state.pred[,j] <- quantile (y.pred.state[,j], c(.25,.5,.75))
}
```

Comparing public opinion estimates to election outcomes

In this example, the estimates of the model come from opinion polls taken immediately before the election, and they can be externally validated by comparing to the actual election outcomes. We can thus treat this as a sort of "laboratory" for testing the accuracy of multilevel models and any other methods that might be used to estimate state-level opinions from national polls.

Figure 14.5 shows the actual election outcome for each state, compared to the model-based estimates of the proportion of Bush supporters. The fit is pretty good, with no strong systematic bias and an average absolute error of only 4.0%.

Comparison to simpler methods

By comparison, Figure 14.6 shows the predictive performance of the estimates based on complete pooling of states (estimating opinion solely based on demographics, thus setting $\alpha_j^{\text{state}} \equiv 0$ for all states) and no pooling (corresponding to completely separate estimates for each state, thus setting $\sigma_{\text{state}} = \sigma_{\text{region}} = \infty$). The complete-pooling model generally shrinks the state estimates too close toward the mean, whereas the no-pooling model does not shrink them enough. To make a numerical comparison, the average absolute error of the state estimates is 4.0% for the multilevel analysis, compared to 5.4% for complete pooling and 10.8% for no pooling.

14.2 Red states and blue states: what's the matter with Connecticut?

Throughout the twentieth century and even before, the Democratic Party in the United States has been viewed as representing the party of the lower classes and thus, by extension, the "average American." More recently, however, a different perspective has taken hold, in which the Democrats represent the elites rather than the masses. These patterns are complicated; on one hand, in recent U.S. presidential elections the Democrats have done best in the richer states of the Northeast and West (often colored blue in electoral maps) while the Republicans have dominated in the poorer "red states" in the South and between the coasts. On the other hand, using census and opinion poll data since 1952, we find that higher-income voters continue to support the Republicans in presidential elections.

We can understand these patterns, first by fitting a sequence of classical regressions and displaying estimates over time (as in Section 4.7), then by fitting some multilevel models:

- Aggregate, by state: to what extent do richer states favor the Democrats?

- Nationally, at the level of the individual voter: to what extent do richer voters support the Republicans?

- Individual voters within states: to what extent do richer voters support the Republicans, within any given state? In other words, how much does context matter?

We fit these models quickly with lmer() and then with Bugs, whose simulations we used to plot and understand the model. Here we describe the model and its estimate without presenting the steps of computation.

Classical regressions of state averages and individuals

Richer states now support the Democrats. We first present the comparison of red and blue states—more formally, regressions of Republican share of the two-party vote on state average per capita income (in tens of thousands of 1996 dollars). Figure 14.7a shows that, since the 1976 election, there has been a steady downward trend in the income coefficient over time. As time has gone on, richer states have increasingly favored the Democrats. For the past twenty years, the same patterns appear when fitting southern and non-southern states separately (Figure 14.7b,c).

Richer voters continue to support the Republicans overall. We fit a logistic regression of reported presidential vote preference ($y_i = 1$ for supporters of the Republican, 0 for the Democrats, and excluding respondents who preferred other candi-

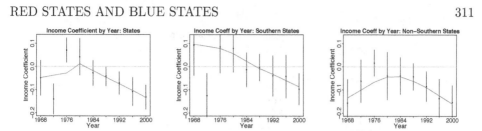

Figure 14.7 *(a) Regression predicting Republican vote share by average income in each state. The model was fit separately for each election year. Estimates and 95% error bars are shown. (b, c) Same model but fit separately to southern and non-southern states each year. Republicans do better in poor states than rich states, especially in recent years.*

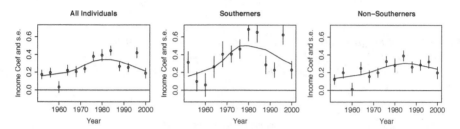

Figure 14.8 *Coefficients for income in logistic regressions of Republican vote, fit to National Election Studies data from each year. The positive coefficients indicate that higher-income voters have consistently supported the Republicans, a pattern that holds both within and outside the South.*

dates or expressed no opinion) on personal income,[1] fit separately to the National Election Study from each presidential election since 1952. Figure 14.8 shows that higher-income people have been consistently more likely to vote Republican. These patterns remain when ethnicity, sex, education, and age are added into the model: after controlling for these other individual-level predictors, the coefficient of income is still consistently positive.

A paradox?   The conflicting patterns of Figures 14.7 and 14.8 have confused many political commentators. How can we understand the pattern of richer states supporting the Democrats, while richer voters support the Republicans? We shall use multilevel modeling to simultaneously study patterns within and between states.

Varying-intercept model of income and vote preference within states

We now focus on the 2000 presidential election using the National Annenberg Election Survey, which, with more than 100,000 respondents, allows accurate estimation of patterns within individual states. We fit a multilevel model that allows income to predict vote preference within each state, while also allowing systematic differences between states:

$$\Pr(y_i = 1) = \text{logit}^{-1}(\alpha_{j[i]} + \beta x_i), \quad \text{for } i = 1, \ldots, n, \tag{14.6}$$

---

[1] The National Election Study uses 1 = 0–16 percentile, 2 = 17–33 percentile, 3 = 34–67 percentile, 4 = 68–95 percentile, 5 = 96–100 percentile. We label these as $-2, -1, 0, 1, 2$, centering at zero (see Section 4.2) so that we can more easily interpret the intercept terms of regressions that include income as a predictor.

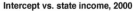

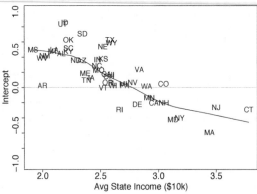

Figure 14.9 *Estimated state intercepts α_j in the varying-intercept logistic regression model (14.6)–(14.7) predicting Republican vote intention given individual income, plotted versus average state income. A nonparametric regression line fitted to the estimates is overlain for convenience.*

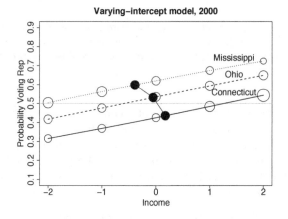

Figure 14.10 *The paradox is no paradox. From the multilevel logistic regression model for the 2000 election: probability of supporting Bush as a function of income category, for a a rich state (Connecticut), a medium state (Ohio), and a poor state (Mississippi). The open circles show the relative proportion (as compared to national averages) of households in each income category in each of the three states, and the solid circles show the average income level and estimated average support for Bush for each state. Within each state, richer people are more likely to vote Republican, but the states with higher income give more support to the Democrats.*

where $j[i]$ indexes the state (from 1 to 50) corresponding to respondent i, x_i is the person's household income (on the five-point scale), and n is the number of respondents in the poll.

We set up a state-level regression for the coefficients α_j, using the state average income level as a group-level predictor, which we label u_j:

$$\alpha_j \sim \text{N}(\gamma_0 + \gamma_1 u_j, \sigma_\alpha^2), \quad \text{for } j = 1, \ldots, 50. \tag{14.7}$$

Figure 14.9 shows the estimated state intercepts α_j, plotted versus average state income. There is a negative correlation between intercept and state income, which

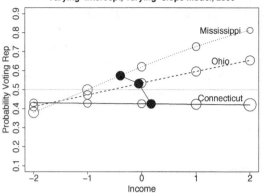

Figure 14.11 *From the multilevel logistic regression with varying intercepts and slopes for the 2000 election: probability of supporting Bush as a function of income category, for a a rich state (Connecticut), a medium state (Ohio), and a poor state (Mississippi). The open circles show the relative proportion (as compared to national averages) of households in each income category in each of the three states, and the solid circles show the average income level and estimated average support for Bush for each state. Income is a very strong predictor of vote preference in Mississippi, a weaker predictor in Ohio, and does not predict vote choice at all in Connecticut. See Figure 14.12 for estimated slopes in all 50 states, and compare to Figure 14.10, in which the state slopes are constrained to be equal.*

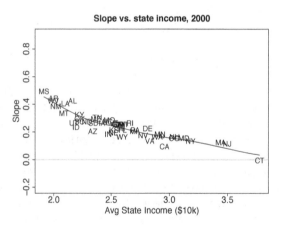

Figure 14.12 *Estimated coefficient for income within state plotted versus average state income, for the varying-intercept, varying-slope multilevel model (14.8)–(14.9) fit to the Annenberg survey data from 2000. A nonparametric regression line fitted to the estimates is overlain for convenience.*

tells us that, after adjusting for individual income, voters in richer states tend to support Democrats.

To understand the model as a whole, we display in Figure 14.10 the estimated logistic regression line, $\text{logit}^{-1}(\alpha_j + \beta x)$, for three states j: Connecticut (the richest state), Ohio (a state in the middle of the income distribution), and Mississippi (the poorest state). The graph shows a statistical resolution of the red-blue paradox. Within each state, income is positively correlated with Republican vote choice, but average income varies by state. For each of the three states in the plot, the open circles show the relative proportion of households in each income category (as compared to national averages), and the solid circle shows the average income level and estimated average support for Bush in the state. The Bush-supporting states have more lower-income people, and as a result there is a negative correlation between average state income and state support for Bush, even amid the positive slope for each state. The poor people in "red" (Republican-leaning) states tend to be Democrats; the rich people in "blue" (Democratic-leaning) states tend to be Republicans. Income matters; also geography matters. Individual income is a positive predictor, and state average income is a negative predictor, of Republican presidential vote support.

Varying-intercept, varying-slope model

As Figure 14.10 shows, income and state are both predictive of vote preference. It is thus natural to consider their interaction, which in a multilevel context is a varying-intercept, varying-slope model:

$$\Pr(y_i\!=\!1) = \text{logit}^{-1}(\alpha_{j[i]} + \beta_{j[i]}x_i), \quad \text{for } i = 1, \ldots, n, \tag{14.8}$$

where, as in (14.6), x_i is respondent i's income (on the -2 to $+2$ scale). The state-level intercepts and slopes that are themselves modeled given average state incomes u_j:

$$
\begin{aligned}
\alpha_j &= \gamma_0^\alpha + \gamma_1^\alpha u_j + \epsilon_j^\alpha, \quad \text{for } j = 1, \ldots, 50 \\
\beta_j &= \gamma_0^\beta + \gamma_1^\beta u_j + \epsilon_j^\beta, \quad \text{for } j = 1, \ldots, 50,
\end{aligned}
\tag{14.9}
$$

with errors $\epsilon_j^\alpha, \epsilon_j^\beta$ having mean 0, variances $\sigma_\alpha^2, \sigma_\beta^2$, and correlation ρ, all estimated from data. By including average income as a state-level predictor, we are not requiring the intercepts and slopes to vary linearly with income—the error terns ϵ_j allow for deviation from the model—but rather are allowing the model to find such linear relations to the extent they are supported by the data.

From this new model, we indeed find strong variation among states in the role of income in predicting vote preferences. Figure 14.11 recreates Figure 14.10 with the estimated varying intercepts and slopes. As before, we see generally positive slopes within states and a negative slope between states. What is new, though, is a systematic pattern of the within-state slopes, with the steepest slope in the poorest state—Mississippi—and the shallowest slope in the richest state—Connecticut.

Figure 14.12 shows the estimated slopes for all 50 states and reveals a clear pattern, with high coefficients—steep slopes—in poor states and low coefficients in rich states. Income matters more in "red America" than in "blue America." The varying-intercept, varying-slope multilevel model has been a direct approach for us to discover these patterns.

14.3 Item-response and ideal-point models

We could have introduced these in Chapter 6 in the context of classical generalized linear models, but item-response and ideal-point models are always applied to data with multilevel structure, typically non-nested, for example with measurements associated with persons and test items, or judges and cases. As with the example of the previous section, we present the models here, deferring computation until the presentation of Bugs in Part 2B of this book.

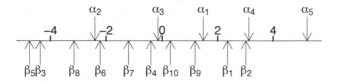

Figure 14.13 *Illustration of the logistic item-response (Rasch) model,* $\Pr(y_i = 1) = \mathrm{logit}^{-1}(\alpha_{j[i]} - \beta_{k[i]})$, *for an example with 5 persons j (with abilities α_j) and 10 items k (with difficulties β_k). If your ability α is greater than the difficulty β of an item, then you have a better-than-even chance of getting that item correct. This graph also illustrates the nonidentifiability in the model: the probabilities depend only on the* relative *positions of the ability and difficulty parameters; thus, a constant could be added to all the α_j's and all the β_k's, and the model would be unchanged. One way to resolve this nonidentifiability is to constrain the α_j's to have mean 0. Another solution is to give the α_j's a distribution with mean fixed at 0.*

The basic model with ability and difficulty parameters

A standard model for success or failure in testing situations is the logistic item-response model, also called the Rasch model. Suppose J persons are given a test with K items, with $y_{jk} = 1$ if the response is correct. Then the logistic model can be written as

$$\Pr(y_{jk}=1) = \mathrm{logit}^{-1}(\alpha_j - \beta_k), \qquad (14.10)$$

with parameters:

- α_j: the *ability* of person j
- β_k: the *difficulty* of item k.

In general, not every person is given every item, so it is convenient to index the individual responses as $i = 1, \ldots, n$, with each response i associated with a person $j[i]$ and item $k[i]$. Thus model (14.10) becomes

$$\Pr(y_i=1) = \mathrm{logit}^{-1}(\alpha_{j[i]} - \beta_{k[i]}). \qquad (14.11)$$

Figure 14.13 illustrates the model as it might be estimated for 5 persons with abilities α_j, and 10 items with difficulties β_k. In this particular example, questions 5, 3, and 8 are easy questions (relative to the abilities of the persons in the study), and all persons except person 2 are expected to answer more than half the items correctly. More precise probabilities can be calculated using the logistic distribution: for example, α_2 is 2.4 higher than β_5, so the probability that person 2 correctly answers item 5 is $\mathrm{logit}^{-1}(2.4) = 0.92$, or 92%.

Identifiability problems

This model is not identified, whether written as (14.10) or as (14.11), because a constant can be added to all the abilities α_j and all the difficulties β_k, and the predictions of the model will not change. The probabilities depend only on the *relative* positions of the ability and difficulty parameters. For example, in Figure 14.13, the scale could go from -104 to -96 rather than -4 to 4, and the model would be unchanged—a difference of 1 on the original scale is still a difference of 1 on the shifted scale.

From the standpoint of classical logistic regression, this nonidentifiability is a simple case of collinearity and can be resolved by constraining the estimated parameters in some way: for example, setting $\alpha_1 = 0$ (that is, using person 1 as a "baseline"), setting $\beta_1 = 0$ (so that a particular item is the comparison point), constraining the α_j's to sum to 0, or constraining the β_j's to sum to 0. In a multilevel model, such constraints are unnecessary, as we discuss next.

Multilevel model

The natural multilevel model for (14.11) assigns normal distributions to the ability and difficulty parameters:

$$\alpha_j \sim N(\mu_\alpha, \sigma_\alpha^2), \text{ for } j = 1, \dots, J$$
$$\beta_k \sim N(\mu_\beta, \sigma_\beta^2), \text{ for } k = 1, \dots, K.$$

This model is nonidentified for the reasons discussed above: now it is μ_α and μ_β that are not identified, because a constant can be added to each without changing the predictions. The simplest way to identify the multilevel model is set μ_α to 0, or to set μ_β to 0 (but not both).

As usual, we can add group-level predictors. In this case, the "groups" are the persons and items:

$$\alpha_j \sim N(X_j^\alpha \gamma_\alpha, \sigma_\alpha^2), \text{ for } j = 1, \dots, J$$
$$\beta_k \sim N(X_j^\beta, \sigma_\beta^2), \text{ for } k = 1, \dots, K.$$

In an educational testing example, the person-level predictors X^α could include age, sex, and previous test scores, and the item-level predictors X^β could include a prior measure of item difficulty (perhaps the average score for that item from a previous administration of the test).

Defining the model using redundant parameters

Another way to identify the model is by allowing the parameters α and β to "float" and then defining new quantities that are well identified. The new quantities can be defined, for example, by rescaling based on the mean of the α_j's:

$$\alpha_j^{\text{adj}} = \alpha_j - \bar{\alpha}, \text{ for } j = 1, \dots, J$$
$$\beta_k^{\text{adj}} = \beta_k - \bar{\alpha}, \text{ for } k = 1, \dots, K. \tag{14.12}$$

The new ability parameters α_j^{adj} and difficulty parameters β_k^{adj} are well defined, and they work in place of α and β in the original model:

$$\Pr(y_i = 1) = \text{logit}^{-1}(\alpha_{j[i]}^{\text{adj}} - \beta_{k[i]}^{\text{adj}}).$$

This holds because we subtracted the same constant from the α's and β's in (14.12). For example, it would *not* work to subtract $\bar{\alpha}$ from the α_j's and $\bar{\beta}$ from the β_k's because then we would lose our ability to distinguish the position of the parameters relative to each other.

Adding a discrimination parameter

The item-response model can be generalized by allowing the slope of the logistic regression to vary by item:

$$\Pr(y_i = 1) = \text{logit}^{-1}(\gamma_{k[i]}(\alpha_{j[i]} - \beta_{k[i]})). \tag{14.13}$$

In this new model, γ_k is called the *discrimination* of item k: if $\gamma_k = 0$, then the item does not "discriminate" at all ($\Pr(y_i = 1) = 0.5$ for any person), whereas high

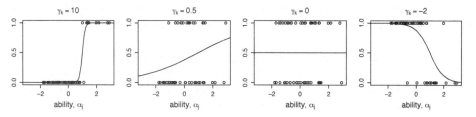

Figure 14.14 *Curves and simulated data from the logistic item-response (Rasch) model for items k with "difficulty" parameter $\beta_k = 1$ and high, low, zero, and negative "discrimination" parameters γ_k.*

values of γ_k correspond to strong relation between ability and the probability of getting a correct response. Figure 14.14 illustrates.

In educational testing, it is generally desirable for items k to have high values of γ_k, because the responses to these items can better "discriminate" between high and low abilities (see the left plot in Figure 14.14). The ideal test would have several items, each with high γ_k, and with difficulties β_k that span the range of the abilities of the persons being tested. Items with γ_k near zero do not do a good job at discriminating between abilities (see the center two plots in Figure 14.14), and negative values of γ_k correspond to items where low-ability persons do better. Such items typically represent mistakes in the construction of the test.

Including the discrimination parameter creates additional identifiability problems which we will discuss in the context of an example in the next section.

An ideal-point model for Supreme Court voting

Ideal-point modeling is an application of item-response models to a setting where what is being measured is not "ability" of individuals and "difficulty" of items, but rather positions of individuals and items on some scale of values.

We illustrate with a study of voting records of U.S. Supreme Court justices, using all the Court's decisions since 1954. Each vote i is associated with a justice $j[i]$ and a case $k[i]$, with an outcome y_i that equals 1 if the justice voted "yes" on the case and 0 if "no." In this particular example, the votes have been coded so that a "yes" response ($y_i = 1$) is intended to correspond to the politically "conservative" outcome, with "no" ($y_i = 0$) corresponding to a "liberal" vote.

As with the item-response models discussed above, the data are modeled with a logistic regression, with the probability of voting "yes" depending on the "ideal point" α_j for each justice, the "position" β_k for each case, and a "discrimination parameter" γ_k for each case, following the three-parameter logistic model (14.13).

The positions on this scale (equivalent to the α's and β's on Figure 14.14) represent whatever dimension is best able to explain the voting patterns. For the Supreme Court, we represent it as an ideological dimension, with liberal justices having positions on the left side of the scale (negative α_j's) and conservatives being on the right side (positive α_j's).

For any given justice j and case k, the difference between α_j and β_k indicates the relative positions of the justice and the case—if a justice's ideal point is near a case's position, then the case could go either way, but if the ideal point is far from the position, then the justice's vote is highly predictable. The discrimination parameter γ_k captures the importance of the positioning in determining the justices' votes: if $\gamma_k = 0$, the votes on case k are purely random; and if γ_k is very large (in absolute value), then the relative positioning of justice and case wholly determines the outcome. Changing the sign of γ changes which justices are expected to vote yes and which to vote no.

Model (14.13) has two indeterminacies: an additive *aliasing* in α and β (that is, a situation in which values of α and β can be changed while keeping the model's predictions unchanged), and a multiplicative aliasing in all three parameters. The additive aliasing occurs because a constant can be added to all the α's and all the β's, leaving the model predictions (and thus the likelihood) unchanged. The multiplicative aliasing arises when multiplying the γ's by a constant and dividing the α's and β's by that same constant. We can resolve both these indeterminacies by constraining the α_j's to have mean 0 and standard deviation 1 or, in a multilevel context, by giving the α_j a $N(0,1)$ distribution. In contrast the parameters β and γ are unconstrained (or, in a multilevel context, have $N(\mu_\beta, \sigma_\beta^2)$ and $N(\mu_\gamma, \sigma_\gamma^2)$ distributions whose means and variances are estimated from the data, as part of a multilevel model).

Even after constraining the distribution of the position parameters α_j, one indeterminacy remains in model (14.13): a reflection invariance associated with multiplying all the γ_k's, α_j's, and β_k's by -1. If no additional constraints are assigned to this model, this aliasing will cause a bimodal likelihood and posterior distribution. It is desirable to select just one of these modes for our inferences. (Among other problems, if we include both modes, then each parameter will have two maximum likelihood estimates and a posterior mean of 0.)

We first briefly discuss two simple and natural ways of resolving the aliasing. The first approach is to constrain the γ's to all have positive signs. This might seem to make sense, since the outcomes have been precoded so that positive y_i's correspond to conservative votes. However, we do not use this approach because it relies too strongly on the precoding, which, even if it is generally reasonable, is not perfect. We would prefer to estimate the ideological direction of each vote from the data and then compare to the precoding to check that the model makes sense (and to explore any differences found between the estimates and the precoding).

A second approach to resolving the aliasing is to choose one of the α's, β's, or γ's, and restrict its sign or choose two and constrain their relative position. For example, we could constrain α_j to be negative for the extremely liberal William Douglas, or constrain α_j to be positive for the extremely conservative Antonin Scalia. Or, we could constrain Douglas's α_j to be less than Scalia's α_j.

Only a single constraint is necessary to resolve the two modes; if possible, however, it should be a clear-cut division. One can imagine a general procedure that would be able to find such divisions based on the data, but in practice it is simpler to constrain using prior information such as the identification of extremely liberal and conservative judges in this example. (Not all choices of constraints would work. For example, if we were to constrain $\alpha_j > 0$ for a merely moderately conservative judge such as Sandra Day O'Connor, this could split the likelihood surface across both modes, rather than cleanly selecting a single mode.)

The alternative approach we actually use in this example is to encode the additional information in the form of a group-level regression predictor, whose coefficient we constrain to be positive. Various case-level and justice-level predictors can be added to model (14.13), but the simplest is an indicator that equals 1 for Scalia, -1 for Douglas, and 0 for all other justices. We set up a multilevel model for the justices' ideal points,

$$\alpha_j = \delta x_j + \text{error}_j \tag{14.14}$$

where x_j is this Scalia/Douglas predictor. Constraining the regression coefficient

$\delta > 0$ identifies the model (by aligning the positive direction with the difference between these two extreme justices) but in a flexible way that allows us to estimate our full model.

Two-dimensional item-response or ideal-point models

In a two-dimensional item-response model, the task of getting an item correct requires a combination of two "skills," which can be represented for each person j as a two-dimensional "ability" vector $(\alpha_j^{(1)}, \alpha_j^{(2)})$. (For example, on a high school general aptitude test, the two dimensions might correspond to verbal and mathematical ability.) The two-dimensional "difficulty" parameter $(\beta_k^{(1)}, \beta_k^{(2)})$ represents the thresholds required to perform well on the task, and the discrimination parameters $\gamma_k^{(1)}, \gamma_k^{(2)}$ indicate the relevance of each of the two skills to task k.

Success on the two skills can be combined in a variety of ways. For example, in a "conjunctive" model, both skills are required to perform the task correctly; thus,

$$\text{conjunctive model:} \quad \Pr(y_i = 1) \;=\; \text{logit}^{-1}\left[\gamma_{k[i]}^{(1)}\left(\alpha_{j[i]}^{(1)} - \beta_{k[i]}^{(1)}\right)\right]$$
$$\times \text{logit}^{-1}\left[\gamma_{k[i]}^{(2)}\left(\alpha_{j[i]}^{(2)} - \beta_{k[i]}^{(2)}\right)\right].$$

In a "disjunctive" model, either skill is sufficient to perform the task:

$$\text{disjunctive model:} \quad 1 - \Pr(y_i = 1) \;=\; \left(1 - \text{logit}^{-1}\left[\gamma_{k[i]}^{(1)}\left(\alpha_{j[i]}^{(1)} - \beta_{k[i]}^{(1)}\right)\right]\right)$$
$$\times \left(1 - \text{logit}^{-1}\left[\gamma_{k[i]}^{(2)}\left(\alpha_{j[i]}^{(2)} - \beta_{k[i]}^{(2)}\right)\right]\right).$$

Perhaps the most straightforward model is additive on the logistic scale:

$$\text{additive model:} \quad \Pr(y_i = 1) \;=\; \text{logit}^{-1}\left[\gamma_{k[i]}^{(1)}\left(\alpha_{j[i]}^{(1)} - \beta_{k[i]}^{(1)}\right)\right.$$
$$\left. + \gamma_{k[i]}^{(2)}\left(\alpha_{j[i]}^{(2)} - \beta_{k[i]}^{(2)}\right)\right].$$

In the "ideal-point" formulation of these models, α_j represents the ideal point of justice j in two dimensions (for example, a left-right dimension for economic issues, and an authoritarian-libertarian dimension on social issues), β_k is the indifference point for case k in these dimensions, the signs of the two components of γ_k give the direction of a Yes vote in terms of the two issues, and the absolute values of $\gamma_k^{(1)}, \gamma_k^{(2)}$ indicate the importance of each issue in determining the vote.

Other generalizations

As formulated so far, the probabilities in the item-response and ideal-point models range from 0 to 1 and are symmetric about 0.5 (see Figure 14.14). Real data do not necessarily look like this. One simple way to generalize the model is to limit the probabilities to a fixed range:

$$\Pr(y_i = 1) = \pi_1 + (1 - \pi_0 - \pi_1)\text{logit}^{-1}[\gamma_{k[i]}(\alpha_{j[i]} - \beta_{k[i]})].$$

In this model, every person has an immediate probability of π_1 of success and π_0 of failure, with the logistic regression model applying to the remaining $(1 - \pi_0 - \pi_1)$ of outcomes. For example, suppose we are modeling responses to a multiple-choice exam, and $\pi_1 = 0.25$ and $\pi_0 = 0.05$. We could interpret this as a 25% chance of getting an item correct by guessing, along with a 5% chance of getting an item wrong by a careless mistake.

Another way to generalize item response and ideal point models is to go beyond the logistic distribution, for example using a robit model as described in Section 6.6 that allows for occasional mispredictions.

14.4 Non-nested overdispersed model for death sentence reversals

So far in this chapter we have presented logistic regression for binary data points y_i that can equal 0 or 1. The model can also be used for proportions, in which each data point y_i equals the number of "successes" out of n_i chances. For example, Section 6.3 describes data on death penalty reversals, in which i indexes state-years (for example, Alabama in 1983), n_i is the number of death sentences given out in that particular state in that particular year, and y_i is the number of these death sentences that were reversed by a higher court. We now describe how we added multilevel structure to this model.

Non-nested model for state and year coefficients

The death penalty model had several predictors in X, including measures of the frequency that the death sentence was imposed, the backlog of capital cases in the appeals courts, the level of political pressure on judges, and other variables at the state-year level.

In addition, we included indicators for the years from 1973 to 1995 and the 34 states (all of those in this time span that had death penalty laws). The regression model with all these predictors can be written as

$$
\begin{aligned}
y_i &\sim \text{Bin}(n_i, p_i) \\
p_i &= \text{logit}^{-1}(X_i\beta + \alpha_{j[i]} + \gamma_{t[i]}),
\end{aligned} \tag{14.15}
$$

where j indexes states and t indexes years. We complete the multilevel model with distributions for the state and year coefficients,

$$
\begin{aligned}
\alpha_j &\sim \text{N}(0, \sigma_\alpha^2) \\
\gamma_t &\sim \text{N}(a + bt, \sigma_\gamma^2).
\end{aligned}
$$

The coefficients for year include a linear time trend to capture the overall increase in reversal rates during the period under study. The model for the γ_t's also includes an intercept, and so we do not need to include a constant term in the model for the α_j's or in the matrix X of individual-level predictors in (14.15).

In this particular example, we are not particularly interested in the coefficients for individual states or years; rather, we want to include these sources of variability into the model in order to get appropriate uncertainty estimates for the coefficients of interest, β.

Multilevel overdispersed binomial regression

Testing for overdispersion. Model (14.15) is inappropriate for the death penalty data because the data are overdispersed, as discussed in Section 6.3. To measure the overdispersion, we compute the standardized residuals, $z_i = (y_i - p_i)/\sqrt{p_i(1 - p_i)/n_i}$ with p_i as defined in (14.15). Under the binomial model, the residuals should have mean 0 and standard deviation 1, and so $\sum_i z_i^2$ should look like a random draw from a χ^2 distribution with degrees of freedom equal to 520 (the number of state-years in the data).

Testing for overdispersion in a classical binomial regression is described in Section

6.3, where the z_i's are computed based on estimated probabilities $\hat{p}_i$, and $\sum_i z_i^2$ is compared to a χ^2 distribution with degrees of freedom adjusted for the number of coefficients estimated in the model.

Beta-binomial model. There are two natural overdispersed generalizations of the multilevel binomial regression (14.15). The first approach uses the beta-binomial distribution:

$$y_i \sim \text{beta-binomial}(n_i, p_i, \omega),$$

where $\omega \geq 1$ is the overdispersion parameter (and the model with $\omega = 1$ reduces to the binomial).

Binomial-normal model. The other direct way to construct an overdispersed binomial distribution is to add normal errors on the logistic scale, keeping the binomial model but adding a data-level error ξ_i to the linear predictor in (14.15):

$$p_i = \text{logit}^{-1}(X_i\beta + \alpha_{j[i]} + \gamma_{t[i]} + \xi_i),$$

with these errors having their own normal distribution:

$$\xi_i \sim \text{N}(0, \sigma_\xi^2).$$

The resulting model reduces to the binomial when $\sigma_\xi = 0$; otherwise it is overdispersed.

With moderate sample sizes, it is typically difficult to distinguish between the beta-binomial and binomial-normal models, and the choice between them is one of convenience. The beta-binomial model adds only one new parameter and so can be easier to fit; however, the binomial-normal model has the advantage that the new error term ξ_i is on the same scale as the group-level predictors, α_j and γ_t, which can make the fitted model easier to understand.

14.5 Bibliographic note

Multilevel logistic regression has a long history in the statistical and applied literature which we do not attempt to trace here: the basic ideas are the same as in multilevel linear models (see references in Sections 12.10 and 13.8) but with complications arising from the discreteness of the data and the nonlinearity of some of the computational steps.

The example of state-level opinions from national polls comes from Gelman and Little (1997) and Park, Gelman, and Bafumi (2004). The analysis of income and voting comes from Gelman, Shor, et al. (2005); see also Wright (1989), Ansolabehere, Rodden, and Snyder (2005), and McCarty, Poole, and Rosenthal (2005) for related work. Figure 14.10, which simultaneously displays patterns within and between groups, is related to the "B-K plot" (discussed by Wainer, 2002, and named after Baker and Kramer, 2001).

The multilevel framework for item-response and ideal-point models appears in Bafumi, Gelman, and Park (2005). See Lord and Novick (1968) and van der Linden and Hambleton (1997) for more on item-response models, and Poole and Rosenthal (1997), Jackman (2001), and Martin and Quinn (2002a) for more on ideal-point models. Loken (2004) discusses identifiability problems in models with aliasing.

The death sentencing example comes from Gelman, Liebman, et al. (2004). See Donohue and Wolfers (2006) for an overview of some of the research literature on death sentencing.

14.6 Exercises

1. The folder **nes** contains the survey data of presidential preference and income for the 1992 election analyzed in Section 5.1, along with other variables including sex, ethnicity, education, party identification, political ideology, and state.

   (a) Fit a logistic regression predicting support for Bush given all these inputs except state. Consider how to include these as regression predictors and also consider possible interactions.

   (b) Now formulate a model predicting support for Bush given the same inputs but allowing the intercept to vary over state. Fit using lmer() and discuss your results.

   (c) Create graphs of the probability of choosing Bush given the linear predictor associated with your model separately for each of eight states as in Figure 14.2.

2. The well-switching data described in Section 5.4 are in the folder **arsenic**.

   (a) Formulate a multilevel logistic regression model predicting the probability of switching using log distance (to nearest safe well) and arsenic level and allowing intercepts to vary across villages. Fit this model using lmer() and discuss the results.

   (b) Extend the model in (b) to allow the coefficient on arsenic to vary across village, as well. Fit this model using lmer() and discuss the results.

   (c) Create graphs of the probability of switching wells as a function of arsenic level for eight of the villages.

   (d) Compare the fit of the models in (a) and (b).

3. Three-level logistic regression: the folder **rodents** contains data on rodents in a sample of New York City apartments.

   (a) Build a varying intercept logistic regression model (varying over buildings) to predict the presence of rodents (the variable **rodent2** in the dataset) given indicators for the ethnic groups (**race**) as well as other potentially relevant predictors describing the apartment and building. Fit this model using lmer() and interpret the coefficients at both levels.

   (b) Now extend the model in (b) to allow variation across buildings within community district and then across community districts. Also include predictors describing the community districts. Fit this model using lmer() and interpret the coefficients at all levels.

   (c) Compare the fit of the models in (a) and (b).

4. Item-response model: the folder **exam** contains data on students' success or failure (item correct or incorrect) on a number of test items. Write the notation for an item-response model for the ability of each student and level of difficulty of each item.

5. Multilevel logistic regression with non-nested groupings: the folder **speed.dating** contains data from an experiment on a few hundred students that randomly assigned each participant to 10 short dates with participants of the opposite sex (Fisman et al., 2006). For each date, each person recorded several subjective numerical ratings of the other person (attractiveness, compatibility, and some

other characteristics) and also wrote down whether he or she would like to meet the other person again. Label

$$y_{ij} = \begin{cases} 1 & \text{if person } i \text{ is interested in seeing person } j \text{ again} \\ 0 & \text{otherwise} \end{cases}$$

and $r_{ij1}, \ldots, r_{ij6}$ as person i's numerical ratings of person j on the dimensions of attractiveness, compatibility, and so forth.

(a) Fit a classical logistic regression predicting $\Pr(y_{ij} = 1)$ given person i's 6 ratings of person j. Discuss the importance of attractiveness, compatibility, and so forth in this predictive model.

(b) Expand this model to allow varying intercepts for the persons making the evaluation; that is, some people are more likely than others to want to meet someone again. Discuss the fitted model.

(c) Expand further to allow varying intercepts for the persons being rated. Discuss the fitted model.

6. Varying-intercept, varying-slope logistic regression: continuing with the speed-dating example from the previous exercise, you will now fit some models that allow the coefficients for attractiveness, compatibility, and the other attributes to vary by person.

(a) Fit a no-pooling model: for each person i, fit a logistic regression to the data y_{ij} for the 10 persons j whom he or she rated, using as predictors the 6 ratings $r_{ij1}, \ldots, r_{ij6}$. (Hint: with 10 data points and 6 predictors, this model is difficult to fit. You will need to simplify it in some way to get reasonable fits.)

(b) Fit a multilevel model, allowing the intercept and the coefficients for the 6 ratings to vary by the rater i.

(c) Compare the inferences from the multilevel model in (b) to the no-pooling model in (a) and the complete-pooling model from part (a) of the previous exercise.

## CHAPTER 15

Multilevel generalized linear models

As with linear and logistic regressions, generalized linear models can be fit to multilevel structures by including coefficients for group indicators and then adding group-level models. We illustrate in this chapter with three examples from our recent applied research: an overdispersed Poisson model for police stops, a multinomial logistic model for storable voting, and an overdispersed Poisson model for social networks.

15.1 Overdispersed Poisson regression: police stops and ethnicity

We return to the New York City police example introduced in Sections 1.2 and 6.2, where we formulated the problem as an overdispersed Poisson regression, and here we generalize to a multilevel model. In order to compare ethnic groups while controlling for precinct-level variation, we perform multilevel analyses using the city's 75 precincts. Allowing precinct-level effects is consistent with theories of policing such as the "broken windows" model that emphasize local, neighborhood-level strategies. Because it is possible that the patterns are systematically different in neighborhoods with different ethnic compositions, we divide the precincts into three categories in terms of their black population: precincts that were less than 10% black, 10%–40% black, and more than 40% black. We also account for variation in stop rates between the precincts within each group. Each of the three categories represents roughly one-third of the precincts in the city, and we perform separate analyses for each set.

Overdispersion as a variance component

As discussed in Chapter 6, data that are fit by a generalized linear model are *overdispersed* if the data-level variance is higher than would be predicted by the model. Binomial and Poisson regression models are subject to overdispersion because these models do not have variance parameters to capture the variation in the data. This was illustrated at the end of Section 6.2 for the Poisson model of police stops.

As discussed at the end of Section 14.4, overdispersion can be directly modeled using a data-level variance component in a multilevel model. For example, we can extend the classical Poisson regression of Section 6.2,

$$\text{Poisson regression:} \quad y_i \sim \text{Poisson}\left(u_i e^{X_i \beta}\right),$$

to the multilevel model,

$$\text{overdispersed Poisson regression:} \quad y_i \ \sim \ \text{Poisson}\left(u_i e^{X_i \beta + \epsilon_i}\right)$$
$$\epsilon_i \ \sim \ \text{N}(0, \sigma_\epsilon^2).$$

The new hyperparameter σ_ϵ measures the amount of overdispersion, with $\sigma_\epsilon = 0$ corresponding to classical Poisson regression. We shall use this model for the police stops.

Section 15.3 illustrates the use of the negative binomial model, which is a different overdispersed extension of the Poisson.

Multilevel Poisson regression model

For each ethnic group $e = 1, 2, 3$ and precinct p, we model the number of stops y_{ep} using an overdispersed Poisson regression with indicators for ethnic groups, a multilevel model for precincts, and using n_{ep}, the number of arrests recorded by the Department of Criminal Justice Services (DCJS) for that ethnic group in that precinct in the previous year (multiplied by $15/12$ to scale to a 15-month period), as a baseline, so that $\log(\frac{15}{12} n_{ep})$ is an offset:

$$
\begin{aligned}
y_{ep} &\sim \text{Poisson}\left(\tfrac{15}{12} n_{ep} e^{\mu + \alpha_e + \beta_p + \epsilon_{ep}}\right) \\
\alpha_e &\sim \text{N}(0, \sigma_\alpha^2) \\
\beta_p &\sim \text{N}(0, \sigma_\beta^2) \\
\epsilon_{ep} &\sim \text{N}(0, \sigma_\epsilon^2),
\end{aligned}
\tag{15.1}
$$

where the coefficients α_e control for ethnic groups, the β_p's adjust for variation among precincts, and the ϵ_{ep}'s allow for overdispersion (see Chapter 6). The parameter σ_β represents variation in the rates of stops among precincts, and σ_ϵ represents variation in the data beyond that explained by the Poisson model. We are not particularly interested in the other variance parameter, σ_α; instead we work with the individual coefficients, α_e.

Constraining a batch of coefficients to sum to 0. When comparing ethnic groups, we can look at the ethnicity coefficients relative to their mean:

$$
\alpha_e^{\text{adj}} = \alpha_e - \bar{\alpha}, \text{ for } e = 1, 2, 3. \tag{15.2}
$$

We examine the exponentiated coefficients $\exp(\alpha_e^{\text{adj}})$, which represent relative rates of stops compared to arrests, after controlling for precinct. Having done this, we also adjust the intercept of the model accordingly:

$$
\mu^{\text{adj}} = \mu + \bar{\alpha}. \tag{15.3}
$$

Now $\mu^{\text{adj}} + \alpha_e^{\text{adj}} = \mu + \alpha_e$ for each ethnic group e, and so we can use μ^{adj} and α^{adj} in place of μ and α without changing the model for the data.

In multilevel modeling, it makes sense to fit the full model (15.1) and then define the constrained parameters of interest as in (15.2) and (15.3), rather than trying to fit a model in which the original α parameters are constrained. We discuss this issue further in Section 19.4.

Separately fitting the model to different subsets of the data. By comparing to arrest rates, we can also separately analyze stops associated with different sorts of crimes. We do a separate comparison for each of four types of offenses ("suspected charges" as characterized on the official form): violent crimes, weapons offenses, property crimes, and drug crimes. For each, we model the number of stops y_{ep} by ethnic group e and precinct p for that crime type, using as a baseline the DCJS arrest rates n_{ep} for that crime type.

We thus estimate model (15.1) for twelve separate subsets of the data, corresponding to the four crime types and the three categories of precincts (less than 10% black population, 10–40% black, and more than 40% black). An alternative approach would be to fit a single model to all the data with interactions between crime types and precinct categories, and the other predictors in the model. It is simpler, however, to just fit the model separately to each of the twelve subsets.

Proportion black in precinct	Parameter	Crime type			
		Violent	Weapons	Property	Drug
< 10%	intercept, μ^{adj}	−0.85 (0.07)	0.13 (0.07)	−0.58 (0.21)	−1.62 (0.16)
	α_1^{adj} [blacks]	0.40 (0.06)	0.16 (0.05)	−0.32 (0.06)	−0.08 (0.09)
	α_2^{adj} [hispanics]	0.13 (0.06)	0.12 (0.04)	0.32 (0.06)	0.17 (0.10)
	α_3^{adj} [whites]	−0.53 (0.06)	−0.28 (0.05)	0.00 (0.06)	−0.08 (0.09)
	σ_β	0.33 (0.08)	0.38 (0.08)	1.19 (0.20)	0.87 (0.16)
	σ_ϵ	0.30 (0.04)	0.23 (0.04)	0.32 (0.04)	0.50 (0.07)
10–40%	intercept, μ^{adj}	−0.97 (0.07)	0.42 (0.07)	−0.89 (0.16)	−1.87 (0.13)
	α_1^{adj} [blacks]	0.38 (0.04)	0.24 (0.04)	−0.16 (0.06)	−0.05 (0.05)
	α_2^{adj} [hispanics]	0.08 (0.04)	0.13 (0.04)	0.25 (0.06)	0.12 (0.06)
	α_3^{adj} [whites]	−0.46 (0.04)	−0.36 (0.04)	−0.08 (0.06)	−0.07 (0.05)
	σ_β	0.49 (0.07)	0.47 (0.07)	1.21 (0.17)	0.90 (0.13)
	σ_ϵ	0.24 (0.03)	0.24 (0.03)	0.38 (0.04)	0.32 (0.04)
> 40%	intercept, μ^{adj}	−1.58 (0.10)	0.29 (0.11)	−1.15 (0.19)	−2.62 (0.12)
	α_1^{adj} [blacks]	0.44 (0.06)	0.30 (0.07)	−0.03 (0.07)	0.09 (0.06)
	α_2^{adj} [hispanics]	0.11 (0.06)	0.14 (0.07)	0.04 (0.07)	0.09 (0.07)
	α_3^{adj} [whites]	−0.55 (0.08)	−0.44 (0.08)	−0.01 (0.07)	−0.18 (0.09)
	σ_β	0.48 (0.10)	0.47 (0.11)	0.96 (0.18)	0.54 (0.11)
	σ_ϵ	0.24 (0.05)	0.37 (0.05)	0.42 (0.07)	0.28 (0.06)

Figure 15.1 *Estimates and standard errors for the intercept μ^{adj}, ethnicity parameters α_e^{adj}, and the precinct-level and precinct-by-ethnicity-level variance parameters σ_β and σ_ϵ, for the multilevel Poisson regression model (15.1), fit separately to three categories of precinct and four crime types. The estimates of $e^{\mu+\alpha_e}$ are displayed graphically in Figure 15.2, and alternative model specifications are shown in Figure 15.5.*
It would be preferable to display these results graphically. We show them in tabular form here to give a sense of the inferences that result from the 12 multilevel models that were fit to these data.

Figure 15.1 shows the estimates from model (15.1) fit to each of four crime types in each of three categories of precinct. The standard-deviation parameters σ_β and σ_ϵ are substantial,[1] indicating the relevance of multilevel modeling for these data.

The parameters of most interest are the rates of stop (compared to previous year's arrests) for each ethnic group, $e^{\mu+\alpha_e}$, for $e = 1, 2, 3$. We display these graphically in Figure 15.2. Stops for violent crimes and weapons offenses were the most controversial aspect of the stop-and-frisk policy (and represent more than two-thirds of the stops), but for completeness we display all four categories of crime here.

Figure 15.2 shows that, for the most frequent categories of stops—those associated with violent crimes and weapons offenses—blacks and hispanics were much more likely to be stopped than whites, in all categories of precincts. For violent crimes, blacks and hispanics were stopped 2.5 times and 1.9 times as often as whites, respectively, and for weapons crimes, blacks and hispanics were stopped 1.8 times and 1.6 times as often as whites. In the less common categories of stop, whites were slightly more often stopped for property crimes and more often stopped for drug crimes, in proportion to their previous year's arrests in any given precinct.

---

[1] Recall that these effects are all on the logarithmic scale, so that an effect of 0.3, for example, corresponds to a multiplicative effect of exp(0.3) = 1.35, or a 35% increase in the probability of being stopped.

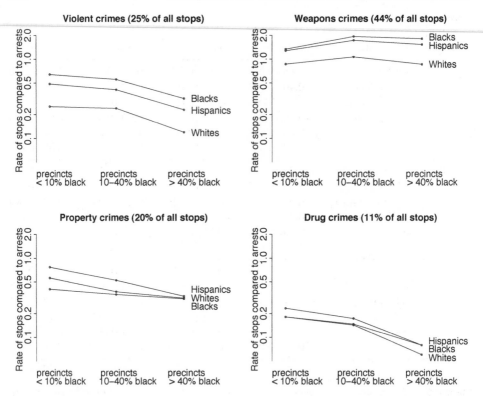

Figure 15.2 *Estimated rates $e^{\mu+\alpha_e}$ at which people of different ethnic groups were stopped for different categories of crime, as estimated from multilevel regressions (15.1) using previous year's arrests as a baseline and controlling for differences between precincts. Separate analyses were done for the precincts that had less than 10%, 10%–40%, and more than 40% black population. For the most common stops—violent crimes and weapons offenses— blacks and hispanics were stopped about twice as often as whites. Rates are plotted on a logarithmic scale. Numerical estimates and standard errors appear in Figure 15.1.*

Alternative model specifications

In addition to fitting model (15.1) as described above, we consider two forms of alternative specifications: first, fitting the same model but changing the batching of precincts; and, second, altering the role played in the model by the previous year's arrests. We compare the fits under these alternative models to assess the sensitivity of our findings to the details of model specification.

Modeling variability across precincts

The batching of precincts into three categories is convenient and makes sense— neighborhoods with different levels of minority populations differ in many ways, and fitting the model separately to each group of precincts is a way to include contextual effects. However, there is an arbitrariness to the division. We explore this by portioning the precincts into different numbers of categories and seeing how the model estimates change.

Including precinct-level predictors. Another approach to controlling for systematic variation among precincts is to include precinct-level predictors, which can be included along with the individual precinct-level effects in the multilevel model.

Parameter	Crime type			
	Violent	Weapons	Property	Drug
intercept, μ^{adj}	-0.66 (0.08)	0.08 (0.11)	-0.14 (0.24)	-0.98 (0.17)
α_1^{adj} [blacks]	0.41 (0.03)	0.24 (0.03)	-0.19 (0.04)	-0.02 (0.04)
α_2^{adj} [hispanics]	0.10 (0.03)	0.12 (0.03)	0.23 (0.04)	0.15 (0.04)
α_3^{adj} [whites]	-0.51 (0.03)	-0.36 (0.03)	-0.05 (0.04)	-0.13 (0.04)
ζ_1 [coeff for prop. black]	-1.22 (0.18)	0.10 (0.19)	-1.11 (0.45)	-1.71 (0.31)
ζ_2 [coef for prop. hispanic]	-0.33 (0.23)	0.71 (0.27)	-1.50 (0.57)	-1.89 (0.41)
σ_β	0.40 (0.04)	0.43 (0.04)	1.04 (0.09)	0.68 (0.06)
σ_ϵ	0.25 (0.02)	0.27 (0.02)	0.37 (0.03)	0.37 (0.03)

Figure 15.3 *Estimates and standard errors for the parameters of model (15.4) that includes proportion black and hispanic as precinct-level predictors, fit to all 75 precincts. The results for the parameters of interest, α_e^{adj}, are similar to those obtained by fitting the basic model separately to each of three categories of precincts, as displayed in Figures 15.1 and 15.2. As before, the model is fit separately to the data from four different crime types.*

As discussed earlier, the precinct-level information that is of greatest interest, and also that has greatest potential to affect our results, is the ethnic breakdown of the population. Thus we consider as regression predictors the proportion black and hispanic in the precinct, replacing model (15.1) by

$$y_{ep} \sim \mathrm{Poisson}\left(\tfrac{15}{12}n_{ep}e^{\mu+\alpha_e+\zeta_1 z_{1p}+\zeta_2 z_{2p}+\beta_p+\epsilon_{ep}}\right), \tag{15.4}$$

where z_{1p} and z_{2p} represent the proportion of the people in precinct p who are black and hispanic, respectively. We also considered variants of model (15.4) including the quadratic terms, z_{1p}^2, z_{2p}^2, and $z_{1p}z_{2p}$, to examine sensitivity to nonlinearity.

Figure 15.3 shows the results from model (15.4), which is fit to all 75 precincts but controls for the proportion black and proportion hispanic in precincts. The inferences are similar to those obtained from the main analysis presented earlier. Including quadratic terms and interactions in the precinct-level model (15.4), and including the precinct-level predictors in the models fit to each of the three subsets of the data, similarly had little effect on the parameters of interest, α_e^{adj}.

Changing the number of precinct categories. Figure 15.4 displays the estimated rates of stops for violent crimes, compared to the previous year's arrests, for each of the three ethnic groups, for analyses dividing the precincts into 5, 10, and 15 categories ordered by percent black population in precinct. For simplicity, we only give results for violent crimes; these are typical of the alternative analyses for all four crime types. For each of the three graphs in Figure 15.4, the model was separately estimated for each batch of precincts, and these estimates are connected in a line for each ethnic group. Compared to the upper-left plot in Figure 15.2, which shows the results from dividing the precincts into three categories, we see that dividing into more groups adds noise to the estimation but does not change the overall pattern of differences between the groups.

Modeling the relation of stops to previous year's arrests

We also consider different ways of using the number of DCJS arrests n_{ep} in the previous year, which plays the role of a baseline (or offset, in generalized linear models terminology) in model (15.1).

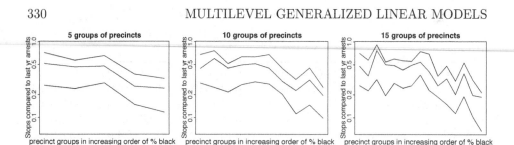

Figure 15.4 *Estimated rates $e^{\mu+\alpha_e}$ at which people of different ethnic groups were stopped for violent crimes, as estimated from models dividing precincts into 5, 10, and 15 categories. For each graph, the top, middle, and lower lines correspond to blacks, hispanics, and whites, respectively. These plots show the same general patterns as the model with 3 categories (the upper-left graph in Figure 15.2) but with increasing levels of noise.*

Using a linear predictor instead of an offset. Including the past arrest rate as an offset makes sense because we are interested in the rate of stops per crime, and we are using past arrests as a proxy for crime rate and for police expectations about the demographics of perpetrators. However, another option is to include the logarithm of the number of past arrests as a linear predictor instead:

$$y_{ep} \sim \text{Poisson}\left(\tfrac{15}{12}e^{\gamma \log n_{ep}+\mu+\alpha_e+\beta_p+\epsilon_{ep}}\right). \tag{15.5}$$

Model (15.5) reduces to the offset model (15.1) if $\gamma = 1$. We can thus fit (15.5) and see if the inferences for α_e^{adj} change compared to the earlier model that implicitly fixes γ to 1.

Two-stage models of arrest and stop rates. We can take this idea further by modeling past arrests as a proxy rather than the actual crime rate. We try this in two ways, for each labeling the true crime rate for each ethnicity in each precinct as θ_{ep}, with separate multilevel Poisson regressions for this year's stops and last year's arrests (as always, including the factor $\tfrac{15}{12}$ to account for our 15 months of stop data). In the first formulation, we model last year's arrests as Poisson distributed with mean θ:

$$
\begin{aligned}
y_{ep} &\sim \text{Poisson}\left(\tfrac{15}{12}\theta_{ep}e^{\mu+\alpha_e+\beta_p+\epsilon_{ep}}\right) \\
n_{ep} &\sim \text{Poisson}(\theta_{ep}) \\
\log\theta_{ep} &= \log N_{ep} + \tilde{\alpha}_e + \tilde{\beta}_p + \tilde{\epsilon}_{ep}.
\end{aligned} \tag{15.6}
$$

Here we are using N_{ep}, the population of ethnic group e in precinct p, as a baseline for the model of crime frequencies. The second-level error terms $\tilde{\beta}$ and $\tilde{\epsilon}$ are given normal hyperprior distributions as with model (15.1).

Our second two-stage model is similar to (15.6) but moving the new error term $\tilde{\epsilon}$ to the model for n_{ep}:

$$
\begin{aligned}
y_{ep} &\sim \text{Poisson}\left(\tfrac{15}{12}\theta_{ep}e^{\mu+\alpha_e+\beta_p+\epsilon_{ep}}\right) \\
n_{ep} &\sim \text{Poisson}(\theta_{ep}e^{\tilde{\epsilon}_{ep}}) \\
\log\theta_{ep} &= \log N_{ep} + \tilde{\alpha}_e + \tilde{\beta}_p.
\end{aligned} \tag{15.7}
$$

Under this model, arrest rates n_{ep} are equal to the underlying crime rates, θ_{ep}, on average, but with overdispersion compared to the Poisson error distribution.

Figure 15.5 displays parameter estimates from the models that differently incorporate the previous year's arrest rates n_{ep}. For conciseness we display results for violent crimes only, for simplicity including all 75 precincts in the models. (Similar results are obtained when fitting the model separately in each of three categories

Parameter	offset (15.1)	Model for previous year's arrests		
		regression (15.5)	2-stage (15.6)	2-stage (15.7)
intercept, μ^{adj}	−1.08 (0.06)	−0.94 (0.16)	−1.07 (0.06)	−1.13 (0.07)
α_1^{adj} [blacks]	0.40 (0.03)	0.41 (0.03)	0.40 (0.03)	0.42 (0.08)
α_2^{adj} [hispanics]	0.10 (0.03)	0.10 (0.03)	0.10 (0.03)	0.14 (0.09)
α_3^{adj} [whites]	−0.50 (0.03)	−0.51 (0.03)	−0.50 (0.03)	−0.56 (0.09)
γ [coef for $\log n_{ep}$]		0.97 (0.03)		
σ_β	0.51 (0.05)	0.51 (0.05)	0.51 (0.05)	0.27 (0.12)
σ_ϵ	0.26 (0.02)	0.26 (0.02)	0.24 (0.02)	0.67 (0.04)

Figure 15.5 *Estimates and standard errors for parameters under model (15.1) and three alternative specifications for the previous year's arrests n_{ep}: treating $\log(n_{ep})$ as a predictor in the Poisson regression model (15.5), and the two-stage models (15.6) and (15.7). For simplicity, results are displayed for violent crimes only, for the model fit to all 75 precincts. The three α_e^{adj} parameters are nearly identical under all four models, with the specification affecting only the intercept.*

of precincts, and for the other crime types.) The first two columns of Figure 15.5 show the result from our main model (15.1) and the alternative model (15.5), which includes $\log n_{ep}$ as a regression predictor. The two models differ only in that the first restricts γ to be 1, but as we can see, γ is estimated very close to 1 in the regression formulation, and the coefficients α_e^{adj} are essentially unchanged. (The intercept changes a bit because $\log n_{ep}$ does not have a mean of 0.)

The last two columns in Figure 15.5 show the estimates from the two-stage regression models (15.6) and (15.7). The models differ in their estimates of the variance parameters σ_β and σ_ϵ, but the estimates of the key parameters α_e^{adj} are essentially the same as in the original model.

We also performed analyses including indicators for the month of arrest. Rates of stops were roughly constant over the 15-month period and did not add anything informative to the comparison of ethnic groups.

15.2 Ordered categorical regression: storable votes

At the end of Section 6.5, we described data on "storable votes" from a study in experimental economics. Each student in the experiment played a voting game 30 times, in each play receiving an input x between 1 and 100 and then giving a response of 1, 2, or 3. As described in Section 6.5, we fit an ordered multinomial logistic regression to these outcomes, fitting the model separately for each student.

We now return to these data and fit a multilevel model. With only 30 data points on each student, there is some uncertainty in the estimate for the parameters for each student, and a multilevel model should allow more precise inferences—at least if the students are similar to each other. (As discussed in Section 12.7, multilevel modeling is most effective when the parameters in a batch are similar to each other.)

In our multilevel model for the storable votes study, we simply expand the model (6.11) allowing each of the parameters to vary by student j:

$$c_{j\,1.5} \sim \text{N}(\mu_{1.5}, \sigma_{1.5}^2)$$
$$c_{j\,2.5} \sim \text{N}(\mu_{2.5}, \sigma_{2.5}^2)$$
$$\log \sigma_j \sim \text{N}(\mu_{\log \sigma}, \sigma_{\log \sigma}^2),$$

with the hyperparameters $\sigma_{1.5}, \sigma_{2.5}, \mu_{\log \sigma}, \sigma_{\log \sigma}$ estimated from the data. The un-

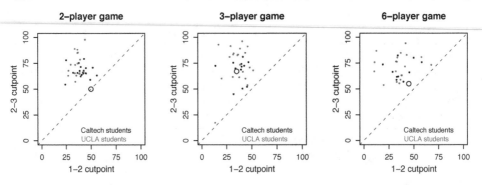

Figure 15.6 *Estimated cutpoints $c_{j\,1.5}, c_{j\,2.5}$ for students j in three experiments on storable voting. Each set of parameters was estimated using a multilevel ordered logistic regression model. The dots all fall above the diagonal line because the 2–3 cutpoint is restricted to be at least as high as the 1–2 cutpoint. The large circle on each graph corresponds to a theoretically optimal cutpoint for each game. Deviations of actual strategies from the theoretical optimum are of interest.*

pooled model, estimating the parameters separately for each student, corresponds to hyperparameters set to $\sigma_{1.5} = \infty, \sigma_{1.5} = \infty, \sigma_{\log \sigma} = \infty$. (The log transformation is used for σ_j so that its distribution is restricted to be positive.)

Having fit the multilevel model, we can now examine how the parameters vary among the students in the experiment. The parameters $c_{1.5}$ and $c_{2.5}$ correspond to the cutpoints for voting 1, 2, 3; variation in these two parameters thus corresponds to variation in the monotonic strategies used by the students in the voting game. The parameter σ represents the variation in the model, with higher values indicating a less purely monotonic strategy.

Figure 15.6 shows estimated cutpoints $c_{j\,1.5}, c_{j\,2.5}$ for students j in three different storable voting experiments. The general distribution of the cutpoints is similar in the three games, despite the different positions of the theoretically optimal cutpoints (indicated by a solid circle in each plot). This suggests that students are following a range of approximately monotonic strategies that are determined more by the general structure of the game than by detailed strategic reasoning. For the practical implementation of storable votes, this finding is somewhat encouraging, in that the "players" are using storable votes to express their preferences on individual issues without feeling the need to manipulate the voting game.

15.3 Non-nested negative-binomial model of structure in social networks

Understanding the structure of social networks, and the social processes that form them, is a central concern of sociology for both theoretical and practical reasons. Networks have been found to have important implications for social mobility, getting a job, the dynamics of fads and fashion, attitude formation, and the spread of infectious disease.

This section discusses how we used an overdispersed Poisson regression model to learn about social structure. We fit the model to a random-sample survey of Americans who were asked,[2] "How many X's do you know?" for a variety of characteristics X, defined by name (Michael, Christina, Nicole, ...), occupation (postal worker,

---

[2] The respondents were told, "For the purposes of this study, the definition of knowing someone is that you know them and they know you by sight or by name, that you could contact them,

pilot, gun dealer, ...), ethnicity (Native American), or experience (prisoner, auto accident victim, ...). For a complete list of the groups, see Figure 15.10 on page 338.

Background

The original goals of the survey were (1) to estimate the distribution of individuals' network size, defined to be the number of acquaintances, in U.S. population and (2) to estimate the sizes of certain subpopulations, especially those that are hard to count using regular survey results.

Before describing our regression modeling, we explain how these two estimates can be roughly computed from the data. First, to estimate the social network size of a single individual, one can use his or her total response for a set of subpopulations with known sizes, and then scale up using the sizes of these groups in the population.

To illustrate, suppose you know 2 persons named Nicole. At the time of the survey, there were 358,000 Nicoles out of 280 million Americans. Thus, your 2 Nicoles represent a fraction $\frac{2}{358,000}$ of all the Nicoles. Extrapolating to the entire country yields an estimate of $\frac{2}{358,000} \cdot (280 \text{ million}) = 1560$ people known by you. A more precise estimate can be obtained by averaging these estimates using a range of different groups. This is only a crude inference since it assumes that everyone has equal propensity to know someone from each group. However, as an estimation procedure, it has the advantage of not requiring a respondent to recall his or her entire network, which typically numbers in the hundreds.

The second use for which this survey was designed is to estimate the size of certain hard-to-count populations. To do this, one can combine the estimated network size information with the responses to the questions about how many people the respondents know in the hard-to-count population.

For example, the survey respondents know, on average, 0.63 homeless people. If it is estimated that the average network size is 750, then homeless people represent a fraction of $\frac{0.63}{750}$ of an average person's social network. The total number of homeless people in the country can then be estimated as $\frac{0.63}{750} \cdot (280 \text{ million}) = 0.24$ million. This estimate relies on idealized assumptions (most notably, that homeless persons have the same social network size, on average, as Americans as a whole) but can be used as a starting point for estimating the sizes of groups that are difficult to measure directly.

Our regression model performs more precise versions of these estimates but, more interestingly, uses overdispersion in the data to reveal information about social structure in the acquaintanceship network. We use the variation in response data to study the heterogeneity of relative propensities for people to form ties to people in specific groups.

---

that they live within the United States, and that there has been some contact (either in person, by telephone or mail) in the past two years."

In addition, the data have some minor problems. For the fewer than 0.4% of responses that were missing, we followed the usual practice with this sort of unbalanced data of assuming an ignorable model (that is, constructing the likelihood using the observed data). Sometimes responses were categorized, and then we use the central value in the bin (for example, imputing 7.5 for the response "5–10"). In addition, to correct for some responses that were suspiciously large (for example, a person claiming to know more than 50 Michaels), we truncate all responses at 30. (As a sensitivity analysis, we tried changing the truncation point to 50; this had essentially no effect on our results.) We also inspected the data using scatterplots of responses, which revealed a respondent who was coded as knowing 7 persons of every category. We removed this case from the dataset.

Modeling the data

For respondent $i = 1, \ldots, 1370$ and subpopulations $k = 1, \ldots, 32$ (see Figure 15.10 for the list of groups asked about in the survey), we use the notation y_{ik} for the number of persons in group k known by person i. We consider three increasingly complicated models for these data.

Most simply, the *Erdos-Renyi* model is derived from the assumption that any person in the population is equally likely to know any other person, so that social links occur completely at random. A consequence of this assumption is that y_{ik} should have a Poisson distribution with mean proportional to the size of subpopulation k.

More generally, one could consider a *null model* in which individuals i have varying levels of gregariousness or popularity, so that the expected number of persons in group k known by person i will be proportional to this gregariousness parameter, which we label a_i. Departure from this model—patterns not simply explained by differing group sizes or individual popularities—can be viewed as evidence of structured social acquaintance networks.

As we shall see, the null model (and its special case, the Erdos-Renyi model) fails to account for much of social reality, including the "How many X's do you know?" survey data. In fact, some individuals are much more likely to know people of some groups. To capture this aspect of social structure, we set up a new model in which, for any individual, the relative propensities to know people from different groups vary. We call this the *overdispersed model* since this variation results in overdispersion in the count data y_{ik}.

Figure 15.7 shows some of the data—the distributions of responses to the questions, "How many people named Nicole do you know?" and "How many Jaycees do you know?" along with the expected distributions under the Erdos-Renyi model, our null model, and our overdispersed model. We chose these two groups to plot because they are close in average number known (0.9 Nicoles, 1.2 Jaycees) but have much different distributions. The distribution for Jaycees has much more variation, with more zero responses and more responses in the upper tail.[3]

Comparing the models, the Erdos-Renyi model implies a Poisson distribution for the responses to each question, whereas the other models allow for more dispersion. The distributions under the null model are more dispersed to reflect that social network sizes vary greatly among individuals. The distributions under the overdispersed model are even more spread out—especially for the Jaycees—reflecting estimated variation in relative propensities for people to know members of the Jaycees. As we shall see, both these sources of variation—variation in social network sizes and variations in relative propensities to form ties to specific groups—can be estimated from the data in the survey.

The three models can be written as follows in statistical notation as $y_{ik} \sim$ Poisson(λ_{ik}), with increasingly general forms for λ_{ik}:

$$\text{Erdos-Renyi model:} \quad \lambda_{ik} = a b_k$$
$$\text{our null model:} \quad \lambda_{ik} = a_i b_k$$
$$\text{our overdispersed model:} \quad \lambda_{ik} = a_i b_k g_{ik}.$$

The null model goes beyond the Erdos-Renyi model by allowing the gregariousness parameters to differ between individuals (a_i) and prevalence parameters

---

[3] "Jaycees" are members of the Junior Chamber of Commerce, a community organization of people between the ages of 21 and 39. Because the Jaycees are a social organization, it makes sense that not everyone has the same propensity to know one—people who are in the social circle of one Jaycee are particularly likely to know others.

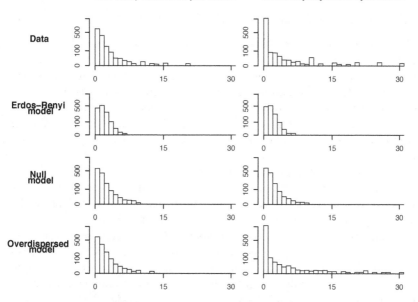

Figure 15.7 *Distributions of responses to "How many persons do you know named Nicole?" and "How any Jaycees do you know?" from survey data and from random simulations under three fitted models: the Erdos-Renyi model (completely random links), our null model (some people more gregarious than others, but uniform relative propensities for people to form ties to all groups), and our overdispersed model (variation in gregariousness and variation in propensities to form ties to different groups). The four models are listed in order of increasing dispersion than the one above, with the overdispersed model fitting the data reasonably well. The propensities to form ties to Jaycees show much more variation than the propensities to form ties to Nicoles, and hence the Jaycees counts are much more overdispersed. (The data also show minor idiosyncrasies such as small peaks at the responses 10, 15, 20, and 25. All values greater than 30 have been truncated at 30.)*

between groups (b_k). The overdispersed model generalized further by allowing different individuals to differ in their relative propensities to form ties to people in specific groups (g_{ik}). When fitting the overdispersed model, we will not attempt to estimate all the individual g_{ik}'s; rather, we estimate certain properties of their distributions.

The overdispersed model

Overdispersion in these data can arise if the relative propensity for knowing someone in prison, for example, varies from respondent to respondent. We can write this in the generalized linear model framework as

$$y_{ik} \sim \text{Poisson}(e^{\alpha_i + \beta_k + \gamma_{ik}}), \tag{15.8}$$

where each $\gamma_{ik} = \log(g_{ik}) \equiv 0$ in the null model. For each subpopulation k, we let the multiplicative factors $g_{ik} = e^{\gamma_{ik}}$ follow a gamma distribution with a value of 1 for the mean and a value of $1/(\omega_k - 1)$ for the shape parameter.[4] This distribution

---

[4] If we wanted, we could allow the mean of the gamma distribution to vary also; however, this would be redundant with a location shift in β_k; see model (15.8). The mean of the gamma distribution for the $e^{\gamma_{ik}}$'s cannot be identified separately from β_k, which we are already estimating from data.

is convenient because then the γ's can be integrated out of model (15.8) to yield

$$y_{ik} \sim \text{negative-binomial}(\text{mean} = e^{\alpha_i + \beta_k}, \text{overdispersion} = \omega_k). \qquad (15.9)$$

(The usual parametrization of this distribution is $y \sim$ Negative-binomial(A, B), but for this example it is more convenient to express in terms of the mean $\lambda = A/B$ and overdispersion $\omega = 1 + 1/B$.) Setting $\omega_k = 1$ corresponds to setting the shape parameter in the gamma distribution to ∞, which in turn implies that the g_{ik}'s have zero variance, reducing to the null model with no overdispersion. Higher values of ω_k correspond to overdispersion—that is, more variation in the distribution of connections involving group k than would be expected under the Poisson regression, as would be expected if there is variation among respondents in the relative propensity to know someone in group k.

Our primary goal in fitting model (15.9) is to estimate the overdispersions ω_k and thus learn about biases that exist in the formation of social networks. As a byproduct, we also estimate the gregariousness parameters α_i and the group prevalence parameters β_k. As we have discussed, $a_i = e^{\alpha_i}$ represents the expected number of persons known by respondent i, and $b_k = e^{\beta_k}$ is the proportion of subgroup k in the social network, counting each link in the network in proportion to the probability that it will be recalled by a random respondent in the survey.

We estimate the α's, β's, and ω's with a multilevel model. The respondent parameters α_i are assumed to follow a normal distribution with unknown mean μ_α and standard deviation σ_α. We similarly fit the subgroup parameters β_k with a normal distribution $\text{N}(\mu_\beta, \sigma_\beta^2)$, with these hyperparameters also estimated from the data. For simplicity, we assign independent Uniform(0,1) prior distributions to the overdispersion parameters on the inverse scale: $p(1/\omega_k) \propto 1$. (The overdispersions ω_k are constrained to the range $(1, \infty)$, and so it is convenient to put a model on the inverses $1/\omega_k$, which fall in the interval $(0, 1)$.)

Nonidentifiability

The model as given is not fully identified. Any constant C can be added to all the α_i's and subtracted from all the β_k's, and the likelihood will remain unchanged (since it depends on these parameters only through sums of the form $\alpha_i + \beta_k$). If we also add C to μ_α and subtract C from μ_β, then the prior density also is unchanged as well. It would be possible to identify the model by anchoring it at some arbitrary point—for example, setting μ_α to zero—but we prefer to let all the parameters float, since including this redundancy can allow the iterative computations to converge more quickly (a point we discuss more generally in Section 19.4).

We choose a constant C so that the parameters e^{β_k} correspond to the proportion of the entire social network associated with subpopulation k. We perform this renormalization using the known population sizes of the named subgroups (the number of Michaels, Nicoles, and so forth, in the population), which can be obtained from the U.S. Census. The choice of C is somewhat elaborate, including adjustments for the rare and common male and female names, and we do not present the details here.

Computation

We would like to fit the model in Bugs, but with more than 1400 parameters (an α_j for each of the 1370 survey respondents, a β_k for each of the 32 subgroups, and some hyperparameters), it runs too slowly. Instead, we use a more efficient program called

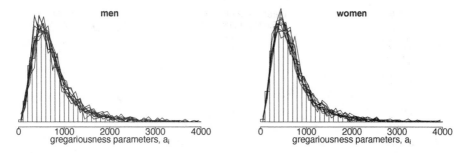

Figure 15.8 *Estimated distributions of "gregariousness" or expected number of acquaintances, $a_i = e^{\alpha_i}$ from the fitted model. Men and women have similar distributions (with medians of about 610 and means about 750), with a great deal of variation among persons. The overlain lines are posterior simulation draws indicating inferential uncertainty in the histograms.*

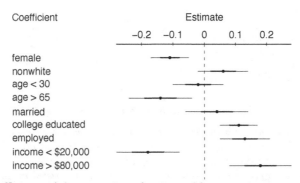

Figure 15.9 *Coefficients of the regression of estimated log gregariousness parameters α_i on personal characteristics. Because the regression is on the logarithmic scale, the coefficients (with the exception of the constant term) can be interpreted as proportional differences: thus, with all else held constant women have social network sizes 11% smaller than men, persons over 65 have social network sizes 14% lower than others, and so forth. The R^2 of the model is only 10%, indicating that these predictors explain very little of the variation in gregariousness in the population.*

Umacs (Universal Markov Chain sampler) which requires the posterior distribution to be specified as an R function. Details appear in Section 18.7.

We fit the model first using all the data and then separately for the male and female respondents (582 men and 784 women, with 4 individuals excluded due to missing gender information). Fitting the models separately for men and women makes sense since many of the subpopulations under study are single-sex groups. As we shall see, men tend to know more men and women tend to know more women, and more subtle sex-linked patterns also occur.

The distribution of social network sizes a_i

Figure 15.8 displays estimated distributions of the gregariousness parameters $a_i = e^{\alpha_i}$ for the survey respondents, showing separate histograms of the posterior simulations from the model estimated separately to the men and the women.

The spread in each of the histograms of Figure 15.8 almost entirely represents population variability. The model allows us to estimate the individual a_i's to within a coefficient of variation of about $\pm 25\%$. When taken together this allows us to

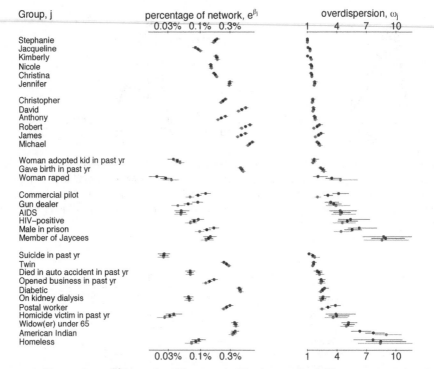

Figure 15.10 *For each group k in the "How many X's do you know?" survey, we plot the estimate (and 95% interval) of b_k and ω_k. The estimates and uncertainty lines are clustered in groups of three; for each group, the top, middle, and bottom dots/lines correspond to men, all respondents, and women, respectively. The groups are listed in categories—female names, male names, female, male (or primarily male), and mixed-sex groups—and in increasing average overdispersion within each category.*

estimate the distribution very precisely. This precision can be seen in the solid lines that are overlaid on Figure 15.8 and represent inferential uncertainty.

Figure 15.9 presents a simple regression analysis estimating some of the factors predictive of $\alpha_i = \log(a_i)$, using questions asked of the respondents in the survey. These explanatory factors are relatively unimportant in explaining social network size: the regression summarized in Figure 15.9 has an R^2 of only 10%. The largest effects are that persons with a college education, a job outside the home, and high incomes know more people, and persons over 65 and those having low incomes know fewer people. These factors all have effects in the range of 10%–20%.

Relative sizes b_k of subpopulations

We now consider the parameters describing the 32 subpopulations. The left panels of Figure 15.10 show the 32 subpopulations k and the estimates of $b_k = e^{\beta_k}$, the proportion of links in the network that go to a member of group k. The right panel displays the estimated overdispersions ω_k. The sample size is large enough that the 95% error bars are tiny for the β_k's and reasonably small for the ω_k's as well. (It is a general property of statistical estimation that mean parameters, such as the β's in this example, are easier to estimate than dispersion parameters such as the ω's.) The figure also displays the separate estimates from the men and women.

Considering the β's first, the clearest pattern in Figure 15.10 is that respondents

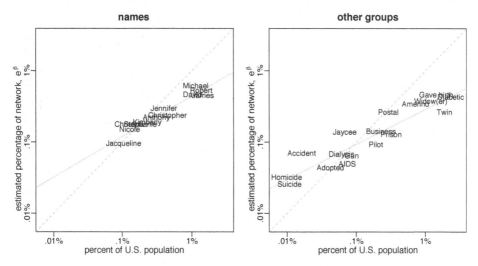

Figure 15.11 *Log-log plot of estimated prevalence of groups in the population (as estimated from the "How many X's do you know?" survey) plotted versus actual group size (as determined from public sources). Names and other groups are plotted separately, on a common scale, with fitted regression lines shown. The solid lines have slopes 0.53 and 0.42, compared to a theoretical slope of 1 (as indicated by the dotted lines) that would be expected if all groups were equally popular, on average, and equally recalled by respondents.*

of each sex tend to know more people in groups of their own sex. We can also see that the 95% intervals are wider for groups with lower β's, which makes sense because the data are discrete and, for these groups, the counts y_{ik} are smaller and provide less information.

Another pattern in the estimated b_k's is the way that they scale with the size of group k. One would expect an approximate linear relation between the number of people in group k and our estimate for b_k: that is, on a graph of $\log b_k$ versus $\log(\text{group size})$, we would expect the groups to fall roughly along a line with slope 1. As can be seen in Figure 15.11, however, this is not the case. Rather, the estimated prevalence increases approximately with square root of population size, a pattern that is particularly clean for the names. This relation has also been observed by earlier researchers.

Discrepancies from the linear relation can be explained by difference in average degrees (for example, as members of a social organization, Jaycees would be expected to know more people than average, so their b_k should be larger than another group of equal numbers), inconsistency in definitions (for example, what is the definition of an American Indian?), and ease or difficulty of recall (for example, a friend might be a twin without you knowing it, whereas you would probably know whether she gave birth in the past year).

This still leaves unanswered the question of why square root (that is, a slope of $1/2$ in the log-log plot) rather than linear (a slope of 1). It is easier to recall rare persons and events, whereas more people in more common categories are easily forgotten. You will probably remember every Ulysses you ever met, but it can be difficult to recall all the Michaels and Roberts you know even now. The recall process for rarer names reaches deeper into one's personal network of acquaintances.

Another pattern in Figure 15.11 is that the line for the names is higher than for the other groups. We suppose that is because, for a given group size, it is easier

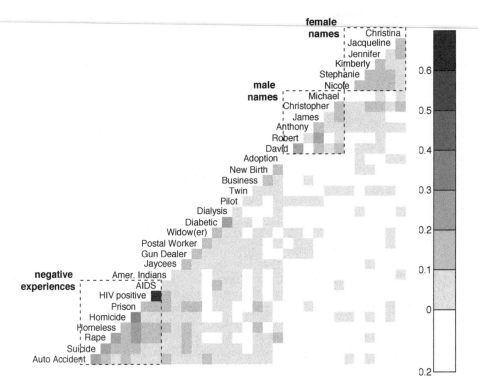

Figure 15.12 *Correlations of the residuals r_{ik} among the survey respondents: people who know more HIV-positive persons know more AIDS patients, etc. The groups other than the names are listed based on a clustering algorithm that maximizes correlations between nearby groups.*

to recall names than characteristics. After all, you know the name of almost all your acquaintances, but you could easily be unaware that a friend has diabetes, for example.

Overdispersion parameters ω_k for subpopulations

Recall that we introduced the overdispersed model to attempt to estimate the variability in respondents' relative propensities to form ties to members of different groups. For groups where $\omega_k = 1$, we can conclude that there is no variation in these relative propensities, so that persons in group k appear to be randomly distributed in the social network. However, for groups where ω_k is much greater than 1, the null model is a poor fit to the data, and persons in group k do not appear to be uniformly distributed in the social network. Rather, overdispersion implies that the relative propensity to know persons of group k varies in the general population.

The right panel of Figure 15.10 displays the estimated overdispersions ω_k, and they are striking. First, we observe that the names have overdispersions of between 1 and 2—that is, indicating very little variation in relative propensities. In contrast, the other groups have a wide range of overdispersions, with the highest values for Jaycees and American Indians (two groups with dense internal networks) and homeless persons, who are both geographically and socially localized.

These results are consistent with our general understanding and also potentially reveal patterns that would not be apparent without this analysis. For example, it

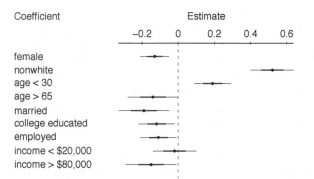

Figure 15.13 *Coefficients of the regression of residuals for the "How many males in federal prison do you know?" question on personal characteristics. Being male, nonwhite, young, unmarried, etc., are associated with knowing more people than expected in federal prison. However, the R^2 of the regression is only 11%, indicting that most of the variation in the data is not captured by these predictors.*

is no surprise that there is high variation in the propensity to known someone who is homeless, but it is perhaps surprising that AIDS patients are less overdispersed than HIV-positive persons, or that new business owners are no more overdispersed than new mothers.

Analysis of residuals

Further features of these data can be studied using residuals from the overdispersed model. A natural object of study is correlation: for example, do people who know more Anthonys tend to know more gun dealers (after controlling for the fact that social network sizes differ, so that anyone who knows more X's will tend to know more Y's)? For each survey response y_{ik}, we can define the standardized residual as

$$\text{residual: } r_{ik} = \sqrt{y_{ik}} - \sqrt{a_i b_k}, \tag{15.10}$$

the excess people known after accounting for individual and group parameters. (It is standard to compute residuals of count data on the square root scale to stabilize the variance.)

For each pair of groups k_1, k_2, we can compute the correlation of their vectors of residuals; Figure 15.12 displays the matrix of these correlations. Care must be taken when interrupting the figure. At first, it may appear that the correlations are quite small. However, this is in some sense a natural result of our model. That is, if the correlations where all positive for a specific group, then the popularity b of that group would increase.

Several patterns can be seen in Figure 15.12. First, there is a slight positive correlation within male and female names. Second, perhaps more interesting sociologically, there is a positive correlation between the categories that can be considered negative experiences—homicide, suicide, rape, died in a car accident, homelessness, and being in prison. That is, someone with a higher relative propensity to know someone with one bad experience is also likely to have a higher propensity to know someone who had a different bad experience.

Instead of correlating the residuals, we could have examined the correlations of the raw data. However, these would be more difficult to interpret because we would find positive correlations everywhere, for the uninteresting reason that some respondents know many more people than others, so that if you know more of any category of person, you are likely to know more in just about any other category.

One can also model the residuals based on individual-level predictors. For example, Figure 15.13 shows the estimated coefficients of a regression model fit to the residuals of the null model for the "How many males do you know in state or federal prison?" question. It is no surprise that being male, nonwhite, young, unmarried, less educated, unemployed, and so forth are associated with knowing more men than expected in state or federal prison. However, the R^2 of the regression model is only 11%.

As with the correlation analysis, by performing this regression on the residuals and not the raw data, we are able to focus on the relative number of prisoners known, without being distracted by the total network size of each respondent (which we have separately analyzed in Figure 15.9).

15.4 Bibliographic note

Examples of hierarchical generalized linear modeling in the literature include Novick, Lewis, and Jackson (1973), Clayton and Kaldor (1987), Karim and Zeger (1992), Barry et al. (2003), and many others. Johnson (1996, 1997) applies multilevel ordered discrete regression models to student grades.

The New York City police analysis comes from Gelman, Fagan, and Kiss (2006). Some references on neighborhoods and crime include Wilson and Kelling (1982), Skogan (1990), and Sampson, Raudenbush, and Earls (1997). The storable votes experiment and model are described in Casella, Gelman, and Palfrey (2006). The social network example comes from Zheng, Salganik, and Gelman (2006), with the analysis based on survey data of Killworth et al. (1998) and McCarty et al. (2000). For more on statistical models for network data, see Newman (2003), Hoff, Raftery, and Handcock (2002), and Hoff (2003, 2005).

15.5 Exercises

1. Multilevel ordered logit: using the National Election Study data from the year 2000 (data available in the folder **nes**), set up an ordered logistic regression predicting the response to the question on vote intention (0 = Gore, 1 = no opinion or other, 2 = Bush), given the predictors shown in Figure 5.4 on page 84, and with varying intercepts for states. (You will fit the model using Bugs in Exercise 17.10.)

2. Using the same data as the previous exercise:

   (a) Formulate a model to predict party identification (which is on a five-point scale) using ideology and demographics with a multilevel ordered categorical model allowing both the intercept and the coefficient on ideology to vary over state.

   (b) Fit the model using `lmer()` and discuss your results.

3. Multinomial choice models: fit and evaluate a multilevel model to the Academy Awards data from Exercise 6.11.

Part 2B: Fitting multilevel models

We next explain how to fit multilevel models in Bugs, as called from R. We illustrate with several examples and discuss some general issues in model fitting and tricks that can help us estimate multilevel models using less computer time. We also present the basics of Bayesian inference (as a generalization of the least squares and maximum likelihood methods used for classical regression), which is the approach used in problems such as multilevel models with potentially large numbers of parameters.

Appendix C discusses some software that is available to quickly and approximately fit multilevel models. We recommend using Bugs for its flexibility in modeling; however, these simpler approaches can be useful to get started, explore models quickly, and check results.

## CHAPTER 16

Multilevel modeling in Bugs and R: the basics

In this chapter we introduce the fitting of multilevel models in Bugs as run from R. Following a brief introduction to Bayesian inference in Section 16.2, we fit a varying-intercept multilevel regression, walking through each step of the model. The computations in this chapter parallel Chapter 12 on basic multilevel models. Chapter 17 presents computations for the more advanced linear and generalized linear models of Chapters 12–15.

16.1 Why you should learn Bugs

As illustrated in the preceding chapters, we can quickly and easily fit many multilevel linear and generalized linear models using the `lmer()` function in R. Functions such as `lmer()`, which use point estimates of variance parameters, are useful but can run into problems. When the number of groups is small or the multilevel model is complicated (with many varying intercepts, slopes, and non-nested components), there just might not be enough information to estimate variance parameters precisely. At that point, we can get more reasonable inferences using a Bayesian approach that averages over the uncertainty in all the parameters of the model.

We recommend the following strategy for multilevel modeling:

1. Start by fitting classical regressions using the `lm()` and `glm()` functions in R. Display and understand these fits as discussed in Part 1 of this book.

2. Set up multilevel models—that is, allow intercepts and slopes to vary, using non-nested groupings if appropriate—and fit using `lmer()`, displaying as discussed in most of the examples of Part 2A.

3. As described in this part of the book, fit fully Bayesian multilevel models, using Bugs to obtain simulations representing inferential uncertainty about all the parameters in a model. Use these simulations to summarize uncertainty about coefficients, predictions, and other quantities of interest.

4. Finally, for some large or complicated models, it is necessary to go one step further and do some programming in R, as illustrated in Section 18.7.

For some analyses, it will not be necessary to go through all four steps. In fact, as illustrated in the first part of this book, often step 1 is enough.

In multilevel settings, the speed and convenience of `lmer()` allow us to try many specifications in building a model, and then the flexibility of Bugs gives us a chance to understand any particular model more fully. Also, as we discuss in Section 16.10, Bugs has an open-ended format that allows models to be expanded more generally than can be done using standard multilevel modeling packages.

16.2 Bayesian inference and prior distributions

The challenge in fitting a multilevel model is estimating the data-level regression (including the coefficients for all the group indicators) along with the group-level model. The most direct way of doing this is through *Bayesian inference*, a statistical method that treats the group-level model as "prior information" in estimating the individual-level coefficients. Chapter 18 discusses Bayesian inference as a generalization of least squares and maximum likelihood estimation. Here we briefly characterize the prior distributions that are commonly used in multilevel models.

In a Bayesian framework, all parameters must have prior distributions, and in the models of this book the prior distributions almost all fall into one of two categories: group-level models and noninformative uniform distributions. The group-level models themselves are either normal distributions (whose mean and standard deviation are themselves typically given noninformative prior distributions) or linear regressions (whose coefficients and error standard deviations are again typically modeled noninformatively).

Classical regression

Classical regression and generalized linear models represent a special case of multilevel modeling in which there is no group-level model—in Bayesian terms, no prior information. For the classical regression model, $y_i = X_i\beta + \epsilon_i$, with independent errors $\epsilon_i \sim N(0, \sigma_y^2)$, the corresponding prior distribution is a uniform distribution on the entire range $(-\infty, \infty)$ for each of the components of β, and uniform on $(0, \infty)$ for σ_y as well.[1] That is, the classical model ascribes no structure to the parameters. Similarly, a classical logistic regression, $\Pr(y_i = 1) = \text{logit}^{-1}(X_i\beta)$, has a uniform prior distribution on the components of β.

Simplest varying-intercept model

The simplest multilevel regression is a varying-intercept model with normally distributed individual and group-level errors: $y_i \sim N(\alpha_{j[i]} + \beta x_i, \sigma_y^2)$ and $\alpha_j \sim N(\mu_\alpha, \sigma_\alpha^2)$. The normal distribution for the α_j's can be thought of as a prior distribution for these intercepts. The parameters of this prior distribution, μ_α and σ_α, are called *hyperparameters* and are themselves estimated from the data. In Bayesian inference, all the hyperparameters, along with the other unmodeled parameters (in this case, β and σ_y) also need a prior distribution which, as in classical regression, is typically set to a uniform distribution.[2] The complete prior distribution can be written in probability notation as $p(\alpha, \beta, \mu_\alpha, \sigma_y, \sigma_\alpha) \propto \prod_{j=1}^{J} N(\alpha_j | \mu_\alpha, \sigma_\alpha^2)$—that is, independent normal distributions for the α_j's so that their probability densities are multiplied to create the joint prior density.

Varying-intercept, varying-slope model

The more complicated model, $y_i = \alpha_{j[i]} + \beta_{j[i]} x_i + \epsilon_i$, has $2J$ modeled parameters, which are modeled as J pairs, (α_j, β_j). Their "prior distribution" is the bivariate normal distribution (13.1) on page 279, once again with independent uniform prior distributions on the hyperparameters.

---

[1] Technically, the classical least squares results are reproduced with a prior distribution that is uniform on $\log \sigma_y$, that is, the prior distribution $p(\log \sigma_y) \propto 1$, which is equivalent to $p(\sigma_y) \propto 1/\sigma_y$. The distinction between $p(\sigma_y) \propto 1$ and $p(\sigma_y) \propto 1/\sigma_y$ matters little in practice, and for convenience we work with the simpler uniform distribution on σ_y itself. See Gelman et al. (2003) for further discussion of such points.

[2] The inverse-gamma distribution is often used as a prior distribution for variance parameters; however, this model creates problems for variance components near zero, and so we prefer the uniform or, if more information is necessary, the half-t model; see Section 19.6.

Multilevel model with group-level predictors: exchangeability and prior distributions

We typically do not assign a model to coefficients of group-level predictors, or of individual-level predictors that do not vary by group. That is, in Bayesian terminology, we assign noninformative uniform prior distributions to these coefficients. More interestingly, a group-level regression induces different prior distributions on the group coefficients.

Consider a simple varying-intercept model with one predictor at the individual level and one at the group level:

$$y_i = \alpha_{j[i]} + \beta x_i + \epsilon_i, \quad \epsilon_i \sim N(0, \sigma_y^2), \text{ for } i = 1, \dots, n$$

$$\alpha_j = \gamma_0 + \gamma_1 u_j + \eta_j, \quad \eta_j \sim N(0, \sigma_\alpha^2), \text{ for } j = 1, \dots, J. \quad (16.1)$$

The first equation, for the y_i's, is called the *data model* or the *likelihood* (see Sections 18.1–18.2). The second equation, for the α_j's, is called the *group-level model* or the *prior model*. (The choice of name does not matter except that it draws attention either to the grouped structure of the data or to the fact that the parameters α_j are given a probability model.)

The α_j's in (16.1) have different prior distributions. For any particular group j, its α_j has a prior distribution with mean $\hat{\alpha}_j = \gamma_0 + \gamma_1 u_j$ and standard deviation σ_α. (As noted, this prior distribution depends on unknown parameters $\gamma_0, \gamma_1, \sigma_\alpha$, which themselves are estimated from the data and have noninformative uniform prior distributions.) These prior estimates of the α_j's differ because they differ in the values of their predictor u_j; the α_j's thus are not "exchangeable" and have different prior distributions.

An equivalent way to think of this model is as an exchangeable prior distribution on the group-level errors, η_j. From this perspective, the α_j's are determined by the group-level predictors u_j and the η_j's, which are assigned a common prior distribution with mean 0 and standard deviation σ_α. This distribution represents the possible values of η_j in a hypothetical population from which the given J groups have been sampled.

Thus, a prior distribution in a multilevel model can be thought of in two ways: as models that represent a "prior," or group-level, estimate for each of the α_j's; or as a single model that represents the distribution of the group-level errors, η_j. When formulating models with group-level predictors in Bugs, the former approach is usually more effective (it avoids the step of explicitly defining the η_j's, thus reducing the number of variables in the Bugs model and speeding computation).

Noninformative prior distributions

Noninformative prior distributions are intended to allow Bayesian inference for parameters about which not much is known beyond the data included in the analysis at hand. Various justifications and interpretations of noninformative priors have been proposed over the years, including invariance, maximum entropy, and agreement with classical estimators. In our work, we consider noninformative prior distributions to be "reference models" to be used as a standard of comparison or starting point in place of the proper, informative prior distributions that would be appropriate for a full Bayesian analysis.

We view any noninformative prior distribution as inherently provisional—after the model has been fit, one should look at the posterior distribution and see if it makes sense. If the posterior distribution does not make sense, this implies that additional prior knowledge is available that has not been included in the model, and that contradicts the assumptions of the prior distribution (or some other part of the model) that has been used. It is then appropriate to go back and alter the model to be more consistent with this external knowledge.

16.3 Fitting and understanding a varying-intercept multilevel model using R and Bugs

We introduce Bugs by stepping through the varying-intercept model for log radon levels described in Section 12.3. We first fit classical complete-pooling and no-pooling models in R using `lm()`, then perform a quick multilevel fit using `lmer()` as described in Section 12.4, then fit the multilevel model in Bugs, as called from R.

Setting up the data in R

We begin by loading in the data: radon measurements and floors of measurement for 919 homes sampled from the 85 counties of Minnesota. (The dataset also contains several other measurements at the house level that we do not use in our model.) Because it makes sense to assume multiplicative effects, we want to work with the logarithms of radon levels; however, some of the radon measurements have been recorded as 0.0 picoCuries per liter. We make a simple correction by rounding these up to 0.1 before taking logs.

R code
```
srrs2 <- read.table ("srrs2.dat", header=TRUE, sep=",")
mn <- srrs2$state=="MN"
radon <- srrs2$activity[mn]
y <- log (ifelse (radon==0, .1, radon))
n <- length(radon)
x <- srrs2$floor[mn]                    # 0 for basement, 1 for first floor
```

County indicators. We must do some manipulations in R to code the counties from 1 to 85:

R code
```
srrs2.fips <- srrs2$stfips*1000 + srrs2$cntyfips
county.name <- as.vector(srrs2$county[mn])
uniq.name <- unique(county.name)
J <- length(uniq.name)
county <- rep (NA, J)
for (i in 1:J){
    county[county.name==uniq.name[i]] <- i
}
```

There may very well be a better way to create this sort of index variable; this is just how we did it in one particular problem.

Classical complete-pooling regression in R

We begin with simple classical regression, ignoring the county indicators (that is, complete pooling):

R code
```
lm.pooled <- lm (y ~ x)
display (lm.pooled)
```

which yields

R output
```
            coef.est coef.se
(Intercept) 1.33     0.03
x           -0.61     0.07
  n = 919, k = 2
  residual sd = 0.82, R-Squared = 0.07
```

Classical no-pooling regression in R

Including a constant term and 84 county indicators. Another alternative is to include all 85 indicators in the model—actually, just 84 since we already have a constant term:

```
lm.unpooled.0 <- lm (formula = y ~ x + factor(county))
```
R code

which yields

```
                coef.est coef.se
(Intercept)       0.84     0.38
x                -0.72     0.07
factor(county)2   0.03     0.39
factor(county)3   0.69     0.58
  . . .
factor(county)85  0.35     0.66
  n = 919, k = 86
  residual sd = 0.76, R-Squared = 0.29
```
R output

This is the no-pooling regression. Here, county 1 has become the reference condition. Thus, for example, the log radon levels in county 2 are 0.03 higher, and so unlogged radon levels are approximately 3% higher in county 2, on average, than those in county 1, after controlling for the floor of measurement. The model including county indicators fits quite a bit better than the previous regression (the residual standard deviation has declined from 0.82 to 0.76, and R^2 has increased from 7% to 29%)—but this is no surprise since we have added 84 predictors. The estimates for the individual counties in this new model are highly uncertain (for example, counties 2, 3, and 85 shown above are not statistically significantly different from the default county 1).

Including 85 county indicators with no constant term. For making predictions about individual counties, it it slightly more convenient to fit this model without a constant term, so that each county has its own intercept:

```
lm.unpooled <- lm (formula = y ~ x + factor(county) - 1)
```
R code

(adding "−1" to the linear model formula removes the constant term), to yield

```
                coef.est coef.se
x                -0.72     0.07
factor(county)1   0.84     0.38
factor(county)2   0.87     0.10
  . . .
factor(county)85  1.19     0.53
  n = 919, k = 86
  residual sd = 0.76, R-Squared = 0.77
```
R output

These estimates are consistent with the previous parameterization—for example, the estimate for county 1 is 0.84 with a standard error of 0.38, which is identical to the inference for the intercept from the previous model. The estimate for county 2 is 0.87, which equals the intercept from the previous model, plus 0.03, which was the estimate for county 2, when county 1 was a baseline. The standard error for county 2 has declined because the uncertainty about the intercept for county 2 is less than the uncertainty about the difference between counties 1 and 2. Moving to the bottom of the table, the residual standard deviation is unchanged, as is appropriate given that this is just a shifting of the constant term within an existing model. Oddly, however, the R^2 has increased from 29% to 77%—this is because the lm() function calculates explained variance differently for models with no constant term, an issue that does not concern us. (For our purposes, the correct R^2 for this model is 29%; see Section 21.5.)

Setting up a multilevel regression model in Bugs

We set up the following Bugs code for a multilevel model for the radon problem, saving it in the file radon.1.bug (in the same working directory that we are using for our R analyses). Section 16.4 explains this model in detail.

Bugs code
```
model {
  for (i in 1:n){
    y[i] ~ dnorm (y.hat[i], tau.y)
    y.hat[i] <- a[county[i]] + b*x[i]
  }
  b ~ dnorm (0, .0001)
  tau.y <- pow(sigma.y, -2)
  sigma.y ~ dunif (0, 100)
  for (j in 1:J){
    a[j] ~ dnorm (mu.a, tau.a)
  }
  mu.a ~ dnorm (0, .0001)
  tau.a <- pow(sigma.a, -2)
  sigma.a ~ dunif (0, 100)
}
```

Calling Bugs from R

Our R environment has already been set up to be ready to call Bugs (see Appendix C). We execute the following R code to set up the data, initial values, and parameters to save for the Bugs run:

R code
```
radon.data <- list ("n", "J", "y", "county", "x")
radon.inits <- function (){
  list (a=rnorm(J), b=rnorm(1), mu.a=rnorm(1),
      sigma.y=runif(1), sigma.a=runif(1))}
radon.parameters <- c ("a", "b", "mu.a", "sigma.y", "sigma.a")
```

Once again, these details will be explained in Section 16.4. The R code continues with a call to Bugs in "debug" mode:

R code
```
radon.1 <- bugs (radon.data, radon.inits, radon.parameters,
  "radon.1.bug", n.chains=3, n.iter=10, debug=TRUE)
```

Here, we have run Bugs for just 10 iterations in each chain. We can look at the output in the Bugs window. When we close the Bugs window, R resumes. In the R window, we can type

R code
```
plot (radon.1)
print (radon.1)
```

and inferences for a, b, mu.a, sigma.y, sigma.a (the parameters included in the radon.parameters vector that was passed to Bugs) are displayed in a graphics window and in the R console. Having ascertained that the program will run, we now run it longer:

R code
```
radon.1 <- bugs (radon.data, radon.inits, radon.parameters,
  "radon.1.bug", n.chains=3, n.iter=500)
```

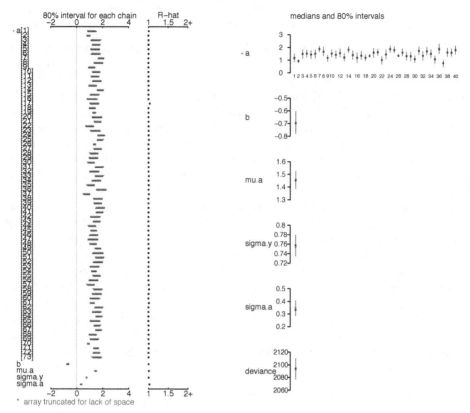

Figure 16.1 *Summary of Bugs simulations for the multilevel regression with varying intercepts and constant slope, fit to the Minnesota radon data. R-hat is near 1 for all parameters, indicating approximate convergence. The intervals for the county intercepts* α_j *indicate the uncertainty in these estimates (see also the error bars in Figure 12.3a on page 256).*

A Bugs window opens, and after about a minute, it closes and the R window becomes active again. Again, we can `plot` and `print` the fitted Bugs object, yielding the display shown in Figure 16.1, and the following text in the R window:

```
Inference for Bugs model at "radon.1.bug"                              R output
 3 chains, each with 500 iterations (first 250 discarded)
 n.sims = 750 iterations saved
          mean   sd   2.5%    25%    50%     75%   97.5% Rhat n.eff
a[1]       1.2  0.3    0.7    1.0    1.2     1.3     1.7    1   230
a[2]       0.9  0.1    0.7    0.9    0.9     1.0     1.1    1   750
 . . .
a[85]      1.4  0.3    0.8    1.2    1.4     1.6     1.9    1   750
b         -0.7  0.1   -0.8   -0.7   -0.7    -0.6    -0.5    1   750
mu.a       1.5  0.1    1.3    1.4    1.5     1.5     1.6    1   240
sigma.y    0.8  0.0    0.7    0.7    0.8     0.8     0.8    1   710
sigma.a    0.3  0.0    0.3    0.3    0.3     0.4     0.4    1    95
deviance 2093.4 12.5 2069.0 2085.0 2093.0 2101.7 2119.0    1   510

pD = 77.7 and DIC = 2171 (using the rule, pD = var(deviance)/2)
DIC is an estimate of expected predictive error (lower deviance is better).
```

```
For each parameter, n.eff is a crude measure of effective sample size,
and Rhat is the potential scale reduction factor (at convergence, Rhat=1).
```

The first seven columns of numbers give inferences for the model parameters. For example, α_1 has a mean estimate of 1.2 and a standard error of 0.3. The median estimate of α_1 is 1.2, with a 50% uncertainty interval of $[1.0, 1.3]$ and a 95% interval of $[0.7, 1.7]$. Moving to the bottom of the table, the 50% interval for the floor coefficient, β, is $[-0.7, -0.6]$, and the average intercept, μ_α, is estimated at 1.5 (see Figure 12.4 on page 257). The within-county standard deviation σ_y is estimated to be much larger than the between-county standard deviation σ_α. (This can also be seen in Figure 12.4—the scatter between points within each county is much larger than the county-to-county variation among the estimated regression lines.)

At the bottom of the table, we see p_D, the estimated effective number of parameters in the model, and DIC, the deviance information criterion, an estimate of predictive error. We return to these in Section 24.3.

Finally, we consider the rightmost columns of the output from the Bugs fit. Rhat gives information about convergence of the algorithm. At convergence, the numbers in this column should equal 1; before convergence, it should be larger than 1. If Rhat is less than 1.1 for all parameters, then we judge the algorithm to have approximately converged, in the sense that the parallel chains have mixed well (see Section 18.6 for more context on this). The final column, n.eff, is the "effective sample size" of the simulations, also discussed in Section 18.6.

Summarizing classical and multilevel inferences graphically

We can use R to summarize our inferences obtained from Bugs. For example, to display individual-level regressions as in Figure 12.4 on page 257, we first choose the counties to display, construct jittered data, and compute the range of the data (so that all eight counties will be displayed on a common scale):

R code
```
display8 <- c (36, 1, 35, 21, 14, 71, 61, 70)   # choose 8 counties
x.jitter <- x + runif(n,-.05,.05)
x.range <- range (x.jitter)
y.range <- range (y[!is.na(match(county,display8))])
```

We then pull out the appropriate parameter estimates from the classical fits:

R code
```
a.pooled <- coef(lm.pooled)[1]           # complete-pooling intercept
b.pooled <- coef(lm.pooled)[2]           # complete-pooling slope
a.nopooled <- coef(lm.unpooled)[2:(J+1)] # no-pooling intercepts
b.nopooled <- coef(lm.unpooled)[1]       # no-pooling slope
```

and summarize the parameters in the fitted multilevel model by their median estimates, first attaching the Bugs object and then computing the medians, component by component:

R code
```
attach.bugs (radon.1)
a.multilevel <- rep (NA, J)
for (j in 1:J){
  a.multilevel[j] <- median (a[,j])
}
b.multilevel <- median (b)
```

The computation for a is more complicated;[3] because a is a vector of length J, its n.sims simulations are saved as a matrix with dimensions n.sims $\times$ J.

We can now make the graphs in Figure 12.4:

```
par (mfrow=c(2,4))                                                    R code
for (j in display8){
  plot (x.jitter[county==j], y[county==j], xlim=c(-.05,1.05),
    ylim=y.range, xlab="floor", ylab="log radon level", main=uniq[j])
  curve (a.pooled + b.pooled*x, lwd=.5, lty=2, col="gray10", add=TRUE)
  curve (a.nopooled[j] + b.nopooled*x, lwd=.5, col="gray10", add=TRUE)
  curve (a.multilevel[j] + b.multilevel*x, lwd=1, col="black", add=TRUE)
}
```

(This can be compared with the code on page 261 for making this graph using the point estimates from lmer().)

To display the estimates and uncertainties versus sample size as in Figure 12.3b on page 256, we first set up the sample size variable,

```
sample.size <- as.vector (table (county))                            R code
sample.size.jitter <- sample.size*exp(runif(J,-.1,.1))
```

and then plot the estimates and standard errors:

```
plot (sample.size.jitter, a.multilevel, xlab="sample size in county j",   R code
  ylim=range(y), ylab=expression (paste ("intercept, ", alpha[j],
  "   (multilevel inference)")), pch=20, log="x")
for (j in 1:J){
  lines (rep(sample.size.jitter[j],2), median(a[,j])+c(-1,1)*sd(a[,j]))
}
abline (a.pooled, 0, lwd=.5)
```

The last line of code above places a thin horizontal line at the complete-pooling estimate, as can be seen in Figure 12.3b.

16.4 Step by step through a Bugs model, as called from R

A Bugs model must include a specification for every data point, every group-level parameter, and every hyperparameter. We illustrate here for the radon model shown in the previous section. For help in programming models in Bugs in general, open the Bugs window and click on Help, then Examples. Chapter 19 discusses some methods for running Bugs faster and more reliably.

The individual-level model

The varying-intercept model on page 350 starts with a probability distribution for each data point; hence the looping from 1 to n:

```
model {                                                              Bugs code
  for (i in 1:n){
    y[i] ~ dnorm (y.hat[i], tau.y)
    y.hat[i] <- a[county[i]] + b*x[i]
  }
```

---

[3] The three lines computing a.multilevel could be compressed into the single command, a.multilevel <- apply (a, 2, median), but we find it clearer to compute the components one at a time.

The Bugs code is abstracted in several steps from the model

$$y_i \sim N(\alpha_{j[i]} + \beta x_i, \sigma_y^2). \tag{16.2}$$

First, Bugs does not allow composite expressions in its distribution specifications (we cannot write y[i] ~ dnorm (a[county[i]] + b*x[i], tau.y) because the first argument to dnorm is too complex), and so we split (16.2) into two lines:

$$\begin{aligned}
y_i &\sim& N(\hat{y}_i, \sigma_y^2) \\
\hat{y}_i &=& \alpha_{j[i]} + \beta x_i.
\end{aligned} \tag{16.3}$$

In the notation of our variables in R, this second line becomes

$$\hat{y}_i = \text{a}_{\text{county}[i]} + \text{b} \cdot \text{x}_i. \tag{16.4}$$

(In the program we use a, b rather than α, β to make the code easier to follow.)

Second, as noted above, we replace the subscripts $j[i]$ in the model with the variable county[i] in R and Bugs, thus freeing j to be a looping index.

Finally, Bugs parameterizes normal distributions in terms of the inverse-variance, $\tau = 1/\sigma^2$, a point we shall return to shortly.

The group-level model

The next step is to model the group-level parameters. For our example here, these are the county intercepts α_j, which have a common mean μ_α and standard deviation σ_α:

$$\alpha_j \sim N(\mu_\alpha, \sigma_\alpha^2).$$

This is expressed almost identically in Bugs:

Bugs code
```
for (j in 1:J){
    a[j] ~ dnorm (mu.a, tau.a)
}
```

The only difference from the preceding statistical formula is the use of the inverse-variance parameter $\tau_\alpha = 1/\sigma_\alpha^2$.

Prior distributions

Every parameter in a Bugs model must be given either an assignment (as is done for the temporary parameters y.hat[i] defined within the data model) or a distribution. The parameters a[j] were given a distribution as part of the group-level model, but this still leaves b and tau.y from the data model and mu.a and tau.a from the group-level model to be defined.

The specifications for these parameters are called *prior distributions* because they must be specified before the model is fit to data. In the radon example we follow common practice and use noninformative prior distributions:

Bugs code
```
b ~ dnorm (0, .0001)
mu.a ~ dnorm (0, .0001)
```

The regression coefficients μ_α and β are each given normal prior distributions with mean 0 and standard deviation 100 (thus, they each have inverse-variance $1/100^2 = 10^{-4}$). This states, roughly, that we expect these coefficients to be in the range $(-100, 100)$, and if the estimates are in this range, the prior distribution is providing very little information in the inference.

```
tau.y <- pow(sigma.y, -2)
sigma.y ~ dunif (0, 100)
tau.a <- pow(sigma.a, -2)
sigma.a ~ dunif (0, 100)
```

Bugs code

We define the inverse-variances τ_y and τ_α in terms of the standard-deviation parameters, σ_y and σ_α, which are each given uniform prior distributions on the range $(0, 100)$.

Scale of prior distributions

Constraining the absolute values of the parameters to be less than 100 is not a serious restriction—the model is on the log scale, and there is no way we will see effects as extreme as -100 or 100 on the log scale, which would correspond to multiplicative effects of e^{-100} or e^{100}.

Here are two examples where the prior distributions with scale 100 would *not* be noninformative:

- In the regression of earnings (in dollars) on height, the coefficient estimate is 1300 (see model (4.1) on page 53). Fitting the model with a $N(0, 100^2)$ prior distribution (in Bugs, `dnorm(0,.0001)`) would pull the coefficient toward zero, completely inappropriately.

- In the model on page 88 of the probability of switching wells, given distance to the nearest safe well (in 100-meter units), the logistic regression coefficient of distance is -0.62. A normal prior distribution with mean 0 and standard deviation 100 would be no problem here. If, however, distance were measured in 100-kilometer units, its coefficient would become -620, and its estimate would be strongly and inappropriately affected by the prior distribution with scale 100.

Noninformative prior distributions

To summarize the above discussion: for a prior distribution to be noninformative, its range of uncertainty should be clearly wider than the range of reasonable values of the parameters. Our starting point is regression in which the outcome y and the predictors x have variation that is of the order of magnitude of 1. Simple examples are binary variables (these are 0 or 1 by definition), subjective measurement scales (for example, 1–5, or 1–10, or -3 to $+3$), and proportions. In other cases, it makes sense to transform predictors to a more reasonable scale—for example, taking a 0–100 score and dividing by 100 so the range is from 0 to 1, or taking the logarithm of earnings or height. One of the advantages of logarithmic and logistic regressions is that these automatically put the outcomes on scales for which typical changes are 0.1, or 1, but not 10 or 100. As long as the predictors are also on a reasonable scale, one would not expect to see coefficients much higher than 10 in absolute value, and so prior distributions with scale 100 are noninformative.

At this point one might ask, why not simply set a prior distribution with mean 0 and a huge standard deviation such as 100,000 (this would be `dnorm(0,1.E-10)` in Bugs) to completely ensure noninformativeness? We do not do this for two reasons. First, Bugs can be computationally unstable when parameters have extremely wide ranges. It is safer to keep the values near 1. (This is why we typically use $N(0, 1)$ and Uniform$(0, 1)$ distributions for initial values, as we shall discuss.)

The second reason for avoiding extremely wide prior distributions is that we do not actually want to work with coefficients that are orders of magnitude away from zero. Part of this is for ease of interpretation (just as we transformed to `dist100` in the arsenic example on page 88). This can also be interpreted as a form of model checking—we set up a model so that parameters should not be much greater than 1 in absolute value; if they are, this indicates some aspect of the problem that we do not understand.

Data, initial values, and parameters

To return to the implementation of the radon example, we set up and call Bugs using the following sequence of commands in R:

R code
```
radon.data <- list ("n", "J", "y", "county", "x")
radon.inits <- function (){
  list (a=rnorm(J), b=rnorm(1), mu.a=rnorm(1),
        sigma.y=runif(1), sigma.a=runif(1))}
radon.parameters <- c ("a", "b", "mu.a", "sigma.y", "sigma.a")
radon.1 <- bugs (radon.data, radon.inits, radon.parameters,
  "radon.1.bug", n.chains=3, n.iter=500)
```

The first argument to the `bugs()` function lists the data—including outcomes, inputs, and indexing parameters such as n and J—that are written to a file to be read by Bugs.

The second argument to `bugs()` is a function that returns a list of the starting values for the algorithm. Within the list are random-number generators—for example, `rnorm(J)` is a vector of length J of random numbers from the $N(0, 1)$ distribution, and these random numbers are assigned to α to start off the Bugs iterations. In this example, we follow our usual practice and assign random numbers from normal distributions for all the parameters—except those constrained to be positive (here, σ_y and σ_α), to which we assign uniformly distributed random numbers (which, by default in R, fall in the range $[0, 1]$). For more details on the random number functions, type `?rnorm` and `?runif` in R.

We generally supply initial values (using random numbers) for all the parameters in the model. (When initial values are not specified, Bugs generates them itself; however, Bugs often crashes when using its self-generated initial values.)

The third argument to the `bugs()` function is a vector of the names of the parameters that we want to save from the Bugs run. For example, the vector of α_j parameters is represented by `"a"`.

Number of sequences and number of iterations

Bugs uses an iterative algorithm that runs several Markov chains in parallel, each starting with some list of initial values and endlessly wandering through a distribution of parameter estimates. We would like to run the algorithm until the simulations from separate initial values converge to a common distribution, as Figure 16.2 illustrates. Specifying initial values using random distributions (as described above) ensures that different chains start at different points.

We assess convergence by checking whether the distributions of the different simulated chains mix; we thus need to simulate *at least 2 chains*. We also need to run the simulations "long enough," although it is generally difficult to know ahead of time how long is necessary. The `bugs()` function is set up to run for `n.iter` iterations and discard the first half of each chain (to lose the influence of the starting values). Thus, in the example presented here, Bugs ran `n.chains` = 3 sequences, each for `n.iter` = 500 iterations, with the first 250 from each sequence discarded.

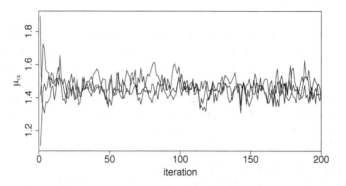

Figure 16.2 *Illustration of convergence for the parameter μ_α from the Bugs simulations for the radon model. The three chains start at different (random) points and, for the first 50 iterations or so, have not completely mixed. By 200 iterations, the chains have mixed fairly well. Inference is based on the last halves of the simulated sequences. Compare to Figure 16.3, which shows poor mixing.*

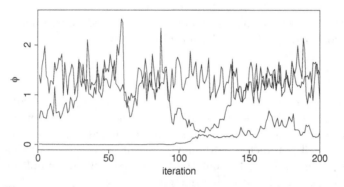

Figure 16.3 *Illustration of poor convergence for the parameter ϕ from a slowly converging Bugs simulation. Compare to Figure 16.2, which shows rapid mixing.*

We offer the following general advice on how long to run the simulations.

1. When first setting up a model, set `n.iter` to a low value such as 10 or 50—the model probably will not run correctly at first, so there is no need to waste time waiting for computations.

2. If the simulations have not reached approximate convergence (see Section 16.4), run longer—perhaps 500 or 1000 iterations—so that running Bugs takes between a few seconds and a few minutes.

3. If your Bugs run is taking a long time (more than a few minutes for the examples of the size presented in this book, or longer for larger datasets or more elaborate models), and the simulations are still far from convergence, then play around with your model to get it to converge faster; see Chapter 19 for more on this.

4. Another useful trick to speed computations, especially when in the exploratory model-fitting stage, is to work with a subset of your data—perhaps half, or a fifth, or a tenth. For example, analyze the radon data from a sample of 20 counties, rather than the full set of 85. Bugs will run faster with smaller datasets and fewer parameters.

Summary and convergence

From a Bugs run, you will see means, standard deviations, and quantiles for all the parameters that are saved. You also get, for each parameter, a convergence statistic, $\hat{R}$, and an effective number of independent simulation draws, n_{eff}. We typically monitor convergence using $\hat{R}$, which we call the *potential scale reduction factor*—for each parameter, the possible reduction in the width of its confidence interval, were the simulations to be run forever. Our usual practice is to run simulations until $\hat{R}$ is no greater than 1.1 for all parameters.

For example, here was the output after `n.iter` $= 50$ iterations:

R output
```
Inference for Bugs model at "radon.1.bug"
 3 chains, each with 50 iterations (first 25 discarded)
 n.sims = 75 iterations saved
          mean   sd   2.5%   25%   50%   75%  97.5% Rhat n.eff
 . . .
 mu.a     1.1   0.1   1.4   1.5   1.6   1.7   1.8  1.7    6
 . . .
```

$\hat{R} = 1.7$ indicates that the simulations were still far from convergence.

After `n.iter` $= 500$ iterations, however, we achieved approximate convergence with the radon model, with $\hat{R} < 1.1$ for all parameters.

$\hat{R}$ and n_{eff}

For each parameter that is saved, $\hat{R}$ is, approximately, the square root of the variance of the mixture of all the chains, divided by the average within-chain variance. If $\hat{R}$ is much greater than 1, the chains have not mixed well. We usually wait until $\hat{R} \leq 1.1$ for all parameters, although, if simulations are proceeding slowly, we might work provisionally with simulations that have still not completely mixed, for example, with $\hat{R} = 1.5$ for some parameters.

Printing a Bugs object also reports n_{eff}, the "effective number of simulation draws." If the simulation draws were independent, then n_{eff} would be the number of saved draws, which is $n_{\text{chains}} \cdot n_{\text{iter}}/2$ (dividing by 2 because our programs automatically discard the first half of the simulations from each chain). Actual Markov chain simulations tend to be autocorrelated and so the effective number of simulation draws is smaller. We usually like to have n_{eff} to be at least 100 for typical estimates and confidence intervals.

Accessing the simulations

We can use the simulations for predictions and uncertainty intervals for any functions of parameters, as with the propagation of error in classical regressions in Chapters 2 and 3. To access the simulations, we must first `attach` them in R. In the example above, we saved the Bugs output into the R object `radon.1` (see page 356), and we can load in the relevant information with the command

R code
```
attach.bugs (radon.1)
```

Each variable that was saved in the Bugs computation now lives as an R object, with its 750 simulation draws (3 chains × 500 iterations × last half of the iterations are saved = 750). Each of the scalar parameters β, μ_α, σ_y, and σ_α is represented by a vector of length 750, and the vector parameter α is saved as a 750×85 matrix. Extending this, a 10×20 matrix parameter would be saved as a $750 \times 10 \times 20$ array, and so forth.

We can access the parameters directly. For example, a 90% interval for β would be computed by

```
quantile (b, c(0.05,0.95))
```
R code

For another example, what is the probability that average radon levels (after controlling for floor-of-measurement effects) are higher in county 36 (Lac Qui Parle) than in county 26 (Hennepin)?

```
mean (a[,36] > a[,26])
```
R code

Fitted values, residuals, and other calculations

We can calculate fitted values and residuals from the multilevel model:

```
y.hat <- a.multilevel[county] + b.multilevel*x
y.resid <- y - y.hat
```
R code

and then plot them:

```
plot (y.hat, y.resid)
```
R code

Alternatively, we can add `y.hat` to the vector of parameters to save in the Bugs call, and then access the simulations of `y.hat` after the Bugs run and the call to `attach.bugs`.

We can also perform numerical calculations, such as the predictions described in Section 12.8 or anything that might be of interest. For example, what is the distribution of the difference in absolute (not log) radon level in a house with no basement in county 36 (Lac Qui Parle), compared to a house with no basement in county 26 (Hennepin)?

```
lqp.radon <- rep (NA, n.sims)
hennepin.radon <- rep (NA, n.sims)
for (s in 1:n.sims){
  lqp.radon[s] <- exp (rnorm (1, a[s,36] + b[s], sigma.y[s]))
  hennepin.radon[s] <- exp (rnorm (1, a[s,26] + b[s], sigma.y[s]))
}
radon.diff <- lqp.radon - hennepin.radon
hist (radon.diff)
print (mean(radon.diff))
print (sd(radon.diff))
```
R code

The expected difference comes to 2.0 picoCuries per liter, with a standard deviation of 4.6 and a wide range of uncertainty. Here we have compared two randomly selected houses, not the two county averages. If we wanted inference for the difference between the two county averages, we could simply take `exp(a[,36]+b)` - `exp(a[,26]+b)`.

We further discuss multilevel predictions in Section 16.6.

16.5 Adding individual- and group-level predictors

Classical complete-pooling and no-pooling regressions

Classical regressions and generalized linear models can be fit easily enough using R, but it can sometimes be useful also to estimate them using Bugs—often as a step toward fitting more complicated models. We illustrate with the radon example.

Complete pooling. The complete-pooling model is a simple linear regression of log radon on basement status and can be written in Bugs as

Bugs code
```
model {
  for (i in 1:n){
    y[i] ~ dnorm (y.hat[i], tau.y)
    y.hat[i] <- a + b*x[i]
  }
  a ~ dnorm (0, .0001)
  b ~ dnorm (0, .0001)
  tau.y <- pow(sigma.y, -2)
  sigma.y ~ dunif (0, 100)
}
```

No pooling. The no-pooling model can be fit in two ways: either by fitting the above regression separately to the data in each county (thus, running a loop in R for the 85 counties), or else by allowing the intercept α to vary but with a noninformative prior distribution for each α_j (so that this is still a classical regression):

Bugs code
```
model {
  for (i in 1:n){
    y[i] ~ dnorm (y.hat[i], tau.y)
    y.hat[i] <- a[county[i]] + b*x[i]
  }
  b ~ dnorm (0, .0001)
  tau.y <- pow(sigma.y, -2)
  sigma.y ~ dunif (0, 100)

  for (j in 1:J){
    a[j] ~ dnorm (0, .0001)
  }
}
```

Classical regression with multiple predictors

The Bugs model can easily include multiple predictors in y.hat. For example, we can add an indicator for whether the measurement was taken in winter (when windows are closed, trapping radon indoors):

Bugs code
```
        y.hat[i] <- a + b.x*x[i] + b.winter*winter[i]
```

and add an interaction:

Bugs code
```
        y.hat[i] <- a + b.x*x[i] + b.winter*winter[i] +
                    b.x.winter*x[i]*winter[i]
```

In each case, we would set up these new coefficients with dnorm(0,.0001) prior distributions.

As the number of predictors increases, it can be simpler to set up a vector β of regression coefficients:

Bugs code
```
        y.hat[i] <- a + b[1]*x[i] + b[2]*winter[i] + b[3]*x[i]*winter[i]
```

and then assign these noninformative prior distributions:

Bugs code
```
        for (k in 1:K){
          b[k] ~ dnorm (0, .0001)
        }
```

with K added to the list of data in the call of **bugs()** from R.

Vector-matrix notation in Bugs. One can go further by creating a matrix of predictors in R:

```
X <- cbind (x, winter, x*winter)                                R code
K <- ncol (X)
```

and then in the Bugs model, using the inner-product function:

```
    y.hat[i] <- a + inprod(b[],X[i,])                           Bugs code
```

Finally, one could include the intercept in the list of β's, first including a constant term in the predictor matrix:

```
ones <- rep (1, n)                                              R code
X <- cbind (ones, x, winter, x*winter)
K <- ncol (X)
```

and then simplify the expression for y.hat in the Bugs model:

```
    y.hat[i] <- inprod(b[],X[i,])                               Bugs code
```

with the coefficients being b[1], ..., b[4].

In a varying-intercept model, it can be convenient to keep the intercept α separate from the other coefficients β. However, in model with a varying intercept and several varying slopes, it can make sense to use the unified notation including all of them in a matrix B, as we discuss in Sections 17.1 and 17.2.

Multilevel model with a group-level predictor

Here is the Bugs code for model (12.15) on page 266, which includes a group-level predictor, u_j (the county-level uranium measure in the radon example):

```
model {                                                         Bugs code
  for (i in 1:n){
    y[i] ~ dnorm (y.hat[i], tau.y)
    y.hat[i] <- a[county[i]] + b*x[i]
  }
  b ~ dnorm (0, .0001)
  tau.y <- pow(sigma.y, -2)
  sigma.y ~ dunif (0, 100)

  for (j in 1:J){
    a[j] ~ dnorm (a.hat[j], tau.a)
    a.hat[j] <- g.0 + g.1*u[j]
  }
  g.0 ~ dnorm (0, .0001)
  g.1 ~ dnorm (0, .0001)
  tau.a <- pow(sigma.a, -2)
  sigma.a ~ dunif (0, 100)
}
```

16.6 Predictions for new observations and new groups

We can simulate predictions in two ways: directly in Bugs, by adding additional units or groups to the model, or in R, by drawing simulations based on the model fit to existing data. We demonstrate both approaches for the radon example, in each case demonstrating how to forecast for new houses in existing counties and for new houses in new counties.

Predicting a new unit in an existing group using Bugs

Bugs automatically makes predictions for modeled data with NA values. Thus, to predict the log radon level for a new house in county 26 with no basement, we merely need to extend the dataset by one point. We first save the current dataset to an external file and then extend it:

R code
```
save ("n", "y", "county", "x", file="radon.data")
n <- n + 1
y <- c (y, NA)
county <- c (county, 26)
x <- c (x, 1)
```

For convenience, we then add a line at the end of the Bugs model to flag the predicted measurement:

Bugs code
```
y.tilde <- y[n]
```

and rename this file as a new model, radon2a.bug. We now fit this Bugs model, after alerting it to save the inferences for the additional data point:

R code
```
radon.parameters <- c (radon.parameters, "y.tilde")
radon.2a <- bugs (radon.data, radon.inits, radon.parameters,
  "radon.2a.bug", n.chains=3, n.iter=500)
```

The prediction, $\tilde{y}$, now appears if radon.2a is printed or plotted, or we can access it directly; for example,

R code
```
attach.bugs (radon.2a)
quantile (exp (y.tilde), c(.25,.75))
```

gives a 50% confidence interval for the (unlogged) radon level in this new house.

We can similarly make predictions for any number of new houses by adding additional NA's to the end of the data vector. It is necessary to specify the predictors for these new houses; if you set county or x to NA for any of the data, the Bugs model will not run. Bugs allows missing data in modeled, but not unmodeled, data (a distinction we discuss further in Section 16.8).

Predicting a new unit in a new group using Bugs

The same approach can be used to make predictions for houses in unmeasured counties (ignoring for the moment that this particular survey included all the counties in Minnesota, or else considering an unmeasured county in a similar neighboring state). We merely need to extend the number of counties in the hypothetical dataset and specify the group-level predictor u_j (in this case, the county uranium measurement) for this new county. For simplicity we here consider a hypothetical new county with uranium measurement equal to the average of the 85 existing counties:

R code
```
u.tilde <- mean (u)
```

We now load back the original data, save everything, and extend to a new house in a new county:

R code
```
load ("radon.data")
save ("n", "y", "county", "x", "J", "u", file="radon.data")
n <- n + 1
y <- c (y, NA)
county <- c (county, J+1)
x <- c (x, 1)
J <- J + 1
u <- c (u, u.tilde)
```

We can then run the model as before, again using the simulations of $\tilde{y}$ to summarize the uncertainty about the radon level in this new house, this time in a new county.

Prediction using R

The other approach to prediction is to use R to simulate from the distribution of the new data, conditional on the estimated parameters from the model. Section 12.8 laid out how to do this using estimates from `lmer()`; here we do the same thing using Bugs output. The only difference from before is that instead of working with functions such as `coef()` and `sim()` applied to objects created from `lmer()`, we work directly with simulated parameters after attaching Bugs fits.

For a new house with no basement in county 26:

```
attach.bugs (radon.2)                                               R code
y.tilde <- rnorm (n.sims, a[,26] + b*1, sigma.y)
```

This creates a vector of n_{sims} simulations of $\tilde{y}$. The only tricky thing here is that we need to use matrix notation for `a` (its 26th column contains the simulations for a_{26}), but we can write `b` and `sigma.y` directly, since as scalar parameters these are saved as vectors of simulations. The simplest way to understand this is to perform the calculations on your own computer, running Bugs for just 10 iterations so that the saved objects are small and can be understood by simply typing `a`, `b`, and so forth in the R console.

Continuing, prediction for a new house with no basement in a new county with uranium level $\tilde{u}$ requires simulation first of the new $\tilde{\alpha}_j$, then of the radon measurement in the house within this county:

```
a.tilde <- rnorm (n.sims, g.0 + g.1*u.tilde, sigma.a)               R code
y.tilde <- rnorm (n.sims, a.tilde + b*1, sigma.y)
```

16.7 Fake-data simulation

As discussed in Sections 8.1–8.2, a good way to understand a model-fitting procedure is by simulating and then fitting a model to fake data:

1. Specify a reasonable "true" value for each of the parameters in the model. Label the vector of specified parameters as θ^{true}; these values should be reasonable and be consistent with the model.

2. Simulate a fake dataset y^{fake} using the model itself along with the assumed θ^{true}.

3. Fit the model to the fake data, and check that the inferences for the parameters θ are consistent with the "true" θ^{true}.

We illustrated this process for classical regression with numerical checks in Section 8.1 and graphical checks in Section 8.2. Here we demonstrate fake-data checking for the varying-intercept radon model (12.15) on page 266 with floor and uranium as individual- and county-level predictors, respectively.

Specifying "true" parameter values

For a classical regression, one must simply specify the coefficients β and residual standard deviation σ to begin a fake-data simulation. Multilevel modeling is more complicated: one must first specify the hyperparameters, then simulate the modeled parameters from the group-level distributions.

Specifying the unmodeled parameters. We start by specifying β^{true}, γ_0^{true}, γ_1^{true}, σ_y^{true}, and $\sigma_\alpha^{\text{true}}$: these are the parameters that do not vary by group, and they get the simulation started.

We do *not* want to choose round numbers such as 0 or 1 for the parameters, since these can mask potential programming errors. For example, a variance is specified as σ rather than σ^2 would not show up as an inconsistency if σ were set to 1.

There are two natural ways to set the parameters. The first approach is just to pick values that seem reasonable; for example, $\beta^{\text{true}} = -0.5$, $\gamma_0^{\text{true}} = 0.5$, $\gamma_1^{\text{true}} = 2.0$, $\sigma_y^{\text{true}} = 3.0$, $\sigma_\alpha^{\text{true}} = 1.5$. (Variation among groups is usually less than variation among individuals within a group, and so it makes sense to set $\sigma_\alpha^{\text{true}}$ to a smaller value than σ_y^{true}. The model is on the log scale, so it would not make sense to choose numbers such as 500 or -400 that are high in absolute value; see the discussion on page 355 on the scaling of prior distributions.)

The other way to choose the parameters is to use estimates from a fitted model. For this example, we could set $\beta^{\text{true}} = -0.7$, $\gamma_0^{\text{true}} = 1.5$, $\gamma_1^{\text{true}} = 0.7$, $\sigma_y^{\text{true}} = 0.8$, $\sigma_\alpha^{\text{true}} = 0.2$, which are the median estimates of the parameters from model (12.15) as fitted in Bugs (see code at the very end of Section 16.5). Supposing we have labeled this `bugs` object as `radon.2`, then we can set the "true" values in R:

R code
```
attach.bugs (radon.2)
b.true <- median (b)
g.0.true <- median (g.0)
g.1.true <- median (g.1)
sigma.y.true <- median (sigma.y)
sigma.a.true <- median (sigma.a)
```

Simulating the varying coefficients. We now simulate the α_j's from the group-level model given the "true" parameters using a loop:[4]

R code
```
a.true <- rep (NA, J)
for (j in 1:J){
    a.true[j] <- rnorm (1, g.0.true + g.1.true*u[j], sigma.a.true)
}
```

Simulating fake data

We can now simulate the dataset y^{fake}:[5]

R code
```
y.fake <- rep (NA, n)
for (i in 1:n){
    y.fake[i] <- rnorm (1, a.true[county[i]] + b.true*x[i], sigma.y.true)
}
```

Inference and comparison to "true" values

We can now fit the model in Bugs using the fake data. The fitting procedure is the same except that we must pass `y.fake` rather than `y` to Bugs, which we can do by explicitly specifying the data to be passed:[6]

---

[4] Or, more compactly but perhaps less clearly, in vector form:
```
a.true <- rnorm (J, g.0.true + g.1.true*u, sigma.a.true).
```
[5] Again, an experienced R programmer would use the vector form:
```
y.fake <- rnorm (n, a.true[county] + b.true*x, sigma.y.true).
```
[6] Alternatively, we could use the existing data object after saving the real data and renaming the fake data:
```
y.save <- y; y <- y.fake.
```

```
radon.data <- list (n=n, J=J, y=y.fake, county=county, x=x, u=u)
```
R code

We can then specify the rest of the inputs, run Bugs, and save it into a new R object:

```
radon.inits <- function (){
    list (a=rnorm(J), b=rnorm(1), g.0=rnorm(1), g.1=rnorm(1),
        sigma.y=runif(1), sigma.a=runif(1))}
radon.parameters <- c ("a", "b", "g.0", "g.1", "sigma.y", "sigma.a")
radon.2.fake <- bugs (radon.data, radon.inits, radon.parameters,
    "radon.2.bug", n.chains=3, n.iter=500)
```
R code

We are now ready to compare the inferences to the true parameter values. To start, we can display the fitted model (`print(radon.2.fake)`) and compare inferences to the true values. Approximately half the 50% intervals and approximately 95% of the 95% intervals should contain the true values; about half of the median estimates should be above the true parameter values and about half should be below. In our example, the output looks like:

	mean	sd	2.5%	25%	50%	75%	97.5%	Rhat	n.eff
a[1]	1.0	0.2	0.7	0.9	1.0	1.1	1.4	1.0	280
a[2]	0.9	0.1	0.7	0.8	0.9	0.9	1.0	1.0	190
. . .									
a[85]	1.8	0.2	1.4	1.7	1.8	1.9	2.2	1.0	470
b	-0.6	0.1	-0.7	-0.6	-0.6	-0.6	-0.5	1.0	750
g.0	1.5	0.0	1.4	1.4	1.5	1.5	1.5	1.0	140
g.1	0.8	0.1	0.6	0.7	0.8	0.9	1.0	1.0	68
sigma.y	0.8	0.0	0.7	0.7	0.8	0.8	0.8	1.0	240
sigma.a	0.2	0.0	0.1	0.2	0.2	0.2	0.3	1.1	36

R output

In the particular simulation we ran, the 85 values of α^{true} were 1.2, 0.9, ..., 1.8 (as we can see by simply typing `a.true` in the R console), and the "true" values of the other parameters, are $\beta^{\text{true}} = -0.7$, $\gamma_0^{\text{true}} = 1.5$, $\gamma_1^{\text{true}} = 0.7$, $\sigma_y^{\text{true}} = 0.8$, $\sigma_\alpha^{\text{true}} = 0.2$, as noted earlier. About half of these fall within the 50% intervals, as predicted.

Checking coverage of 50% intervals

For a more formal comparison, we can measure the coverage of the intervals. For example, to check the coverage of the 50% interval for α_1:

```
attach.bugs (radon.2.fake)
a.true[1] > quantile (a[,1], .25) & a.true[1] < quantile (a[,1], .75)
```
R code

which, for our particular simulation, yields the value FALSE. We can write a loop to check the coverage for all 85 α_j's:[7]

```
cover.50 <- rep (NA, J)
for (j in 1:J){
    cover.50[j] <- a.true[j] > quantile (a, .25) &
                   a.true[j] < quantile (a, .75)
}
mean (cover.50)
```
R code

which comes to 0.51 in our example—well within the margin of error for a random simulation.

Other numerical and graphical checks are also possible, following the principles of Chapter 8.

---

[7] Again, we could write the calculation more compactly in vectorized form as
```
cover.50 <- a.true > quantile (a, .25) & a.true < quantile (a, .75).
```

Category	List of objects
Modeled data	y
Unmodeled data	n, J, county, x, u
Modeled parameters	a
Unmodeled parameters	b, g.0, g.1, sigma.y, sigma.a
Derived quantities	y.hat, tau.y, a.hat, tau.a
Looping indexes	i, j

Figure 16.4 *Classes of objects in the Bugs model and R code of Section 16.8. Data are specified in the list of data sent to Bugs,* parameters *are nondata objects that are given distributions ("~" statements in Bugs),* derived quantities *are defined deterministically ("<-" statements in Bugs), and* looping indexes *are defined in* for *loops in the Bugs model.*

16.8 The principles of modeling in Bugs

Data, parameters, and derived quantities

Every object in a Bugs model is data, parameter, or derived quantity. Data are specified in the bugs() call, parameters are modeled, and derived quantities are given assignments. Some but not all data are modeled.

We illustrate with the varying-intercept Bugs model on page 361, stored in the file radon3.bug:

Bugs code
```
model {
  for (i in 1:n){
    y[i] ~ dnorm (y.hat[i], tau.y)
    y.hat[i] <- a[county[i]] + b*x[i]
  }
  b ~ dnorm (0, .0001)
  tau.y <- pow(sigma.y, -2)
  sigma.y ~ dunif (0, 100)

  for (j in 1:J){
    a[j] ~ dnorm (a.hat[j], tau.a)
    a.hat[j] <- g.0 + g.1*u[j]
  }
  g.0 ~ dnorm (0, .0001)
  g.1 ~ dnorm (0, .0001)
  tau.a <- pow(sigma.a, -2)
  sigma.a ~ dunif (0, 100)
}
```

as called from R as follows:

R code
```
radon.data <- list ("n", "J", "y", "county", "x", "u")
radon.inits <- function (){
  list (a=rnorm(J), b=rnorm(1), g.0=rnorm(1), g.1=rnorm(1),
    sigma.y=runif(1), sigma.a=runif(1))}
radon.parameters <- c ("a", "b", "sigma.y", "sigma.a")
radon.3 <- bugs (radon.data, radon.inits, radon.parameters,
  "radon.3.bug", n.chains=3, n.iter=500)
```

This model has the following classes of objects, which we summarize in Figure 16.4.

- The *data* are the objects that are specified by the data input to the bugs() call:

- *Modeled data*: These are the data objects that are assigned probability distributions (that is, they are to the left of a "∼" in a line of Bugs code). In our example, the only modeled data are the components of y.

- *Unmodeled data*: These are the data that are not assigned any distribution in the Bugs code. The unmodeled data objects in our example are n, J, county, x, u.

• Next come the *parameters*, which are assigned probability distributions (that is, they are to the left of a "∼" in a line of Bugs code) but are not specified as data in the bugs() call:

- *Modeled parameters*: We label parameters as "modeled" if they are assigned informative prior distributions that depend on hyperparameters (which typically are themselves estimated from the data). The only modeled parameters in our example are the elements of a, whose distribution depends on a.hat and tau.a.

- *Unmodeled parameters*: These are the parameters with noninformative prior distributions (typically dnorm(0,.0001) or dunif(0,100)); in our example, these are b, g.0, g.1, sigma.y, sigma.a. Strictly speaking, what we call "unmodeled parameters" are actually modeled with wide, "noninformative" prior distributions. What is important here is that their distributions in the Bugs model are specified not based on other parameters in the model but rather based on constants such as .0001.

• *Derived quantities*: These are objects that are defined deterministically (that is, with "<-" in the Bugs code); in our example: y.hat, tau.y, a.hat, tau.a.

• *Looping indexes*: These are integers (in our example, i and j) that are defined in for loops in the Bugs model.

To figure out whether an object is a parameter or a derived quantity, it can be helpful to scan down the code to see how it is defined. For example, reading the model on page 361 from the top, it is not clear at first whether tau.y is a parameter or a derived quantity. (We know it is not data since it is not specified in the data list supplied in the call to bugs().) But reading through the model, we see the line tau.y <- ..., so we know it is a derived quantity.

Missing data

Modeled data can have elements with NA values, in which case these elements are implicitly treated as parameters by the Bugs model—that is, they are estimated stochastically along with the other uncertain quantities in the model. Bugs will not accept *unmodeled* data (for example, the regression predictors x and u in our example) with NA values, because these objects in the model have no distributions specified and thus cannot be estimated stochastically. If predictors have missing data, they must either be imputed before being entered into the Bugs model, or they must themselves be modeled.

Changing what is included in the data

Any combination of data, parameters, or derived quantities can be saved as parameters in the bugs() call. But only parameters and missing data, not observed data and not derived quantities, can be given initial values.

Conversely, the status of the objects in a given Bugs model can change by changing the corresponding call in R. For example, we can call the model without specifying y:

R code
```
radon.data <- list ("n", "J", "county", "x", "u")
```

and the components of y become modeled parameters—they are specified by distributions in the Bugs model and not included as data. Bugs then draws from the "prior predictive distribution"—the model unconditional on y. Alternatively, if b, g.0, g.1, sigma.y, sigma.a are included in the data list but y is not, Bugs will perform "forward simulation" for y given these specified parameters.

We cannot remove the other objects—n, J, county, x, u—from the data list, because these are not specified in the Bugs model. For example, there is no line in the model of the form, n ∼ ... If we did want to remove any of these data objects, we would need to include them in the model, either as parameters (defined by ∼) or derived quantities (defined by <-).

We can, however, specify the values of parameters in the model. For example, suppose we wanted to fit the model with the variance parameters set to known values, for example, $\sigma_y = 0.7$ and $\sigma_\alpha = 0.4$. We can simply define them in the R code and include them in the data:

R code
```
sigma.y <- .7
sigma.alpha <- .4
radon.data <- list("n","J","y","county","x","u","sigma.y","sigma.alpha")
```

and remove them from the inits() function.

Each object can be defined at most once

Every object is modeled or defined at most once in a Bugs model (except that certain transformations can be done by declaring a variable twice). For example, the example we have been discussing includes the lines

Bugs code
```
for (i in 1:n){
  y[i] ~ dnorm (y.hat[i], tau.y)
  y.hat[i] <- a[county[i]] + b*x[i]
}
```

It might seem more compact to express this as

Bugs code
```
for (i in 1:n){
  y[i] ~ dnorm (y.hat, tau.y)
  y.hat <- a[county[i]] + b*x[i]
}
```

thus "overwriting" the intermediate quantity y.hat at each step of the loop. Such code would work in R but is not acceptable in Bugs. The reason is that lines of Bugs code are *specifications* of a model, not *instructions* to be executed. In particular, these lines define the single variable y.hat multiple times, which is not allowed in Bugs.

For another example, we have been coding the inverse-variance in terms of the standard-deviation parameter:

Bugs code
```
tau.y <- pow(sigma.y, -2)
sigma.y ~ dunif (0, 100)
```

It might seem natural to write this transformation as

```
sigma.y <- sqrt(1/tau.y)                              Bugs code
sigma.y ~ dunif (0, 100)
```

but this will not work: in this model, `sigma.y` is defined twice and `tau.y` is defined not at all. We must use the earlier formulation in which `tau.y` (which is specified in the data model) is defined in terms of `sigma.y`, which is then given its own prior distribution.

16.9 Practical issues of implementation

In almost any application, a good starting point is to run simple classical models in R (for example, complete-pooling and no-pooling regressions) and then replicate them in Bugs, checking that the estimates and standard errors are approximately unchanged. Then gradually complexify the model, adding multilevel structures, group-level predictors, varying slopes, non-nested structures, and so forth, as appropriate.

We suggest a simple start for both statistical and computational reasons. Even in classical regression, it is a good idea to include the most important predictors first and then see what happens when further predictors and interactions are added. Multilevel models can be even more difficult to understand, and so it makes sense to build up gradually. In addition, it is usually a mistake in Bugs to program a complicated model all at once; it typically will not run, and then you have to go back to simpler models anyway until you can get the program working.

If a model does not run, you can use the `debug=TRUE` option in the call to `bugs()`. Then the Bugs window will stay open and you might be able to figure out what's going wrong, as we discuss in Section 19.1.

How many chains and how long to run?

We usually run a small number of chains such as 3. This is governed by the `n.chains` argument of the `bugs()` function.

In deciding how long to run the simulations, we balance the goals of speed and convergence. We start by running Bugs for only a few iterations in debug mode until we can get our script to run without crashing. Once it works, we will do a fairly short run—for example, `n.iter` = 100 or 500. At this point:

- If approximate convergence has been reached ($\widehat{R} < 1.1$ for all parameters), we stop.

- If the sequences seem close to convergence (for example, $\widehat{R} < 1.5$ for all parameters), then we repeat, running longer (for example, 1000 or 2000 iterations).

- If our Bugs run takes more than a few minutes, and the sequences are still far from convergence, we step back and consider our options, which include:

  - altering the Bugs model to run more efficiently (see the tips in Chapter 17),
  - fitting the Bugs model to a sample of the data (see Section 19.2),
  - fitting a simpler model.

In some settings with complicated models, it may be necessary to run Bugs for a huge number of iterations, but in the model-building stage, we generally recommend *against* the "brute force" approach of simply running for 50,000 or 100,000 iterations. Even if this tactic yields convergence, it is typically not a good long-run solution, since it ensures long waits for fitting the inevitable alterations of the model (for example, from adding new predictors).

Initial values

The `bugs()` function takes an `inits` argument, which is a function that must be written for creating starting points. It is not too important exactly what these starting points are, as long as they are dispersed (so the different chains start at different points; see Figure 16.2 on page 357) and reasonable. If parameters get started at values such as 10^{-4} or 10^6, Bugs can drift out of range and crash.

It is convenient to specify initial values as a function using random numbers, so that running it for several chains will automatically give different starting points. Typically we start parameters from random $N(0, 1)$ distributions, unless they are restricted to be positive, in which case we typically use random numbers that are uniformly distributed between 0 and 1 (the Uniform$(0, 1)$ distribution). Vector and matrix parameters must be set up as vectors and matrices; for example, if a is a scalar parameter, b is a vector of length J, C is a $J \times K$ matrix, and σ is a scalar parameter constrained to be positive:

R code
```
inits <- function() {list (a=rnorm(1), b=rnorm(J),
    C=array(rnorm(J*K), c(J,K)), sigma=runif(1))}
```

Here we have used the defaults of the `rnorm()` and `runif()` functions. If instead, for example, we want to use $N(0, 2^2)$ and Uniform$(0.1, 10)$ distributions, we can write

R code
```
inits <- function() {list (a=rnorm(1,0,2), b=rnorm(J,0,2),
    C=array(rnorm(J*K,0,2), c(J,K)), sigma=runif(1,.1,10))}
```

16.10 Open-ended modeling in Bugs

This book focuses on the most standard models, beginning with linear regression and then adding various complications one step at a time. From this perspective, Bugs is useful because it accounts for uncertainty in all parameters when fitting multilevel models. However, a quite different advantage of Bugs is its modular structure, which allows us to fit models of nearly arbitrary complexity.

We demonstrate here by considering a hypothetical study of a new teaching method applied in J different classrooms containing a total of n students. Our data for this example will be the treatment indicator T (defined at the school level) and, for each student, a pre-treatment assessment, x (on a 1–10 scale, say) and a post-treatment test score, y (on a 0–100 scale).

The minimal model for such data is a hierarchical regression with varying intercepts for schools:

Bugs code
```
model {
  for (i in 1:n){
    y[i] ~ dnorm (y.hat[i], tau.y)
    y.hat[i] <- a[school[i]] + b*x[i]
  }
  b ~ dnorm (0, .0001)
  tau.y <- pow(sigma.y, -2)
  sigma.y ~ dunif (0, 100)

  for (j in 1:J){
    a[j] ~ dnorm (a.hat[j], tau.a)
    a.hat[j] <- g.0 + g.1*T[j]
  }
  g.0 ~ dnorm (0, .0001)
```

```
    g.1 ~ dnorm (0, .0001)
    tau.a <- pow(sigma.a, -2)
    sigma.a ~ dunif (0, 100)
  }
```

The natural extension here is to allow the coefficient β for the pre-treatment assessment to vary by group. This is a varying-intercept, varying-slope model, the implementation of which we shall discuss in Sections 17.1–17.2. Here, however, we shall consider some less standard extensions, to demonstrate the flexibility of Bugs.

Nonlinear and nonadditive models

The relation between inputs and regression predictor need not be linear. From classical regression we are already familiar with the inclusion of interactions and transformed inputs as additional predictors, for example, altering the expression for $\hat{y}$:

```
    y.hat[i] <- a[school[i]] + b[1]*x[i] + b[2]*pow(x[i],2) +          Bugs code
      b[3]*T[school[i]]*x[i]
```

Another option is to define transformed variables in R and then include them as predictors in the Bugs model; for example, for the squared term

```
    x.sq <- x^2                                                        R code
```

and then, in the Bugs model:

```
    y.hat[i] <- a[school[i]] + b[1]*x[i] + b[2]*x.sq[i]                 Bugs code
```

Nonlinear functions of data and parameters. More interesting are models that cannot simply be expressed as regressions. For example, suppose we wanted to fit a model with diminishing returns for the pre-treatment assessment, such as $y = \alpha - \beta e^{-\gamma x}$. We can simultaneously estimate the linear parameters α, β and the nonlinear γ in a Bugs model:

```
    y.hat[i] <- a[school[i]] + b*exp(-g*x[i])                          Bugs code
```

and also add a noninformative prior distribution for γ. The parameters β and γ could also be allowed to vary by group. More complicated expressions are also possible, for example,

```
    y.hat[i] <- a + b[1]*exp(-g[1]*x1[i]) + b[2]*exp(-g[2]*x2[i])      Bugs code
```

or

```
    y.hat[i] <- (a + b*x1[i])/(1 + g*x2[i])                            Bugs code
```

or whatever. The point here is not to try an endless variety of models but to be able to fit models that might be suggested by theoretical considerations, and to have the flexibility to alter functional forms as appropriate.

Unequal variances

Perhaps the data-level variance should be different for students in the treated and control groups. (The variance in the treatment group could be higher, for example, if the treatment worked very well on some students and poorly on others. Or, in the other direction, the treated students could show lower variance if the effect of the treatment is to pull all students up to a common level of expertise.)

Different variances for treated and control units.   We can allow for either of these possibilities by changing the data distribution to

Bugs code
```
y[i] ~ dnorm (y.hat[i], tau.y[T[school[i]+1]])
```

(Adding 1 to T allows the index to take on the values 1 and 2.) The specification of `tau.y` in the model is then expanded to

Bugs code
```
for (k in 1:2){
    tau.y[k] <- pow(sigma.y[k], -2)
    sigma.y[k] ~ dunif (0, 100)
}
```

Different variance within each group.   Similarly, we can simply let the data variance vary by school by changing the data model to,

Bugs code
```
y[i] ~ dnorm (y.hat[i], tau.y[school[i]])
```

and then specifying the distribution of `tau.y` within the `for (j in 1:J)` loop:

Bugs code
```
tau.y[j] <- pow(sigma.y[j], -2)
sigma.y[j] ~ dunif (0, 100)
```

Modeling the varying variances.   Once we have a parameter that varies by group, it makes sense to model it, for example using a lognormal distribution:

Bugs code
```
tau.y[j] <- pow(sigma.y[j], -2)
sigma.y[j] ~ dlnorm (mu.lsigma.y, tau.lsigma.y)
```

This extra stage of modeling allows us to better adjust for unequal variances when the sample size within groups is small (so that within-group variances cannot be precisely estimated individually). We also must specify noninformative prior distributions for these hyperparameters, outside the `for (j in 1:j)` loop:

Bugs code
```
mu.lsigma.y ~ dnorm (0, .0001)
tau.lsigma.y <- pow(sigma.lsigma.y, -2)
sigma.lsigma.y ~ dunif (0, 100)
```

Variances that also differ systematically between treatment and control.   We can extend the hierarchical model for the variances to include treatment as a group-level predictor for the variance model:

Bugs code
```
tau.y[j] <- pow(sigma.y[j], -2)
sigma.y[j] ~ dlnorm (log.sigma.hat[j], sigma.log.sigma.y)
log.sigma.hat[j] <- d.0 + d.1*T[j]
```

and again specifying noninformative prior distributions outside the loop.

Other distributional forms

There is no need to restrict ourselves to the normal distribution. For example, the t distribution allows for occasional extreme values. Here is how it can be set up in Bugs with degrees of freedom ν_y estimated from the data:

Bugs code
```
y[i] ~ dt (y.hat[i], tau.y, nu.y)
```

and then, outside the loop, we must put in a prior distribution for ν_y. Bugs restricts this degrees-of-freedom parameter to be at least 2, so it is convenient to assign a uniform distribution on its inverse:

Bugs code
```
nu.y <- 1/nu.inv.y
nu.inv.y ~ dunif (0, .5)
```

The t model could similarly be applied to the group-level model for the α_j's as well.

16.11 Bibliographic note

Details on Bugs are in Appendix C. For fitting models in Bugs, the two volumes of examples in the online help are a good starting point, and the textbooks by Lancaster (2004) and Congdon (2001, 2003) are also helpful because they use Bugs for all their examples. Gelman and Rubin (1992) and Brooks and Gelman (1998) discuss the use of multiple chains to monitor the convergence of iterative simulations. Kass et al. (1998) present a lively discussion of practical issues in implementing iterative simulation.

16.12 Exercises

1. Elements of a Bugs model: list the elements of the model on page 370 by category: modeled data, unmodeled data, modeled parameters, unmodeled parameters, derived quantities, and looping indexes (as in Figure 16.4).

2. Find all the errors in the following Bugs model:

```
model {                                                              Bugs code
  for (i in 1:n){
    y[i] ~ dnorm (a[state[i]] + theta*treat[i] + b*hispanic, tau.y)
  }
    theta ~ dnorm (0, .0001)
    b ~ dnorm (0, 1000)
  for (j in 1:J){
    a[j] ~ rnorm (mu.a, tau.a^2)
  }
  mu.a ~ dnorm (0, .0001)
  tau.a <- pow (sigma.a, -2)
  sigma.a ~ dunif (0, 100)
  tau.y <- pow (sigma.y, -2)
  sigma.y <- dunif (0, 100)
}
```

3. Using the data in folder **cd4** regarding CD4 percentages for young children with HIV, we shall revisit Exercise 12.2.

   (a) Use Bugs to fit the model in Exercise 12.2(a). Interpret the results.
   (b) Use Bugs to fit the model in Exercise 12.2(b). Interpret the results.
   (c) How do the results from these models compare to the fits from `lmer()`?
   (d) Summarize the results graphically as in Section 16.3.

4. Repeat the predictions described in Exercise 12.3 using the output from the Bugs fits from Exercise 16.3 instead.

5. Scaling of coefficients: again using the data in folder **cd4**, fit the model you formulated in Exercise 12.2(b), just as you did in Exercise 16.3(b). What happens if you rescale time so that it is in units of days rather than years? How does this influence your prior distributions and starting values?

6. Convergence of iterative simulation: return to the beauty and teaching evaluations example introduced in Exercise 3.5 and revisited in Exercises 12.6 and 13.1.

   (a) Write a varying-intercept model for these data with no group-level predictors. Fit this model using Bugs but allow for only 10 iterations. What do the $\hat{R}$ values look like?

(b) Now fit the model again allowing for enough iterations to achieve convergence. How many iterations were required? Interpret the results from this model.

(c) Write a varying-intercept model that you would like to fit to these data that includes three group-level predictors. Fit this model using Bugs. How many iterations were required for convergence for this model? Interpret the results of the model.

(d) Create fake-data simulations to check the fit of the models in (b) and (c).

7. This exercise will use the data you found for Exercise 4.7. This time, rather than repeating the same analysis across each year, or country (or whatever group the data vary across), fit a multilevel model using Bugs instead. Compare the results to those obtained in your earlier analysis.

8. Impact of the prior distribution: you will use Bugs to fit several versions of the varying-intercept model to the radon data using floor as a house-level predictor and uranium as a county-level predictor.

(a) How do the inferences change if you assign normal prior distributions with mean 5 and standard deviation 1000 to the coefficients for floor and uranium.

(b) How do the inferences change if you switch to normal prior distributions with mean 0 and standard deviation 0.1?

(c) Now try normal prior distributions with mean 5 and standard deviation 1.

(d) Now try t prior distributions with mean 5, standard deviation 1, and 4 degrees of freedom.

(e) Now try Uniform$(-100,100)$ prior distributions, then Uniform$(-1,1)$ prior distributions.

(f) Discuss the impact of the prior distributions on the inferences.

## CHAPTER 17

Fitting multilevel linear and generalized linear models in Bugs and R

This chapter presents Bugs code for some of the multilevel models from Chapters 13–15, including varying intercepts and slopes, non-nested groupings, and multilevel versions of logistic regression and other generalized linear models.

17.1 Varying-intercept, varying-slope models

Simple model with no correlation between intercepts and slopes

We start with the varying-intercept, varying-slope radon model of Section 13.1, temporarily simplifying by ignoring the possible correlation between intercepts and slopes—that is, we model the intercepts and slopes as independent. Ignoring the multilevel correlation is inappropriate but can be helpful in getting started with the Bugs modeling.

```
model {
  for (i in 1:n){
    y[i] ~ dnorm (y.hat[i], tau.y)
    y.hat[i] <- a[county[i]] + b[county[i]]*x[i]
  }
  tau.y <- pow(sigma.y, -2)
  sigma.y ~ dunif (0, 100)

  for (j in 1:J){
    a[j] ~ dnorm (a.hat[j], tau.a)
    b[j] ~ dnorm (b.hat[j], tau.b)
    a.hat[j] <- mu.a
    b.hat[j] <- mu.b
  }
  mu.a ~ dnorm (0, .0001)
  mu.b ~ dnorm (0, .0001)
  tau.a <- pow(sigma.a, -2)
  tau.b <- pow(sigma.b, -2)
  sigma.a ~ dunif (0, 100)
  sigma.b ~ dunif (0, 100)
}
```

Bugs code

This model looks a little more complicated than it needs to, in that we could have simply inserted mu.a and mu.b into the expressions for a[j] and b[j]. We include the intermediate quantities a.hat[j] and b.hat[j] because they become useful when the model includes correlations and group-level predictors, as we discuss in Section 17.2.

Modeling the correlation

We can write model (13.1) on page 279—which allows a correlation ρ between the varying intercepts and slopes—as follows:

Bugs code
```
model {
  for (i in 1:n){
    y[i] ~ dnorm (y.hat[i], tau.y)
    y.hat[i] <- a[county[i]] + b[county[i]]*x[i]
  }
  tau.y <- pow(sigma.y, -2)
  sigma.y ~ dunif (0, 100)

  for (j in 1:J){
    a[j] <- B[j,1]
    b[j] <- B[j,2]
    B[j,1:2] ~ dmnorm (B.hat[j,], Tau.B[,])
    B.hat[j,1] <- mu.a
    B.hat[j,2] <- mu.b
  }
  mu.a ~ dnorm (0, .0001)
  mu.b ~ dnorm (0, .0001)

  Tau.B[1:2,1:2] <- inverse(Sigma.B[,])
  Sigma.B[1,1] <- pow(sigma.a, 2)
  sigma.a ~ dunif (0, 100)
  Sigma.B[2,2] <- pow(sigma.b, 2)
  sigma.b ~ dunif (0, 100)
  Sigma.B[1,2] <- rho*sigma.a*sigma.b
  Sigma.B[2,1] <- Sigma.B[1,2]
  rho ~ dunif (-1, 1)
}
```

Here we are using capital letters (B, Tau, Sigma) for matrix parameters and lower-case for vectors and scalars.

Scaled inverse-Wishart model

Another approach is to model the covariance matrix for the intercepts and slopes directly using the scaled inverse-Wishart distribution described at the end of Section 13.3. A useful trick with the scaling is to define the coefficients α_j, β_j in terms of multiplicative factors, ξ^α, ξ^β, with unscaled parameters $\alpha_j^{\mathrm{raw}}, \beta_j^{\mathrm{raw}}$ having the inverse-Wishart distribution. We first give the Bugs model, then explain it:

Bugs code
```
model {
  for (i in 1:n){
    y[i] ~ dnorm (y.hat[i], tau.y)
    y.hat[i] <- a[county[i]] + b[county[i]]*x[i]
  }
  tau.y <- pow(sigma.y, -2)
  sigma.y ~ dunif (0, 100)

  for (j in 1:J){
    a[j] <- xi.a*B.raw[j,1]
    b[j] <- xi.b*B.raw[j,2]
    B.raw[j,1:2] ~ dmnorm (B.raw.hat[j,], Tau.B.raw[,])
    B.raw.hat[j,1] <- mu.a.raw
```

```
    B.raw.hat[j,2] <- mu.b.raw
}
mu.a <- xi.a*mu.a.raw
mu.b <- xi.b*mu.b.raw
mu.a.raw ~ dnorm (0, .0001)
mu.b.raw ~ dnorm (0, .0001)

xi.a ~ dunif (0, 100)
xi.b ~ dunif (0, 100)

Tau.B.raw[1:2,1:2] ~ dwish (W[,], df)
df <- 3
Sigma.B.raw[1:2,1:2] <- inverse(Tau.B.raw[,])
sigma.a <- xi.a*sqrt(Sigma.B.raw[1,1])
sigma.b <- xi.b*sqrt(Sigma.B.raw[2,2])
rho <- Sigma.B.raw[1,2]/sqrt(Sigma.B.raw[1,1]*Sigma.B.raw[2,2])
}
```

The quantities that must be initialized—the parameters, in the terminology of Figure 16.4—include B^{raw}, the ξ's, the μ^{raw}'s, and T_B^{raw}. But we are actually interested in the derived quantities α, β, the μ's, the σ's, and ρ.

In the specification of the Wishart distribution for the inverse-variance Tau.B.raw, the matrix W is the prior scale, which we set to the identity matrix (in R, we write W <- diag(2)), and the degrees of freedom, which we set to 1 more than the dimension of the matrix (to induce a uniform prior distribution on ρ, as discussed near the end of Section 13.3). The call to the above model in R then looks something like

```
W <- diag (2)                                                   R code
radon.data <- list ("n", "J", "y", "county", "x", "W")
radon.inits <- function (){
  list (B.raw=array(rnorm(2*J),c(J,2)), mu.a.raw=rnorm(1),
    mu.b.raw=rnorm(1), sigma.y=runif(1), Tau.B.raw=rwish(3,diag(2)),
    xi.a=runif(1), xi.b=runif(1))}
radon.parameters <- c ("a", "b", "mu.a", "mu.b", "sigma.y",
  "sigma.a", "sigma.b", "rho")
M1 <- bugs (radon.data, radon.inits, radon.parameters,
  "wishart1.bug", n.chains=3, n.iter=2000)
```

The advantage of this model, as compared to the simple model of σ_α, σ_β, and ρ presented on page 376, is that it generalizes more easily to models with more than two varying coefficients, as we discuss next. In addition, the extra parameters ξ_α, ξ_β can actually allow the computations to converge faster, an issue to which we shall return in Section 19.5 in the context of using redundant multiplicative parameters to speed computations for multilevel models.

Modeling multiple varying coefficients

As discussed in Section 13.3, when more than two coefficients are varying (for example, a varying intercept and two or more varying slopes), it is difficult to model the group-level correlations directly because of constraints involved in the requirement that the covariance matrix be positive definite. With more than two varying coefficients, it is simplest to just use the scaled inverse-Wishart model described above, using a full matrix notation to allow for an arbitrary number K of coefficients that vary by group.

Bugs code
```
model {
  for (i in 1:n){
    y[i] ~ dnorm (y.hat[i], tau.y)
    y.hat[i] <- inprod(B[county[i],],X[i,])
  }
  tau.y <- pow(sigma.y, -2)
  sigma.y ~ dunif (0, 100)

  for (j in 1:J){
    for (k in 1:K){
      B[j,k] <- xi[k]*B.raw[j,k]
    }
    B.raw[j,1:K] ~ dmnorm (mu.raw[], Tau.B.raw[,])
  }
  for (k in 1:K){
    mu[k] <- xi[k]*mu.raw[k]
    mu.raw[k] ~ dnorm (0, .0001)
    xi[k] ~ dunif (0, 100)
  }
  Tau.B.raw[1:K,1:K] ~ dwish (W[,], df)
  df <- K+1
  Sigma.B.raw[1:K,1:K] <- inverse(Tau.B.raw[,])
  for (k in 1:K){
    for (k.prime in 1:K){
      rho.B[k,k.prime] <- Sigma.B.raw[k,k.prime]/
        sqrt(Sigma.B.raw[k,k]*Sigma.B.raw[k.prime,k.prime])
    }
    sigma.B[k] <- abs(xi[k])*sqrt(Sigma.B.raw[k,k])
  }
}
```

And here is how the model could be called in R, assuming that the predictors (including the constant term) have already been bundled into a $n \times K$ matrix X:

R code
```
W <- diag (K)
data <- list ("n", "J", "K", "y", "county", "X", "W")
inits <- function (){
  list (B.raw=array(rnorm(J*K),c(J,K)), mu.raw=rnorm(K),
        sigma.y=runif(1), Tau.B.raw=rwish(K+1,diag(K)), xi=runif(K))}
parameters <- c ("B", "mu", "sigma.y", "sigma.B", "rho.B")
M2 <- bugs (radon.data, radon.inits, radon.parameters,
  "wishart2.bug", n.chains=3, n.iter=2000)
```

This reduces to our earlier model when $K = 2$.

Adding unmodeled individual-level coefficients

The above model allows all the coefficients B to vary. In practice it can be convenient to leave some regression coefficients unmodeled and let others vary by group. The simplest way to do this in the Bugs model is to add a vector β^0 of unmodeled coefficients for a matrix X^0 of predictors; the model is then $y_i = X_i^0 \beta^0 + X_i B_{j[i]} + \epsilon_i$, with the fourth line of the Bugs model being written as

Bugs code
```
y.hat[i] <- inprod(b.0[],X.0[i,]) + inprod(B[county[i],],X[i,])
```

or, in one of the simpler models with only two varying coefficients,

Bugs code
```
y.hat[i] <- inprod(b.0[],X.0[i,]) + a[county[i]] + b[county[i]]*x[i]
```

In either case, if we define K^0 as the number of unmodeled coefficients (that is, the number of columns of X^0), we can specify a noninformative prior distribution at the end of the Bugs model:

```
for (k in 1:K.0){
  b.0[k] ~ dnorm (0, .0001)
}
```
Bugs code

17.2 Varying intercepts and slopes with group-level predictors

We can add group-level predictors to the varying-intercept, varying-slope Bugs models of the previous section by replacing mu.a and mu.b by group-level regressions.

Simplest varying-intercept, varying-slope model. For example, we can add a group-level predictor u to the very first model of this chapter by replacing the expressions for a.hat[j] and b.hat[j] with

```
a.hat[j] <- g.a.0 + g.a.1*u[j]
b.hat[j] <- g.b.0 + g.b.1*u[j]
```
Bugs code

and then removing the prior distributions for mu.a and mu.b and replacing with dnorm (0, .0001) prior distributions for each of g.a.0, g.a.1, g.b.0, and g.b.1.

Model in which intercepts and slopes are combined into a matrix, B. A similar operation works with the model on page 376 also, except that a.hat[j], b.hat[j] become B.hat[j,1], B.hat[j,2].

Scaled inverse-Wishart model. In the scaled model on page 376, we add a group-level predictor by replacing mu.a.raw with g.a.0.raw + g.a.1.raw*u[j] and similarly for mu.b.raw. We continue by multiplying the raw parameters for a and b by xi.a and xi.b, respectively to get g.a.0, g.a.1, g.b.0, g.b.1.

Multiple group-level predictors

If we have several group-level predictors (expressed as a matrix U with J rows, one for each group), we use notation such as

```
a.hat[j] <- inprod (g.a[], U[j,])
b.hat[j] <- inprod (g.b[], U[j,])
```
Bugs code

with a loop setting up the prior distribution for the elements of g.a and g.b.

Multiple varying coefficients with multiple group-level predictors

In a more general notation, we can express both the individual-level and group-level regression using matrices, generalizing the model on page 377 to:

```
model {
  for (i in 1:n){
    y[i] ~ dnorm (y.hat[i], tau.y)
    y.hat[i] <- inprod(B[county[i],],X[i,])
  }
  tau.y <- pow(sigma.y, -2)
  sigma.y ~ dunif (0, 100)

  for (k in 1:K){
    for (j in 1:J){
```
Bugs code

```
          B[j,k] <- xi[k]*B.raw[j,k]
        }
        xi[k] ~ dunif (0, 100)
      }
      for (j in 1:J){
        B.raw[j,1:K] ~ dmnorm (B.raw.hat[j,], Tau.B.raw[,])
        for (k in 1:K){
          B.raw.hat[j,k] <- inprod(G.raw[k,],U[j,])
        }
      }
      for (k in 1:K){
        for (l in 1:L){
          G[k,l] <- xi[k]*G.raw[k,l]
          G.raw[k,l] ~ dnorm (0, .0001)
        }
      }
      Tau.B.raw[1:K,1:K] ~ dwish (W[,], df)
      df <- K+1
      Sigma.B.raw[1:K,1:K] <- inverse(Tau.B.raw[,])
      for (k in 1:K){
        for (k.prime in 1:K){
          rho.B[k,k.prime] <- Sigma.B.raw[k,k.prime]/
            sqrt(Sigma.B.raw[k,k]*Sigma.B.raw[k.prime,k.prime])
        }
        sigma.B[k] <- abs(xi[k])*sqrt(Sigma.B.raw[k,k])
      }
    }
```

Adding unmodeled individual-level coefficients

One can further extend these models by adding a matrix of predictors X^0 with unmodeled coefficients β^0 as described at the end of Section 17.1. In Bugs, this is done by adding inprod(b.0[],X.0[i,]) to the expression for y.hat[i].

17.3 Non-nested models

Here is the Bugs code for model (13.9) on page 289, in which the data are grouped in two different ways (by treatment and airport in the flight simulator example):

Bugs code
```
model {
  for (i in 1:n){
    y[i] ~ dnorm (y.hat[i], tau.y)
    y.hat[i] <- mu + gamma[treatment[i]] + delta[airport[i]]
  }
  mu ~ dnorm (0, .0001)
  tau.y <- pow(sigma.y, -2)
  sigma.y ~ dunif (0, 100)

  for (j in 1:n.treatment){
    gamma[j] ~ dnorm (0, tau.gamma)
  }
  tau.gamma <- pow(sigma.gamma, -2)
  sigma.gamma ~ dunif (0, 100)

  for (k in 1:n.airport){
```

```
      delta[k] ~ dnorm (0, tau.delta)
    }
    tau.delta <- pow(sigma.delta, -2)
    sigma.delta ~ dunif (0, 100)
  }
```

With non-nested models, there is a choice of where to put the intercept or constant term: here we have included it as μ in the data model, with zero means for the group effects, but another option would be to include a mean parameter for the γ_j's or δ_k's. Including a constant term in more than one place would lead to nonidentifiability. As we discuss in Section 19.4, it can be useful for both conceptual and computational reasons to include redundant mean parameters and then reparameterize the model to regain identifiability.

17.4 Multilevel logistic regression

Multilevel generalized linear models have the same structure as linear models, altering only the data model. To illustrate, we show the logistic regression for state-level opinions from Section 14.1. The model as written in Bugs looks more complicated than it really is, because we must explicitly specify distributions for all the parameters and we must transform from standard-deviation parameters σ to inverse-variances τ. Here is the model:

```
model {                                                         Bugs code
  for (i in 1:n){
    y[i] ~ dbin (p.bound[i], 1)
    p.bound[i] <- max(0, min(1, p[i]))
    logit(p[i]) <- Xbeta[i]
    Xbeta[i] <- b.0 + b.female*female[i] + b.black*black[i] +
      b.female.black*female[i]*black[i] + b.age[age[i]] + b.edu[edu[i]] +
      b.age.edu[age[i],edu[i]] + b.state[state[i]]
  }
  b.0 ~ dnorm (0, .0001)
  b.female ~ dnorm (0, .0001)
  b.black ~ dnorm (0, .0001)
  b.female.black ~ dnorm (0, .0001)

  for (j in 1:n.age) {b.age[j] ~ dnorm (0, tau.age)}
  for (j in 1:n.edu) {b.edu[j] ~ dnorm (0, tau.edu)}
  for (j in 1:n.age) {for (k in 1:n.edu){
    b.age.edu[j,k] ~ dnorm (0, tau.age.edu)}}
  for (j in 1:n.state) {
    b.state[j] ~ dnorm (b.state.hat[j], tau.state)
    b.state.hat[j] <- b.region[region[j]] + b.v.prev*v.prev[j]}
  b.v.prev ~ dnorm (0, .0001)
  for (j in 1:n.region) {b.region[j] ~ dnorm (0, tau.region)}

  tau.age <- pow(sigma.age, -2)
  tau.edu <- pow(sigma.edu, -2)
  tau.age.edu <- pow(sigma.age.edu, -2)
  tau.state <- pow(sigma.state, -2)
  tau.region <- pow(sigma.region, -2)

  sigma.age ~ dunif (0, 100)
  sigma.edu ~ dunif (0, 100)
```

```
      sigma.age.edu ~ dunif (0, 100)
      sigma.state ~ dunif (0, 100)
      sigma.region ~ dunif (0, 100)
  }
```

The "logistic regression" part of this model is at the beginning. The data distribution, dbin, is the *binomial distribution* with $N = 1$, meaning that $y_i = 1$ with probability p_i and 0 otherwise. The quantity p.bound is defined to restrict the probability to lie between 0 and 1, a trick we use to keep Bugs from crashing. The model continues with a usual multilevel formulation, with the only new feature being the nested loops for the matrix parameter b.age.edu. The model is actually slow to converge under this parameterization, but in improving it we shall keep this basic structure (see Section 19.4 for details).

17.5 Multilevel Poisson regression

Poisson models are straightforward to code in Bugs. Here we show model (15.1) from page 326 for the analysis of the police stops. This model includes crossed multilevel predictors and overdispersion:

Bugs code
```
model {
  for (i in 1:n){
    stops[i] ~ dpois (lambda[i])
    log(lambda[i]) <- offset[i] + mu +
      b.eth[eth[i]] + b.precinct[precinct[i]] + epsilon[i]
    epsilon[i] ~ dnorm (0, tau.epsilon)
  }
  mu ~ dnorm (0, .0001)
  mu.adj <- mu + mean(b.eth[]) + mean(b.precinct[])
  tau.epsilon <- pow(sigma.epsilon, -2)
  sigma.epsilon ~ dunif (0, 100)

  for (j in 1:n.eth){
    b.eth[j] ~ dnorm (0, tau.eth)
    b.eth.adj[j] <- b.eth[j] - mean(b.eth[])
  }
  tau.eth <- pow(sigma.eth, -2)
  sigma.eth ~ dunif (0, 100)

  for (j in 1:n.precinct){
    b.precinct[j] ~ dnorm (0, tau.precinct)
    b.precinct.adj[j] <- b.precinct[j] - mean(b.precinct[])
  }
  tau.precinct <- pow(sigma.precinct, -2)
  sigma.precinct ~ dunif (0, 100)
}
```

As in (15.1), the ϵ_i's represent overdispersion, and σ_ϵ is a measure of the amount of overdispersion in the data. We computed the offset term in R (as $\log(\frac{15}{12}n)$) and included it in the data in the bugs() call. (Alternatively, we could have computed the offset directly within the initial loop of the Bugs model.) Finally, the adjusted parameters mu.adj and b.eth.adj correspond to μ' and α'_e in equations (15.2) and (15.3). For completeness we have adjusted the precinct intercepts to sum to zero also. (This adjustment is less important because there are 75 of them, so their mean will be close to zero in any case.)

17.6 Multilevel ordered categorical regression

Sections 6.5 and 15.2 describe an ordered categorical logistic regression. Here we show the model as written in Bugs. The data-level model is adapted from an example from the online Bugs manual, and the rest of the Bugs code describes the particular model we fit to the storable-votes data. In this model, dcat is the Bugs notation for the distribution defined by probabilities p[i,1], p[i,2], p[i,3] of observation y_i falling in each of the 3 categories, and these probabilities p are defined in terms of the cumulative probabilities Q, which follow a logistic regression.

```
model {                                                          Bugs code
  for (i in 1:n){
    y[i] ~ dcat(P[i,])
    P[i,1] <- 1 - Q[i,1]
    for (i.cut in 2:n.cut){
      P[i,i.cut] <- Q[i,i.cut-1] - Q[i,i.cut]
    }
    P[i,n.cut+1] <- Q[i,n.cut]
    for (i.cut in 1:n.cut){
      logit(Q[i,i.cut]) <- z[i,i.cut]
      Z[i,i.cut] <- (x[i] - C[player[i],i.cut])/s[player[i]]
    }
  }
  for (i.player in 1:n.player){
    C[i.player,1] ~ dnorm (mu.c[1], tau.c[1])I(0,C[i.player,2])
    C[i.player,2] ~ dnorm (mu.c[2], tau.c[2])I(C[i.player,1],100)
    s[i.player] ~ dlnorm (mu.log.s, tau.log.s)I(1,100)
  }
  for (i.cut in 1:n.cut){
    mu.c[i.cut] ~ dnorm (0, 1.E-6)
    tau.c[i.cut] <- pow(sigma.c[i.cut], -2)
    sigma.c[i.cut] ~ dunif (0, 1000)
  }
  mu.log.s ~ dnorm (0, .0001)
  tau.log.s <- pow(sigma.log.s, -2)
  sigma.log.s ~ dunif (0, 1000)
}
```

We continue with the notation in which matrices (in this case, P, Q, Z, and C) are written as capital letters, with lowercase used for vectors and scalars.

The above model uses bounds (the I(,) notation) for two purposes. First, the bounds c[i.player,1] and c[i.player,2] constrain these two parameters for each player to fall between 0 and 100 (see the discussion following model (6.11) on page 121), and to be in increasing order, which is necessary so that the probability for each category is nonnegative.

The second use of bounds in the Bugs model is to constrain each s[i.player] to fall between 1 and 100, which we do for purely computational reasons. When we allowed these parameters to float freely, Bugs would crash. We suspect that Bugs was crashing because the ratios calculated for the logistic distribution were too extreme, and constraining the parameters s stopped this problem. Another way to constrain this in Bugs would be to parameterize in terms of $1/s$.

17.7 Latent-data parameterizations of generalized linear models

Logistic regression

As discussed in Section 5.3, logistic regression can be expressed directly or using latent parameters. We can similarly implement both formulations in Bugs.

Section 17.4 shows an example of the direct parameterization. The equivalent latent-data version begins as follows:

Bugs code
```
model {
  for (i in 1:n){
    z.lo[i] <- -100*equals(y[i],0)
    z.hi[i] <- 100*equals(y[i],1)
    z[i] ~ dlogis (Xbeta[i], 1) I(z.lo[i], z.hi[i])
```

After this, the two models are the same. For each data point, we have simply defined the latent variable z_i and restricted it to be positive if $y_i = 1$ and negative if $y_i = 0$. (For a logistic model, restricting z_i to the range $(0, 100)$ is essentially equivalent to restricting to $(0, \infty)$.)

In calling the model from R, it is helpful to initialize the z_i's along with the other parameters in the model. We want to use random numbers, but they must respect the restriction that z be positive if and only if $y = 1$. We can do this by adding the following, within the list *inside* the inits() function:

R code
```
z=runif(n)*ifelse(y==1,1,-1)
```

This gives initial values for z in the range $(0, 1)$ if $y = 1$, or the range $(-1, 0)$ if $y = 0$.

Robit regression

Section 6.6 describes a robust alternative to logistic regression constructed by replacing the logistic latent-data distribution with the t. We can implement robit regression in Bugs in three steps. First, within the data model (in the for (i in 1:n) loop, we use the t distribution:

Bugs code
```
z[i] ~ dt (z.hat[i], tau.z, df) I(z.lo[i], z.hi[i])
```

Second, outside the data loop, we scale the t distribution as indicated in (6.15) on page 125, so that its variance equals 1 for any value of the degrees-of-freedom parameter:

Bugs code
```
tau.z <- df/(df-2)
```

Third, we give the inverse-variance parameter a uniform prior distribution from 0 to 0.5, which has the effect of restricting the degrees of freedom to be at least 2 (a constraint required by Bugs):

Bugs code
```
df <- 1/df.inv
df.inv ~ dunif (0, .5)
```

In addition, we must initialize z within the inits() function as described previously for latent-data logistic regression. The probit model can be implemented in a similar way.

17.8 Bibliographic note

The Gibbs sampler and Metropolis algorithm were first presented by Metropolis et al. (1953), for applications in statistical physics. Tanner and Wong (1987), Gelfand and Smith (1990), and Gelfand et al. (1990) demonstrated the application of these computational ideas to general statistical models, including hierarchical linear regression. Zeger and Karim (1991), Karim and Zeger (1992), Albert and Chib (1993), Dellaportas and Smith (1993), and others developed Gibbs sampler and Metropolis algorithms for hierarchical generalized linear models. Chen, Shao, and Ibrahim (2000) and Liu (2002) review more advanced work on iterative simulation.

The augmented-data formulation of multilevel regression appears in Lindley and Smith (1972) and Hodges (1998). Gelman et al. (2003) derive the algebra of Bayesian multilevel modeling and discuss partial pooling and the Gibbs sampler in detail; chapters 11 and 15 of that book discuss computation for multilevel models. The general references on multilevel modeling (see the note at the end of Chapter 12) are relevant here. See appendix C of Gelman et al. (2003) for more on programming the Gibbs sampler and related simulation algorithms in R.

17.9 Exercises

1. Parameterizing varying-intercept, varying-slope models: the folder **nes** contains data from the National Election Study surveys. Set up a model for party identification (as a continuous outcome, as in Section 4.7), given the predictors shown in Figure 4.6 on page 74, and also allowing the intercept and the coefficient for ideology to vary by state. You will fit various versions of this model (using data from the year 2000) in Bugs.

   (a) Fit the model with no correlation between the intercepts and the slopes (the coefficients for ideology).

   (b) Fit the model with a uniform prior distribution on the correlation between the intercepts and slopes.

   (c) Fit the scaled inverse-Wishart model.

   (d) Compare the inferences from the three models.

2. Understanding and summarizing varying-intercept, varying-slope models: continue the example from the previous exercise.

   (a) Add, as a group-level predictor, the average income by state. Discuss how the parameter estimates change when this predictor has been added.

   (b) Graph the fitted model as in Figures 13.1–13.2 on page 281.

3. Multiple varying coefficients:

   (a) Repeat the previous exercise, this time allowing the coefficient for income to vary by state as well.

   (b) Repeat, allowing all the coefficients to vary by state.

   (c) Summarize your inferences from these models graphically.

4. Fitting non-nested models: use Bugs to fit the model for Olympic judging from Exercises 13.3.

5. Models with unequal variances: use Bugs to fit the model you set up for the age-guessing data in Exercise 13.4.

6. Multilevel logistic regression: use Bugs to fit the model for the well-switching data from Exercise 14.2.

7. Fitting a choice model: set up a model for the arsenic decision problem, as described near the end of Section 6.8, modeling a distribution for the parameters (a_i, b_i, c_i) in the population.

8. Elements of a Bugs model:

   (a) List the elements of the model on page 379 by category: modeled data, unmodeled data, modeled parameters, unmodeled parameters, derived quantities, and looping indexes (as in Figure 16.4).

   (b) Do the same for the model on page 381.

   (c) Do the same for the model on page 383.

9. Ordered logit: consider the ordered logistic regression for vote intention in Exercise 15.1.

   (a) Fit a classical ordered logistic regression (without the coefficients for states) using `lmer()`.

   (b) Fit the classical ordered logistic regression using Bugs. Compare to the estimate from (a).

10. Multilevel ordered logit:

   (a) Take the previous exercise and allow the intercepts to vary by state.

   (b) Plot the fitted model from (a), along with the model from Exercise 17.9(a), for a set of eight states (as in Figure 14.2 on page 307, but for this three-category model) and discuss how they differ.

   (c) Discuss whether it is worth fitting this model, as compared to a logistic model that includes only two categories, discarding the respondents who express no opinion or support other candidates.

11. Multilevel ordered logit, probit, and robit models: the folder `storable` has data from the storable-votes experiment described in Sections 6.5, 15.2, and 17.6.

   (a) Fit an ordered logistic regression for the three categories, with a different intercept for each person.

   (b) Fit the same model but using the probit link.

   (c) Fit the same model but using the robit link.

   (d) Compare the estimates from the three models. The coefficients might not be comparable, so you have to plot the fitted models as in Figure 6.4 on page 121.

12. Multivariate outcomes: the folder `beta.blockers` contains data from a meta-analysis of 22 clinical trials of beta-blockers for reducing mortality after myocardial infarction (from Yusuf et al., 1985; see also Carlin, 1992, and Gelman et al., 2003, chapter 5).

   (a) Set up and fit a logistic regression model to estimate the effect of the treatment on the probability of death, allowing different death rates and different treatment effects for the different studies. Summarize by the estimated treatment average effect and its standard error.

   (b) Make a graph or graphs displaying the data and fitted model.

## CHAPTER 18

Likelihood and Bayesian inference and computation

Most of this book concerns the interpretation of regression models, with the understanding that they can be fit to data fairly automatically using R and Bugs. However, it can be useful to understand some of the theory behind the model fitting, partly to connect to the usual presentation of these models in statistics and econometrics.

This chapter outlines some of the basic ideas of likelihood and Bayesian inference and computation, focusing on their application to multilevel regression. One point of this material is to connect multilevel modeling to classical regression; another is to give enough insight into the computation to allow you to understand some of the practical computational tips presented in the next chapter.

18.1 Least squares and maximum likelihood estimation

We first present the algebra for classical regression inference, which is then generalized when moving to multilevel modeling. We present the formulas here without derivation; see the references listed at the end of the chapter for more.

Least squares

The classical linear regression model is $y_i = X_i\beta + \epsilon_i$, where y and ϵ are (column) vectors of length n, X is a $n \times k$ matrix, and β is a vector of length k. The vector β of coefficients is estimated so as to minimize the errors ϵ_i. If the number of data points n exceeds the number of predictors[1] k, it is not generally possible to find a β that gives a perfect fit (that would be $y_i = X_i\beta$, with no error, for all data points $i = 1, \ldots, n$), and the usual estimation goal is to choose the estimate $\hat{\beta}$ that minimizes the sum of the squares of the residuals $r_i = y_i - X_i\hat{\beta}$. (We distinguish between the *residuals* $r_i = y_i - X_i\hat{\beta}$ and the *errors* $\epsilon_i = y_i - X_i\beta$.) The sum of squared residuals is $SS = \sum_{i=1}^{n}(y_i - X_i\hat{\beta})^2$; the $\hat{\beta}$ that minimizes it is called the *least squares estimate* and can be written in matrix notation as

$$\hat{\beta} = (X^t X)^{-1} X^t y. \tag{18.1}$$

We rarely work with this expression directly, since it can be computed directly in the computer (for example, using the `lm()` command in R).

The errors ϵ come from a distribution with mean 0 and variance σ^2. This standard deviation can be estimated from the residuals, as

$$\hat{\sigma}^2 = \frac{1}{n-k} SS = \frac{1}{n-k} \sum_{i=1}^{n}(y_i - X_i\hat{\beta})^2, \tag{18.2}$$

with $n-k$ rather than $n-1$ in the denominator to adjust for the estimation of the

---

[1] The constant term, if present in the model, counts as one of the predictors; see Section 3.4.

k-dimensional parameter β. (Since β is estimated to minimize the sum of squared residuals, SS will be, on average, lower by a factor of $\frac{n-k}{n}$ than the sum of squared errors.)

Maximum likelihood

As just described, least squares estimation assumes linearity of the model and independence of the errors. If we further assume that the errors are normally distributed, so that $y_i \sim N(X_i\beta, \sigma^2)$ for each i, the least squares estimate $\hat{\beta}$ is also the maximum likelihood estimate. The *likelihood* of a regression model is defined as the probability of the data given the parameters and inputs; thus, in this example,

$$p(y|\beta, \sigma, X) = \prod_{i=1}^{n} N(y_i | X_i\beta, \sigma^2), \qquad (18.3)$$

where $N(\cdot|\cdot, \cdot)$ represents the normal probability density function, $N(y|m, \sigma^2) = \frac{1}{\sqrt{2\pi}\sigma} \exp\left(-\frac{1}{2}\left(\frac{y-m}{\sigma}\right)^2\right)$. The model can also be written in vector-matrix notation as $y \sim N(X\beta, \sigma^2 I_n)$, where I_n is the n-dimensional identity matrix. Giving a diagonal covariance matrix to this multivariate normal distribution implies independence of the errors.

Expression (18.3) is a special case of the general expression for the likelihood of n independent measurements given a vector parameter θ and predictors X:

$$p(y|\theta, X) = \prod_{i=1}^{n} p(y_i | \theta, X_i). \qquad (18.4)$$

The maximum likelihood estimate is the vector θ for which this expression is maximized, given data X, y. (In classical least squares regression, θ corresponds to the vector of coefficients β, along with the error scale, σ.) In general, we shall use the notation $p(y|\theta)$ for the likelihood as a function of parameter vector θ, with the dependence on the predictors X implicit.

The likelihood can then be written as

$$p(y|\beta, \sigma, X) = N(y|X\beta, \sigma^2 I_n). \qquad (18.5)$$

Using the standard notation for the multivariate normal distribution with mean vector m and covariance matrix Σ, this becomes

$$N(y|m, \Sigma) = (2\pi)^{-n/2} |\Sigma|^{-1/2} \exp\left(-\frac{1}{2}(y-m)^t \Sigma^{-1}(y-m)\right).$$

Expressions (18.3) and (18.5) are equivalent and are useful at different times when considering generalizations of the model.

A careful study of (18.3) or (18.5) reveals that maximizing the likelihood is equivalent to minimizing the sum of squared residuals; hence the least squares estimate $\hat{\beta}$ can be viewed as a maximum likelihood estimate under the normal model.

There is a small twist in fitting regression models, in that the maximum likelihood estimate of σ is $\sqrt{\frac{1}{n}\sum_{i=1}^{n}(y_i - X_i\hat{\beta})^2}$, with $\frac{1}{n}$ instead of $\frac{1}{n-k}$. The estimate with $\frac{1}{n-k}$ is generally preferred: the maximum likelihood estimate of (β, σ) simply takes the closest fit and needs to be adjusted to account for the fitting of k regression coefficients.

Weighted least squares

The least squares estimate counts all n data points equally in minimizing the sum of squares. If some data are considered more important than others, this can be captured in the estimation by minimizing a *weighted* sum of squares, $WSS = \sum_{i=1}^{n} w_i(y_i - X_i\hat{\beta})^2$, so that points i with larger weights w_i count more in the optimization. The weighted least squares estimate is

$$\hat{\beta}^{\text{WLS}} = (X^t W X)^{-1} X^t W y, \tag{18.6}$$

where W is the diagonal matrix whose elements are the weights w_i.

Weighted least squares is equivalent to maximum likelihood estimation of β in the normal regression model

$$y_i \sim \text{N}(X_i\beta, \sigma^2/w_i), \tag{18.7}$$

with independent errors with variances inversely proportional to the weights. Points with high weights have low error variances and are thus expected to lie closer to the fitted regression function.

Weighted least squares can be further generalized to fit data with correlated errors; if the data are fit by the model $y \sim \text{N}(X\beta, \Sigma)$, then the maximum likelihood estimate is $\hat{\beta} = (X^t\Sigma^{-1}X)^{-1}X^t\Sigma^{-1}y$ and minimizes the expression $(y - X\beta)^t\Sigma^{-1}(y - X\beta)$, which can be seen as a generalization of the "sum of squares" concept.

Generalized linear models

Classical linear regression can be motivated in a purely algorithmic fashion (as "least squares") or as maximum likelihood inference under a normal model. With generalized linear models, the algorithmic justification is usually set aside, and maximum likelihood is the starting point. We illustrate with the two most important examples.

Logistic regression.   For binary logistic regression with data $y_i = 0$ or 1, the likelihood is

$$p(y|\beta, X) = \prod_{i=1}^{n} \left\{ \begin{array}{ll} \text{logit}^{-1}(X_i\beta) & \text{if } y_i = 1 \\ 1 - \text{logit}^{-1}(X_i\beta) & \text{if } y_i = 0, \end{array} \right.$$

which can be written more compactly, but equivalently, as

$$p(y|\beta, X) = \prod_{i=1}^{n} \left(\text{logit}^{-1}(X_i\beta) \right)^{y_i} \left(1 - \text{logit}^{-1}(X_i\beta) \right)^{1-y_i}.$$

To find the β that maximizes this expression, we can compute the derivative $dp(y|\beta, X)/d\beta$ of the likelihood (or, more conveniently, the derivative of the logarithm of the likelihood), set this derivative equal to 0, and solve for β. There is no closed-form solution, but the maximum likelihood estimate can be found using an iteratively weighted least squares algorithm, each step having the form of a weighted least squares computation, with the weights changing at each step.

Not just a computational trick, iteratively weighted least squares can be understood statistically as a series of steps approximating the logistic regression likelihood by a normal regression model applied to transformed data. We shall not discuss this further here, however. We only mentioned the algorithm to give a sense of how likelihood functions are used in classical estimation.

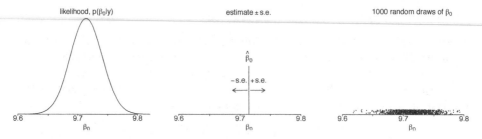

Figure 18.1 *(a) Likelihood function for the parameter β_0 in the trivial linear regression $y = \beta_0 + error$, of log earnings y_i. (b) Mode of the likelihood function and range indicating ± 1 standard error as computed from the inverse-second-derivative-matrix of the log likelihood at the mode. (c) 1000 random simulation draws from the normal distribution with this mean and standard deviation, representing the distribution of uncertainty in the inference for β_0. The simulations have been vertically jittered to make them visible. (For this one-dimensional problem it would be better to display the simulations as a histogram; we use a dotplot here for compatability with the scatterplot of the two-dimensional simulations in Figures 18.2–18.3.)*

Poisson regression.   For Poisson regression (6.3), the likelihood is

$$p(y|\beta, X, u) = \prod_{i=1}^{n} \text{Poisson}\left(y_i \big| u_i e^{X_i \beta}\right),$$

where each factor has the Poisson probability density function: $\text{Poisson}(y|m) = \frac{1}{y!} m^y e^{-m}$.

18.2 Uncertainty estimates using the likelihood surface

In maximum likelihood estimation, the likelihood function can be viewed as a "hill" with $\hat{\beta}$ identifying the location of the top of the hill—that is, the mode of the likelihood function. We illustrate with two simple regression examples.

One-parameter example: linear regression with just a constant term

Figure 18.1 demonstrates likelihood estimation for the simple problem of regression with only a constant term; that is, inference for β_0 in the model $y_i = \beta_0 + \epsilon_i$, $i = 1, \ldots, n$, for the earnings data from Chapter 2. In this example, β_0 corresponds to the average log earnings in the population represented by the survey. For simplicity, we shall assume that σ, the standard deviation of the errors ϵ_i, is known and equal to the sample standard deviation of the data.

   Figure 18.1a plots the likelihood function for β_0. The peak of the function is the maximum likelihood estimate, which in this case is simply $\bar{y}$, the average log earnings reported in the sample. The range of the likelihood function tells us that it would be extremely unlikely for these data to occur if the true β_0 were as low as 9.6 or as high as 9.8. Figure 18.1b shows the maximum likelihood estimate ± 1 standard error, and Figure 18.1c displays 1000 random draws from the normal distribution representing uncertainty in β_0.

Two-parameter example: linear regression with two coefficients

Figure 18.2 illustrates the slightly more complicated case of a linear regression model with two coefficients (corresponding to a constant term and a linear predic-

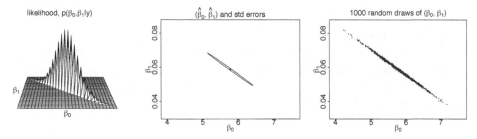

Figure 18.2 *(a) Likelihood function for the parameters β_0, β_1 in the linear regression $y = \beta_0 + \beta_1 x + error$, of log earnings, y_i, on heights, x_i. (The spiky pattern of the three-dimensional plot is an artifact of the extreme correlation of the distribution.) (b) Mode of the likelihood function (that is, the maximum likelihood estimate $(\hat{\beta}_0, \hat{\beta}_1)$) and ellipse summarizing the inverse-second-derivative-matrix of the log likelihood at the mode. (c) 1000 random simulation draws from the normal distribution centered at $(\hat{\beta}_0, \hat{\beta}_1)$ with variance matrix equal to the inverse of the negative second derivative of the log likelihood.*

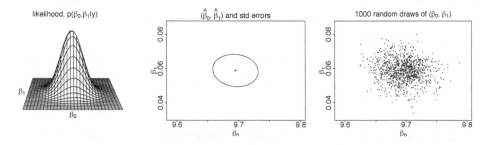

Figure 18.3 *(a) Likelihood function, (b) mode and uncertainty ellipse, and (c) 1000 simulation draws of the regression coefficients for the model $y = \beta_0 + \beta_1 z + error$, of log earnings, y_i, on mean-centered heights, $z_i = x_i - \bar{x}$. The inferences for the parameters β_0, β_1 are now independent. Compare to Figure 18.2.*

tor). (Strictly speaking this model has three parameters—β_0, β_1, and σ—but for simplicity we display the likelihood of β_0, β_1 conditional on the estimated $\hat{\sigma}$.)

Figure 18.2a shows the likelihood as a function of (β_0, β_1). The area with highest likelihood surrounding the peak can be represented by an ellipse as is shown in Figure 18.2b. Figure 18.2c displays 1000 random draws from the normal distribution with covariance matrix represented by this ellipse. The shape of the uncertainty ellipse, or equivalently the correlation of the simulation draws, tells us something about the information available about the two parameters. For example, the data are consistent with β_0 being anywhere between 4.5 and 7.2, and with β_1 being anywhere between 0.04 and 0.08. However, the inferences for these two parameters are correlated: if β_0 is 4.5, then β_1 must be near 0.08, and if β_0 is 7, then β_1 must be near 0.04. To understand this inferential correlation, see Figure 4.1 on page 54: the regression line must go through the cloud of points, which is far from the y-axis. Lines of higher slope (for which β_1 is higher) intersect the y-axis at a lower value (and thus have lower values of β_0), and vice versa.

It is convenient to reparameterize the model so that the inferences for the intercept and slope coefficients are uncorrelated. We can do this by replacing the predictor x_i by its mean-centered values, $z_i = x_i - \bar{x}$—that is, height relative to the average height in the sample. Figure 18.3 shows the likelihood function and simulations for the coefficients in the regression of $y = \beta_0 + \beta_1 z + error$.

Nonidentified parameters and the likelihood function

In maximum likelihood inference, parameters in a model are nonidentified if they can be changed without affecting the likelihood. Continuing with the "hill" analogy, nonidentifiability corresponds to a "ridge" in the likelihood—that is, a direction in parameter space in which the likelihood is flat. This occurs, for example, when predictors in a classical regression are collinear.

Summarizing uncertainty about β and σ using the variance matrix from a fitted regression

We summarize the fit of a model $y = X\beta + \epsilon$ by a least squares estimate $\hat{\beta} = (X^tX)^{-1}X^ty$ and a *variance matrix* (or *covariance matrix*) of estimation,

$$V_\beta = (X^tX)^{-1}\sigma^2. \tag{18.8}$$

We represent the uncertainty in the estimated β vector using the normal distribution with mean $\hat{\beta}$ and variance matrix V_β. Figures 18.1–18.3 show examples of the estimated $\mathrm{N}(\hat{\beta}, V_\beta)$ distribution in one and two dimensions.

Expression (18.8) depends on the unknown σ^2, which we can estimate most simply with $\hat{\sigma}^2$ from (18.2) on page 387. To better capture uncertainty, we first compute $\hat{\sigma}^2$ and then sample $\sigma^2 = \hat{\sigma}^2(n-k)/X^2_{n-k}$, where X^2_{n-k} represents a random draw from the χ^2 distribution with $n-k$ degrees of freedom. These steps are performed by the `sim()` function we have written in R, as we describe next.

18.3 Bayesian inference for classical and multilevel regression

Bayesian inference for classical regression

In Bayesian inference, the likelihood is multiplied by a prior distribution, and inferences are typically summarized by random draws from this product, the *posterior distribution*.

The simplest form of Bayesian inference uses a uniform prior distribution, so that the posterior distribution is the same as the likelihood function (when considered as a function of the parameters), as pictured, for example, in the left graphs in Figures 18.1–18.3. The random draws shown in the rightmost graphs in these figures correspond to random draws from the posterior distribution, assuming a uniform prior distribution. In this way, informal Bayesian inference is represented as discussed in Section 7.2, using the simulations obtained from the `sim()` function in R (which draws from the normal distribution with mean $\hat{\beta}$ and standard deviation V_β). This is basically a convenient way to summarize classical regression, especially for propagating uncertainty for predictions.

Informative prior distributions in a single-level regression

Bayesian inference can also be used to add numerical information to a regression model. Usually we shall do this using a multilevel model, but we illustrate here with the simpler case of a specified prior distribution—the regression of log earnings on height, shown in Figures 18.2 and 18.3. Suppose we believed that β_1 was probably between 0 and 0.05—that is, a predicted difference of between 0 and 5% in earnings per inch of height. We could code this as a normal prior distribution with mean 2.5% and standard deviation 2.5%, that is, $\beta_1 \sim \mathrm{N}(0.025, 0.025^2)$.

Mathematically, this prior distribution can be incorporated into the regression by treating it as an additional "data point" of 0.025, measured directly on β_2, with a standard deviation of 0.025. This in turn can be computed using a weighted regression of an augmented data vector y_* on an augmented predictor matrix X_* with augmented weight vector w_*. These are defined as follows:

$$
y_* = \begin{pmatrix} y_1 \\ y_2 \\ \vdots \\ y_n \\ 0.025 \end{pmatrix}, \quad X_* = \begin{pmatrix} 1 & x_1 \\ 1 & x_2 \\ \vdots & \vdots \\ 1 & x_n \\ 0 & 1 \end{pmatrix}, \quad w_* = \begin{pmatrix} 1 \\ 1 \\ \vdots \\ 1 \\ \sigma_y^2/0.025^2 \end{pmatrix}. \tag{18.9}
$$

We have added the prior information as a new data point and given it a weight of the data variance (which can be estimated from the classical regression) divided by the prior variance. This weighting makes sense:

- If $\sigma_y > 0.025$, then the prior distribution is *more* informative than any data point, and so the prior "data point" will be given a high weight.

- If $\sigma_y = 0.025$, then the prior distribution has the same information as one data point and so is given equal weight.

- If $\sigma_y < 0.025$, then the prior distribution has *less* information than a single data point and so gets a lower weight.

We rarely use formulation (18.9) directly, but similar ideas apply with multilevel models, in which the group-level model for parameters α_j and β_j can be interpreted as prior information.

Collinearity and Bayesian regression

In the matrix-algebra language of (18.1) and (18.6), the Bayesian estimate—the least squares estimate of the augmented regression based on (18.9)—contains the expression $X_*^t \text{Diag}(w_*)X_*$, which is simply $X^t X$ with an added term corresponding to the prior information. With collinear predictors, the original $X^t X$ is noninvertible, but the new $X_*^t \text{Diag}(w_*)X_*$ might be invertible, depending on the structure of the new information.

For example, in a classical varying-intercept model, we would include only $J-1$ group indicators as predictors, because a regression that included the constant term along with indicators for all J groups would be collinear and nonidentifiable. But the multilevel model has the effect of adding a prior distribution for the J coefficients of the group indicators, thus adding a term to $X^t X$ which makes the new matrix invertible, even with the constant term included as well.

Simple multilevel model with no predictors

Bayesian inference achieves partial pooling for multilevel models by treating the group-level model as defining prior distributions for varying intercepts and slopes.

We begin by working through the algebra for the radon model with no individual- or group-level predictors, simply measurements within counties:

$$
\begin{aligned}
y_i &\sim \text{N}(\alpha_{j[i]}, \sigma_y^2) \text{ for } i = 1, \ldots, n \\
\alpha_j &\sim \text{N}(\mu_\alpha, \sigma_\alpha^2) \text{ for } j = 1, \ldots, J.
\end{aligned} \tag{18.10}
$$

We label the number of houses in county j as n_j (so that Lac Qui Parle County has $n_j = 2$, Aitkin County has $n_j = 4$, and so forth; see Figure 12.2 on page 255).

Sample size in group, n_j	Estimate, $\hat{\alpha}_j$
$n_j = 0$	$\hat{\alpha}_j = \mu_\alpha$ (complete pooling)
$n_j < \sigma_y^2/\sigma_\alpha^2$	$\hat{\alpha}_j$ closer to μ_α
$n_j = \sigma_y^2/\sigma_\alpha^2$	$\hat{\alpha}_j = \frac{1}{2}\bar{y}_j + \frac{1}{2}\mu_\alpha$
$n_j > \sigma_y^2/\sigma_\alpha^2$	$\hat{\alpha}_j$ closer to $\bar{y}_j$
$n_j = \infty$	$\hat{\alpha}_j = \bar{y}_j$ (no pooling)

Figure 18.4 *Summary of partial pooling of multilevel estimates as a function of group size.*

Complete-pooling and no-pooling estimates. As usual, we begin with the classical estimates. In complete pooling, all counties are considered to be equivalent, so that $\alpha_1 = \alpha_2 = \cdots = \alpha_J = \mu_\alpha$, and the model reduces to $y_i \sim N(\mu_\alpha, \sigma_y^2)$ for all measurements y. The estimate of μ_α, and thus of all the individual α_j's, is then simply $\bar{y}$, the average of the n measurements in the data.

In the no-pooling model, each county is estimated alone, so that each α_j is estimated by $\bar{y}_j$, the average of the measurements in county j.

Multilevel inference if the hyperparameters were known. The multilevel model (18.10) has *data-level regression coefficients* $\alpha_1, \ldots, \alpha_J$ and *hyperparameters* μ_α, σ_y, and σ_α. In multilevel estimation, we perform inference for both sets of parameters. To explain how to do this, we first work out the inferences for each set of parameters separately.

The key step of multilevel inference is estimation of the data-level regression coefficients given the data and hyperparameters—that is, acting as if the hyperparameters were known. As discussed in the regression context in Section 12.2, the estimate of each α_j will be a compromise between $\bar{y}_j$ and μ_α, the unpooled estimate in county j and the average over all the counties.

Given the hyperparameters, the inferences for the α_j's follow independent normal distributions, which we can write as

$$\alpha_j | y, \mu_\alpha, \sigma_y, \sigma_\alpha \sim N(\hat{\alpha}_j, V_j), \quad \text{for } j = 1, \ldots, J, \tag{18.11}$$

where the estimate and variance of estimation are

$$\hat{\alpha}_j = \frac{\frac{n_j}{\sigma_y^2}\bar{y}_j + \frac{1}{\sigma_\alpha^2}\mu_\alpha}{\frac{n_j}{\sigma_y^2} + \frac{1}{\sigma_\alpha^2}}, \quad V_j = \frac{1}{\frac{n_j}{\sigma_y^2} + \frac{1}{\sigma_\alpha^2}}. \tag{18.12}$$

The notation "$\alpha_j | y, \mu_\alpha, \sigma_y, \sigma_\alpha \sim$" in (18.11) can be read as, "α_j, given data, μ_α, σ_y, and σ_α, has the distribution ...," indicating that this is the estimate with the hyperparameters assumed known.

The estimate $\hat{\alpha}_j$ in (18.12) can be interpreted as a *weighted average* of $\bar{y}_j$ and μ_α, with relative weights depending on the sample size in the county and the variance at the data and group levels. As shown in Figure 18.4, the key parameter is the variance ratio, $\sigma_y^2/\sigma_\alpha^2$. For counties j for which $n_j = \sigma_y^2/\sigma_\alpha^2$, then the weights in (18.12) are equal, and $\hat{\alpha}_j = \frac{1}{2}\bar{y}_j + \frac{1}{2}\mu_\alpha$. If n_j is greater than the variance ratio, then $\hat{\alpha}_j$ is closer to $\bar{y}_j$; and if n_j is less then the variance ratio, then $\hat{\alpha}_j$ is closer to μ_α.

Crude inference for the hyperparameters. Given the data-level regression coefficients α_j, how can we estimate the hyperparameters, $\sigma_y, \mu_\alpha, \sigma_\alpha$ in the multilevel model (18.10)?

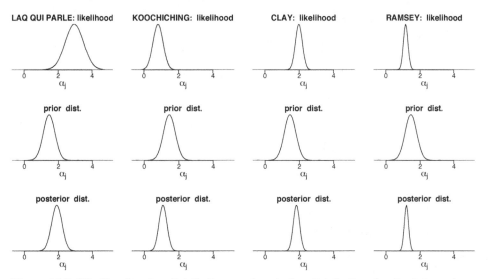

Figure 18.5 *Likelihood, prior distribution, and posterior distribution for the intercept parameter α_j for the simple radon model (with no county-level predictors) in four different counties j in Minnesota with a range of sample sizes in the data. As the sample size in the county increases, the likelihood becomes more informative (see Figure 12.4 on page 257).*

- The natural estimate of the data variance σ_y^2 is simply the residual variance:

$$\hat{\sigma}_y^2 = \frac{1}{n} \sum_{i=1}^{n} (y_i - \alpha_{j[i]})^2. \tag{18.13}$$

- The mean μ_α from the group-level model in (18.10) can be estimated by the average of the county intercepts α_j:

$$\hat{\mu}_\alpha = \frac{1}{J} \sum_{j=1}^{J} \alpha_j, \tag{18.14}$$

with an estimation variance of $\frac{1}{J}\sigma_\alpha^2$.
- The group-level variance σ_α^2 can be estimated by

$$\hat{\sigma}_\alpha^2 = \frac{1}{J} \sum_{j=1}^{J} (\alpha_j - \mu_\alpha)^2. \tag{18.15}$$

Unfortunately, the county parameters α_j are not themselves known, so we cannot directly apply the above formulas. We can, however, use an iterative algorithm that alternately estimates the α_j's and the hyperparameters, as we describe next.

Individual predictors but no group-level predictors

We next consider the varying-intercept model (12.2) from page 256: $y_i \sim N(\alpha_{j[i]} + \beta x_i, \sigma_y^2)$, where $j[i]$ is the county containing house i. The basic varying-intercept model (12.3) is $\alpha_j \sim N(\mu_\alpha, \sigma_\alpha^2)$—that is, a normal prior distribution for each α_j that is common to all counties j. (The hyperparameters $\mu_\alpha, \sigma_\alpha$ must themselves be estimated from the data, but we shall set this issue aside for a moment and just treat them as known.)

The top row of Figure 18.5 shows the likelihood for α_j in four of the counties.

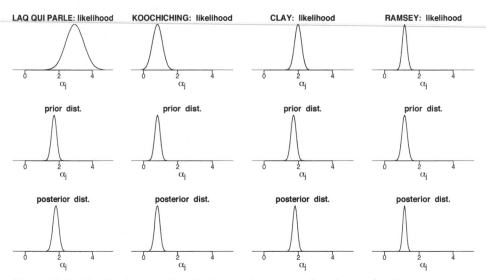

Figure 18.6 *Likelihood, prior distribution, and posterior distribution for the intercept parameter α_j in four counties for the radon model that includes uranium as a county-level predictor. The prior distributions for the counties now differ because of their varying uranium levels (see Figure 12.6 on page 266). Compare to Figure 18.5.*

For each county j, the likelihood indicates the range of values of α_j that are most consistent with the data in that county. The four counties are displayed in increasing order of sample size; the likelihood is more informative as sample size increases. The second row of Figure 18.5 shows the $N(\mu_\alpha, \sigma_\alpha^2)$ prior distribution, which is the same for the four counties. In this context the "prior distributions" do not represent information occurring before the data have been seen; rather, they convey the information about the distribution of the α_j's among the counties, which is relevant for estimating each individual α_j.

The bottom row of Figure 18.5 displays the posterior distributions, which combine the information from the likelihoods and prior distributions. The posterior distribution for each county is centered at a point between the maximum likelihood estimate and the maximum of the prior distribution—a weighted average of likelihood and prior estimates—falling closer to the prior distribution when sample sizes are small and closer to the likelihood when sample sizes are large.

Including group-level predictors

We now move to the radon model including uranium as a county-level predictor. Figure 18.6 displays the likelihood, prior distributions, and posterior distributions for four counties. In this case, the prior distributions for county j is normal with mean $\gamma_0 + \gamma_1 u_j$ and variance σ_α^2. The county uranium levels u_j vary, and so the prior distributions vary also, as can be seen in the second row of Figure 18.6.

Multilevel regression as least squares with augmented data

To understand the matrix algebra of multilevel regression, we continue with the data-augmentation idea illustrated in (18.9) on page 393. Starting with classical weighted least squares with data vector $y = (y_1, \ldots, y_n)$, an $n \times k$ predictor matrix X, and data weights $w = (w_1, \ldots, w_n)$, we define $W_y = \text{Diag}(w_1, \ldots, w_n)$, which is a matrix proportional to the inverse data variances in the model. The model is

$$y \sim N(X\beta, \Sigma_y),$$

where $\Sigma_y = \sigma^2 W_y^{-1}$. The vector of regression coefficients is estimated by weighted least squares as $\hat{\beta}_{\mathrm{wls}} = (X^t W_y X)^{-1} X^t W_y y$, and the corresponding variance matrix is $V_\beta = (X^t W_y X)^{-1} \sigma^2$. We refer to this as "the regression of y on X with weight matrix W_y." (This reduces to classical unweighed regression if the weights are all equal to 1, so that W_y is the identity matrix.)

To fit multilevel models in this framework, we work with the formulation as a large regression model, as in the discussion following (12.10) on page 264. We illustrate with the flight simulator model (13.9) on page 289. Here, β is a vector of length 14: the mean parameter, followed by 5 treatment effects and 8 airport effects. In the notation of (13.9), $\beta = (\alpha_0, \gamma_1, \dots, \gamma_5, \delta_1, \dots, \delta_8)$. We define μ_β and Σ_β as the mean and variance of β in the prior distribution: $\beta \sim N(\mu_\beta, \Sigma_\beta)$. Finally, we define the weight matrix W_β as the inverse-variance of β, scaled by the data variance: $W_\beta = \sigma_y^2 \Sigma_\beta^{-1}$.

For the flight simulator example, μ_β is a vector of 14 zeroes, and W_β is a diagonal matrix with diagonal entries , followed by $\sigma_y^2/\sigma_\gamma^2$ five times, followed by $\sigma_y^2/\sigma_\delta^2$ eight times. The first element of the diagonal of W_β is zero because our model specifies no information about the parameter α_0; the other elements indicate the information in the model about each multilevel parameter, compared to the information in each data point.

The multilevel model can be expressed as a least squares regression of y_* on X_* with weight matrix W_*, where

$$y_* = \begin{pmatrix} y \\ \mu_\beta \end{pmatrix}, \quad X_* = \begin{pmatrix} X \\ I_k \end{pmatrix}, \quad W_* = \begin{pmatrix} W_y & 0 \\ 0 & W_\beta \end{pmatrix}. \tag{18.16}$$

The augmented data correspond to the extra information in the model for β. The augmentation has the effect of partially pooling the least squares estimate of β in the direction of its mean vector μ_β, and can be viewed as a matrix generalization of (18.12).

18.4 Gibbs sampler for multilevel linear models

Gibbs sampling is the name given to a family of iterative algorithms that are used by Bugs ("Bayesian inference using Gibbs sampling") and other programs to fit Bayesian models. The basic idea of Gibbs sampling is to partition the set of unknown parameters and then estimate them one at a time, or one group at a time, with each parameter or group of parameters estimated conditional on all the others. The algorithm is effective because, in a wide range of problems, estimating separate parts of a model is relatively easy, even if it is difficult to see how to estimate all the parameters at once.

Figure 18.7 illustrates the Gibbs sampler for a simple example. In general, the algorithm proceeds as follows:

1. Choose some number n_{chains} of parallel simulation runs (typically a small number such as 3). For each of the chains:

   (a) Start with *initial values* for all the parameters. These should be dispersed (as pictured by the solid squares in Figure 18.7); for convenience we typically use simple random numbers, as discussed in Chapter 16 in the context of implementing models in Bugs.

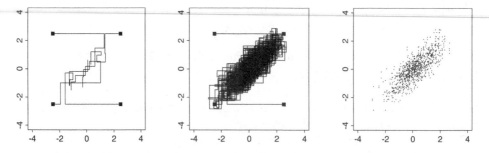

Figure 18.7 *Four independent sequences of the Gibbs sampler for a simple example with two parameters. Initial values of the $n_{chains} = 4$ sequences are indicated by solid squares. (a) First 10 iterations, showing the component-by-component updating of the Gibbs iterations. (b) After 500 iterations, when the sequences have reached approximate convergence. (c) The points from the second halves of the sequences.*

(b) Choose some number n_{iter} of iterations (typically a somewhat large number such as 1000). For each iteration:

Update the parameters, or batches of parameters, one at a time. For each parameter or batch, take a random simulation draw given the data and the current estimate of all the other parameters. (We illustrate with some examples below.)

2. Evaluate the mixing of the simulated chains using the $\hat{R}$ summary, which we have already discussed in Section 16.4 in the context of interpreting the output from Bugs models.

3. If convergence is poor, run longer or alter the model, following the advice in Section 16.9.

The key part of this algorithm is the sequential updating step. Bugs performs it automatically, but here we will show how to compute Gibbs updates "manually" in R for multilevel linear regressions. Our purpose is not to set you up to program these yourself but rather to give enough insight that you can understand roughly how Bugs works, and thus better diagnose and fix problems when Bugs is not working so well.

We present in this section the steps of Gibbs sampling for a series of multilevel linear regressions: first a model with no predictors, then including a predictor at the individual level, then adding one at the group level.

The basic Gibbs sampler structure described here works for multilevel regressions, with the new twist that the regression coefficients can be estimated using an adaptation of classical least squares regression. (Model (18.10) can be considered as a special case of regression with only an intercept and no slope parameters, but this case is so simple that least squares matrix computations were not required.)

Gibbs sampler for a multilevel model with no predictors

We first go through the steps of the Gibbs sampler—mathematically and as programmed in R—for the multilevel model (18.10) with data in groups and no predictors.

The Gibbs sampler starts with initial values for all the parameters and then updates the parameters in turn, giving each a random estimate based on the data and the current guess of the other parameters in the model. For the simple model we are considering here, the Gibbs updating steps are:

1. Update α: For $j = 1, \ldots, J$, compute $\hat{\alpha}_j$ and V_j from (18.12) and then draw α_j from the normal distribution with mean $\hat{\alpha}_j$ and variance V_j.

2. Update μ_α: Compute $\hat{\mu}_\alpha$ from (18.14) and then draw μ_α from the normal distribution with mean $\hat{\mu}_\alpha$ and variance σ_α^2 / J.

3. Update σ_y: Compute $\hat{\sigma}_y^2$ from (18.13) and then draw $\sigma_y^2 = \hat{\sigma}_y^2 / X_{n-1}^2$, where X_{n-1}^2 is a random draw from a χ^2 distribution with $n - 1$ degrees of freedom.

4. Update σ_α: Compute $\hat{\sigma}_\alpha^2$ from (18.15) and then draw $\sigma_\alpha^2 = \hat{\sigma}_\alpha^2 / X_{J-1}^2$, where X_{J-1}^2 is a random draw from a χ^2 distribution with $J - 1$ degrees of freedom.

Each of these steps should seem reasonable; however, the details (such as the χ^2 distributions and their degrees of freedom) are not particularly intuitive and must be derived using probability calculations that are beyond the scope of this book. Each step uses random simulations rather than point estimates so that the procedure captures the inferential uncertainty about the parameters.

Iterating the above four steps produces a "chain" of simulation draws—a sequence of simulations $\alpha_1, \ldots, \alpha_J$; σ_y; μ_α; $\alpha_1, \ldots, \alpha_J$; σ_y; μ_α; and so forth. Looking at any single one of these parameters, we have a sequence of simulations that, if the chain is run long enough, captures the range of uncertainty in the estimation of that parameter. We start several chains with random initial values and then run until the chains have mixed (see Figure 16.2 on page 357).

Programming the Gibbs sampler in R

When Bugs fits model (18.10), it performs a series of computations that are similar to the steps just given. To understand in more detail, we program them here in R. For many applications, we can simply use Bugs, but when computational speed is a concern (for example, with large datasets), or for some complicated models (for example, the social networks model in Section 15.3), it can be necessary to code the Gibbs sampler directly.

We program the Gibbs sampler in three steps: setting up the data, writing functions for the individual parameter updates, and writing a loop for the actual computation. For the radon example, we have already set up the data vector y and the vector of county indexes county, and we are ready to program the parameter updates.

```
a.update <- function() {                                          R code
  a.new <- rep (NA, J)
  for (j in 1:J){
    n.j <- sum (county==j)
    y.bar.j <- mean (y[county==j])
    a.hat.j <- ((n.j/sigma.y^2)*ybar.j + (1/sigma.a^2)*mu.a)/
            (n.j/sigma.y^2 + 1/sigma.a^2)
    V.a.j <- 1/(n.j/sigma.y^2 + 1/sigma.a^2)
    a.new[j] <- rnorm (1, a.hat.j, sqrt(V.a.j))
  }
  return (a.new)
}
mu.a.update <- function() {
  mu.a.new <- rnorm (1, mean(a), sigma.a/sqrt(J))
  return (mu.a.new)
}
sigma.y.update <- function() {
  sigma.y.new <- sqrt(sum((y-a[county])^2)/rchisq(1,n-1))
```

```
    return (sigma.y.new)
  }
  sigma.a.update <- function() {
    sigma.a.new <- sqrt(sum((a-mu.a)^2)/rchisq(1,J-1))
    return (sigma.a.new)
  }
```

These functions have empty argument lists (for example, `sigma.a.update` uses α, μ_α, and J, but these variables are not passed to as arguments in the function call) because we find it convenient to define all variables globally when putting the functions together. Passing functions through argument lists is cleaner in a programming sense but in this context can lead to confusion when models get altered, with parameters added and removed.

Another approach to programming these Gibbs updates would be to write general updating functions for the normal and inverse-χ^2 distributions and to call these by passing arguments through the functions.

In any case, having created the updating functions, we now create the space for three independent chains of length 1000 and give names to the parameters that will be saved in a large array, `sims`, that will contain posterior simulation draws for $\alpha, \mu_\alpha, \sigma_y, \sigma_\alpha$:

R code
```
n.chains <- 3
n.iter <- 1000
sims <- array (NA, c(n.iter, n.chains, J+3))
dimnames (sims) <- list (NULL, NULL,
  c (paste ("a[", 1:J, "]", sep=""), "mu.a", "sigma.y", "sigma.a"))
```

This last bit looks confusing; after running the command in R, it is helpful to type `dimnames(sims)` to see what this name object looks like, and to type `sims[1:5,1,]` to see how the names attach themselves to the `sims` object (or, in this case, the first five steps of the first chain of the `sims` object).

We are now ready to run the Gibbs sampler, first initializing $\mu_\alpha, \sigma_y, \sigma_\alpha$ with random values set crudely based on the range of the data—we need not initialize α because it is updated in the first step in the loop—and then simulating three chains for 1000 iterations each:

R code
```
for (m in 1:n.chains){
  mu.a <- rnorm (1, mean(y), sd(y))
  sigma.y <- runif (1, 0, sd(y))
  sigma.a <- runif (1, 0, sd(y))
  for (t in 1:n.iter){
    a <- a.update ()
    mu.a <- mu.a.update ()
    sigma.y <- sigma.y.update ()
    sigma.a <- sigma.a.update ()
    sims[t,m,] <- c (a, mu.a, sigma.y, sigma.a)
  }
}
```

We then summarize the simulations—view the inferences and check convergence—by using the `as.bugs.array` function to convert them to a Bugs object:

R code
```
sims.bugs <- as.bugs.array (sims)
plot (sims.bugs)
```

Gibbs sampler for a multilevel model with regression predictors

We can easily adapt the above algorithm to include predictors at the individual and group levels. Consider the model

$$y_i \sim N(\alpha_{j[i]} + X_i\beta, \sigma_y^2) \text{ for } i = 1, \ldots, n$$
$$\alpha_j \sim N(U_j\gamma, \sigma_\alpha^2) \text{ for } j = 1, \ldots, J,$$

where X is a matrix of individual-level predictors (without a constant term), U is a matrix of group-level predictors (including a constant term), and β and γ are vectors of coefficients.

After initializing the parameters $\alpha, \beta, \gamma, \sigma_y, \sigma_\alpha$ with random numbers (constraining the σ parameters to be positive), the Gibbs sampler can be implemented as follows:

1. Update α: It is simplest to use the reexpression, $\alpha_j = U_j + \eta_j$; the η_j's are group-level errors that are partially pooled toward their mean of 0. We apply (18.12) to suitably adjusted data y, correcting for individual- and group-level predictors. For each data point, compute $y_i^{\text{temp}} = y_i - X_i\beta - U_{j[i]}\gamma$. Then for $j = 1, \ldots, J$, compute $\hat{\eta}_j$ and V_j from (18.12)—but using y^{temp} in place of y—and draw η_j from the normal distribution with mean $\hat{\eta}_j$ and variance V_j. We complete the updating step by setting each α_j to $U_j\gamma + \eta_j$.

2. Update β: For each data point, compute $y_i^{\text{temp}} = y_i - \alpha_{j[i]}$. Then regress y^{temp} on X to obtain an estimate $\hat{\beta}$ and covariance matrix V_β, inserting σ_y for σ in equation (18.8). Now draw β from the $N(\hat{\beta}, V_\beta)$ distribution.

3. Update γ: Regress α on U (this is a regression with J data points) to obtain an estimate $\hat{\gamma}$ and covariance matrix V_γ, inserting σ_α for σ in (18.8). Now draw γ from the $N(\hat{\gamma}, V_\gamma)$ distribution.

4. Update σ_y: Compute $\hat{\sigma}_y^2 = \frac{1}{n}\sum_{i=1}^{n}(y_i - \alpha_{j[i]} - X_i\beta)^2$ and then draw $\sigma_y^2 = \hat{\sigma}_y^2/X_{n-1}^2$, where X_{n-1}^2 is a random draw from a χ^2 distribution with $n - 1$ degrees of freedom.

5. Update σ_α: Compute $\hat{\sigma}_\alpha^2 = \frac{1}{J}\sum_{i=1}^{J}(\alpha_j - U_j\gamma)^2$ and then draw $\sigma_\alpha^2 = \hat{\sigma}_\alpha^2/X_{J-1}^2$, where X_{J-1}^2 is a random draw from a χ^2 distribution with $J - 1$ degrees of freedom.

This algorithm combines partial pooling (step 1) with classical regression estimation of coefficients (steps 2 and 3) and standard errors (steps 4 and 5). It all works to summarize uncertainty because the parameters are updated iteratively, leading to an inference that includes all aspects of the model.

Programming in R involves writing functions for each of the five steps:

```
a.update <- function() {                                        R code
  y.temp <- y - X%*%b - U[county]%*%g
  eta.new <- rep (NA, J)
  for (j in 1:J){
    n.j <- sum (county==j)
    y.bar.j <- mean (y.temp[county==j])
    eta.hat.j <- ((n.j/sigma.y^2)*y.bar.j/
                  (n.j/sigma.y^2 + 1/sigma.a^2))
    V.eta.j <- 1/(n.j/sigma.y^2 + 1/sigma.a^2)
    eta.new[j] <- rnorm (1, eta.hat.j, sqrt(V.eta.j))
  }
  a.new <- U%*%g + eta.new
  return (a.new)
```

```
}
b.update <- function() {
  y.temp <- y - a[county]
  lm.0 <- lm (y.temp ~ X)
  b.new <- sim (lm.0, n.sims=1)
  return (b.new)
}
g.update <- function() {
  lm.0 <- lm (a ~ U)
  g.new <- sim (lm.0, n.sims=1)
  return (g.new)
}
sigma.y.update <- function() {
  sigma.y.new <- sqrt(sum((y-a[county]-X%*%b)^2)/rchisq(1,n-1))
  return (sigma.y.new)
}
sigma.a.update <- function() {
  sigma.a.new <- sqrt(sum((a-U%*%g)^2)/rchisq(1,J-1))
  return (sigma.a.new)
}
```

(In the calls to lm() in the b.update and g.update functions, we have specified the predictors in matrix form rather than as a formula listing the individual predictor names.)

Now that the updating functions have been written, the Gibbs sampler can be programmed and run as in the example earlier in this section, simply expanding to include the new parameters.

The Gibbs sampler as a general way of working with multilevel models

On page 239 we discussed the simple two-step procedure of first regressing y on X and group indicators to estimate β and the α_j's, then regressing the estimated α_j's on U to estimate γ. Multilevel modeling, in its Gibbs implementation, can be seen as a generalization of two-step regression in which the α_j's are estimated more accurately using partial pooling. Similarly, in more complicated multilevel structures, it often makes sense to program a Gibbs sampler in a way that alternately performs group-level regressions and separate inferences within each group.

If we were to start with a simple no-pooling, complete-pooling, or two-step analysis, and then gradually improve it to account for estimation uncertainty in each step, iterating to allow inferences to be based on the latest estimate for each parameter, then we would end up with a Gibbs sampler.

18.5 Likelihood inference, Bayesian inference, and the Gibbs sampler: the case of censored data

We illustrate some of the ideas of likelihood and Bayesian inference for a *censored-data* model. We begin with a regression of weight (in pounds) on height (in inches), using data from a random sample of Americans. Before fitting, we center the height variable (c.height <- height - mean(height)) so that we can better interpret the intercept as well as the slope of the regression:

R output
```
lm(formula = weight ~ c.height)
            coef.est coef.se
(Intercept)  156.1     0.6
```

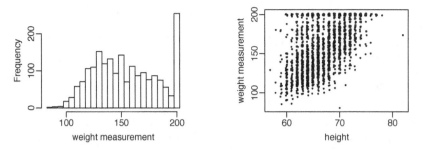

Figure 18.8 *(a) Histogram of weights as recorded on a hypothetical scale that censors measurements at 200; (b) Plot of (jittered values of) measured weight versus height in a sample of adults. The relation between height and weight is clear for low heights but becomes more difficult to follow at the high end, where the censoring becomes more frequent.*

```
c.height        4.9    0.2
  n = 1984, k = 2
  residual sd = 28.6, R-Squared = 0.30
```

Censoring

Now imagine that weights had been measured on a scale that was limited to a maximum of 200, so that any weights greater than 200 were recorded as "200^+" with the superscript indicating the censoring. We artificially perform this censoring on the survey data:

```
C <- 200                                                       R code
censored <- weight >= C
y <- ifelse (censored, C, weight)
```

From here on, we suppose that the censored variable, y, is what was observed, with the true weight only known if $y < 200$. Figure 18.8a shows the measured weights y, and Figure 18.8b shows weight plotted against height.

Naive regression estimate excluding the censored data

The two simple (but wrong) analyses of these data are to ignore the censored measurements, or to include them as measurements of 200. Here is the regression discarding the measurements of 200^+:

```
lm(formula = y ~ c.height, subset = y<200)                    R output
            coef.est coef.se
(Intercept)   148.7     0.5
c.height        3.8     0.1
  n = 1739, k = 2
  residual sd = 20.5, R-Squared = 0.31
```

Both the intercept and slope are too low, which makes sense given Figure 18.8b. This analysis excludes the largest values of weight and thus underestimates the average weight in the population. Also, more data are censored at the high end of heights, so the slope is underestimated too.

Naive regression estimate imputing the censoring point

Another simple but wrong approach is to simply code the 200^+ measurements as 200, which yields the following regression fit:

R output
```
lm(formula = y ~ c.height)
            coef.est coef.se
(Intercept)    153.6     0.5
c.height         4.3     0.1
  n = 1984, k = 2
  residual sd = 23.8, R-Squared = 0.32
```

Once again, this underestimates both the intercept (by using $y = 200$ in cases where we know the true weight is *at least* 200) and also the slope (because more of this bias occurs for taller persons).

Likelihood function accounting for the censoring. A better way to account for the censoring is to include it explicitly in the likelihood function. We write the censoring formally as

$$y_i = \begin{cases} z_i & \text{if } z_i \leq 200 \\ 200^+ & \text{if } z_i > 200, \end{cases} \tag{18.17}$$

with a linear regression for the true weights z_i given heights x_i:

$$y_i \sim N(a + bx_i, \sigma^2). \tag{18.18}$$

For the uncensored data points, the likelihood is simply the normal distribution, $N(y_i | a + bx_i, \sigma^2)$, as in (18.3) on page 388. For a censored measurement, the likelihood is

$$\Pr(y = 200^+) = \Pr(z_i \geq 200) = \int_{200}^{\infty} N(z_i \,|\, a + bx_i, \sigma^2) = \Phi\left(\frac{a + bx_i - z_i}{\sigma}\right),$$

where Φ is the normal cumulative distribution function (which can be computed using the `pnorm()` function in R).

The likelihood of all the data is then

$$p(y|\beta, \sigma, x) = \prod_{i=1}^{n} p(y_i|\beta, \sigma, x_i), \tag{18.19}$$

where the individual factors of the likelihood are

$$p(y_i|\beta, \sigma, X) = \begin{cases} N(y_i|a + bx_i, \sigma^2) & \text{if } y_i < 200 \\ \Phi((a + bx_i - 200)/\sigma) & \text{if } y_i = 200^+. \end{cases} \tag{18.20}$$

We shall clarify this expression (we hope) by programming it in R.

Maximum likelihood estimate using R

We shall first program the likelihood function in R and then call an optimization routine to find the maximum. In programming the likelihood, it is convenient to express the unknown parameters (in this case, a, b, and σ) as a vector, and then include the data and censoring point as additional arguments to the function. The following function computes the logarithm of the likelihood by evaluating (18.20) one data point at a time and then adding these values (which, on the log scale, is equivalent to the multiplication in (18.19)):

R code
```
Loglik <- function (parameter.vector, x, y, C) {
  a <- parameter.vector[1]
  b <- parameter.vector[2]
  sigma <- parameter.vector[3]
  ll.vec <- ifelse (y<C, dnorm (y, a + b*x, sigma, log=TRUE),
    pnorm ((a + b*x - C)/sigma, log=TRUE))
```

```
    return (sum (ll.vec))
  }
```

We have used the `log=TRUE` options of the `dnorm()` and `pnorm()` functions so that R automatically computes the logarithms of these probabilities. It is more computationally stable to compute probabilities on the log scale and only exponentiate at the end of the calculations.

To find the values of a, b, σ that maximize the log likelihood, we use the `optim()` function in R, which requires initial values (for which we simply use uniformly distributed random numbers) and some specifications:[2]

```
inits <- runif (3)                                                          R code
mle <- optim (inits, Loglik,  lower=c(-Inf,-Inf,1.e-5),
    method="L-BFGS-B", control=list(fnscale=-1), x=c.height,
    y=weight.censored, C=200)
```

We check by typing `print(mle$convergence)` (which should take on the value 0; type `?optim` in R for more details) and then typing

```
mle$par                                                                     R code
```

to find the vector of maximum likelihood estimates, which in this case are 155, 4.8, and 26.5 (corresponding to $\hat{a}$, $\hat{b}$, and $\hat{\sigma}$, respectively).

Fitting the censored-data model using Bugs

Bugs model.   Another way of fitting the censoring model, more consistent with the general approach of this book, is to write it in Bugs. The trick here is to express the model in terms of the true weights, z_i as defined in (18.17), which follow the regression model (18.18). For the censored data (the measurements $y_i = 200^+$, the true weights are unobserved but are constrained to fall in the range $(200, \infty)$. In Bugs, we express this constraint as a lower bound of 200 as follows:

```
model {                                                                     Bugs code
  for (i in 1:n){
    z.lo[i] <- C*equals(y[i],C)
    z[i] ~ dnorm (z.hat[i], tau.y) I(z.lo[i],)
    z.hat[i] <- a + b*x[i]
  }
  a ~ dnorm (0, .0001)
  b ~ dnorm (0, .0001)
  tau.y <- pow(sigma.y, -2)
  sigma.y ~ dunif (0, 100)
}
```

The `I(z.lo[i],)` factor constrains the distribution for $z[i]$ to be above `z.lo[i]`,[3] and this lower bound has been defined using `equals` to equal C (that is, 200) for censored observations and 0 otherwise.[4]

---

[2] In addition to specifying the vector of initial values and the name of the function to be optimized, we need to constrain σ to be positive—this is done by assigning a vector of lower limits to all three parameters, with empty $-\infty$ limits set for a and b. We then must set `method="L-BFGS-B"`, which is the "box constraint" algorithm that allows for bounds on the parameters. Setting `control=list(fnscale=-1)` tells `optim()` to find a maximum, rather than a minimum, of the specified function. Finally, we must specify the values of the other inputs to the `Loglik()` function, which in this case are x, y, and C.

[3] The factor `I(,z.hi[i])` would restrict the distribution to be *lower* than some value `z.hi[i]`, and `I(z.lo[i],z.hi[i])` would constrain to a finite range; see the models in Sections 6.5 and 17.7 for other examples of constrained distributions.

[4] A lower bound of zero is fine given that true weights are always positive. If there were no such natural bound—for example, if we were modeling log(weights)—then one could use an

Fitting the Bugs model from R. To fit the model in Bugs, we must first define z, which equals the weight when observed and is missing otherwise:

R code
```
z <- ifelse (censored, NA, weight.censored)
```

We then set up the bugs() call as usual:

R code
```
data <- list (x=c.height, y=weight.censored, z=z, n=n, C=C)
inits <- function() {
  list (a=rnorm(1), b=rnorm(1), sigma.y=runif(1))}
params <- c ("a", "b", "sigma.y")
censoring.1 <- bugs (data, inits, params, "censoring.bug", n.iter=100)
```

Reproducing some of the output from print(censoring.1):

R output

	mean	sd	2.5%	25%	50%	75%	97.5%	Rhat	n.eff
a	155.3	0.6	154.2	154.9	155.4	155.8	156.4	1	150
b	4.8	0.2	4.5	4.7	4.7	4.9	5.2	1	150
sigma.y	26.5	0.5	25.7	26.1	26.5	26.8	27.3	1	150

This inference is essentially the same as the maximum likelihood estimate—which makes sense, given that the sample size is large and the number of parameters is small—but are clearly different from the naive estimates excluding the censored data ($\hat{b} = 3.8 \pm 0.1$) and ($\hat{b} = 4.3 \pm 0.1$). The censoring model appropriately imputes the missing values, which we know lie above 200.

Gibbs sampler

Yet another way to fit this model is by programming a Gibbs sampler in R, as follows:

1. Impute crude starting values for the missing data—the true weights z_i corresponding to measurements $y = 200^+$.

2. Iterate the following two steps:

   (a) Run a regression of z (including the imputed values) on x and take a random draw from the uncertainty distribution of the parameters a, b, σ.

   (b) Use the estimated a, b, σ to create random imputations of the missing data.

For this problem, there is no real reason to program these steps; as we have just seen, the model is easy to fit in Bugs. This is, however, a good example to illustrate the way the Gibbs sampler handles uncertainty about missing data. We shall give the algebra and R code for each of the above steps.

Crude starting values. We simply impute a random value between C and $2C$ (in our example, 200 and 400 pounds) for each of the missing weights:

R code
```
n.censored <- sum (censored)
z[censored] <- runif (n.censored, C, 2*C)
```

Regression if the exact weights were known. We fit a regression and then draw one simulation value for the parameters a, b, σ:

R code
```
x <- c.height
lm.1 <- lm (z ~ x)
sim.1 <- sim (lm.1, n.sims=1)
a <- sim.1$beta[1]
b <- sim.1$beta[2]
sigma <- sim.1$sigma
```

assignment such as z.lo[i] <- -1.E5 + (1.E5+C)*equals(y[i],C) to set an extremely low bound for uncensored cases.

Imputing the missing values given the fitted regression.   The predictive distribution for any particular censored value i is normal with mean $a + bx_i$ and standard deviation σ, but constrained to be at least 200. We can write an R function to draw from this constrained distribution:

```
rnorm.trunc <- function (n, mu, sigma, lo=-Inf, hi=Inf) {          R code
  p.lo <- pnorm (lo, mu, sigma)
  p.hi <- pnorm (hi, mu, sigma)
  u <- runif (n, p.lo, p.hi)
  return (qnorm (u, mu, sigma))
}
```

This function first locates the constraint points (set by default to $(-\infty, \infty)$ if no constraints are given) in the specified distribution, then draws a sample within these probabilities, and finally transforms back to the original scale. We have written the function to take n independent draws, by analogy to the rnorm() function (type ?rnorm in R for details); this is not the same n that is the length of the data vector y.

We can then use this function to sample the missing z_i's given their predictors x_i:

```
z[censored] <- rnorm.trunc (n.censored, a + b*x[censored], sigma, lo=C)   R code
```

Gibbs sampler: putting it together in a loop.   We can now produce a Gibbs sampler. We first set up a space for 3 chains of 100 iterations each, saving $3 + n_{\text{censored}}$ parameters corresponding to a, b, σ, and the unobserved z_i's:

```
n.chains <- 3                                                     R code
n.iter <- 100
sims <- array (NA, c(n.iter, n.chains, 3 + n.censored))
dimnames (sims) <- list (NULL, NULL,
  c ("a", "b", "sigma", paste ("z[", (1:n)[censored], "]", sep="")))
```

We then program a Gibbs sampler, looping over the 3 chains: each chain starts with random initial values, then a loop through 100 iterations, first updating a, b, σ and then updating the censored components of z, and saving all these parameters at the end of each iteration.

```
for (m in 1:n.chains){                                           R code
  z[censored] <- runif (n.censored, C, 2*C)  # random initial values
  for (t in 1:n.iter){
    lm.1 <- lm (z ~ x)
    sim.1 <- sim (lm.1, n.sims=1)
    a <- sim.1$beta[1]
    b <- sim.1$beta[2]
    sigma <- sim.1$sigma
    z[censored] <- rnorm.trunc (n.censored, a + b*x[censored], sigma, lo=C)
    sims[t,m,] <- c (a, b, sigma, z[censored])
  }
}
```

Finally, we check the convergence:

```
sims.bugs <- as.bugs.array (sims)                                R code
print (sims.bugs)
```

yielding:

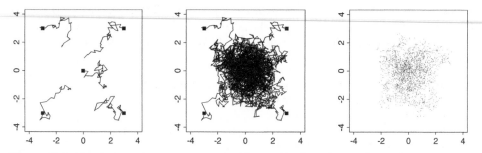

Figure 18.9 *Five independent sequences of a Metropolis algorithm, with overdispersed start-ing points indicated by solid squares. (a) After 50 iterations, the sequences are still far from convergence. (b) After 1000 iterations, with the sequences nearer to convergence. (c) The iterates from the second halves of the sequences, jittered so that steps in which the random walks stood still are not hidden.*

R output

	mean	sd	2.5%	25%	50%	75%	97.5%	Rhat	n.eff
a	155.3	0.6	154.1	154.9	155.3	155.7	156.3	1.0	53
b	4.8	0.2	4.5	4.7	4.8	4.9	5.1	1.0	150
sigma	26.5	0.5	25.7	26.2	26.5	26.9	27.6	1.0	69
z[3]	216.9	13.0	200.7	206.1	215.4	223.7	247.7	1.0	87
z[6]	210.9	9.3	200.3	203.5	208.9	214.9	234.1	1.0	150
z[11]	208.5	7.1	200.3	203.4	206.6	211.7	226.2	1.0	150

. . .

which is essentially identical to the results from the Bugs run (as it should be, given that we are fitting the same model).

Section 25.6 briefly describes a more realistic and complicated example of censoring that arises in a study of reversals of the death penalty, in which cases are censored that are still under consideration by appellate courts.

18.6 Metropolis algorithm for more general Bayesian computation

Moving to even more general models, the Gibbs sampler is a special case of a larger class of *Markov chain simulation algorithms* that can be used to iteratively estimate parameters in any statistical model. Markov chain simulation in general (and the Gibbs sampler in particular) can be thought of as iterative imputation of unknown parameters, or as a random walk through parameter space.

The Gibbs sampler updates the parameters one at a time (or in batches) using their conditional distributions. It can also be efficient to use the *Metropolis algorithm*, which takes a random walk through the space of parameters.

The Gibbs sampler and Metropolis algorithms are special cases of Markov chain simulation (also called *Markov chain Monte Carlo*, or MCMC), a general method based on drawing values of θ from approximate distributions and then correcting those draws to better approximate the target posterior distribution, $p(\theta|y)$. The samples are drawn sequentially, with the distribution of the sampled draws depending on the last value drawn; hence, the draws form a Markov chain. (As defined in probability theory, a *Markov chain* is a sequence of random variables $\theta^{(1)}, \theta^{(2)}, \ldots$, for which, for any t, the distribution of $\theta^{(t)}$ given all previous θ's depends only on the most recent value, $\theta^{(t-1)}$.) The key to the method's success, however, is not the Markov property but rather that the approximate distributions are improved at each step in the simulation, in the sense of converging to the target distribution.

Figure 18.9 illustrates a simple example of a Markov chain simulation—in this case, a Metropolis algorithm in which θ is a vector with only two components, with a bivariate unit normal posterior distribution, $\theta \sim N(0, I)$. First consider Figure 18.9a, which portrays the early stages of the simulation. The space of the figure represents the range of possible values of the multivariate parameter, θ, and each of the five jagged lines represents the early path of a random walk starting near the center or the extremes of the target distribution and jumping through the distribution according to an appropriate sequence of random iterations. Figure 18.9b represents the mature stage of the same Markov chain simulation, in which the simulated random walks have each traced a path throughout the space of θ, with a common stationary distribution that is equal to the target distribution. From a simulation such as 18.9b, we can perform inferences about θ using points from the second halves of the Markov chains we have simulated, as displayed in Figure 18.9c.

It is useful to have some sense of how the Metropolis algorithm works, because it is a key part of Bugs and other programs that perform iterative simulation. For further details on programming the Metropolis algorithm, see Gelman et al. (2003).

18.7 Specifying a log posterior density, Gibbs sampler, and Metropolis algorithm in R

To compute Markov chain simulation, the posterior density function and Gibbs sampler steps must be given. Bugs sets up these specification automatically (determining them from the model file) but for certain applications in which Bugs does not run or is too slow, it is necessary to program the log posterior density and Gibbs sampler steps explicitly.

We illustrate for the overdispersed Poisson regression model for social networks from Section 15.3, in which the large number of parameters (more than 1400) makes Bugs too slow to be practical. Instead, we use Umacs (universal Markov chain sampler), a program under development that performs Gibbs and Metropolis sampling given a specified posterior distribution. We go through the steps here, partly to complete the fitting of the social network model and partly to illustrate Markov chain sampling on a relatively complicated example.

The joint posterior density

The joint posterior density of the model in Section 15.3 can be written as

$$p(\alpha, \beta, \omega, \mu_\alpha, \mu_\beta, \sigma_\alpha, \sigma_\beta | y) \propto \prod_{i=1}^{n} \prod_{k=1}^{K} \binom{y_{ik} + \xi_{ik} - 1}{\xi_{ik} - 1} \left(\frac{1}{\omega_k} \right)^{\xi_{ik}} \left(\frac{\omega_k - 1}{\omega_k} \right)^{y_{ik}}$$

$$\times \prod_{i=1}^{n} N(\alpha_i | \mu_\alpha, \sigma_\alpha^2) \prod_{k=1}^{K} N(\beta_k | \mu_\beta, \sigma_\beta^2) \prod_{k=1}^{K} \omega_k^{-2}, \quad (18.21)$$

where $\xi_{ik} = e^{\alpha_i + \beta_k}/(\omega_k - 1)$, from the definition of the negative binomial distribution. The first factor in the posterior density is the likelihood—the probability density function of data given the parameters—and the remaining factors are the population distributions for each of the α_j's, β_k's, and ω_k's. The prior $p(\omega_k) \propto \omega_k^{-2}$ is equivalent to a uniform prior distribution on $1/\omega$, using the "Jacobian" from probability theory to transform from $1/\omega$ to ω.[5]

In computing, we actually work with the logarithm of the posterior density func-

---

[5] See, for example, Gelman et al. (2003, p. 24) for an explanation of the Jacobian.

tion because then computations are more stable and less likely to result in overflows or underflows than when using the density function itself. We typically only specify the density up to a multiplicative constant (note the proportionality sign in (18.21)) or, equivalently, the log density up to an additive constant—but this is all that is needed for Gibbs/Metropolis calculations.

The simulation algorithm

Our Markov chain simulation for the social network model requires the following steps:

- Gibbs sampler on the hyperparameters $\mu_\alpha, \sigma_\alpha, \mu_\beta, \sigma_\beta$.
- Metropolis jumping for each component of α, β, γ. Jump one vector at once for computational convenience.
- Constraining the components of γ to keep them above 1.
- Renormalization at each step because we are working with an overspecified model. (As discussed in Section 15.3, the model is unchanged if a constant is added to all the components of α and β. This constant thus needs to be specified in some way to stop the simulations from drifting aimlessly.)
- Adaptive Metropolis updating to keep acceptance rates near the target of 44%.
- Simulation of three parallel chains.
- After burn-in: stop adaptation, run awhile, and check convergence.
- Summarize with random simulation draws.

We program the first four of the above items; the others are performed automatically by Umacs.

We obtain posterior simulations using a Gibbs-Metropolis algorithm, iterating the following steps:

1. For each i, update α_i using a Metropolis step with jumping distribution, $\alpha_i^* \sim N(\alpha_i^{(t-1)}, (\text{jumping scale of } \alpha_i)^2)$.
2. For each k, update β_k using a Metropolis step with jumping distribution, $\beta_k^* \sim N(\beta_i^{(t-1)}, (\text{jumping scale of } \beta_k)^2)$.
3. Update $\mu_\alpha \sim N(\hat{\mu}_\alpha, \sigma_\alpha^2/n)$, where $\hat{\mu}_\alpha = \frac{1}{n}\sum_{i=1}^n \alpha_i$.
4. Update $\sigma_\alpha^2 \sim \text{Inv-}\chi^2(n-1, \hat{\sigma}_\alpha^2)$, where $\hat{\sigma}_\alpha^2 = \frac{1}{n}\sum_{i=1}^n (\alpha_i - \mu_\alpha)^2$.
5. Update $\mu_\beta \sim N(\hat{\mu}_\beta, \sigma_\beta^2/n)$, where $\hat{\mu}_\beta = \frac{1}{K}\sum_{k=1}^K \beta_k$.
6. Update $\sigma_\beta^2 \sim \text{Inv-}\chi^2(K-1, \hat{\sigma}_\beta^2)$, where $\hat{\sigma}_\beta^2 = \frac{1}{K}\sum_{k=1}^K (\beta_k - \mu_\beta)^2$.
7. For each k, update ω_k using a Metropolis step with jumping distribution, $\omega_k^* \sim N(\omega_k^{(t-1)}, (\text{jumping scale of } \omega_k)^2)$.
8. Rescale the α's and β's by computing the adjustment term C described on page 336 and adding it to all the α_i's and μ_α and subtracting it from all the β_k's and μ_β.

We construct starting points for the algorithm by fitting a classical Poisson regression (the null model, $y_{ik} \sim \text{Poisson}(\lambda_{ik})$, with $\lambda_{ik} = a_i b_k$) and then estimating the overdispersion for each subpopulation k using the statistic (6.5) on page 114.

Programming in R and Umacs

Setting up the model in Umacs requires several steps, which we include here as illustration of the details required to fully specify multilevel computation in R.

Log likelihood function. We take advantage of the matrix representation of the data $y = (y_{jk})$ to write a function that computes the log likelihood in parallel for all data points at once. In these expressions, y is a 1370×32 matrix; α is a vector of length 1370; β and ω are vectors of length 32, and `data.n=` 1370, the number of survey respondents.

```
f.loglik <- function (y, a, b, o, data.n) {                              R code
  theta.mat <- exp (outer (a, b, "+"))
  O.mat <- outer (rep (1, data.n), o, "*")
  A.mat <- theta.mat/(O.mat-1)      # the "alpha" and "beta" parameters
  B.mat <- 1/(O.mat-1)              # of the negative binomial distribution
  loglik <- lgamma(y+A.mat) - lgamma(A.mat) - lgamma(y+1) +
    (log(B.mat)-log(B.mat+1))*A.mat - log(B.mat+1)*y
  return (loglik)
}
```

The expression for `loglik` is the logarithm of the negative binomial density function.

Log posterior density functions. Our next step is to write functions that compute the log posterior density for each vector of parameters; these log densities are formed by summing the log likelihood by row or column and then adding the log prior distribution. We write different log posterior density functions for each parameter vector in order to make computations more efficient (for example, in updating α, we only need to include factors that depend on this parameter):

```
f.logpost.a <- function() {                                             R code
  loglik <- f.loglik (y, a, b, o, data.n)
  rowSums (loglik, na.rm=TRUE) + dnorm (a, mu.a, sigma.a, log=TRUE)
}
f.logpost.b <- function() {
  loglik <- f.loglik (y, a, b, o, data.n)
  colSums (loglik, na.rm=TRUE) + dnorm (b, mu.b, sigma.b, log=TRUE)
}
f.logpost.o <- function() {
  reject <- !(o>1)            # reject if omega is not greater than 1
  o[reject] <- 2             # set rejected omega's to arbitrary value of 2
  loglik <- f.loglik (y, a, b, o, data.n)
  loglik <- colSums (loglik, na.rm=TRUE) - 2*log(o)
  loglik[reject] <- -Inf    # set loglik to zero for rejected values
  return (loglik)
}
```

We constrain the components of ω to be greater than 1.01 because the model restricts this parameter to be greater than 1, and we want to avoid potential numerical difficulties when any ω_k is exactly 1.

Data and initial values. Next, we load in the data:

```
library ("foreign")                                                     R code
y <- as.matrix (read.dta ("all.dta"))
```

and define initial values for the parameters:

```
a.init       <- function() {rnorm (data.n)}                             R code
b.init       <- function() {rnorm (data.j)}
o.init       <- function() {runif (data.j, 1.01, 50)}
mu.a.init    <- function() {rnorm (1)}
mu.b.init    <- function() {rnorm (1)}
sigma.a.init <- function() {runif (1)}
sigma.b.init <- function() {runif (1)}
```

The initial values (as well as the log posterior densities defined above) are set up as functions with no arguments because Umacs uses the variables in the workspace rather than passing data and parameters back and forth among functions.[6]

Gibbs sampler steps. Having defined the model, data, and inital values, we write the functions for the Gibbs samplers for the hyperparameters:

R code
```
mu.a.update <- function() {
  rnorm (1, mean(a), sigma.a/sqrt(data.n))
}
mu.b.update <- function() {
  rnorm (1, mean(b), sigma.b/sqrt(data.j))
}
sigma.a.update <- function() {
  sqrt (sum((a-mu.a)^2)/rchisq(1, data.n-1))
}
sigma.b.update <- function() {
  sqrt (sum((b-mu.b)^2)/rchisq(1, data.j-1))
}
```

Renormalization step. We next write a function for the renormalization of the α's and β's in terms of the frequencies of the names in the population:

R code
```
renorm.network <- function() {
  const <- log (sum(exp(b[c(2,4,12)]))/.00357) +
    .5*log (sum(exp(b[c(3,7)]))/.00760) -
    .5*log (sum(exp(b[c(6,8,10)]))/.00811)
  a <- a + const
  mu.a <- mu.a + const
  b <- b - const
  mu.b <- mu.b - const
}
```

Setting up the Umacs sampler function. We are now ready to set up the series of steps for Metropolis and Gibbs sampling for the social network model. We update each of the vectors α, β, and ω using the `PSMetropolis()` routine, which stands for "scalar parallel Metropolis"—that is, separately updating each component using Metropolis jumping, automatically tuning these to jump efficiently.[7]

R code
```
s.network <- Sampler (
  y = y,
  data.n = nrow(y),
  data.j = ncol(y),
  a = PSMetropolis (f.logpost.a, a.init),
  b = PSMetropolis (f.logpost.b, b.init),
  o = PSMetropolis (f.logpost.o, o.init),
  mu.a = Gibbs (mu.a.update, mu.a.init),
  mu.b = Gibbs (mu.b.update, mu.b.init),
  sigma.a = Gibbs (sigma.a.update, sigma.a.init),
  sigma.b = Gibbs (sigma.b.update, sigma.b.init),
  renorm.network)
```

This call to `Sampler()` creates a function, `s.network()`, which we can then call to perform the actual sampling.

---

[6] As discussed on page 400, this "global variable" structure makes it easier for us to expand models without having to worry about keeping track of the parameters used in each function call. Other programming strategies are also possible.

[7] Umacs also includes vector Metropolis jumping, in which several components of a vector are altered at once, but in this case the posterior density for each vector of parameters factors into its components, so these components can be efficiently updated in parallel.

Running Umacs and saving the simulations. Finally, we run the sampler for three parallel chains for 2000 iterations, keeping the last 1000. We save the output as a Bugs object and plot it.

```
network.1 <- s.network (n.iter=2000, n.sims=1000, n.chains=3)
network.1.bugs <- as.bugs.array (network.1)
plot (network.1)
```

R code

We can then check convergence (by looking at the values of $\widehat{R}$ in the plot), access the simulations using `attach.bugs(network.1.bugs)`, and make the plots shown in Section 15.3.

18.8 Bibliographic note

For a fuller presentation of our perspective on likelihood and Bayesian data analysis, see Gelman et al. (2003). Other presentations of Bayesian inference include Box and Tiao (1973), Bernardo and Smith (1994), and Carlin and Louis (2001).

For more on prior distributions, see Jeffreys (1961), Jaynes (1983), Box and Tiao (1973), and Meng and Zaslavsky (2002). Many of the concerns in this literature are less urgent in multilevel models, in which most parameters are themselves modeled at the group level—but the prior distribution can still be relevant for the few remaining hyperparameters of any model. Our approach of prior distributions as placeholders or "reference models" follows Bernardo (1979); see also Kass and Wasserman (1996).

Full Bayesian analysis for multilevel models was first performed by Hill (1965), Tiao and Tan (1965, 1966), and Tiao and Box (1967). Important later work includes Lindley and Smith (1972), Efron and Morris (1975), Dempster, Rubin, and Tustakawa (1981), Gelfand and Smith (1990), and Pauler, Wakefield, and Kass (1999).

See Gilks, Richardson, and Spiegelhalter (1996) for more on the Gibbs sampler and the Metropolis algorithm. For an introduction to Bayesian inference for censoring and truncation, see Gelman et al. (2003, section 7.8).

The social network example comes from Zheng, Salganik, and Gelman (2006). Umacs is described by Kerman (2006) and Kerman and Gelman (2006).

18.9 Exercises

1. Linear regression algebra: show that weighted least squares is maximum likelihood estimation for the model (18.7).

2. Bayesian inference: take a multilevel linear model that you have already fit, and make a graph such as in Figure 18.5 or 18.6 showing likelihood, prior distribution, and posterior distribution, in each of several groups.

3. Maximum likelihood estimation: consider the logistic regression you set up in Exercise 5.8(a) for predicting presence of rodents in an apartment given ethnic group.

   (a) Write the likelihood for this model.

   (b) Program the likelihood function in R and use `optim()` to find the maximum likelihood estimate. Check that your estimate is the same as you obtained in Exercise 5.8(a).

4. Censored data: take the data on beauty and teaching evaluations data described in Exercise 3.5 and artificially censor by reporting all course evaluations below 3.0 simply as "*"

(a) Take one of the models from that earlier exercise and write the likelihood function given this mix of observed and censored data.

(b) Find the maximum likelihood estimate in R using the `optim()` function.

(c) Fit the model using Bugs, accounting for the censoring.

(d) Compare the censored-data inferences from the estimates using the complete data.

## CHAPTER 19

Debugging and speeding convergence

Once data and a model have been set up, we face the challenge of debugging or, more generally, building confidence in the model and estimation. The steps of Bugs and R as we have described them are straightforward, but cumulatively they require a bit of effort, both in setting up the model and checking it—adding many lines of code produces many opportunities for typos and confusion. In Section 19.1 we discuss some specific issues in Bugs and general strategies for debugging and confidence building. Another problem that often arises is computational speed, and in Sections 19.2–19.5 we discuss several specific methods to get reliable inferences faster when fitting multilevel models. The chapter concludes with Section 19.6, which is not about computation at all, but rather is a discussion of prior distributions for variance parameters. The section is included here because it discusses models that were inspired by the computational idea described in Section 19.5. It thus illustrates the interplay between computation and modeling which has often been so helpful in multilevel data analysis.

19.1 Debugging and confidence building

Our general approach to finding problems in statistical modeling software is to get various crude models (for example, complete pooling and no pooling, or models with no predictors) to work and then to gradually build up to the model we want to fit. If you set up a complicated model and you can't get it to run—or it will run but its results don't make sense—then either build it from scratch, or strip it down until you can get it to work and make sense. Figure 19.1 illustrates.

It might end up that there was a bug in your original data and model specification, or maybe the model you wanted to fit is not appropriate for your data, and you will move to a more reasonable version. Thus, the debugging often serves a statistical goal too, by motivating the exploration of various approximate and alternative models.

Getting your Bugs program to work

In R, C, and other command-based languages, if a function or script does not work, we have various direct debugging tactics, including executing the code line by line to find where it fails, executing parts of the code separately, and inserting print statements inside the code to display intermediate results. None of these methods work with Bugs, because the lines of a Bugs model are not executed sequentially. Actually, the lines of a Bugs model are not "executed" at all; rather, Bugs parses the entire model and then runs a process.

If there is a problem with the model, or the data, or the initial values, or the simulation itself, Bugs will crash (usually by halting and displaying an error message in the Bugs window—which you will see if you set the `debug=TRUE` option when calling from R—or occasionally by simply not responding, in which case you must close the Bugs window to end the process.

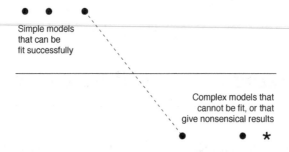

Figure 19.1 *Diagram of advice for debugging. The asterisk on the lower right represents the scenario in which problems arise when trying to fit the desired complex model. The dots on the upper left represent successes at fitting various simple versions, and the dots on the lower right represent failures at fitting various simplifications of the full model. The dotted line represents the idea that the problems can be identified somewhere between the simple models that fit and the complex models that don't.*

Thus, when Bugs fails, we can try to identify the problem from the error message in the Bugs window (as discussed below), but the only general approach is to go back to simpler models that work and then locate the error as indicated in Figure 19.1. Here we briefly list some specific problems that we have encountered in Bugs modeling.

Model not compiling. Common problems in the compilation include: too much inside a distribution argument (for example, `y[i]~dnorm(a+b*x[i],tau)` should be two lines: `y[i]~dnorm(y.hat[i],tau)` and `y.hat[i]<-a+b*x[i]`; see page 354); space in the wrong place in an assignment statement (for example, the expression `tau <- pow (sigma,-2)` should not have a space after "pow"); undefined parameters (that is, parameters used somewhere in the Bugs model but not modeled (with "~") or assigned (with "<-") elsewhere in the model or specified as data); and multiply defined parameters or data (for example, `y.hat<-a+b*x[i]` defined within a loop, thus implicitly creating n different definitions of the same variable `y.hat`). Many of these errors can be detected from the error messages given by Bugs.

Problems with data. Improper data in Bugs include NA's in unmodeled data (see Section 16.8); passing data that are not in the model (this can occur, for example, if a predictor `x` is removed from the model but kept in the data list); passing the wrong data (for example, if a variable name has been changed, or if an R object containing data has been overwritten); passing empty data vectors (for variables that you have forgotten to set up in R); subscripting problems (length or dimension of a data array inconsistent with its use in the Bugs model); and data out of range (for example, passing a negative value to data modeled by a lognormal distribution, or passing a fractional value to data modeled by a binomial distribution).

Problems with initial values. The list of initial values can create problems because of variables that are empty or have missing values (often because of problems in the R code); and if initial values are of range of the model (this can happen, for example, with constrained parameters, as well as more obvious scenarios such as negative variance parameters).

Problems when updating. Bugs sometimes crashes because some of its updating routines are not so robust, especially when in extreme areas of parameter space, such as continuous parameters with very large values or probabilities very close to 0 or 1. One reason we like to give moderate initial values to all the parameters in the model is to avoid these problems. When Bugs updating continues to crash, we can sometimes fix the problem by constraining the parameters, as with the line `p.bound[i]<-max(0,min(1,p[i]))` in the logistic regression model in Section 17.4. Other times Bugs crashes or hangs because of problems in the model that were not detected in the compilation stage, for example, circular definitions (parameter `a` defined in terms of `b`, and vice versa).

Model runs, but results don't make sense. We have already discussed near the end of Section 16.9 that, when Bugs is slow to converge, it can make sense to reparameterize the model to speed the mixing of the Gibbs sampler. We discuss some such techniques later in this chapter. However, it also happens that Bugs converges but to an unreasonable answer, or to a surprising result compared to results from previously fit classical models and simple multilevel models fit using R. Nonsensical inferences commonly arise from problems in the Bugs model, such as the misspecification of a distribution (for example, confusing an inverse-variance parameter τ with a variance or standard-deviation parameter) or when parameters that should have a group-level model are assigned noninformative distributions (for example, `y[i]~dnorm(y.hat[i],tau[i])`, with a separately estimated variance parameter for each observation). Problems also arise when the data sent to the model are not what you thought they were.

Comparisons to simpler models

As illustrated in several of the examples of this book, a good way to understand and build confidence in a multilevel model is to build it up from simpler no-pooling and complete-pooling models, approximate fits using `lmer()`, and simpler versions, for example, excluding some predictors, letting some coefficients vary but not others, and so forth.

Fake-data simulation

Follow the procedure described in Section 16.7: from the estimated model, take the estimates of the unmodeled parameters as the "true values," and from these simulate "true values" of the modeled parameters and then a fake dataset. Estimate the model from the fake data, and check that the estimates for the unmodeled and modeled parameters are close to the true values.

Checking fit of model to data

An important way to build confidence in a model-fitting procedure is to check that the model fits the data. Model-checking tools include residual plots and, more generally, predictive checks where the actual data are compared, numerically or graphically, to replicated data simulated from the fitted model. We discuss this approach further in Chapter 24.

Unexpected difficulties with Bugs

Sometimes a model will fail in Bugs, even though it is only a slight modification of a successful fit. We have encountered difficulties even with pure classical regressions, when there is near-collinearity in the matrix of predictors. Ultimately we believe Bugs will be improved and these problems will no longer arise, but in the meantime we can use work-arounds such as centering predictors about their mean levels and trying different reparameterizations until something works.

19.2 General methods for reducing computational requirements

There are two ways of speeding an iterative algorithm such as the Gibbs sampler: taking less time per iteration or reducing the number of iterations to convergence. Here we briefly discuss some general approaches; then in Sections 19.3–19.5 we consider some specific methods for improving the mixing of Gibbs samplers for multilevel models.

Sampling data to speed computation

Bugs can run slowly with large datasets. Computation can be slow even for moderate-sized datasets when the number of parameters in the model is large, as can easily happen with multilevel models. For example, the polling model in Section 14.1 has more than 70 parameters (a coefficient for each of the 50 states, plus coefficients for the demographic indicators, plus hyperparameters) and about 1500 respondents. The model can be fit in a reasonable time, but when we extended it to handle data from seven different polls, totaling about 10,000 respondents, we had to wait a few minutes for each model fit. In practice, this waiting reduced the number of models we could conveniently fit, thus actually reducing the flexibility of our statistical analysis.

In this sort of setting, it can be effective to *sample* the data, for example, randomly selecting one-tenth of the respondents and analyzing them first. Once a reasonable model has been found, we can go back and take the time to fit it to the entire dataset. Sampling is easy in R; for example,

R code
```
subset <- sample (n, n/10)
n <- length(subset)
y <- y[subset]
X <- X[subset,]
state <- state[subset]
. . .
```

Computation can be sped even more using *cluster sampling*, in which we take a subset of groups, and then a sample of respondents within each group. For example, in the election poll example, we could include half the states, and one-fifth of the data within each sampled state. The advantage of cluster sampling is that it reduces the number of parameters in the model as well as the number of data points. Do not reduce the number of clusters too far, however, or it will make it difficult to estimate the group-level regression coefficients and the group-level variance. (As discussed in Section 12.9, the group-level variance is difficult to estimate if the number of groups is less than 5.)

Thinning output to save memory

Sometimes Bugs must be run a long time to convergence. This happens when the simulation draws are highly autocorrelated, so there is very little information in each draw. In this case, it makes sense to *thin* the Bugs output—that is, to just keep every k^{th} simulation draw and discard most of the rest—to save computation time and memory. We have set up the bugs() function to automatically thin long simulation runs so as to save approximately 1000 simulation draws.

Knowing when to give up

If you have to run Bugs for a long time and still don't reach approximate convergence, then we recommend reformulating the model using the ideas discussed in the rest of this chapter. In the meantime, you can run simpler versions of the model and see if they make sense. Our applied research often proceeds on two tracks, with a series of simple models used to explore data, while we work to get more elaborate models working reliably. Fitting an elaborate model generally becomes more possible as we get a better sense of what answers we should be expecting, based on our simpler data analysis. At the end, the more complicated model can answer questions that we did not even know to ask before we built up to it. An example is the pattern in Figure 14.11 on page 313 of state-by-state variation in the coefficient of income on vote preference, which we were only able to learn about after we succeeded in fitting the varying-intercept, varying-slope model with a group-level predictor.

19.3 Simple linear transformations

We now discuss some methods for reformulating models to make the Gibbs sampler converge in fewer iterations. As discussed in Section 18.6, the Gibbs sampler will run slowly if parameters in the model are highly correlated in their estimation—for example, if α is large, then β must be small and vice versa. This sort of correlation can easily occur between regression intercepts and slopes for predictors that are not centered at zero; see Figure 13.5 on page 288. As illustrated in that example, we improved the model by centering the predictor x before including it in the model.

Centering typically increases the speed of convergence and can be implemented easily in R without requiring any changes to the Bugs model. For example, for the simple regression

$$y_i \sim N(\alpha + \beta_1 X_{i1} + \beta_2 X_{i2}, \sigma_y^2),$$

we can rescale the individual predictors x1 and x2:

```
z1 <- x1 - mean (x1)                                              R code
z2 <- x2 - mean (x2)
```

and then express the Bugs model using the z's rather than x's as predictors. We then have two options: we can simply work with the fitted model in terms of the new, adjusted predictors; or we can adjust the fitted coefficients to work with the original predictors X. The fitted model is

$$
\begin{aligned}
y_i &= \alpha + \beta_1 Z_{i1} + \beta_2 Z_{i2} + \epsilon_i \\
&= \alpha + \beta_1 (X_{i1} - \bar{X}_1) + \beta_2 (X_{i2} - \bar{X}_2) + \epsilon_i \\
&= (\alpha - \beta_1 \bar{X}_1 - \beta_2 \bar{X}_2) + \beta_1 X_{i1} + \beta_2 X_{i2} + \epsilon_i
\end{aligned}
$$

Thus, the intercept on the original scale is simply $\alpha - \beta_1 \bar{X}_1 - \beta_2 \bar{X}_2$, and the slopes are unaffected.

19.4 Redundant parameters and intentionally nonidentifiable models

Another way of speeding the convergence of the Gibbs sampler involves adding redundant coefficients that are collinear with existing predictors in the model.

We have already discussed one connection between multilevel models and identifiability: in a classical regression model, we can include only $J-1$ of the J group-level indicators (with the remaining indicator serving as a baseline), whereas in multilevel regression, we can include coefficients for all J groups, because they are modeled as coming from a common distribution with a finite variance. The group-level model provides information that allows the coefficients to be identified (as expressed mathematically by the augmented data vector y_* in (18.16) on page 397).

Our use of nonidentifiability here is different—it is purely for computational, not modeling, purposes, and is nonidentified even in the hierarchical model.

Redundant mean parameters for a simple nested model

As a simple example, we return to the radon analysis from Chapter 12, simplified to measurements within counties, ignoring all individual-level and group-level predictors. We can write the model as

$$y_i \sim \mathrm{N}(\mu + \eta_{j[i]}, \sigma_y^2), \text{ for } i = 1, \dots, n$$
$$\eta_j \sim \mathrm{N}(0, \sigma_\eta^2), \text{ for } j = 1, \dots, J.$$

In Bugs:

Bugs code
```
model {
    for (i in 1:n){
        y[i] ~ dnorm (y.hat[i], tau.y)
        y.hat[i] <- mu + eta[county[i]]
    }
    mu ~ dnorm (0, .0001)
    tau.y <- pow(sigma.y, -2)
    sigma.y ~ dunif (0, 100)

    for (j in 1:n.county){
        eta[j] ~ dnorm (0, tau.eta)
    }
    tau.eta <- pow(sigma.eta, -2)
    sigma.eta ~ dunif (0, 100)
}
```

With this version of the model, however, the Bugs simulations are slow to converge; see Figure 19.2. It is possible for the simulations to get stuck in a configuration where the entire vector η is far from zero (even though the η_j's are assigned a distribution with mean 0). Ultimately, the simulations will converge to the correct distribution, but we do not want to have to wait. We can speed the convergence by adding a *redundant parameter* for the mean, and then redefining new parameters η centered at an arbitrary group-level mean μ_η, thus replacing the loop for (j in 1:n.county) in the above model by

Bugs code
```
mu.adj <- mu + mean(eta[])
for (j in 1:n.county){
    eta[j] ~ dnorm (mu.eta, tau.eta)
    eta.adj[j] <- eta[j] - mean(eta[])
}
mu.eta ~ dnorm (0, .0001)
}
```

To fit the new model, we must specify initial values to the redundant parameters, μ_η, η_j, but we only need to save the adjusted parameters, $\mu^{\mathrm{adj}}, \eta^{\mathrm{adj}}$:

R code
```
radon.data <- list ("n", "y", "n.county", "county")
radon.inits <- function(){
    list (mu=rnorm(1), mu.eta=rnorm(1), eta=rnorm(n.county),
```

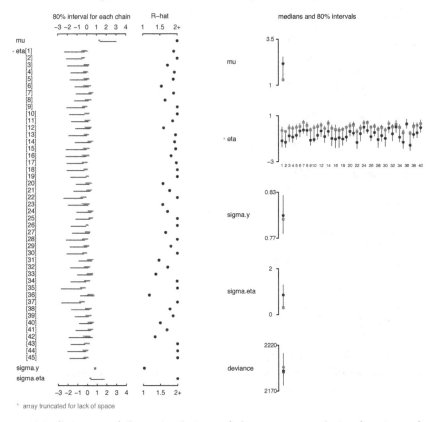

Figure 19.2 *Summary of Bugs simulations of the one-way analysis of variance for the radon data. Convergence is slow, so we used redundant parameters to help the model run more efficiently.*

```
    sigma.y=runif(1), sigma.eta=runif(1))
}
radon.parameters <- c ("mu.adj", "eta.adj", "sigma.y", "sigma.eta")
fit.4 <- bugs (radon.data, radon.inits, radon.parameters,
  "M4.bug", n.chains=3, n.iter=100)
```

Redundant mean parameters for a non-nested model: flight simulator example

We can use the same overparameterization trick for non-nested models. For example, we can express the prior distributions for the flight simulator model (13.9) as

$$\begin{aligned}
\gamma_j &\sim \text{N}(\mu_\gamma, \sigma_\gamma^2), \text{ for } j = 1, \dots, J \\
\delta_k &\sim \text{N}(\mu_\delta, \sigma_\delta^2), \text{ for } k = 1, \dots, K.
\end{aligned} \tag{19.1}$$

The means μ_γ, μ_δ, and the individual raw treatment and airport effects γ_j, δ_k are not separately identified. However, we can define the parameters of interest by centering each batch of group-level coefficients around zero and then redefining the mean. This can all be implemented in Bugs (compare to the model on page 380):

```
model {                                                          Bugs code
  for (i in 1:n){
```

```
      y[i] ~ dnorm (y.hat[i], tau.y)
      y.hat[i] <- mu + g[treatment[i]] + d[airport[i]]
    }
    mu.adj <- mu + mean(g[]) + mean(d[])
    mu ~ dnorm (0, .0001)
    tau.y <- pow(sigma.y, -2)
    sigma.y ~ dunif (0, 100)

    for (j in 1:n.treatment){
      g[j] ~ dnorm (mu.g, tau.g)
      g.adj[j] <- g[j] - mean(g[])
    }
    mu.g ~ dnorm (0, .0001)
    tau.g <- pow(sigma.g, -2)
    sigma.g ~ dunif (0, 100)

    for (k in 1:n.airport){
      d[k] ~ dnorm (mu.d, tau.d)
      d.adj[k] <- d[k] - mean(d[])
    }
    mu.d ~ dnorm (0, .0001)
    tau.d <- pow(sigma.d, -2)
    sigma.d ~ dunif (0, 100)
  }
```

As with the previous example, the redundant location parameters μ_γ and μ_δ can reduce the number of iterations required for the Gibbs sampler to converge for the parameters of interest: μ^{adj}, the γ_j^{adj}'s, and the δ_k^{adj}'s.

Example: multilevel logistic regression for survey responses

For a more elaborate problem in which redundant parameters are useful, we consider the example from Section 14.1 of the probability of a Yes response on a survey, estimated as a function of demographics and state of residence.

We write the model in Bugs by expanding the multilevel logistic regression model on page 381:

Bugs code
```
    mu.adj <- b.0 + mean(b.age[]) + mean(b.edu[]) + mean(b.age.edu[,]) +
      mean(b.state[])
    for (j in 1:n.age){
      b.age[j] ~ dnorm (mu.age, tau.age)
      b.age.adj[j] <- b.age[j] - mean(b.age[])
    }
    for (j in 1:n.edu){
      b.edu[j] ~ dnorm (mu.edu, tau.edu)
      b.edu.adj[j] <- b.edu[j] - mean(b.edu[])
    }
    for (j in 1:n.age) {for (k in 1:n.edu){
      b.age.edu[j,k] ~ dnorm (mu.age.edu, tau.age.edu)
      b.age.edu.adj[j,k] <- b.age.edu[j,k] - mean(b.age.edu[,])
    }
    for (j in 1:n.state){
      b.state[j] ~ dnorm (b.state.hat[j], tau.state)
      b.state.hat[j] <- b.region[region[j]] + b.v.prev*v.prev[j]
    }
    b.v.prev ~ dnorm (0, .0001)
```

```
for (j in 1:n.region){
  b.region[j] ~ dnorm (mu.region, tau.region)
  b.region.adj[j] <- b.region[j] - mean(b.region[])
}
```

We assign noninformative prior distributions to the new hierarchical mean parameters:

```
mu.age ~ dnorm (0, .0001)                                    Bugs code
mu.edu ~ dnorm (0, .0001)
mu.age.edu ~ dnorm (0, .0001)
mu.region ~ dnorm (0, .0001)
```

and retain the rest of the model from page 381. The expression for `mu.adj` does not include `mean(b.region[])` because the parameters `b.region` are not directly included in the expression for `logit(p[i])`; the region coefficients come in through the state coefficients `b.state`.

As with the other examples in this section, the redundant mean parameters, $\mu_{\text{age}}, \mu_{\text{edu}}, \mu_{\text{age.edu}}$, are not separately identified (except by their prior distributions). We will not expect them to converge rapidly in the simulations, and similarly we would expect convergence problems with the multilevel parameters β_{age}, *before* they were centered at these μ's.

A slight complexity arises in that we add a redundant mean to β_{region} but not to β_{state} to be consistent with the regional model that is nested within.

After 500 steps (in the call to `bugs()` from R, we set `n.chains=3` and `n.iter=500`), the β's and σ's have reached approximate convergence.

(A simpler way to write the model would be to replace the μ parameters in the model by zeroes, which would leave all the coefficients identifiable and eliminate the need to center the β's. However, this model can be slower to converge—setting the μ's to zero causes a kind of "gridlock" in the Gibbs sampler with multilevel coefficients.)

"Adjusted" or "raw" parameters

There are two equivalent ways of labeling in a reparameterization. The method we have described so far is to start with an existing model, add parameters to it, then define new "adjusted" parameters that are identified by the data. Thus, for example, a Bugs model would include an unidentified α and an identified α^{adj}, which is what would be saved from the fitting of the model. This approach is convenient in that it begins with an already-working model but is awkward in that the saved parameters do not have the names that were used in the statistical formulation of the model (except in cases such as (15.2) on page 326 for the study of police stops, where we explicitly define adjusted parameters of interest).

An alternative approach is to relabel the parameters in the original model as "raw" and then give the identified parameters the clean names that would be saved in the model fit. Thus, instead of `a.adj[j]<-a[j]-mean(a[])`, the Bugs model would have `a[j]<-a.raw[j]-mean(a.raw[])`. When we know ahead of time that we will be using redundant parameters or models expressed in a compound form (as in the scaled inverse-Wishart model on page 376), we often start right away with the formulation in terms of "raw" parameters to simplify the post-processing of inferences in R.

19.5 Parameter expansion: multiplicative redundant parameters

We next present a slightly more sophisticated trick involving redundant multiplicative as well as additive parameters, an idea that is a response to a particular kind of slow convergence of Gibbs samplers for hierarchical models.

The Gibbs sampler can get stuck

Gibbs samplers for multilevel models can get stuck in the following way. Suppose that a group-level variance parameter σ_α happens to be estimated near zero. Then, in the updating step for $\alpha_1, \ldots, \alpha_J$, these group-level coefficients will be pooled strongly toward their common mean (see Figure 18.4: if σ_α is near zero, then the ratio $\sigma_y^2/\sigma_\alpha^2$ will be large, and so we will be near the top of the table where there is more pooling). So the α_j's will be estimated to be close to each other. But the updating step for σ_α is based on the variance of the α_j's, and so σ_α will be estimated to be near zero, and the algorithm can get stuck. Eventually, the simulations will move through the entire distribution of σ_α, but this can require many iterations.

Solution using parameter expansion

A good way to get the Gibbs sampler unstuck in the above scenario is to be able to rescale the α_j's by multiplying the entire α vector by a constant. We can do this in a multilevel model by adding a redundant multiplicative parameter for each variance component.

For example, we can extend the flight simulator model (13.9) with two new multiplicative parameters, ξ_γ and ξ_δ and then assigning them noninformative prior distributions. The ξ parameters are not identified in the model; their role is to rescale the existing group-level parameters:

	Old	New	
Treatment effects	γ_j	$\xi_\gamma(\gamma_j - \bar\gamma)$	for $j = 1, \ldots, J$
Airport effects	δ_k	$\xi_\delta(\delta_k - \bar\delta)$	for $k = 1, \ldots, K$
Treatment s.d.	σ_γ	$|\xi_\gamma|\,\sigma_\gamma$	
Airport s.d.	σ_δ	$|\xi_\delta|\,\sigma_\delta$	

Implementing all this in Bugs looks elaborate but can be built gradually as an expansion of the model on page 380:

Bugs code
```
model {
  for (i in 1:n){
    y[i] ~ dnorm (y.hat[i], tau.y)
    y.hat[i] <- mu + g[treatment[i]] + d[airport[i]]
  }
  mu ~ dnorm (0, .0001)
  tau.y <- pow(sigma.y, -2)
  sigma.y ~ dunif (0, 100)

  for (j in 1:n.treatment){
    g.raw[j] ~ dnorm (mu.g.raw, tau.g.raw)
    g[j] <- xi.g*(g.raw[j] - mean(g.raw[]))
  }
  xi.g ~ dunif (0, 100)
  mu.g.raw ~ dnorm (0, .0001)
  tau.g.raw <- pow(sigma.g.raw, -2)
```

```
    sigma.g.raw ~ dunif (0, 100)
    sigma.g <- xi.g*sigma.g.raw

    for (k in 1:n.airport){
      d.raw[k] ~ dnorm (mu.d.raw, tau.d.raw)
      d[k] <- xi.d*(d.raw[k] - mean(d.raw[]))
    }
    xi.d ~ dnorm (0, .0001)
    mu.d.raw ~ dnorm (0, .0001)
    tau.d.raw <- pow(sigma.d.raw, -2)
    sigma.d.raw ~ dunif (0, 100)
    sigma.d <- abs(xi.d)*sigma.d.raw
  }
```

Example: multilevel logistic regression for survey responses

We next apply multiplicative redundant parameters to a model with individual- and group-level predictors: the logistic regression for state-level opinions from national polls. In Bugs, we re-express in terms of "raw" parameters and extend the model on page 381 to include the new ξ's:

```
    for (j in 1:n.age){                                          Bugs code
      b.age[j] <- xi.age*(b.age.raw[j] - mean(b.age.raw[]))
      b.age.raw[j] ~ dnorm (0, tau.age.raw)
    }
    for (j in 1:n.edu){
      b.edu[j] <- xi.edu*(b.edu.raw[j] - mean(b.edu.raw[]))
      b.edu.raw[j] ~ dnorm (0, tau.edu.raw)
    }
    for (j in 1:n.age){
      for (k in 1:n.edu){
        b.age.edu[j,k] <- xi.age.edu*(b.age.edu.raw[j,k] -
          mean(b.age.edu.raw[,]))
        b.age.edu.raw[j,k] ~ dnorm (0, tau.age.edu.raw)
      }
    }
    for (j in 1:n.state){
      b.state[j] <- xi.state*(b.state.raw[j] - mean(b.state.raw[]))
      b.state.raw[j] ~ dnorm (b.state.raw.hat[j], tau.state.raw)
      b.state.raw.hat[j] <- b.region.raw[region[j]]+b.v.prev.raw*v.prev[j]
    }
    b.v.prev <- xi.state*b.v.prev
    b.v.prev.raw ~ dnorm (0, .0001)
    for (j in 1:n.region) {
      b.region[j] <- xi.state*b.region.raw[j]
      b.region.raw[j] ~ dnorm (0, tau.region.raw)
    }
    tau.age.raw <- pow(sigma.age.raw, -2)
    tau.edu.raw <- pow(sigma.edu.raw, -2)
    tau.age.edu.raw <- pow(sigma.age.edu.raw, -2)
    tau.state.raw <- pow(sigma.state.raw, -2)
    tau.region.raw <- pow(sigma.region.raw, -2)

    sigma.age.raw ~ dunif (0, 100)
    sigma.edu.raw ~ dunif (0, 100)
    sigma.age.edu.raw ~ dunif (0, 100)
```

```
     sigma.state.raw ~ dunif (0, 100)
     sigma.region.raw ~ dunif (0, 100)

     xi.age ~ dunif (0, 100)
     xi.edu ~ dunif (0, 100)
     xi.age.edu ~ dunif (0, 100)
     xi.state ~ dunif (0, 100)
     sigma.age <- xi.age*sigma.age.raw
     sigma.edu <- xi.edu*sigma.edu.raw
     sigma.age.edu <- xi.age.edu*sigma.age.edu.raw
     sigma.state <- xi.state*sigma.state.raw
     sigma.region <- xi.state*sigma.region.raw          # not "xi.region"
```

This model is awkwardly long, but we have attempted to use parallel structure to write it cleanly.

Application to item-response and ideal-point models

We illustrate a slightly different approach to redundant parameters with the item-response or ideal-point model with a discrimination parameter (see model (14.13) on page 316 and the accompanying discussion). Here, we resolve the additive and multiplicative nonidentifiability by *shifting* the ability and difficulty parameters using the mean of the α_j's and *scaling* the ability, difficulty, and discrimination parameters using the standard deviation of the α_j's. This is equivalent to restricting the abilities to have a mean of 0 and a standard deviation of 1, but the expression using redundant parameters allows for faster convergence of the Gibbs sampler. The Bugs model can be written as

Bugs code
```
model {
   for (i in 1:n){
     y[i] ~ dbin (p.bound[i], 1)
     p.bound[i] <- max(0, min(1, p[i]))
     logit(p[i]) <- g[k[i]]*(a[j[i]] - b[k[i]])
   }
   shift <- mean(a[])
   scale <- sd(a[])

   for (j in 1:J){
     a.raw[j] ~ dnorm (mu.a.raw, tau.a.raw)
     a[j] <- (a.raw[j] - shift)/scale
   }
   mu.a.raw ~ dnorm (0, .0001)
   tau.a.raw <- pow(sigma.a.raw, -2)
   sigma.a.raw ~ dunif (0, 100)

   for (k in 1:K){
     b.raw[k] ~ dnorm (b.hat.raw[j], tau.b.raw)
     b.hat.raw[j] <- b.0.raw + d.raw*x[j]
     b[k] <- (b.raw[k] - shift)/scale
     g.raw[k] ~ dnorm (mu.g.raw, tau.g.raw)
     g[k] <- g.raw[k]*scale
   }
   b.0.raw ~ dnorm (0, .0001)
   mu.g.raw ~ dnorm (0, .0001)
   tau.b.raw <- pow(sigma.b.raw, -2)
   tau.g.raw <- pow(sigma.g.raw, -2)
```

```
    sigma.b.raw ~ dunif (0, 100)
    sigma.g.raw ~ dunif (0, 100)

    d.raw ~ dnorm (0, .0001) I(0,)
    d <- d.raw*scale
}
```

The second-to-last line in this Bugs model constrains the group-level regression coefficient δ to be positive to resolve the reflection invariance in the model, as discussed in the context of model (14.14) on page 318.

The future of redundant parameterization

Bugs (as linked from R) is currently the best general-purpose tool for multilevel modeling, but in its flexibility it cannot do everything well, and so it requires various work-arounds to run effectively in many real problems. In the not-too-distant future, we expect that Bugs or its successor programs will automatically implement reparameterizations internally, so that the user can get the benefits in efficiency without the need to set up the model expansion explicitly.

An analogy could be made to least squares computations fifty years ago, when users had to program and, if necessary, perform steps such as rescaling and pivoting to obtain numerically stable results. Now the least squares routines in linpack, R, and other packages automatically do what is necessary to compute stable least squares estimates for just about any (noncollinear) data.

19.6 Using redundant parameters to create an informative prior distribution for multilevel variance parameters

The prior distribution is sometimes said to represent your knowledge about the parameters before ("prior to") seeing the data. In practice, however, the prior distribution, along with the rest of the model, is set up after seeing data, and so we prefer to think of it as representing any information outside of the data used in the likelihood.

Here we shall consider in depth the choice of prior distribution for group-level variance parameters in multilevel models. For simplicity, we work with a basic two-level normal model of data y_{ij} with group-level coefficients α_j:

$$
\begin{aligned}
y_{ij} &\sim \mathrm{N}(\mu + \alpha_j, \sigma_y^2), \quad i = 1, \ldots, n_j, \quad j = 1, \ldots, J \\
\alpha_j &\sim \mathrm{N}(0, \sigma_\alpha^2), \quad j = 1, \ldots, J.
\end{aligned}
\tag{19.2}
$$

We briefly discuss other hierarchical models at the end of this section.

Model (19.2) has three hyperparameters—μ, σ_y, and σ_α—but here we concern ourselves only with the last of these. Typically, enough data will be available to estimate μ and σ_y that one can use any reasonable noninformative prior distribution—for example, $p(\mu, \sigma_y) \propto 1$ or $p(\mu, \log \sigma_y) \propto 1$.

Various noninformative prior distributions have been suggested in Bayesian literature and software, including an improper uniform density on σ_α and proper distributions such as $\sigma_\alpha^2 \sim$ inverse-gamma$(0.001, 0.001)$. In this section, we explore and make recommendations for prior distributions for σ_α. We find that some purportedly noninformative prior distributions can unduly affect inferences, especially for problems where the number of groups J is small or the group-level standard deviation σ_α is close to zero.

Informative prior distributions

We can construct a rich family of prior distributions by applying a redundant multiplicative reparameterization to model (19.2):

$$y_{ij} \sim \mathrm{N}(\mu + \xi\eta_j, \sigma_y^2)$$
$$\eta_j \sim \mathrm{N}(0, \sigma_\eta^2). \tag{19.3}$$

The parameters α_j in (19.2) correspond to the products $\xi\eta_j$ in (19.3), and the hierarchical standard deviation σ_α in (19.2) corresponds to $|\xi|\sigma_\eta$ in (19.3). The parameters ξ, η, σ_η are not separately identifiable in this model. As discussed in Section 19.5, adding the redundant multiplicative parameter ξ can speed the convergence of the Gibbs sampler.

In addition, this expanded model form allows us to construct a family of prior distributions for the hierarchical variance parameter σ_α by separately assigning prior distributions to both ξ and σ_η, thus implicitly creating a model for $\sigma_\alpha = |\xi|\sigma_\eta$.

For simplicity we restrict ourselves to independent prior distributions on ξ and σ_η, with a normal distribution for ξ and inverse-gamma for σ_η^2. (These are technically known as *conditionally conjugate* prior distributions, for reasons that are not relevant to us here.)

The implicit conditionally conjugate family for σ_α is then the set of distributions corresponding to the absolute value of a normal random variable, divided by the square root of a gamma random variable. That is, σ_α has the distribution of the absolute value of a noncentral-t variate. We call this the *folded noncentral t distribution*, with the "folding" corresponding to the absolute value operator. The noncentral t in this context has three parameters, which can be identified with the mean of the normal distribution for ξ, and the scale and degrees of freedom for σ_η^2. (Without loss of generality, the scale of the normal distribution for ξ can be set to 1 since it cannot be separated from the scale for σ_η.)

The folded noncentral t distribution is not commonly used in statistics, and we find it convenient to understand it through various special and limiting cases. In the limit that the denominator is specified exactly, we have a folded normal distribution; conversely, specifying the numerator exactly yields the square-root-inverse-χ^2 distribution for σ_α.

An appealing two-parameter family of prior distributions is determined by restricting the prior mean of the numerator to zero, so that the folded noncentral t distribution for σ_α becomes simply a half-t—that is, the absolute value of a Student-t distribution centered at zero. We can parameterize this in terms of scale s_α and degrees of freedom ν:

$$p(\sigma_\alpha) \propto \left(1 + \frac{1}{\nu}\left(\frac{\sigma_\alpha}{s_\alpha}\right)^2\right)^{-(\nu+1)/2}$$

This family includes, as special cases, the improper uniform density (if $\nu = -1$) and the proper half-Cauchy, $p(\sigma_\alpha) \propto \left(\sigma_\alpha^2 + s_\alpha^2\right)^{-1}$ (if $\nu = 1$).

Noninformative prior distributions

Uniform prior distributions.   We first consider uniform prior distributions while recognizing that we must be explicit about the scale on which the distribution is defined. Various choices have been proposed for modeling variance parameters. A uniform prior distribution on $\log\sigma_\alpha$ would seem natural—working with the logarithm of a parameter that must be positive—but it results in an improper posterior distribution. An alternative would be to define the prior distribution on a compact set (for example, in the range $[-A, A]$ for some large value of A), but then the posterior distribution would depend strongly on $-A$, the lower bound of the prior support.

The problem arises because the marginal likelihood, $p(y|\sigma_\alpha)$—after integrating over α, μ, σ_y in (19.2)—approaches a finite nonzero value as $\sigma_\alpha \to 0$. Thus, if the prior density for $\log \sigma_\alpha$ is uniform, the posterior distribution will have infinite mass integrating to the limit $\log \sigma_\alpha \to -\infty$. To put it another way, in a hierarchical model the data can never rule out a group-level variance of zero, and so the prior distribution should not put an infinite mass in this area.

Another option is a uniform prior distribution on σ_α itself, which has a finite integral near $\sigma_\alpha = 0$ and thus avoids the above problem. We have generally used this noninformative density in our applied work, but it has a slightly disagreeable miscalibration toward positive values (see footnote 1 on page 433), with its infinite prior mass in the range $\sigma_\alpha \to \infty$. With $J = 1$ or 2 groups, this actually results in an improper posterior density, essentially concluding $\sigma_\alpha = \infty$ and doing no pooling. In a sense this is reasonable behavior, since it would seem difficult from the data alone to decide how much, if any, pooling should be done with data from only one or two groups. However, from a Bayesian perspective it is awkward for the decision to be made ahead of time, as it were, with the data having no say in the matter. In addition, for small J, such as 4 or 5, we worry that the heavy right tail of the posterior distribution would lead to overestimates of σ_α and thus result in pooling that is less than optimal for estimating the individual α_j's.

We can interpret the various improper uniform prior densities as limits of proper distributions. The uniform density on $\log \sigma_\alpha$ is equivalent to $p(\sigma_\alpha) \propto \sigma_\alpha^{-1}$ or $p(\sigma_\alpha^2) \propto \sigma_\alpha^{-2}$, which has the form of an inverse-χ^2 density with 0 degrees of freedom and can be taken as a limit of proper inverse-gamma densities.

The uniform density on σ_α is equivalent to $p(\sigma_\alpha^2) \propto \sigma_\alpha^{-1}$, an inverse-$\chi^2$ density with -1 degrees of freedom. This density cannot easily be seen as a limit of proper inverse-χ^2 densities (since these must have positive degrees of freedom), but it can be interpreted as a limit of the half-t family on σ_α, where the scale approaches ∞ (and any value of ν). Or, in the expanded notation of (19.3), one could assign any prior distribution to σ_η and a normal to ξ, and let the prior variance for ξ approach ∞.

Another noninformative prior distribution sometimes proposed in the Bayesian literature is uniform on σ_α^2. We do not recommend this, as it seems to have the miscalibration toward higher values as described above, but more so, and also requires $J \geq 4$ groups for a proper posterior distribution.

Inverse-gamma(ϵ, ϵ) *prior distributions.* The inverse-gamma(ϵ, ϵ) prior distribution is an attempt at noninformativeness within the conditionally conjugate family, with ϵ set to a low value such as 1 or 0.01 or 0.001 (the latter value being used in some of the examples in Bugs). A difficulty of this prior distribution is that in the limit of $\epsilon \to 0$ it yields an improper posterior density, and thus ϵ must be set to a reasonable value. Unfortunately, for datasets in which low values of σ_α are possible, inferences become very sensitive to ϵ in this model, and the prior distribution hardly looks noninformative, as we illustrate next. So we do not recommend the use of this inverse-gamma model as a noninformative prior distribution.

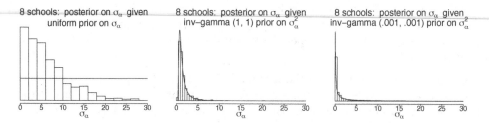

Figure 19.3 *Histograms of posterior simulations of the between-school standard deviation, σ_α, from models with three different prior distributions: (a) uniform prior distribution on σ_α, (b) inverse-gamma$(1,1)$ prior distribution on σ_α^2, (c) inverse-gamma$(0.001, 0.001)$ prior distribution on σ_α^2. Overlain on each is the corresponding prior density function for σ_α. (For models (b) and (c), the density for σ_α is calculated using the gamma density function multiplied by the Jacobian (see the footnote on page 409) of the $1/\sigma_\alpha^2$ transformation.) We prefer the uniform prior distribution shown in the left plot. In the center and right plots, posterior inferences are strongly constrained by the prior distribution. Adapted from Gelman et al. (2003, appendix C).*

Example: educational testing experiments in 8 schools

We demonstrate the properties of some proposed noninformative prior densities with a simple example of data from $J = 8$ educational testing experiments. Here, the parameters $\alpha_1, \ldots, \alpha_8$ represent the relative effects of Scholastic Aptitude Test coaching programs in 8 different schools, and σ_α represents the between-school standard deviations of these effects. The effects are measured as points on the test, which was scored from 200 to 800; thus the largest possible range of effects could be 600 points, with a realistic upper limit on σ_α of 100, say.

Here is the Bugs code for the model with half-Cauchy prior distribution:

Bugs code
```
model {
    for (j in 1:J){                        # J = the number of schools
        y[j] ~ dnorm (theta[j], tau.y[j])   # data model:  the likelihood
        theta[j] <- mu.theta + xi*eta[j]
        tau.y[j] <- pow(sigma.y[j], -2)
    }
    xi ~ dnorm (0, tau.xi)
    tau.xi <- pow(prior.scale, -2)
    for (j in 1:J){
        eta[j] ~ dnorm (0, tau.eta)          # hierarchical model for theta
    }
    tau.eta ~ dgamma (.5, .5)                # chi^2 with 1 d.f.
    sigma.theta <- abs(xi)/sqrt(tau.eta)     # cauchy = normal/sqrt(chi^2)
    mu.theta ~ dnorm (0, .0001)              # noninformative prior on mu
}
```

When running this model from R, we set `prior.scale <- 25`; give y, sigma.y, J, prior.scale as data for the bugs() call; and initialize the parameters eta, xi, mu.theta, tau.eta.

Figure 19.3 shows the posterior distributions for the 8-schools model resulting from three different choices of prior distributions that are intended to be noninformative.

The leftmost histogram of Figure 19.3 displays the posterior distribution for σ_α (as represented by 6000 simulation draws from a model fit using Bugs) for the model with uniform prior density. The data show support for a range of values below $\sigma_\alpha = 20$, with a slight tail after that, reflecting the possibility of larger values, which are difficult to rule out given that the number of groups J is only 8—that is, not much more than the $J = 3$ required to ensure a proper posterior density with finite mass in the right tail.

In contrast, the middle histogram in Figure 19.3 shows the result with an inverse-gamma$(1, 1)$ prior distribution for σ_α^2. This new prior distribution leads to changed inferences. In particular, the posterior mean and median of σ_α are lower and pooling of the α_j's is greater than in the previously fitted model with a uniform prior distribution on σ_α. To understand this, it helps to graph the prior distribution in the range for which the posterior distribution is substantial. The graph shows that the prior distribution is concentrated in the range $[0.5, 5]$, a narrow zone in which the likelihood is close to flat compared to this prior (as we can see because the distribution of the posterior simulations of σ_α closely matches the prior distribution, $p(\sigma_\alpha)$). By comparison, in the left graph, the uniform prior distribution on σ_α seems closer to "noninformative" for this problem, in the sense that it does not appear to be constraining the posterior inference.

Finally, the rightmost histogram in Figure 19.3 shows the corresponding result with an inverse-gamma$(0.001, 0.001)$ prior distribution for σ_α^2. This prior distribution is even more sharply peaked near zero and further distorts posterior inferences, with the problem arising because the marginal likelihood for σ_α remains high near zero.

In this example, we do not consider a uniform prior density on $\log \sigma_\alpha$, which would yield an improper posterior density with a spike at $\sigma_\alpha = 0$, like the rightmost graph in Figure 19.3, but more so. We also do not consider a uniform prior density on σ_α^2, which would yield a posterior distribution similar to the leftmost graph in Figure 19.3, but with a slightly higher right tail.

This example is a gratifying case in which the simplest approach—the uniform prior density on σ_α—seems to perform well. This model is also straightforward to program directly using the Gibbs sampler or in Bugs, using either the basic model (19.2) or using the expanded parameterization (19.3), which converges slightly faster.

The appearance of the histograms and density plots in Figure 19.3 is crucially affected by the choice to plot them on the scale of σ_α. If instead they were plotted on the scale of $\log \sigma_\alpha$, the inverse-gamma$(0.001, 0.001)$ prior density would appear to be the flattest. However, the inverse-gamma(ϵ, ϵ) prior is not at all "noninformative" for this problem since the resulting posterior distribution remains highly sensitive to the choice of ϵ. The hierarchical model likelihood does not constrain $\log \sigma_\alpha$ in the limit $\log \sigma_\alpha \to -\infty$, and so a prior distribution that is noninformative on the log scale will not work.

Weakly informative prior distribution for the 3-schools problem

The uniform prior distribution seems fine for the 8-schools analysis, but problems arise if the number of groups J is much smaller, in which case the data supply little information about the group-level variance, and a noninformative prior distribution can lead to a posterior distribution that is improper or is proper but unrealistically broad.

We demonstrate by reanalyzing the 8-schools example using the data from just 3 of the schools.

Figure 19.4 displays the inferences for σ_α from two different prior distributions. First we continue with the default uniform distribution that worked well with $J = 8$

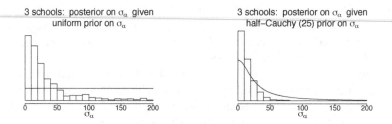

Figure 19.4 *Histograms of posterior simulations of the between-school standard deviation,* σ_α, *from models for the 3-schools data with two different prior distributions on* σ_α*: (a) uniform* $(0, \infty)$*, (b) half-Cauchy with scale 25, set as a weakly informative prior distribution given that* σ_α *was expected to be well below 100. The histograms are not on the same scales. Overlain on each histogram is the corresponding prior density function. With only* $J = 3$ *groups, the noninformative uniform prior distribution is too weak, and the proper Cauchy distribution works better, without appearing to distort inferences in the area of high likelihood.*

(as seen in Figure 19.3). Unfortunately, as the left histogram of Figure 19.4 shows, the resulting posterior distribution for the 3-schools dataset has an extremely long right tail, containing values of σ_α that are too high to be reasonable. This heavy tail is expected since J is so low (if J were any lower, the right tail would have an infinite integral). Using this posterior distribution for σ_α will have the effect of undershrinking the estimates of the school effects α_j. Better estimates should be obtained by including some information about σ_α that will restrict it to a more realistic range.

The right histogram of Figure 19.4 shows the posterior inference for σ_α resulting from a half-Cauchy prior distribution with scale parameter 25. As the line on the graph shows, this prior distribution is close to flat over the plausible range of $\sigma_\alpha <$ 50, falling off gradually beyond this point. We call this prior distribution "weakly informative" on this scale because, even at its tail, it has a gentle slope (unlike, for example, a half-normal distribution) and can let the data dominate if the likelihood is strong in that region. This prior distribution performs well in this example, reflecting the marginal likelihood for σ_α at its low end but removing much of the unrealistic upper tail.

This half-Cauchy prior distribution would also perform well in the 8-schools problem, but it was unnecessary because the default uniform prior gave reasonable results. With only 3 schools, we went to the trouble of using a weakly informative prior, a distribution that was not intended to represent our actual prior state of knowledge about σ_α but rather to constrain the posterior distribution, to an extent allowed by the data.

General comments

Prior distributions for variance parameters.   In fitting hierarchical models, we recommend starting with a noninformative uniform prior density on standard-deviation parameters σ_α. We expect this will generally work well unless the number of groups J is low (below 5, say). If J is low, the uniform prior density tends to lead to high estimates of σ_α, as discussed above. This miscalibration is an unavoidable consequence of the asymmetry in the parameter space, with variance parameters restricted to be positive. Similarly, there are no always-nonnegative classical unbiased estimators of σ_α or σ_α^2 in the hierarchical model.[1]

For a noninformative but proper prior distribution, we recommend approximating the uniform density on σ_α by a uniform on a wide range (for example, uniform from 0 to 100 in the SAT coaching example) or a half-normal centered at 0 with standard deviation set to a high value such as 100. The latter approach is particularly easy to program as a $N(0, 100^2)$ prior distribution for ξ in (19.3).

When more prior information is desired, for instance to restrict σ_α away from very large values, we recommend working within the half-t family of prior distributions, which are more flexible and have better behavior near 0, compared to the inverse-gamma family. A reasonable starting point is the half-Cauchy family, with scale set to a value that is high but not off the scale, for example, 25 in the SAT coaching example.

We do *not* recommend the inverse-gamma(ϵ, ϵ) family of noninformative prior distributions because, as discussed in Sections 3.3 and 4.1, in cases where σ_α is estimated to be near zero, the resulting inferences will be sensitive to ϵ. The setting of near-zero variance parameters is important partly because this is where uncertainty in the variance parameters is particularly relevant for multilevel inference.

Figure 19.3 illustrates the generally robust properties of the uniform prior density on σ_α. Many Bayesians have preferred the inverse-gamma prior family, possibly because its conditional conjugacy suggested clean mathematical properties. However, by writing the hierarchical model in the form (19.3), we see conditional conjugacy in the wider class of half-t distributions on σ_α, which include the uniform and half-Cauchy densities on σ_α (as well as inverse-gamma on σ_α^2) as special cases. From this perspective, the inverse-gamma family has nothing special to offer, and we prefer to work on the scale of the standard deviation parameter σ_α, which is typically directly interpretable in the original model.

Application to other models.   The reasoning in this paper should apply to multilevel models in general. The key idea is that parameters α_j—in general, group-level exchangeable parameters—have a common distribution with some scale parameter, which we label σ_α. In addition, when group-level regression predictors must be estimated, more than $J = 3$ groups may be necessary to estimate σ_α from a noninformative prior distribution, thus requiring at least weakly informative prior distributions for the regression coefficients, the variance parameters, or both.

---

[1] More formally, we can evaluate the inferences using the concept of *calibration* of the posterior mean, the Bayesian analogue to the classical notion of "bias." For any parameter θ, we label the posterior mean as $\hat{\theta} = E(\theta|y)$ and define the *miscalibration* of the posterior mean as $E(\theta|\hat{\theta}, y) - \hat{\theta}$, for any value of $\hat{\theta}$. If the prior distribution is true—that is, if the data are constructed by first drawing θ from $p(\theta)$, then drawing y from $p(y|\theta)$—then the posterior mean is automatically calibrated; that is its miscalibration is 0 for all values of $\hat{\theta}$.

For improper prior distributions, however, things are not so simple, since it is impossible for θ to be drawn from an unnormalized density. To evaluate calibration in this context, it is necessary to posit a "true prior distribution" from which θ is drawn along with the "inferential prior distribution" that is used in the Bayesian inference.

For the hierarchical model discussed here, we can consider the improper uniform density on σ_α as a limit of uniform prior densities on the range $(0, A)$, with $A \to \infty$. For any finite value of A, we can then see that the improper uniform density leads to inferences with a positive miscalibration—that is, overestimates (on average) of σ_α.

We demonstrate this miscalibration in two steps. First, suppose that both the true and inferential prior distributions for σ_α are uniform on $(0, A)$. Then the miscalibration is trivially zero. Now keep the true prior distribution at $U(0, A)$ and let the inferential prior distribution go to $U(0, \infty)$. This will necessarily increase $\hat{\theta}$ for any data y (since we are now averaging over values of θ in the range $[A, \infty)$) without changing the true θ, thus causing the average value of the miscalibration to become positive. Similar issues are discussed by Meng and Zaslavsky (2002).

Further work needs to be done in developing the next level of hierarchical models, in which there are several batches of exchangeable parameters, each with their own variance parameter; we present a simple sort of hierarchical model in Section 22.6 in the context of the analysis of variance.

19.7 Bibliographic note

Gelfand, Sahu, and Carlin (1995), Boscardin (1996), Roberts and Sahu (1997), Gelfand and Sahu (1999), and Sargent, Hodges, Carlin (2000) discuss additive transformations (also called *hierarchical centering*) and related ideas for speeding Gibbs sampler convergence. The properties of redundant multiplicative parameters have been studied by Liu, Rubin, and Wu (1998), Liu and Wu (1999), van Dyk and Meng (2001), and Gelman, Huang, et al. (2006). Gelman et al. (2003, appendix C) present R and Bugs implementations of redundant multiplicative parameterization for a simple multilevel model. Geweke (2004) and Cook, Gelman, and Rubin (2006) present some methods for using fake-data simulation to validate Bayesian software.

The treatment in Section 19.6 of prior distributions for multilevel variance parameters comes from Gelman (2006). The 8-schools example is discussed in detail in Gelman et al. (2003, chapter 5).

19.8 Exercises

1. Fake-data simulation: take an example of a multilevel model from an exercise in one of the previous chapters. Check the fitting using a fake-data simulation:

   (a) Specify the unmodeled parameters (that is, those with noninformative prior distributions in the Bugs model; see Figure 16.4 on page 366), then simulate the modeled parameters, then simulate a fake dataset.

   (b) Fit the model to the fake data and check numerically that the inferences for the unmodeled parameters are consistent with their "true values" you chose in part (a). Check graphically that the inferences for the modeled parameters are consistent with their "true values" you simulated in part (a).

2. Redundant parameterization: take a varying-intercept model from an exercise in one of the previous chapters.

   (a) Fit the model without redundant parameterization.

   (b) Include redundant additive parameters.

   (c) Include redundant additive and multiplicative parameters.

   (d) Check that the different forms of the model give the same inferences. Compare how long (in actual time, not just number of iterations) it takes for each of the simulations to reach approximate convergence.

3. Prior distributions for multilevel variance parameters: consider the varying-intercept radon model with floor of measurement as an individual-level predictor and log uranium as a county-level predictor. Data are in the folder `radon`.

   (a) Fit the model in Bugs using the different prior distributions for the group-level variance parameter discussed in Section 19.6. Compare the inferences for the group-level standard deviation and for a selection of the intercept parameters.

   (b) Repeat (a) but just analyzing the subset of the data corresponding to the first eight counties.

   (c) Repeat just using the data from the first three counties.

Part 3: From data collection to model understanding to model checking

We now go through the steps of understanding and working with multilevel regressions, including designing studies, summarizing inferences, checking the fit of models to data, and imputing missing data.

CHAPTER 20

Sample size and power calculations

20.1 Choices in the design of data collection

Multilevel modeling is typically motivated by features in existing data or the object of study—for example, voters classified by demography and geography, students in schools, multiple measurements on individuals, and so on. Consider all the examples in Part 2 of this book. In some settings, however, multilevel data structures arise by choice from the data collection process. We briefly discuss some of these options here.

Unit sampling or cluster sampling

In a sample survey, data are collected on a set of units in order to learn about a larger population. In unit sampling, the units are selected directly from the population. In cluster sampling, the population is divided into clusters: first a sample of clusters is selected, then data are collected from each of the sampled clusters.

In *one-stage* cluster sampling, complete information is collected within each sampled cluster. For example, a set of classrooms is selected at random from a larger population, and then all the students within each sampled classroom are interviewed. In *two-stage* cluster sampling, a sample is performed within each sampled cluster. For example, a set of classrooms is selected, and then a random sample of ten students within each classroom is selected and interviewed. More complicated sampling designs are possible along these lines, including adaptive designs, stratified cluster sampling, sampling with probability proportional to size, and various combinations and elaborations of these.

Observational studies or experiments with unit-level or group-level treatments

Treatments can be applied (or can be conceptualized as being applied in the case of a purely observational study) at individual or group levels; for example:

- In a medical study, different treatments might be applied to different patients, with patients clustered within hospitals that could be associated with varying intercepts or slopes.

- As discussed in Section 9.3, the Electric Company television show was viewed by classes, not individual students.

- As discussed in Section 11.2, child support enforcement policies are set by states and cities, not individuals.

- In the radon study described in Chapter 12, we can compare houses with and without basements within a county, but we can only study uranium as it varies between counties.

We present a longer list of such designs in the context of experiments in Section 22.4.

Typically, coefficients for factors measured at the individual level can be estimated more accurately than for group-level factors because there will be more individuals than groups; so $1/\sqrt{n}$ is more effective than $1/\sqrt{J}$ at reducing the standard error.

Meta-analysis

The sample size of a study can be increased in several ways:

- Gathering more data of the sort already in the study,

- Including more observations either in a nonclustered setting, as new observations in existing clusters, or new observations in new clusters

- Finding other studies performed under comparable (but not identical) conditions (so new observations in effect are like observations from a new "group").

- Finding other studies on related phenomena (again new observations from a different "group").

For example, in the study of teenage smoking in Section 11.3, these four options could be: (a) surveying more Australian adolescents about their smoking behavior, (b) taking more frequent measurements (for example, asking about smoking behavior every three months instead of every six months), (c) performing a similar survey in other cities or countries, or (d) performing similar studies of other unhealthy behaviors.

The first option is most straightforward—increasing n decreases standard errors in proportion to $1/\sqrt{n}$. The others involve various sorts of multilevel models and are made more effective by collecting appropriate predictors at the individual and group levels. (As discussed in Section 12.3, the more that the variation is explained by external predictors, the more effective the partial pooling will be.) A challenge of multilevel design is to assess the effectiveness of these various strategies for increasing sample size. Finding data from other studies is often more feasible than increasing n in an existing study, but then it is important to either find other studies that are similar, or to be able to model these differences.

Sample size, design, and interactions

Sample size is never large enough. As n increases, we estimate more interactions, which typically are smaller and have relatively larger standard errors than main effects (for example, see the fitted regression on page 63 of log earnings on sex, standardized height, and their interaction). Estimating interactions is similar to comparing coefficients estimated from subsets of the data (for example, the coefficient for height among men, compared to the coefficient among women), thus reducing power because the sample size for each subset is halved, and also the differences themselves may be small. As more data are included in an analysis, it becomes possible to estimate these interactions (or, using multilevel modeling, to include them and partially pool them as appropriate), so this is not a problem. We are just emphasizing that, just as you never have enough money, because perceived needs increase with resources, your inferential needs will increase with your sample size.

20.2 Classical power calculations: general principles, as illustrated by estimates of proportions

Questions of data collection can typically be expressed in terms of estimates and standard errors for quantities of interest. This chapter follows the usual focus on estimating population averages, proportions, and comparisons in sample surveys; or estimating treatment effects in experiments and observational studies. However, the general principles apply for other inferential goals such as prediction and data reduction. The paradigmatic problem of power calculation is the estimation of a parameter θ (for example, a regression coefficient such as would arise in estimating a difference or treatment effect), with the sample size determining the standard error.

Effect sizes and sample sizes

In designing a study to maximize the power of detecting a statistically significant comparison, it is generally better, if possible, to double the effect size θ than to double the sample size n, since standard errors of estimation decrease with the square root of the sample size. This is one reason, for example, why potential toxins are tested on animals at many times their exposure levels in humans; see Exercise 20.3.

Studies are designed in several ways to maximize effect size:

- In drug studies, setting doses as low as ethically possible in the control group and as high as ethically possible in the experimental group.

- To the extent possible, choosing individuals that are likely to respond strongly to the treatment. For example, the Electric Company experiment described in Section 9.3 was performed on poorly performing classes in each grade, for which it was felt there was more room for improvement.

In practice, this advice cannot be followed completely. In the social sciences, it can be difficult to find an intervention with *any* noticeable positive effect, let alone to design one where the effect would be doubled. Also, when treatments in an experiment are set to extreme values, generalizations to more realistic levels can be suspect; in addition, missing data in the control group may be more of a problem if the control treatment is ineffective. Further, treatment effects discovered on a sensitive subgroup may not generalize to the entire population. But, on the whole, conclusive effects on a subgroup are generally preferred to inconclusive but more generalizable results, and so conditions are usually set up to make effects as large as possible.

Power calculations

Before data are collected, it can be useful to estimate the precision of inferences that one expects to achieve with a given sample size, or to estimate the sample size required to attain a certain precision. This goal is typically set in one of two ways:

- Specifying the standard error of a parameter or quantity to be estimated, or

- Specifying the probability that a particular estimate will be "statistically significant," which typically is equivalent to ensuring that its confidence interval will exclude the null value.

In either case, the sample size calculation requires assumptions that typically cannot really be tested until the data have been collected. Sample size calculations are thus inherently hypothetical.

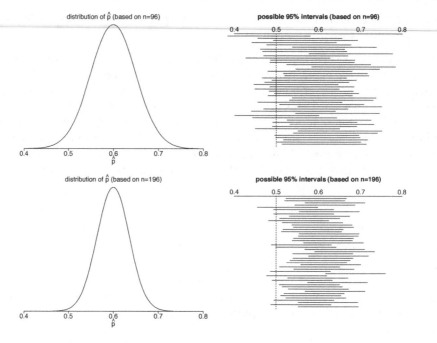

Figure 20.1 *Illustration of simple sample size calculations.*
Top row: (a) distribution of the sample proportion $\hat{p}$ if the true population proportion is $p = 0.6$, based on a sample size of 96; (b) several possible 95% intervals for p based on a sample size of 96. The power is 50%—that is, the probability is 50% that a randomly generated interval will be entirely to the right of the comparison point of 0.5.
Bottom row: corresponding graphs for a sample size of 196. Here the power is 80%.

Sample size to achieve a specified standard error

To understand these two kinds of calculations, consider the simple example of estimating the proportion of the population who support the death penalty (under a particular question wording). Suppose we suspect the population proportion is around 60%. First, consider the goal of estimating the true proportion p to an accuracy (that is, standard error) of no worse than 0.05, or 5 percentage points, from a simple random sample of size n. The standard error of the mean is $\sqrt{p(1-p)/n}$. Substituting the guessed value of 0.6 for p yields a standard error of $\sqrt{0.6 \cdot 0.4/n} = 0.49/\sqrt{n}$, and so we need $0.49/\sqrt{n} \leq 0.05$, or $n \geq 96$. More generally, we do not know p, so we would use a conservative standard error of $\sqrt{0.5 \cdot 0.5/n} = 0.5/\sqrt{n}$, so that $0.5/\sqrt{n} \leq 0.05$, or $n \geq 100$.

Sample size to achieve a specified probability of obtaining statistical significance

Second, suppose we have the goal of demonstrating that more than half the population supports the death penalty—that is, that $p > 1/2$—based on the estimate $\hat{p} = y/n$ from a sample of size n. As above, we shall evaluate this under the hypothesis that the true proportion is $p = 0.60$, using the conservative standard error for $\hat{p}$ of $\sqrt{0.5 \cdot 0.5/n} = 0.5/\sqrt{n}$. The 95% confidence interval for p is $[\hat{p} \pm 1.96 \cdot 0.5/\sqrt{n}]$, and classically we would say we have demonstrated that $p > 1/2$ if the interval lies entirely above $1/2$; that is, if $\hat{p} > 0.5 + 1.96 \cdot 0.5/\sqrt{n}$. The estimate must be at least 1.96 standard errors away from the comparison point of 0.5.

A simple, but not quite correct, calculation, would set $\hat{p}$ to the hypothesized value

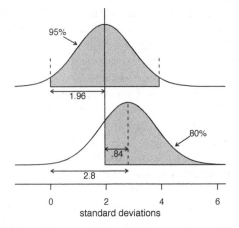

Figure 20.2 *Sketch illustrating that, to obtain 80% power for a 95% confidence interval, the true effect size must be at least 2.8 standard errors from zero (assuming a normal distribution for estimation error). The top curve shows that the estimate must be at least 1.96 standard errors from zero for the 95% interval to be entirely positive. The bottom curve shows the distribution of the parameter estimates that might occur, if the true effect size is 2.8. Under this assumption, there is an 80% probability that the estimate will exceed 1.96. The two curves together show that the lower curve must be centered all the way at 2.8 to get an 80% probability that the 95% interval will be entirely positive.*

of 0.6, so that the requirement is $0.6 > 0.5 + 1.96 \cdot 0.5/\sqrt{n}$, or $n > (1.96 \cdot 0.5/0.1)^2 = 96$. This is mistaken, however, because it confuses the assumption that $p = 0.6$ with the claim that $\hat{p} > 0.6$. In fact, if $p = 0.6$, then $\hat{p}$ depends on the sample, and it has an approximate normal distribution with mean 0.6 and standard deviation $\sqrt{0.6 \cdot 0.4/n} = 0.49/\sqrt{n}$; see Figure 20.1a.

To determine the appropriate sample size, we must specify the desired *power*— that is, the probability that a 95% interval will be entirely above the comparison point of 0.5. Under the assumption that $p = 0.6$, choosing $n = 96$ yields 50% power: there is a 50% chance that $\hat{p}$ will be more than 1.96 standard deviations away from 0.5, and thus a 50% chance that the 95% interval will be entirely greater than 0.5.

The conventional level of power in sample size calculations is 80%: we would like to choose n such that 80% of the possible 95% confidence intervals will not include 0.5. When n is increased, the estimate becomes closer (on average) to the true value, and the width of the confidence interval decreases. Both these effects (decreasing variability of the estimator and narrowing of the confidence interval) can be seen in going from the top half to the bottom half of Figure 20.1.

To find the value of n such that exactly 80% of the estimates will be at least 1.96 standard errors from 0.5, we need

$$0.5 + 1.96 \, \text{s.e.} = 0.6 - 0.84 \, \text{s.e.}$$

Some algebra then yields $(1.96 + 0.84) \, \text{s.e.} = 0.1$. We can then substitute s.e. $= 0.5/\sqrt{n}$ and solve for n.

2.8 standard errors from the comparison point

In summary, to have 80% power, the true value of the parameter must be 2.8 standard errors away from the comparison point: the value 2.8 is 1.96 from the 95% interval, plus 0.84 to reach the 80th percentile of the normal distribution. The bottom row of Figure 20.1 illustrates: with $n = (2.8 \cdot 0.49/0.1)^2 = 196$, and if the true population proportion is $p = 0.6$, there is an 80% chance that the 95% confidence interval will be entirely greater than 0.5, thus conclusively demonstrating that more than half the people support the death penalty.

These calculations are only as good as their assumptions; in particular, one would generally not know the true value of p before doing the study. Nonetheless, power analyses can be useful in giving a sense of the size of effects that one could reasonably expect to demonstrate with a study of given size. For example, a survey of size 196 has 80% power to demonstrate that $p > 0.5$ if the true value is 0.6, and it would easily detect the difference if the true value were 0.7; but if the true p were equal to 0.56, say, then the difference would be only $0.06/(0.5/\sqrt{196}) = 1.6$ standard errors away from zero, and it would be likely that the 95% interval for p would include $1/2$, even in the presence of this true effect. Thus, if the primary goal of the survey were to conclusively detect a difference from 0.5, it would probably not be wise to use a sample of only $n = 196$ unless we suspect the true p is at least 0.6. Such a small survey would "not have the power to" reliably detect differences of less than 0.1.

Estimates of hypothesized proportions

The standard error of a proportion p, if it is estimated from a sample of size n, is $\sqrt{p(1-p)/n}$, which has an upper bound of $0.5/\sqrt{n}$. This upper bound is very close to the actual standard error for a wide range of probabilities p near $1/2$: for example, for $\hat{p} = 0.5$, $\sqrt{0.5 \cdot 0.5} = 0.5$ exactly; for $\hat{p} = 0.6$ or 0.4, $\sqrt{0.6 \cdot 0.4} = 0.49$,; and for $\hat{p} = 0.7$ or 0.3, $\sqrt{0.7 \cdot 0.3} = 0.46$.

If the goal is a specified standard error, then a conservative required sample size is determined by s.e.$= 0.5/\sqrt{n}$, so that $n = (0.5/\text{s.e.})^2$ or, more precisely, $n = p(1-p)/(\text{s.e.})^2$, given a hypothesized p near 0 or 1.

If the goal is 80% power to distinguish p from a specified value p_0, then a conservative required sample size is $n = (2.8 \cdot 0.5/(p - p_0))^2$ or, more precisely, $n = p(1-p)(2.8/(p - p_0))^2$.

Simple comparisons of proportions: equal sample sizes

The standard error of a difference between two proportions is, by a simple probability calculation, $\sqrt{p_1(1-p_1)/n_1 + p_2(1-p_2)/n_2}$, which has an upper bound of $0.5\sqrt{1/n_1 + 1/n_2}$. If we make the restriction $n_1 = n_2 = n/2$ (equal sample sizes in the two groups), the upper bound on the standard error becomes simply $1/\sqrt{n}$. A specified standard error can then be attained with a sample size of $n = 1/(\text{s.e.})^2$.

If the goal is 80% power to distinguish between hypothesized proportions p_1 and p_2 with a study of size n, equally divided between the two groups, a conservative sample size is $n = [2.8/(p_1 - p_2)]^2$ or, more precisely, $n = 2[p_1(1-p_1) + p_2(1-p_2)] \cdot [2.8/(p_1 - p_2)]^2$.

For example, suppose we suspect that the death penalty is 10% more popular in the United States than in Canada, and we plan to conduct surveys in both countries on the topic. If the surveys are of equal sample size, $n/2$, how large must n be so that there is an 80% chance of achieving statistical significance, if the true difference in proportions is 10%? The standard error of $\hat{p}_1 - \hat{p}_2$ is approximately $1/\sqrt{n}$, so for 10% to be 2.8 standard errors from zero, we must have $n > (2.8/0.10)^2 = 784$, or a survey of 392 persons in each country.

Simple comparisons of proportions: unequal sample sizes

In observational epidemiology, it is common to have unequal sample sizes in comparison groups. For example, consider a study in which 20% of units are "cases" and 80% are "controls."

First, consider the goal of estimating the difference between the treatment and control groups, to some specified precision. The standard error of the difference is $\sqrt{p_1(1-p_1)/(0.2n) + p_2(1-p_2)/(0.8n)}$, and this expression has an upper bound of $0.5\sqrt{1/(0.2n) + 1/(0.8n)} = 0.5\sqrt{1/(0.2) + 1/(0.8)}/\sqrt{n} = 1.25/\sqrt{(n)}$. A specified standard error can then be attained with a sample size of $n = (1.25/\text{s.e.})^2$.

Second, suppose we want have sufficient total sample size n to achieve 80% power to detect a difference of 10%, again with 20% of the sample size in one group and 80% in the other. Again, the standard error of $\hat{p}_1 - \hat{p}_2$ is bounded by $1.25/\sqrt{n}$, so for 10% to be 2.8 standard errors from zero, we must have $n > (2.8 \cdot 1.25/0.10)^2 = 1225$, or 245 cases and 980 controls.

20.3 Classical power calculations for continuous outcomes

Sample size calculations proceed much the same way with continuous outcomes, with the added difficulty that the population standard deviation must also be specified along with the hypothesized effect size. We shall illustrate with a proposed experiment adding zinc to the diet of HIV-positive children in South Africa. In various other populations, zinc and other micronutrients have been found to reduce the occurrence of diarrhea, which is associated with immune system problems, as well as to slow the progress of HIV. We first consider the one-sample problem—how large a sample size would we expect to need to measure various outcomes to a specified precision—and then move to two-sample problems comparing treatment to control groups.

Estimates of means

Suppose we are trying to estimate a population mean value θ from data $y_1, \ldots, y_n$, a random sample of size n. The quick estimate of θ is the sample mean, $\bar{y}$, which has a standard error of $\sigma/\sqrt{n}$, where σ is the standard deviation of y in the population. So if the goal is to achieve a specified s.e. for $\bar{y}$, then the sample size must be at least $n = (\sigma/\text{s.e.})^2$.

If the goal is 80% power to distinguish θ from a specified value θ_0, then a conservative required sample size is $n = (2.8\sigma/(\theta - \theta_0))^2$.

Simple comparisons of means

The standard error of $\bar{y}_1 - \bar{y}_2$ is $\sqrt{\sigma_1^2/n_1 + \sigma_2^2/n_2}$. If we make the restriction $n_1 = n_2 = n/2$ (equal sample sizes in the two groups), the standard error becomes simply s.e. $= \sqrt{2(\sigma_1^2 + \sigma_2^2)}/\sqrt{n}$. A specified standard error can then be attained with a sample size of $n = 2(\sigma_1^2 + \sigma_2^2)/(\text{s.e.})^2$. If we further suppose that the variation is the same within each of the groups ($\sigma_1 = \sigma_2 = \sigma$), then s.e. $= 2\sigma/\sqrt{n}$, and the required sample size is $n = (2\sigma/\text{s.e.})^2$.

If the goal is 80% power to detect a difference of Δ, with a study of size n, equally divided between the two groups, then the required sample size is $n = 2(\sigma_1^2 + \sigma_2^2)(2.8/\Delta)^2$. If $\sigma_1 = \sigma_2 = \sigma$, this simplifies to $(5.6\sigma/\Delta)^2$.

For example, consider the effect of zinc supplements on young children's growth. Results of published studies suggest that zinc can improve growth by approximately

Rosado et al. (1997), Mexico	Treatment	Sample size	Avg. # episodes in a year ± s.e.
	placebo	56	1.1 ± 0.2
	iron	54	1.4 ± 0.2
	zinc	54	0.7 ± 0.1
	zinc + iron	55	0.8 ± 0.1

Ruel et al. (1997), Guatemala	Treatment	Sample size	Avg. # episodes per 100 days [95% c.i.]
	placebo	44	8.1 [5.8, 10.2]
	zinc	45	6.3 [4.2, 8.9]

Lira et al. (1998), Brazil	Treatment	Sample size	% days with diarrhea	Prevalence ratio [95% c.i.]
	placebo	66	5%	1
	1 mg zinc	68	5%	1.0 [0.72, 1.4]
	5 mg zinc	71	3%	0.68 [0.49, 0.95]

Muller et al. (2001), West Africa	Treatment	Sample size	# days with diarrhea/ total # days
	placebo	329	997/49021 = 0.020
	zinc	332	869/49086 = 0.018

Figure 20.3 *Results from various experiments studying the effects of zinc supplements on diarrhea in children. We use this information to hypothesize the effect size Δ and within-group standard deviation σ for our planned experiment.*

0.5 standard deviations. That is, $\Delta = 0.5\sigma$ in the our notation. To have 80% power to detect an effect size, it would be sufficient to have a total sample size of $n = (5.6/0.5)^2 = 126$, or $n/2 = 63$ in each group.

Estimating standard deviations using results from previous studies

Sample size calculations for continuous outcomes are based on estimated effect sizes and standard deviations in the population—that is, Δ and σ. Guesses for these parameters can be estimated or deduced from previous studies. We illustrate with the design of a study to estimate the effects of zinc on diarrhea in children. Various experiments have been performed on this topic—Figure 20.3 summarizes the results, which we shall use to get a sense of the sample size required for our study.

We consider the studies reported in Figure 20.3 in order. For Rosado et al. (1997), we shall estimate the effect of zinc by averaging over the iron and no-iron cases, thus an estimated Δ of $\frac{1}{2}(1.1 + 1.4) - \frac{1}{2}(0.7 + 0.8) = 0.5$ episodes in a year, with a standard error of $\sqrt{\frac{1}{4}(0.2^2 + 0.2^2) + \frac{1}{4}(0.1^2 + 0.1^2)} = 0.15$. From this study, it would be reasonable to hypothesize that zinc reduces diarrhea in that population by an average of about 0.3 to 0.7 episodes per year. Next, we can deduce the within-group standard deviations σ using the formula s.e.$= \sigma/\sqrt{n}$; thus the standard deviations are $0.2 \cdot \sqrt{56} = 1.5$ for the placebo group, and similarly for the other three groups are 1.5, 0.7, and 0.7, respectively. (Since the number of episodes is bounded below by zero, it makes sense that when the mean level goes down, the standard deviation decreases also.)

Assuming an effect size of $\Delta = 0.5$ episodes per year and within-group standard deviations of 1.5 and 0.7 for the control and treatment groups, we can evaluate the power of a future study with $n/2$ children in each group. The estimated difference would have a standard error of $\sqrt{1.5^2/(n/2) + 0.7^2/(n/2)} = 2.4/\sqrt{n}$, and so for the effect size to be at least 2.8 standard errors away from zero (and thus to have 80% power to attain statistical significance), n would have to be at least $(2.8 \cdot 2.4/0.5)^2 = 180$ persons in the two groups.

Now turning to the Ruel et al. (1997) study, we first see that rates of diarrhea—for control and treated children both—are much higher than in the previous study: 8 episodes per hundred days, which corresponds to 30 episodes per year, more than 20 times the rate in the earlier group. We are clearly dealing with much different populations here. In any case, we can divide the confidence interval widths by 4 to get standard errors—thus, 1.1 for the placebo group and 1.2 for the treated group—yielding an estimated treatment effect of 1.8 with standard error 1.6, which is consistent with a treatment effect of nearly zero or as high as about 4 episodes per 100 days. When compared to the average observed rate in the control group, the estimated treatment effect from this study is about half that of the Rosado et al. (1997) experiment: $1.8/8.1 = 0.22$, compared to $0.5/1.15 = 0.43$, which suggests a higher sample size might be required. However, the wide confidence bounds of the Ruel et al. (1997) study make it consistent with the larger effect size.

Next, Lira et al. (1998) report the average percent of days with diarrhea of children in the control and two treatment groups corresponding to a low (1 mg) or high (5 mg) dose of zinc. We shall consider only the 5 mg condition as this is closer to the treatment we are considering in our experiment. The estimated effect of the treatment is to multiply the number of days with diarrhea by 68%—that is, a reduction of 32%, which again is consistent with the approximate 40% decrease found in the first study. To make a power calculation, we first convert the confidence interval $[0.49, 0.95]$ for this multiplicative effect to the logarithmic scale—thus, an additive effect of $[-0.71, -0.05]$ on the logarithm—then divide by 4 to get an estimated standard error of 0.16 on this scale. The estimated effect of 0.68 is -0.38 on the log scale, thus 2.4 standard errors away from zero. For this effect size to be 2.8 standard errors from zero, we would need to increase the sample size by a factor of $(2.8/2.4)^2 = 1.4$, thus moving from approximately 70 children to approximately 100 in each of the two groups.

Finally, Muller et al. (2001) compare the proportion of days with diarrhea, which declined from 2.03% in the controls to 1.77% among children who received zinc. Unfortunately, no standard error is reported for this 13% decrease, and it is not possible to compute it from the information in the article. However, the estimates of within-group variation σ from the other studies would lead us to conclude that we would need a very large sample size to be likely to reach statistical significance, if the true effect size were only 10%. For example, from the Lira et al. (1998) study, we estimate a sample size of 100 in each group is needed to detect an effect of 32%; thus to detect a true effect of 13% we would need a sample size of $100(0.32/0.13)^2 = 600$.

These calculations are necessarily speculative; for example, to detect an effect of 10% (instead of 13%), the required sample size would be $100(0.32/0.10)^2 = 1000$ per group, a huge change considering the very small change in hypothesized treatment effects. Thus, it would be misleading to think of these as "required sample sizes." Rather, these calculations tell us how large the effects are that we could expect to have a good chance of discovering, given any specified sample size.

The first two studies in Figure 20.3 report the frequency of episodes, whereas the last two studies give the proportion of days with diarrhea, which is proportional to the frequency of episodes multiplied by the average duration of each episode. Other data (not shown here) show no effect of zinc on average duration, and so we treat all four studies as estimating the effects on frequency of episodes.

In conclusion, a sample size of about 100 per treatment group should give adequate power to detect an effect of zinc on diarrhea, if its true effect is to reduce the frequency, on average, by 30%–50% compared to no treatment. A sample size of 200 per group would have the same power to detect effects a factor $\sqrt{2}$ smaller, that is, effects in the 20%–35% range.

Including more regression predictors

Now suppose we are comparing treatment and control groups with additional pre-treatment data available on the children (for example, age, height, weight, and health status at the start of the experiment). These can be included in a regression. For simplicity, we consider a model with no interactions—that is, with coefficients for the treatment indicator and the other inputs—in which case, the treatment coefficient represents the causal effect, the comparison between the two groups after controlling for pre-treatment differences.

Sample size calculations for this new study are exactly as before, except that the within-group standard deviation σ is replaced by the residual standard deviation of the regression. This can be hypothesized in its own right or in terms of the added predictive power of the pre-treatment data. For example, if we hypothesize a within-group standard deviation of 0.2, then a residual standard deviation of 0.14 would imply that half the variance within any group is explained by the regression model, which would actually be pretty good.

Adding predictors tends to decrease the residual standard deviation and thus reduce the required sample size for any specified level of precision or power.

Estimation of regression coefficients more generally

More generally, sample sizes for regression coefficients and other estimands can be calculated using the rule that standard errors are proportional to $1/\sqrt{n}$; thus, if inferences exist under a current sample size, effect sizes can be estimated and standard errors extrapolated for other hypothetical samples.

We illustrate with the example of the survey earnings and height discussed in Chapter 4. The coefficient for the sex-earnings interaction in model (4.2) on page 63 is plausible (a positive interaction, implying that an extra inch of height is worth 0.7% more for men than for women), but it is not statistically significant—the standard error is 1.9%, yielding a 95% interval of $[-3.1, 4.5]$, which contains zero.

Simple sample size and power calculations.   How large a sample size would be needed for the coefficient on the interaction to be statistically significant? A simple calculation uses the fact that standard errors are proportional to $1/\sqrt{n}$. For a point estimate of 0.7% to achieve statistical significance, it would need a standard error of 0.35%, which would require the sample size to be increased by a factor of $(1.9\%/0.35\%)^2 = 29$. The original survey had a sample of 1192; this implies a required sample size of $29 \cdot 1192 = 35,000$.

To extend this to a power calculation, we suppose that the true β for the interaction is equal to 0.7% and that the standard error is as we have just calculated. With a standard error of 0.35%, the estimate from the regression would then be statistically significant only if $\hat{\beta} > 0.7\%$ (or, strictly speaking, if $\hat{\beta} < -0.7\%$, but that latter possibility is highly unlikely given our assumptions). If the true coefficient is β, we would expect the estimate from the regression to possibly take on values in the range $\beta \pm 0.35\%$ (that is what is meant by "a standard error of 0.35%"), and thus if β truly equals 0.7%, we would expect $\hat{\beta}$ to exceed 0.7%, and thus achieve statistical significance, with a probability of 1/2—that is, 50% power. To get 80% power, we need the true β to be 2.8 standard errors from zero, so that there is an 80% probability that $\hat{\beta}$ is at least 2 standard errors from zero. If $\beta = 0.7\%$, then its standard error would have to be no greater than $0.7\%/2.8 = 0.25\%$, so that the survey would need a sample size of $(1.9\%/0.25\%)^2 \cdot 1192 = 70{,}000$.

This power calculation is only provisional, however, because it makes the very strong assumption that the β is equal to 0.7%, the estimate that we happened to obtain from our survey. But the estimate from the regression is $0.7\% \pm 1.9\%$, which implies that these data are consistent with a low, zero, or even negative value of the true β (or, in the other direction, a true value that is greater than the point estimate of 0.7%). If the true β is actually less than 0.7%, then even a sample size of 70,000 will be insufficient for 80% power.

This is not to say the power calculation is useless but just to point out that, even when done correctly, it is based on an assumption that is inherently untestable from the available data (hence the need for a larger study). So we should not necessarily expect statistical significance from a proposed study, even if the sample size has been calculated correctly.

20.4 Multilevel power calculation for cluster sampling

With multilevel data structures and models, power calculations become more complicated because there is the option to set the sample size at each level. In a cluster sampling design, one can choose the number of clusters to sample and the number of units to sample within each cluster. In a longitudinal study, one can choose the number of persons to study and the frequency of measurement of each person. Options become even more involved for more complicated designs, such as those involving treatments at different levels. We illustrate here with examples of quick calculations for a survey and an experiment and then in Section 20.5 discuss a general approach for power calculations using simulations.

Standard deviation of the mean of clustered data

Consider a survey in which it is desired to estimate the average value of y in some population, and data are collected from J equally sized clusters selected at random from a larger population, with m units measured from each sampled cluster, so that the total sample size is $n = Jm$.[1] In this symmetric design, the estimate for the population total is simply the sample mean, $\bar{y}$. If the number of clusters in the population is large compared to J, and the number of units within each cluster is large compared to m, then the standard error of $\bar{y}$ is

$$\text{standard error of } \bar{y} = \sqrt{\sigma_y^2/n + \sigma_\alpha^2/J}. \qquad (20.1)$$

---

[1] In the usual notation for survey sampling, one might use a and A for the number of clusters in the sample and population, respectively. Here we use the capital letter J to indicate the number of selected clusters to be consistent with our general multilevel-modeling notation of J for the number of groups in the data.

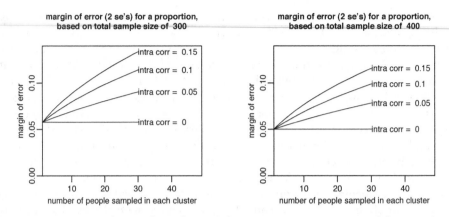

Figure 20.4 *Margin of error for inferences for a proportion as estimated from a cluster sample, as a function of cluster size and intraclass correlation, for two different proposed values of total sample size. The lines on the graphs do* not *represent a fitted model; they are based on analytical calculations using the variance formulas for cluster sampling.*

(The separate variance parameters σ_y^2 and σ_α^2, needed for the power calculations, can be estimated from the cluster-sampled data using a multilevel model.)

This formula can also be rewritten as

$$\text{standard error of } \bar{y} = \sqrt{\frac{\sigma_{\text{total}}^2}{Jm}[1 + (m-1)\text{ICC}]}, \tag{20.2}$$

where σ_{total} represents the standard deviation of all the data (mixing all the groups; thus $\sigma_{\text{total}}^2 = \sigma_y^2 + \sigma_\alpha^2$ for this simple model), and ICC is the *intraclass correlation*,

$$\text{intraclass correlation: } \text{ICC} = \frac{\sigma_\alpha^2}{\sigma_\alpha^2 + \sigma_y^2}, \tag{20.3}$$

the fraction of total variation in the data that is accounted for by between-group variation. The intraclass correlation can also be thought of as the correlation among units within the same group. Formulas (20.1) and (20.2) provide some intuition regarding the extent to which clustering can affect our standard errors. The greater the correlation among units within a group (that is, the bigger ICC is) the greater the impact on the standard error. If there is no intraclass correlation (that is, ICC = 0) the standard error of $\bar{y}$ is simply $\sigma_{\text{total}}/\sqrt{n}$.

Example of a sample size calculation for cluster sampling

We illustrate sample size calculations for cluster sampling with a design for a proposed study of residents of New York City. The investigators were planning to study approximately 300 or 400 persons sampled for convenience from 10 or 20 U.S. Census tracts, and they wanted to get a sense of how much error the clustering was introducing into the estimation. The number of census tracts in the city and the population of each tract are large enough that (20.1) was a reasonable approximation.

Figure 20.4 shows the margin of error for $\bar{y}$ from this formula, as a function of the sample size within clusters, for several values of the intraclass correlation. When the correlation is zero, the clustering is irrelevant and the margin of error only depends on the total sample size, n. For positive values of intraclass correlation (so that people within a census tract are somewhat similar to each other, on average), the standard error increases as the number of clusters decreases with fixed sample size. For the higher values of intraclass correlation shown in the graphs, it seems that it would be best to choose enough clusters so that no more than 20 persons are selected within each cluster.

But why, in Figure 20.4, do we think that interclass correlations between 0 and 15% are plausible? To start with, for binary data, the denominator of (20.3) can be reasonably approximated by 0.25 (since $p(1 - p) \approx 0.25$ if p is not too close to 0 or 1). Now suppose that the clusters themselves differ in some particular average outcome with a standard error of 0.2—this is a large value of σ_α, with, for example, the percentages of Yes responses in some clusters as low as 0.3 and in others as high as 0.7. The resulting intraclass correlation is $0.2^2/0.25 = 0.16$. If, instead, $\sigma_\alpha = 0.1$ (so that, for example, the average percentage of Yes in clusters varies from approximately 0.4 to 0.6), the intraclass correlation is 0.04. Thus, it seems reasonable to consider correlations ranging from 0 to 5% to 15% as in Figure 20.4.

20.5 Multilevel power calculation using fake-data simulation

Figure 20.5a shows measurements of the immune system (CD4 percentage, transformed to the square root scale to better fit an additive model) taken over a two-year period on a set of HIV-positive children who were not given zinc. The observed noisy time series can be fitted reasonably well by a varying-intercept, varying-slope model of the form, $y_{jt} \sim N(\alpha_j + \beta_j t, \sigma_y^2)$, where j indexes children, t indexes time, and the data variance represents a combination of measurement errors, short-term variation in CD4 levels, and departures from a linear trend within each child. This model can also be written more generally as $y_i \sim N(\alpha_{j[i]} + \beta_{j[i]} t_i, \sigma_y^2)$, where i indexes measurements taken at time t_i on person $j[i]$. Here is the result of the quick model fit:

```
lmer(formula = y ~ time + (1 + time | person))                         R output
             coef.est coef.se
(Intercept)   4.8      0.2
time         -0.5      0.1
Error terms:
 Groups    Name        Std.Dev. Corr
 person    (Intercept) 1.3
           time        0.7      0.1
 Residual              0.7
# of obs: 369, groups: person, 83
```

Of most interest are the time trends β_j, whose average is estimated at -0.5 with a standard deviation of 0.7 (we thus estimate that most, but not all, of the children have declining CD4 levels during this period). The above display also gives us estimates for the intercepts and the residual standard deviation.

We then fit the model in Bugs to get random simulations of all the parameters. The last three panels of Figure 20.5 show the results: the estimated trend line for each child, a random draw of the set of 83 trend lines, and a random replicated dataset (following the principles of Section 8.3) with measurements at the time points observed for the actual data. The replicated dataset looks generally like the actual data, suggesting that the linear-trend-plus-error model is a reasonable fit.

Modeling a hypothetical treatment effect

We shall use these results to perform a power calculation for a proposed new study of dietary zinc. We would like the study to be large enough that the probability is

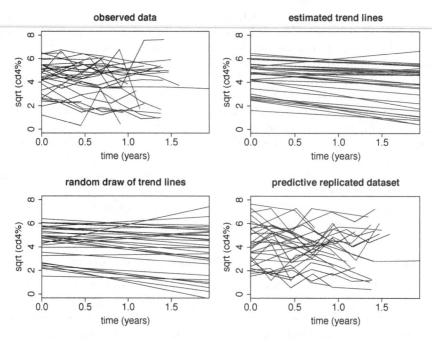

Figure 20.5 *(a) Progression of CD4 percentage over time (on the square root scale) for 83 untreated children j in the HIV study; (b) individual trend lines $\hat{\alpha}_j + \hat{\beta}_j t$ (posterior mean estimates from multilevel model); (c) a single posterior draw from the set of individual trend lines $\alpha_j + \beta_j t$; (d) a replicated dataset $(\tilde{y}_{jt})$ simulated from the posterior predictive distribution.*

at least 80% that the average estimated treatment effect is statistically significant at the 95% level.

A hypothesized model of treatment effects. To set up this power calculation we need to make assumptions about the true treatment effect and also specify all the other parameters that characterize the study. Our analysis of the HIV-positive children who did not receive zinc found an average decline in CD4 (on the square root scale) of 0.5 per year. We shall suppose in our power calculation that the true effect of the treatment is to reduce this average decline to zero.

We now set up a model for the hypothetical treatment and control data. So far, we have fitted a model to "controls," but that model can be used to motivate hypotheses for effects of treatments applied after the initial measurement ($t = 0$). To start with, the parameters α_j, β_j cleanly separate into an intercept that is unaffected by the treatment (and can thus be interpreted as an unobserved unit-level characteristic) and a slope β_j that is potentially affected. A model of linear trends can then be written as

$$y_i \sim \text{N}(\alpha_{j[i]} + \beta_{j[i]} t_i, \sigma_y^2), \text{ for } i = 1, \ldots, n$$

$$\begin{pmatrix} \alpha_j \\ \beta_j \end{pmatrix} \sim \text{N}\left(\begin{pmatrix} \gamma_0^\alpha \\ \gamma_0^\beta + \gamma_1^\beta z_j \end{pmatrix}, \begin{pmatrix} \sigma_\alpha^2 & \rho\sigma_\alpha\sigma_\beta \\ \rho\sigma_\alpha\sigma_\beta & \sigma_\beta^2 \end{pmatrix} \right), \text{ for } j = 1, \ldots, J,$$

where

$$z_j = \begin{cases} 1 & \text{if child } i \text{ received the treatment} \\ 0 & \text{otherwise.} \end{cases}$$

The treatment z_j affects the slope β_j but not the intercept α_j because the treatment can have no effect at time zero. As noted, we shall suppose γ_0^β, the slope for controls, to be -0.5, with a treatment effect of $\gamma_1^\beta = 0.5$. We complete the model by setting the other parameters to their estimated values from the control data: $\mu_\alpha = 4.8$, $\sigma_\alpha = 1.3$, $\sigma_y = 0.7$, $\sigma_\beta = 0.7$. For simplicity, we shall set ρ, the correlation between intercepts and slopes, to zero, although it was estimated at 0.1 from the actual data.

Design of the study. The next step in the power analysis is to specify the design of the study. We shall assume that J HIV-positive children will be randomly assigned into two treatments, with $J/2$ receiving regular care and $J/2$ receiving zinc supplements as well. We further assume that the children's CD4 percentages are measured every two months over a year (that is, seven measurements per child). We will now determine the J required for 80% power, if the true treatment effect is 0.5, as assumed above.

Quick power calculation for classical regression

We first consider a classical analysis, in which a separate linear regression is fitted for each child: $y_{jt} = \alpha_j + \beta_j t + \text{error}$. The trend estimates $\hat{\beta}_j$ would then be averaged for the children in the control and treatment groups, with the difference between the group mean trends being an estimated treatment effect. For simplicity, we assume the model is fitted separately for each child—that is, simple least squares, not a multilevel model.

This problem then has the structure of a simple classical sample size calculation, with the least squares estimate $\hat{\beta}_j$ being the single "data point" for each child j and an assumed effect size $\Delta = 0.5$. We must merely estimate σ, the standard deviation of the $\hat{\beta}_j$'s within each group, and we can determine the required total sample size as $J = (2 \cdot 2.8\sigma/\Delta)^2$.

If $\hat{\beta}_j$ were a perfect estimate of the child's trend parameter, then σ would simply be the standard deviation of the β_j's, or 0.7 from the assumptions we have made. However, we must also add the variance of estimation, which in this case (from the formula for least squares estimation with a single predictor) is $\frac{1}{\sqrt{(-3/6)^2 + (-2/6)^2 + \cdots + (3/6)^2}}\sigma_y = 1.13\sigma_y = 0.8$ (based on the estimate of $\sigma_y = 0.7$ from our multilevel model earlier). The total standard deviation of $\hat{\beta}_j$ is then $\sqrt{\sigma_\beta^2 + 1.13^2\sigma_y^2} = \sqrt{0.7^2 + 0.8^2} = 1.1$. The sample size required for 80% power to find a statistically significant difference in trends between the two groups is then $J = (2 \cdot 2.8 \cdot 1.1/0.5)^2 = 150$ children total (that is, 75 per group).

This sample size calculation is based on the assumption that the treatment would, on average, eliminate the observed decline in CD4 percentage. If instead we were to hypothesize that the treatment would cut the decline in half, the required sample size would quadruple, to a total of 600 children.

Power calculation for multilevel estimate using fake-data simulation

Power calculations for any model can be performed by simulation. This involves repeatedly simulating data from the hypothetical distribution that we expect our sampled data to come from (once we perform the intended study) and then fitting a multilevel model to each dataset. This can be computer-intensive, and practical compromises are sometimes needed so that the simulation can be performed in a reasonable time. Full simulation using Bugs is slow because it involves nested loops (100 or 1000 sets of fake data; for each, the looping of a Gibbs sampler required to

fit a model in Bugs). Instead, we fit the model to each fake dataset quickly using
lmer(). We illustrate with the zinc treatment example.

Simulating the hypothetical data. The first step is to write a function in R that
will generate data from the distribution assumed for the control children (based on
our empirical evidence) and the distribution for the treated children (based on our
assumptions about how their change in CD4 count might be different were they
treated). This function generates data from a sample of J children (half treated,
half controls), each measured K times during a 1-year period.

R code
```
CD4.fake <- function (J, K){
  time <- rep (seq(0,1,length=K), J)      # K measurements during the year
  person <- rep (1:J, each=K)             # person ID's
  treatment <- sample (rep (0:1, J/2))
  treatment1 <- treatment[person]
#                                          # hyperparameters:
  mu.a.true <- 4.8                        #   more generally, these could
  g.0.true <-  -.5                        #   be specified as additional
  g.1.true <-   .5                        #   arguments to the function
  sigma.y.true <-  .7
  sigma.a.true <- 1.3
  sigma.b.true <-  .7
#                                          # person-level parameters
  a.true <- rnorm (J, mu.a.true, sigma.a.true)
  b.true <- rnorm (J, g.0.true + g.1.true*treatment, sigma.b.true)
#                                          # data
  y <- rnorm (J*K, a.true[person] + b.true[person]*time, sigma.y.true)
  return (data.frame (y, time, person, treatment1))
}
```

The function returns a data frame with the simulated measurements along with
the input variables needed to fit a model to the data and estimate the average
treatment effect, γ_1. We save treatment as a data-level predictor (which we call
treatment1) because this is how it must be entered into lmer().

Fitting the model and checking the power. Next we can embed the fake-data simulation CD4.fake() in a loop to simulate 1000 sets of fake data; for each, we fit the
model and obtain confidence intervals for the parameter of interest:

R code
```
CD4.power <- function (J, K, n.sims=1000){
  signif <- rep (NA, n.sims)
  for (s in 1:n.sims){
    fake <- CD4.fake (J, K)
    lme.power <- lmer (y ~ time + time:treatment1 +
      (1 + time | person), data=fake)
    theta.hat <- fixef(lme.power)["time:treatment1"]
    theta.se <- se.fixef(lme.power)["time:treatment1"]
    signif[s] <- (theta.hat - 2*theta.se) > 0      # returns TRUE or FALSE
  }
  power <- mean (signif)                            # proportion of TRUE
  return (power)
}
```

This function has several features that might need explaining:

- The function definition sets the number of simulations to the default value of
  1000. So if CD4.power() is called without specifying the n.sims argument, it
  will automatically run 1000 simulations.

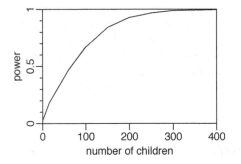

Figure 20.6 *Power (that is, the probability that estimated treatment effect is statistically significantly positive) as a function of number of children, J, for the hypothetical zinc study, as computed using fake-data simulation with multilevel inference performed by* `lmer()`. *The simulations are based on particular assumptions about the treatment effect and the variation among children and among measurements within children. We also have assumed K = 7 measurements for each child during the year of the study, a constraint determined by the practicalities of the experiment. Reading off the curve, 80% power is achieved at approximately J = 130.*

- The `lmer()` call includes the interaction `time:treatment1` and the main effect `time` but *not* the main effect `treatment1`. This allows the treatment to affect the slope but not the intercept, which is appropriate since the treatment is performed after time 0.

- The data frame `fake` is specified as an argument to `lmer()` so that the analysis knows what dataset to use.

- We assume the estimated treatment effect of the hypothetical study is statistically significantly positive if the lower bound of its 95% interval exceeds zero.

- The function returns the proportion of the 1000 simulations where the result is statistically significant; thus, the power (as computed via simulation) for a study with J children measured at K equally spaced times during the year.

Putting it all together to compute power as a function of sample size. Finally, we put the above simulation in a loop and compute the power at several different values of J, running from 20 to 400, and plot a curve displaying power as a function of sample size; the result is shown in Figure 20.6. Our quick estimate based on classical regression was that 80% power is achieved with $J = 150$ children (75 in each treatment group) also applies to the multilevel model in this case. The classical computation works in this case because the treatment is at the group level (in this example, persons are the groups, and CD4 measurements are the units) and the planned study is balanced.

At the two extremes:

- The power is 0.025 in the limit $J \to 0$. With a small enough sample, the treatment effect estimate is essentially random, and so there is a 2.5% chance that it is more than 2 standard errors above zero.

- Under the assumption that the true effect is positive, the power is 1 in the limit $J \to \infty$, at which point there are enough data to estimate the treatment effect perfectly.

Using simulation for power analyses allows for greater flexibility in study design. For instance, besides simply calculating how power changes as sample size increases, we might also have investigated a different kind of change in study design such as changes in the percentage of study participants allocated to treatment versus control groups. This aspect of study design might be particularly relevant if treatment participants are more costly than control participants, for instance (see Exercise 20.6). Another design feature that could be varied is the number of measurements per person, and the simulation can also include missing data, nonlinearity, unequal variance, and other generalizations of the model.

20.6 Bibliographic note

Scott and Smith (1969), Cochran (1977), Goldstein and Silver (1989), and Lohr (1999) are standard and useful references for models used in survey sampling, and Groves et al. (2004) goes over the practical aspects of survey design. Montgomery (1986) and Box, Hunter, and Hunter (2005) review the statistical aspects of experimental design; Trochim (2001) is a more introductory treatment with useful practical advice on research methods.

Hoenig and Heisey (2001) and Lenth (2001) provide some general warnings and advice on sample size and power calculations. Design issues and power calculations for multilevel studies are discussed by Snijders and Bosker (1993), Raudenbush (1997), Snijders, Bosker, and Guldemond (1999), Raudenbush and Xiaofeng (2000), and Raudenbush and Bryk (2002).

20.7 Exercises

1. Sample size calculations for estimating proportions:

   (a) How large a sample survey would be required to estimate, to within a standard error of $\pm 3\%$, the proportion of the U.S. population who support the death penalty?

   (b) About 14% of the U.S. population is Latino. How large would a national sample of Americans have to be in order to estimate, to within a standard error of $\pm 3\%$, the proportion of Latinos in U.S. who support the death penalty?

   (c) How large would a national sample of Americans have to be in order to estimate, to within a standard error of $\pm 1\%$, the proportion who are Latino?

2. Consider an election with two major candidates, A and B, and a minor candidate, C, who are believed to have support of approximately 45%, 35%, and 20% in the population. A poll is to be conducted with the goal of estimating the difference in support between candidates A and B. How large a sample would you estimate is needed to estimate this difference to within a standard error of 5%? (Hint: consider an outcome variable that is coded as $+1$, -1, and 0 for supporters of A, B, and C, respectively.)

3. Effect size and sample size: consider a toxin that can be tested on animals at different doses. Suppose a typical exposure level for humans is 1 (in some units), and at this level the toxin is hypothesized to introduce a risk of 0.01% of death per person.

   (a) Consider different animal studies, each time assuming a linearity in the dose-response relation (that is, 0.01% risk of death per animal per unit of the toxin), with doses of 1, 100, and 10,000. At each of these exposure levels, what sample size is needed to have 80% power of detecting the effect?

   (b) This time assume that response is a logged function of dose and redo the calculations in (a).

4. Cluster sampling with equal-sized clusters: a survey is being planned with the goal of interviewing n people in some number J of clusters. For simplicity, assume simple random sampling of clusters and a simple random sample of size n/J (appropriately rounded) within each sampled cluster.

   Consider inferences for the proportion of Yes responses in the population for some question of interest. The estimate will be simply the average response for the n people in the sample. Suppose that the true proportion of Yes responses is not too far from 0.5 and that the standard deviation among the mean responses of clusters is 0.1.

   (a) Suppose the total sample size is $n = 1000$. What is the standard error for the sample average if $J = 1000$? What if $J = 100, 10, 1$?

   (b) Suppose the cost of the survey is $50 per interview, plus $500 per cluster. Further suppose that the goal is to estimate the proportion of Yes responses in the population with a standard error of no more than 2%. What values of n and J will achieve this at the lowest cost?

5. Simulation for power analysis: the folder `electric.company` contains data from the Electric Company experiment analyzed in Chapter 9. Suppose you wanted to perform a new experiment under similar conditions, but for simplicity just for second-graders, with the goal of having 80% power to find a statistically significant result (at the 95% level) in grade 2.

   (a) State clearly the assumptions you are making for your power calculations. (Hint: you can set the numerical values for these assumptions based on the analysis of the existing Electric Company data.)

   (b) Suppose that the new data will be analyzed by simply comparing the average scores for the treated classrooms to the average scores for the controls. How many classrooms would be needed for 80% power?

   (c) Repeat (b), but supposing that the new data will be analyzed by comparing the average gain scores for the treated classrooms to the average gain scores of the controls.

   (d) Repeat (b), but supposing that the new data will be analyzed by regression, controlling for pre-test scores as well as the treatment indicator.

6. Optimal design:

   (a) Suppose that the zinc study described in Section 20.5 would cost $150 for each treated child and $100 for each control. Under the assumptions given in that section, determine the number of control and treated children needed to attain 80% power at minimal total cost. You will need to set up a loop of simulations as illustrated for the example in the text. Assume that the number of measurements per child is fixed at $K = 7$ (that is, measuring every two months for a year).

   (b) Make a generalization of Figure 20.6 with several lines corresponding to different values of the design parameter K, the number of measurements for each child.

## CHAPTER 21

Understanding and summarizing the fitted models

Now that we can fit multilevel models, we should consider how to understand and summarize the parameters (and important transformations of these parameters) thus estimated.

Inferences from classical regression are typically summarized by a table of coefficient estimates and standard errors, sometimes with additional information on residuals and statistical significance (see, for example, the R output on page 39). With multilevel models, however, the sheer number of parameters adds a challenge to interpretation. The coefficient list in a multilevel model can be arbitrarily long (for example, the radon analysis has 85 county-level coefficients for the varying-intercept model, or 170 coefficients if the slope is allowed to vary also), and it is unrealistic to expect even the person who fit the model to be able to interpret each number separately. We prefer graphical displays such as the generic `plot` of a Bugs object or plots of fitted multilevel models such as displayed in the examples in Part 2A of this book.

Our general plan is to follow the same structures when plotting as when modeling. Thus, we plot data with data-level regressions (as in Figure 12.5 on page 266), and estimated group coefficients with group-level regressions (as in Figure 12.6). More complicated plots can be appropriate for non-nested models (for example, Figure 13.10 on page 291 and Figure 13.12 on page 293). More conventional plots of parameter estimates and standard errors (such as Figure 14.1 on page 306) can be helpful in multilevel models too. It is also sometimes feasible to display between-group and within-group information on the same graph (see Figure 14.11 on page 313). Finally, specific models can inspire new ideas for graphs, as in Figure 15.4 on page 330.

21.1 Uncertainty and variability

Uncertainty reflects lack of complete knowledge about a parameter; *variability* refers to underlying differences among individuals or groups. As sample size goes to infinity, uncertainty goes to zero (more precisely, in a well-designed study, uncertainty about key parameters goes to zero), but variability will always be present.

Distinguishing between uncertainty and variability in a multilevel model

For example, consider the varying-intercept radon model, $y_i \sim N(\alpha_{j[i]} + \beta x_i, \sigma_y^2)$. Here is a subset of the Bugs inferences for the parameters (more details appear on page 351):

	mean	sd
a[1]	1.2	0.3
a[2]	0.9	0.1

. . .

R output

```
a[85]         1.4  0.3
b            -0.7  0.1
mu.a          1.5  0.1
sigma.y       0.8  0.05
sigma.a       0.3  0.05
```

The numbers in the last column of this table show the *uncertainty* about each parameter. For example, we estimate the intercept for county 1 to be 1.2 with an uncertainty of 0.3. The parameters σ_y and σ_α represent *variability* among houses within a county and variability among counties. Thus, the unexplained variation of radon levels of houses within a county (after controlling for floor of measurement), is estimated at 0.8, and the unexplained variation among county-average radon levels is estimated at 0.3.

As we get more data within *existing* counties, uncertainty about individual α_j's would be expected to decline. If data from *more* counties were added, uncertainty about the group-level parameters μ_α and σ_α would be expected to decline. But under either (or both) of these scenarios, there is no reason to expect the estimates of σ_y and σ_α to change.

A varying-intercept, varying-slope example

For another example, consider the model from page 449 of variation in time trends of CD4 counts in a sample of children. The model is $y_i \sim N(\alpha_{j[i]} + \beta_{j[i]}t_i, \sigma_y^2)$. The lmer display includes both uncertainty and variation, and they can easily be confused:

R output
```
lmer(formula = y ~ time + (1 + time | person))
                coef.est coef.se
(Intercept)  4.8         0.2
time        -0.5         0.1
Error terms:
 Groups    Name        Std.Dev. Corr
 person    (Intercept) 1.3
           time        0.7      0.1
 Residual              0.7
# of obs: 369, groups: person, 83
```

There is no confusion about the coefficient estimates: 4.8 and -0.5 are the estimates of the average α_j, β_j in the population, so the "average person," in some sense, has a time trend of $4.8 - 0.5t$. But care must be taken in interpreting standard errors and standard deviations. We will focus here on the time trend (the slope in the model) because it is of more substantive interest. The *standard error* of the time coefficient is 0.1, which tells us the precision of the "-0.5" estimate: we are essentially certain that the population-average slope is negative, that is, that CD4 counts are on average declining in this population. The estimated *standard deviation* of the slopes is 0.7, which implies that the individual slopes vary greatly, with some being very negative and some being moderately positive. In this example, both the uncertainty in the mean and the variation in the population are important.

Variability can be interesting in itself

In some cases, the magnitude of variation can be substantively relevant. For example, Kane et al. (2006) estimate the "value added" by teachers in elementary and middle schools in New York City. They fit regression models of students' test scores, given previous test scores, background information on students, classes, and schools, and varying intercepts for teachers.

The models were actually fit with classical regression using an indicator variable for each teacher, and then the estimates were post-processed to distinguish inferential variability and consistent teacher effects, with the latter identified based on correlations between the estimated teacher effects in successive years and in different classes taught by the same teacher within a year. For our purposes we can imagine that a multilevel varying-intercept model was fit. We have no particular interest in thousands of individual teacher effects, but we do care about the group-level coefficients (in this case, the "groups" are the teachers, and the predictors describe teacher certification and experience) and the residual standard deviation of teacher effects.

The analysis found the teacher characteristics to be weak predictors of value added—in particular, teachers with formal certifications and better college grades did essentially the same (as measured by the test scores of the students they taught), on average, compared to seemingly less-well-qualified teachers. In contrast, the unexplained teacher-level variation was large, with students of the best teachers scoring about 0.2 to 0.4 better than students of the worst teachers (after controlling for previous test scores and other covariates). This unexplained variation in teacher effects (on a scale in which 1 unit represents the standard deviation of scores of all students within a grade) is moderately large, and it led the researchers to conclude that "policies that enable districts to attract and retain high quality teachers (or screen-out less effective teachers) have potentially large benefits for student achievement."

21.2 Superpopulation and finite-population variances

Each factor in a multilevel model corresponds to a set of linear parameters or coefficients. Label the coefficients in a particular factor as $\alpha_1, \ldots, \alpha_J$, where J is the number of groups or categories of this factor. The variation among the α_j's can be summarized in two ways:

- The *superpopulation* standard deviation σ_α, which represents the variation among the modeled probability distribution from which the α_j's were drawn, is relevant for determining the uncertainty about the value of a new group not in the original set of J.

- The *finite-population* standard deviation s_α of the particular J values α_j describes variation within the existing data.

Example of a non-nested model

We illustrate with the simple model (13.9) for the flight simulator example introduced in Section 13.5. The model has three (non-nested) levels:

$$
\begin{aligned}
y_i &\sim \ \mathrm{N}(\gamma_{j[i]} + \delta_{k[i]}, \ \sigma_y^2), \text{ for } i = 1, \ldots, n \\
\gamma_j &\sim \ \mathrm{N}(\mu_\gamma, \sigma_\gamma^2), \text{ for } j = 1, \ldots, J \\
\delta_k &\sim \ \mathrm{N}(\mu_\delta, \sigma_\delta^2), \text{ for } k = 1, \ldots, K.
\end{aligned}
$$

with diffuse normal prior distributions specified for μ_γ and μ_δ. When we specify the model this way, μ_γ and μ_δ are not separately identified; however we can compute summaries of interest using the ideas of redundant additive parameterization as described in Section 19.4.

The superpopulation standard deviations at each level are simply $\sigma_\gamma, \sigma_\delta, \sigma_y$. The finite-population standard deviations are

$$s_\gamma = \sqrt{\frac{1}{J-1}\sum_{j=1}^{J}(\gamma_j - \bar{\gamma})^2}$$

$$s_\delta = \sqrt{\frac{1}{K-1}\sum_{k=1}^{K}(\delta_k - \bar{\delta})^2}$$

$$s_y = \sqrt{\frac{1}{n-1}\sum_{i=1}^{n}(\epsilon_i - \bar{\epsilon})^2}. \tag{21.1}$$

In defining s_y, we work with the data-level errors

$$\epsilon_i = y_i - \hat{y}_i = y_i - (\mu + \gamma_{j[i]} + \delta_{k[i]}).$$

Computation

In Bugs, the superpopulation standard deviations $\sigma_y, \sigma_\gamma, \sigma_\delta$ are already defined in the model (see the model at the beginning of Section 17.3). To define the finite-population standard deviations, we add the lines

Bugs code
```
for (i in 1:n){
  e.y[i] <- y[i] - y.hat[i]
}
s.y <- sd(e.y[])

s.g <- sd(g[])
s.d <- sd(d[])
```

When group-level predictors are present, they must be subtracted before computing the standard deviations as defined above. Suppose, for example, the airport coefficients γ_j had a group-level model with two predictors, u_1 and u_2:

Bugs code
```
for (i in 1:J){
  g[j] ~ dnorm (g.hat[j], tau.g)
  g.hat[j] <- a.0 + a.1*u.1[j] + a.2*u[2]
}
tau.g <- pow(sigma.g, -2)
```

Then `sigma.gamma` is the superpopulation standard deviation, and the corresponding finite-population standard deviation is defined by

Bugs code
```
for (j in 1:J){
  e.g[j] <- g[j] - g.hat[j]
}
s.g <- sd(e.g[])
```

When using the convenient $\gamma_j \sim N(\hat{\gamma}_j, \sigma_\gamma^2)$ notation, we can simply subtract $\hat{\gamma}_j$ to define the group-level error; we need not worry about how $\hat{\gamma}$ is constructed. (For example, the model for γ could itself include group-level parameters.)

The finite-population standard deviation is estimated more precisely than the superpopulation standard deviations

The superpopulation and finite-population standard deviations are *not* two different statistical "estimators" of a common quantity; rather, they are two different

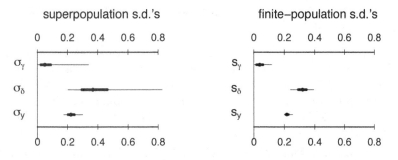

Figure 21.1 *Median estimates, 50% intervals, and 95% intervals for (a) superpopulation and (b) finite-population standard deviations of the treatment-level, airport-level, and data-level errors in the flight simulator example. The two sorts of standard-deviation parameters have essentially the same estimates, but the finite-population quantities are estimated much more precisely.*

quantities that can both be estimated from the multilevel model. We can get a point estimate and uncertainty intervals for both. In general, the point estimates of σ and s will be similar to each other, but s will have less uncertainty than σ. That is, the variation is more precisely estimated for the finite population than the superpopulation. This makes sense because we have more information about the units we observe than the full population from which they are sampled.

More generally, consider a hypothetical group-level model,

$$\alpha_j = U_j\gamma + \eta_j, \quad \eta_j \sim \mathrm{N}(0, \sigma_\alpha^2), \quad \text{for } j = 1, \ldots, J.$$

The superpopulation standard deviation is simply σ_α, and the finite-population standard deviation is $s_\alpha = \mathrm{sd}_{j=1}^J \eta_j = \sqrt{\frac{1}{J-1} \sum_{j=1}^J (\eta_j - \bar{\eta})^2}$.

Example with only two groups.   Both measures of variation are important. To see how they differ, consider the extreme case in which $J = 2$ and there are no group-level predictors (so that, in classical estimation, there would be a constraint such as $\alpha_1 + \alpha_2 = 0$) and a large amount of data are available in both groups. In that case, almost nothing can be said about the superpopulation standard deviation, but the finite-population standard deviation can be still be estimated well if the sample size within groups is large. The two parameters α_1 and α_2 will be estimated accurately (in either a classical or a multilevel model) and so will $s_\alpha^2 = \frac{1}{2-1}((\alpha_1 - \bar{\alpha})^2 + (\alpha_2 - \bar{\alpha})^2) = (\alpha_1 - \alpha_2)^2/2$. The superpopulation variance σ_α^2, on the other hand, is only being estimated by a measurement from a distribution that is proportional to a χ^2 with 1 degree of freedom. We know much about the two parameters α_1, α_2 but can say little about others from their batch.

Fixed and random effects

As we discussed in Section 11.4, we believe that much of the statistical literature on fixed and random effects can be fruitfully reexpressed in terms of finite-population and superpopulation inferences. In some contexts (for example, collecting data on the 50 states of the United States), the finite population seems more meaningful; whereas in others (for example, subject-level effects in a psychological experiment), interest clearly lies in the superpopulation.

Chapter 22 applies finite-population standard deviations to the analysis of variance.

Finite-population standard deviations for regression coefficients that are not part of a multilevel model

A key idea in analysis of variance, as we discuss in Chapter 22, is to summarize each factor in a regression model by its finite-population standard deviation. As we have seen above, this is typically straightforward for batches of varying coefficients. For an unmodeled coefficients we can define finite-population standard deviations as the standard deviation of the coefficients in the population of predicted values.

We illustrate with some examples:

- *Binary predictor.* Suppose sex is included in a model as an indicator x that equals 1 for men and 0 for women. Label the proportion of men in the sample is λ. If the coefficient for x is β, then the finite-population standard deviation is simply the standard deviation of a random variable that equals β with probability λ and 0 with probability $1-\lambda$; this comes to $\sqrt{\lambda(1-\lambda)}|\beta|$.

- *Unmodeled categorical predictor.* Suppose age is included using three indicator variables corresponding to the categories 18–29, 30–44, and 45–64 (with the final category, 65+, being the baseline), and label the proportion of the sample in each category as $\lambda_1, \lambda_2, \lambda_3$, and $\lambda_4 = 1 - \lambda_1 - \lambda_2 - \lambda_3$. If the coefficients for the three indicators are $\beta_1, \beta_2, \beta_3$, and (implicitly) $\beta_4 = 0$, then the finite-sample standard deviation is the standard deviation of a random variable that takes on these four values with these four probabilities. This can be easily computed in R from the vectors λ and β:

R code
```
b.mean <- sum (b*lambda)
b.sd <- sqrt (sum (lambda*(b-b.mean)^2))
```

Alternatively, we can perform the calculation directly using indexing:

R code
```
b.sd <- sd (b[x])
```

assuming `x` is the index variable for age that takes on the values 1, 2, 3, 4.

- *Continuous linear predictor.* Suppose age is simply included as a linear predictor x with coefficient β. The corresponding finite-population standard deviation is simply the standard deviation of x in the sample, multiplied by the absolute value of β.

- *Continuous nonlinear predictor.* Suppose age is included as a continuous predictor x with linear and quadratic coefficients, β_1 and β_2. The finite-population standard deviation is then the standard deviation of $\beta_1 + \beta_2 x$ in the data, and can simply be calculated in R as `sd(b[1]+b[2]*x)`, using the data vector `x`.

These examples do not cover all cases. For example, it is not clear how to define the standard deviation of variables that are also included as interactions.

21.3 Contrasts and comparisons of multilevel coefficients

A key difference between classical and multilevel models is the treatment of categorical or discrete inputs. As discussed in Section 11.4, in a classical regression it is possible to include group indicators or group-level predictors, but it is difficult to do both. (It is possible to perform two-level regression, first including group indicators and then regressing the estimated group indicators on group-level predictors—but this approach runs into difficulties when within-group sample sizes are small.)

Contrasts: including an input both numerically and categorically

Fortunately, including group indicators and also group-level predictors is straightforward and often useful in multilevel modeling. The only challenge comes in the interpretation of the group-level coefficients. We illustrate with an example from education.

Advanced Placement (AP) tests are taken by students in high school and are used by colleges to place students in introductory courses. As part of a study of the relevance of AP tests to the college curriculum, we performed a regression predicting the grades in introductory calculus classes for students who had taken the AP Calculus exam. The outcome of the regression was course grade,[1] and the inputs were AP score (which took on the discrete values 1, 2, 2.5, 3, 3.5, 4, 4.5, 5, 5.5, and 6.5; we label these as $u_j, j = 1, \ldots, 10$)[2] and Scholastic Aptitude Test (SAT) score (which we treated as a continuous predictor).

There are two natural ways to code AP score as regression predictors: as a single continuous predictor taking on numerical values u_j from 1 through 6.5, or as 10 indicators corresponding to its 10 different levels. Either of these approaches could make sense: on one hand, the AP grades are ordered, and it is reasonable, at least as a starting point, to consider the jump from a score of 1 to 2, or 2 to 3, and so forth, as corresponding to roughly equal jumps in course grade. On the other hand, nonlinear patterns are possible: for example, perhaps scores of 4 and 5 are both high enough to assure a good grade in the calculus class, so that little benefit would be seen beyond a score of 4.

Classical regression. In a classical regression, we can include AP score as a continuous predictor (and also, for example, include the square of AP score to model nonlinearity)—or we could include indicators for four of the five AP scores, for example, treating a score of 1 as the baseline category. But it would not be possible to do both, because the linear trend is collinear with the set of indicators.

That is, we can model the coefficients for the AP scores, $\alpha_j, j = 1, \ldots, 5$, as $\alpha_j = \gamma_0 + \gamma_1 u_j$ (the linear model), or $\alpha_j = \gamma_0 + \gamma_1 u_j + \gamma_2 u_j^2$ (the quadratic model), or simply as 10 different α_j's—but there is no way in a classical model to include the linear or quadratic trend *and* the separate coefficients for each level.

Multilevel regression. In a multilevel model we can include AP both as a continuous predictor and as indicators: $\alpha_j \sim N(\gamma_0 + \gamma_1 u_j, \sigma_\alpha^2)$ or, equivalently,

$$\alpha_j = \gamma_0 + \gamma_1 u_j + \eta_j, \tag{21.2}$$

with group-level errors $\eta_j \sim N(0, \sigma_\alpha^2)$. This model allows the α_j's to have arbitrary levels, but with the group-level model (21.2) pulling the estimates toward linearity to the extent supported by the data.

The two classical models (linearity at one extreme or 10 separate coefficients at the other extreme) are special cases of the multilevel model, corresponding to $\sigma_\alpha = 0$ or ∞, respectively.

For another example of this sort of multilevel model, see Figure 13.12 on page 293. We further discuss contrasts in the context of the analysis of variance in Section 22.5.

---

[1] Grades of A+, A, A-, B+, ... were coded as 4.3, 4.0, 3.7, 3.3, ... For simplicity we treated the grades as continuous measurements.

[2] The AP Calculus test had two different forms, each on its own 1–5 scale. We performed a preliminary analysis to equate the two tests and put them on a common scale by adding 1.5 to scores on the more difficult test form.

The finite-population group-level coefficient. A challenge remains in interpreting the group-level slope, γ_1 in model (21.2). The difficulty is in the partial nonidentifiability between γ_1 and the η_j's—more precisely, between γ_1 and whatever linear trend there is in the η_j's. This is only a partial nonidentifiability, because the η_j's are pooled toward zero by the multilevel model—but with only 10 groups, it creates practical problems.

In this particular example, the estimated slope among the 10 groups was clearly positive—α_j increased with j, roughly linearly with the score u_j. But the inference for the group-level slope γ_1 had a wide standard error—the point estimate of γ_1 was positive but its 95% confidence interval actually included zero and negative values. The problem arose because there was a possibility, with only $J = 10$ groups, that the finite-sample regression of the η_j's was strongly positive or negative.

Thinking about the problem, we realized that what we really wanted was the *finite-population* slope, that is, the linear regression of the α_j's on the u_j's for these 10 groups. This finite-population quantity will be estimated more precisely than the superpopulation slope γ_1.

Having fit the model in Bugs, one can quickly compute a 95% interval, for example, for the finite-population slope in R:

R code
```
attach.bugs (AP.fit)
finite.slope <- rep (NA, n.sims)
for (s in 1:n.sims){
  finite.pop <- lm (a[s,] ~ u)
  finite.slope[s] <- coef(finite.pop)["u"]
}
quantile (finite.slope, c(.025,.975))
```

This can be compared, for example, to the 95% interval for γ_1, which will be wider. Performing this regression on the multilevel coefficients is equivalent to constraining the group-level errors η_j in (21.2) to have a mean of 0 and a slope of 0. The step can thus be thought of as a generalization of the additive redundant parameterization of Section 19.4.

Inferences for multilevel parameters defined relative to their finite-population average

Sometimes each individual coefficient is not of intrinsic interest but we are interested in comparisons among the coefficients. Moreover, defining predictors relative to each other yields much more stable estimates when sample sizes are small. We here discuss both the finite-population and superpopulation contrasts that can be estimated using multilevel models.

Consider the flight simulator data in Section 13.5, where the airport coefficients δ_k come from a distribution with common mean μ_δ. The left panel of Figure 21.2 shows the estimates and standard errors for these parameters. These reflect superpopulation contrasts because they measure the position of each airport relative to the population mean μ_δ.

Alternatively, we can define the varying airport intercepts relative to the mean of these coefficients in our sample, computing an adjusted coefficient $\delta_k^{\text{adj}} = \delta_k - \bar{\delta}$ for each airport k. This sort of mean shift is motivated in Section 19.4 for computational purposes—to help the Gibbs sampler converge faster—but it also gives us more precise inferences about the relative values of the airport coefficients for this sample. The right panel of Figure 21.2 shows the estimates and standard errors for the mean-centered airport coefficients. The point estimates are in the same place but

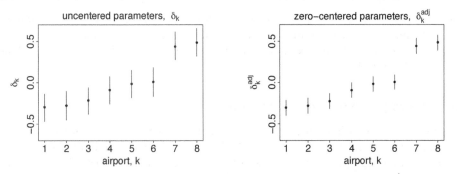

Figure 21.2 *Estimates and standard errors for the airport coefficients in the flight simulator example: (a) parameters with population mean subtracted out, $\delta_k - \mu_\delta$; (b) parameters centered around the sample mean $\delta_k^{\mathrm{adj}} = \delta_k - \bar{\delta}$. The point estimates are the same for the two sets of parameters, but the centered versions have smaller, finite-sample standard errors. The airports have been ordered in increasing average response (see Figure 13.8 on page 290).*

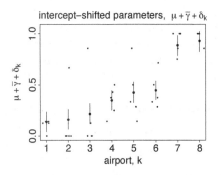

Figure 21.3 *Estimates and standard errors for the airport coefficients in the flight simulator example, after adding in the average values of the other terms in the regression model, thus moving to the scale of the data. The small dots in the graphs show the data. (It makes sense that many of the data fall outside the standard-error bounds, which represent inferential uncertainty about the airport coefficients but do not include the data-level error that is in the model.)*

the finite-population standard errors are smaller than the superpopulation standard errors used in the $\delta_k - \mu_\delta$ contrast.

Adding the average value of the other predictors back in.   Another way of summarizing the differences across airports is to add the average values of the other terms in the regression, to create coefficients that are on the scale of the original data. For the flight simulator example, the other terms in the regression model are the mean level μ and the treatment effects γ_j:

$$y_i \sim \mathrm{N}(\mu + \gamma_{j[i]} + \delta_{k[i]}, \sigma_y^2).$$

The average value of the other coefficients is μ plus the average of the γ_j's, and so the shifted-to-the-data-scale airport coefficients are $\mu + \bar{\gamma} + \delta_k$ for the airports $k = 1, \ldots, K$. Figure 21.3 shows the inference for these shifted parameters, with the data displayed also. In essence, then, these parameters represent the predicted success rates for each airport for the average treatment. (The plot reveals some potential poor fit of the linear model as fit to the data y_i, which in fact are success rates that are bounded between 0 and 1. But here we are simply focusing on the mechanics of displaying and understanding the coefficients in a non-nested multilevel model.)

This is the method we used to display the fitted model and data in the various plots in Chapters 12–15.

21.4 Average predictive comparisons

Section 5.7 discusses how to calculate average predictive differences for logistic regressions. The challenge is that the model is on the logistic scale but we would like to interpret changes on the probability scale. Further complications arise with multilevel models, and we discuss the general issue of average predictive comparisons here.

Notation and basic definition of predictive comparisons

Assume you have fit a probability model predicting a numerical outcome y based on vectors x of inputs and θ of parameters. As in Section 3.4, we denote the separate sources of information in x that are used in the predictive model as *inputs* or *input variables*, as distinguished from the *linear predictors* that form the columns of the design matrix of a linear or generalized linear model. For example, consider a logistic regression of some individual outcome on age, sex, an age × sex interaction, and age^2. There are two inputs (age and sex) but five linear predictors (including the constant term).

In any model, we consider the scalar inputs one at a time, using the notation

$$u : \quad \text{the input of interest,}$$
$$v : \quad \text{all the other inputs.}$$

Thus, $x = (u, v)$. We focus on the expected predictive difference in y per unit difference in the input of interest, u, with v (the other components of x) held constant:

$$b_u(u^{(\text{lo})}, u^{(\text{hi})}, v, \theta) = \frac{\mathrm{E}(y | u^{(\text{hi})}, v, \theta) - \mathrm{E}(y | u^{(\text{lo})}, v, \theta)}{u^{(\text{hi})} - u^{(\text{lo})}} \tag{21.3}$$

and the average predictive difference per unit of u:

$$B_u(u^{(\text{lo})}, u^{(\text{hi})}) = \frac{1}{n} \sum_{i=1}^{n} b_u(u^{(\text{lo})}, u^{(\text{hi})}, v_i, \theta). \tag{21.4}$$

We assume that $\mathrm{E}(y | x, \theta)$ is a known function (such as the inverse-logit) that can be computed directly. For a linear model with no interactions, b_u does not depend on $u^{(1)}, u^{(2)}$, or v, and is simply the regression coefficient associated with u. More generally, however, b_u varies as a function of these inputs, and it can be useful to summarize the predicted difference with some sort of weighted average as in (21.4). In practice, one must also average over θ (or plug in a point estimate).

Expressions (21.3) and (21.4) are similar to (5.8) and (5.9) on page 103 except in dividing by $u^{(\text{hi})} - u^{(\text{lo})}$. For some settings the total difference is relevant; other times we are interested in the difference per unit of u.[3] As we shall see, options also arise in the definition and estimation of average predictive changes with continuous inputs u. This section presents one particular set of choices and illustrates with an example.

---

[3] We use the notation δ for predictive differences and Δ for their averages, and the notation b, B for ratio comparisons of the form (21.3) and (21.4), by analogy to regression coefficients β.

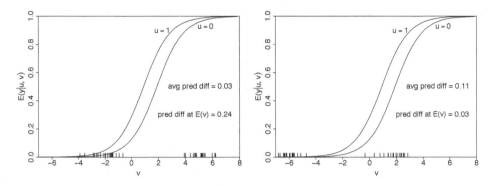

Figure 21.4 *Hypothetical examples illustrating the need for averaging over the distribution of v (rather than simply working with a central value) in computing the predictive comparison. Each graph shows a hypothesized logistic regression model for an outcome y given a binary input of interest, u, and a continuous input variable, v. The vertical lines on the x-axis indicate the values of v in the hypothetical dataset. (a) In the left plot, data v are concentrated near the ends of the predictive range. Hence the average predictive comparison is small. In contrast, $E(v)$ is near the center of the range; hence the predictive comparison at this average value is large, even though this is not appropriate for any of the data points individually. (b) Conversely, in the right plot, the average predictive comparison is reasonably large, but this would not be seen if the predictive comparison were evaluated at the average value of v.*

Problems with evaluating predictive comparisons at a central value

A rough approach that is sometimes used is to evaluate $E(y|u, v)$ at a central value v_0—perhaps the mean or the median of the data—and then estimate predictive comparisons by holding v constant at this value. Evaluating differences about a central value can work well in practice, but one can run into problems when the space of inputs is very spread out (in which case no single central value can be representative) or if many of the inputs are binary or bimodal (in which case the concept of a "central value" is less meaningful). In addition, this approach is hard to automate since it requires choices about how to set up the range for each input variable. In fact, our research in this area was motivated by practical difficulties that can arise in trying to implement this central-value approach.

We illustrate some challenges in defining predictive comparisons with a simple hypothetical example of a logistic regression model of data y on a binary input of interest, u, and a continuous input variable, v. The curves in each plot of Figure 21.4 show the assumed predictive relationship. In this example, u has a constant effect on the logit scale but, on the scale of $E(y)$, the predictive comparison b_u (as defined in (21.3), with $u^{(\text{lo})} = 0$ and $u^{(\text{hi})} = 1$) is high for v in the middle of the plotted range and low at the extremes. As a result, the average predictive comparison B_u, defined in (21.4), depends on the distribution of the other input, v.

The two plots in Figure 21.4 show two examples in which the average predictive comparison differs from the predictive comparison evaluated at a central value. In the first plot, the data are at the extremes, so the average predictive comparison is small—but the predictive comparison evaluated at $E(v)$ is misleadingly large. The predictive comparison in y corresponding to a difference in u is small, because switching u from 0 to 1 typically has the effect of switching $E(y|u, v)$ from, say, 0.02 to 0.05 or from 0.96 to 0.99. In contrast, the predictive comparison if evaluated at

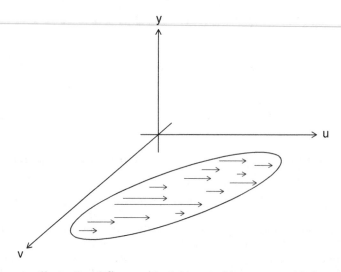

Figure 21.5 *Diagram illustrating differences in the input of interest u, with the other inputs v held constant. The ellipse in (u, v)-space represents the joint distribution p(u, v), and as the arrows indicate, we wish to consider differences in u in the region of support of this distribution.*

the mean of the data is large, where switching u from 0 to 1 switches $E(y|u, v)$ from 0.36 to 0.60.

The second plot in Figure 21.4 shows a similar example, but where the centrally computed predictive comparison is too low compared to the population-average predictive comparison. Here, the centrally located value of v is already near the edge of the curve, at which point a difference in u corresponds to only a small difference in $E(y|u, v)$ of only 0.03. In comparison, the average predictive comparison, averaging over the data locations, has the larger value of 0.11, which appropriately reflects that many of the sample data are in the range where a difference in u can correspond to a large difference in $E(y)$.

General approach to defining population predictive comparisons

The basic predictive comparison b_u defined in (21.3) depends in general on $u^{(lo)}$ and $u^{(hi)}$ (the beginning and end points of the hypothesized difference in the input of interest), v (the values of the other inputs), and θ (the parameters of the model). We define the *average predictive comparison* b_u as the mean value of b_u over some specified distribution of the inputs and parameters. We apply this idea to various sorts of inputs u, starting with numerical input variables, including continuous and binary inputs as special cases, and moving to unordered categorical variables, interactions, and constraints.

It turns out that the form of the input of interest u, not the form of data y or other predictors v, is crucial in deciding how to define average predictive comparisons. In any application, we would compute the average predictive comparison of each of the inputs to a model one at a time—that is, treating each component of x in turn as the "input of interest." This is often a goal of regression modeling: estimating the predictive comparison of each input with all other inputs held constant.

Averaging over $u^{(lo)}, u^{(hi)}$, and v in this way is equivalent to counting all pairs of transitions of $(u^{(lo)}, v^{(1)})$ to $(u^{(hi)}, v^{(2)})$ in which $v^{(1)} = v^{(2)}$—that is, differences in u with v held constant. Figure 21.5 illustrates. We average over θ using simulations from the fitted model (as obtained, for example, in Bugs), which in a Bayesian context is the posterior distribution, or classically could be a point estimate or an uncertainty distribution defined by simulations (as discussed at the end of the previous section). The distributions of (u, v) and θ are independent because we are working in a regression context in which θ represents the parameters of the model for y conditional on u and v.

In the special case in which u is a binary input, the average predictive comparison is a simple average of the differences $(\mathrm{E}(y|u^{(\mathrm{hi})}, v, \theta) - \mathrm{E}(y|u^{(\mathrm{lo})}, v, \theta))$. More generally, the average predictive comparison has the form of a ratio of averages.

Models with interactions

These definitions automatically apply to models with interactions. The key is that u represents a single input, and $x = (u, v)$ represents the vector of inputs to the predictive model. As discussed throughout this book, the vector of inputs is not in general the same as the vector of linear predictors. For example, in the simple example at the beginning of this section, sex is included on its own and also interacted with age. When defining the predictive comparison for sex, we must alter this input wherever it occurs in the model—that is, both the "sex" predictor and the "sex × age" predictor must be changed. For another example, the constant term in a regression is *not* an input in our sense and has no corresponding predictive comparison, since it can take on only one possible value.

From a computational perspective, it is important that the model be coded in terms of its separate inputs. Thus, to compute predictive comparisons, it is not enough simply to specify the design matrix of a regression model; one must be able to evaluate $\mathrm{E}(y)$ as a function of the original inputs.

Inputs that are not always active

A model will sometimes have inputs that are involved in prediction for only some of the data. For example, consider an experiment in which some units are given the control (no treatment) and others are given the treatment, in doses ranging from 10 to 20 (on some meaningful scale). Suppose the data are fit by a generalized linear model with treatment indicator, dose, and some pre-treatment measurements as predictors.

Now consider how to define the average predictive comparison for dose. One approach is to consider treatment and dose to be a single input with value 0 for control units and the dose for treated units. This will not be appropriate, however, if we are particularly interested in the effect of dose in the range 10 to 20, conditional on treatment. We can define the predictive comparison for dose, restricting all integrals over v to the subspace in which dose is defined (in this case, the treated units).

We can formally define the average predictive comparison for a partially active input u by introducing a function $\zeta_u(v)$ that equals 1 wherever u is defined and 0 elsewhere. Then all the earlier definitions hold, as long as we insert the factor $\zeta_u(v)$ in all the integrals.

Average predictive comparisons for multilevel models

In a linear multilevel model, one might use the standard deviation of a batch of coefficients as a measure of their importance for predicting the outcome variable. But in a nonlinear model, this does not directly transfer to the scale of measurement.

For the purpose of defining average predictive comparisons, we can treat a batch of K parameters $\phi_k, k = 1, \ldots, K$, as an unordered categorical input variable with K levels $u^{(k)}$. The essence of a variance components model is that the parameters ϕ_k in a batch are considered to have been drawn from a continuous population distribution. In the following example we consider vector ϕ_k's.

Example: a multilevel logistic regression of prison sentences

We illustrate average predictive comparisons in multilevel models with a model of the geographic variation of the severity of prison sentences in the United States. A study was performed of the sentences of 8446 convicted felons in 39 of the 75 most populous counties in the United States during May 1998. The outcome variable for this study was "sentence severity," defined as $y_{ij} = 1$ if the offender i in county j received a prison sentence, or 0 for a jail or noncustodial sentence (considered to be much less severe than prison). A multilevel logistic regression model was fit with 12 individual-level variables from the State Court Processing Statistics program of the Bureau of Justice Statistics, linked to six county-level variables using the Federal Information Processing Standards code. Information collected in this program includes demographic characteristics, criminal history, details of pretrial processing, disposition, and sentencing of felony defendants.

Under the model, the outcomes are independent with probabilities

$$\Pr(y_i = 1) = \text{logit}^{-1}\left(X_i G_{j[i]}\eta + X_i\alpha_{j[i]}\right), \qquad (21.5)$$

where X_i represents a vector of measurements on K individual-level variables and G_j is a $K \times M$ block-diagonal matrix of measurements on L county-level variables. In particular, interactions between individual- and county-level variables account for dependence of individual-level comparisons across counties, so that $M = KL$ if all county-level variables are used to explain these individual-level comparisons. The coefficients η in (21.5) represent main effects and interactions of the predictors and are constant across counties, and the unmodeled coefficients in the vector α have a j-subscript and represent varying coefficients across counties, or can be viewed as interactions between the predictors X and the county indicators j.

Applying average predictive comparisons to a single model

Fitting this two-level multilevel model is straightforward; however, the presence of individual-county interactions and varying county coefficients complicates the interpretation of the parameters η and α_j. In contrast, average predictive comparisons provide a clear indication of the overall contribution of each variable to the probability of receiving a prison sentence (rather than a jail or noncustodial sentence).

Figure 21.6 displays the average predictive comparison for each variable in the model, together with a average predictive comparison for the county indicators. Horizontal bars indicate ± 1 standard error for each average predictive comparison, calculated as described earlier in this section. Due to computational limitations, we based all calculations on a randomly drawn subset of $L = 100$ posterior samples, with $n = 4500$ data points for the binary inputs, $n = 450$ for the continuous inputs,

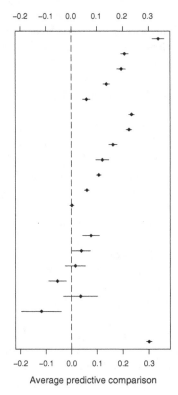

Figure 21.6 *Estimated average predictive comparisons* B_u *for the probability of a prison sentence (rather than a jail or noncustodial sentence), for each input variable u in the prison example. Horizontal lines show* ±1 *standard-error bounds. The first set of inputs, with initial letters I, are at the level of the individual offender; the second set, with initial letters C, are county-level inputs; and the last corresponds to the differences associated with varying the county indicator, keeping all other inputs constant. Many of the individual predictors are associated with large differences and county itself predicts a fair amount of variability, but the county-level variables are associated with relatively small differences (recall that these average predictive comparisons correspond to differences of one standard deviation in each input).*

and $n = 4000$ for the varying county coefficients. Results varied little on repeating the calculations with different random subsets.

Individual-level variables in Figure 21.6 are denoted with an initial "I," and county-level variables are denoted with an initial "C." The five individual-level variables for "most serious conviction charge" (ICVIOL1, ICTRAF, ICVIOL2, ICPROP, and ICDRUG) are relative to a reference category of weapons, driving-related, and other public order offenses. The 12 individual-level variables and 2 of the county-level variables are binary, and the remaining 4 county-level variables are continuous. Finally, the county indicators are a batch of parameters in a multilevel model.

The individual-level predictor associated with the largest difference in the probability of receiving a prison sentence is ICVIOL1 (murder, rape, or robbery), with an estimated average predictive comparison of 0.38 (and standard error 0.03). That is, the expected difference in the probability of receiving a prison sentence between a randomly chosen individual in the population charged with murder, rape, or robbery and a similar individual charged with a reference category offense is 0.38.

Other charges are less highly associated with a prison sentence, with decreasing probability: drug trafficking (with an estimated average predictive comparison of 0.21), then assault (0.19), then property offenses (0.14), and finally drug possession offenses (0.05). Other individual-level variables can be interpreted similarly, as can the two binary county-level variables (which compare individuals in southern and non-southern counties, and individuals in counties with and without state sentencing guidelines).

Predictive comparisons for the continuous input variables correspond to changes of 1 standard deviation in each, one at a time. Standard deviations for the four variables (CCONS, CCRIME, CBLPCT, and CUNEMP) are 13%, 220 per 10,000, 12%, and 1.8%, respectively. The positive average predictive comparisons for CCONS and CCRIME suggest that, comparing otherwise-similar cases, those in counties with higher conservative populations or higher crime rates have slightly higher probabilities of receiving a prison sentence. Conversely, the negative average predictive comparison for CUNEMP suggests lessened sentence severity in high-unemployment counties, with all other inputs fixed. Taking into account all other factors, the proportion of the county's population that is African American (CBLPCT) has little bearing by itself on sentence severity. (However, this factor does play a role in reducing or increasing the contributions of various individual-level variables, as can be seen in more detailed analysis.)

The average predictive comparison for the county indicators differs from the other average predictive comparisons in this example in that it considers just the magnitude, rather than the sign, of the comparisons. This is because "county" is the only unordered categorical variable in the model. To understand its average predictive comparison, consider the probabilities of receiving a prison sentence for two individuals who are identical in all respects except that they are in different counties. So, the two individuals will share the same values for individual-level variables but have different values for county-level variables (and county indicators). The county average predictive comparison of 0.37 represents the root mean square of the difference in the probability of receiving a prison sentence between a randomly chosen individual in one county and a similar individual in another county.

Using average predictive comparisons to compare models

The multilevel logistic regression model (21.5) has varying coefficients for the within-county intercepts as well as for each individual predictor. We also fit a multilevel model with varying intercepts only, as well as a non-multilevel complete pooling model that ignores the multilevel nature of the data and excludes county indicators. Whereas the regression coefficients have different interpretations for the different models, predictive comparisons allow for direct comparison.

Figure 21.7 displays the average predictive comparison for each variable across all three models. The average predictive comparisons and standard errors are very similar across the two multilevel models, perhaps suggesting that the additional variation for each individual predictor may be redundant. The individual-level comparisons are also very similar for the non-multilevel model. However, the county-level comparisons tend to be smaller in magnitude and have smaller standard errors for the non-multilevel model; the average predictive comparison for unemployment, CUNEMP, even has the opposite sign. The non-multilevel model did not fit this dataset well, and the average predictive comparisons displayed here suggest that while individual-level comparisons may be robust to model misspecification of this nature, higher-level comparisons can clearly be adversely affected.

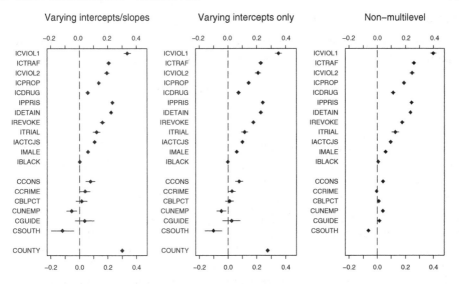

Figure 21.7 *Estimated average predictive comparisons for the probability of a prison sentence for each input variable across three models in the prison example. Horizontal lines show ±1 standard-error bounds. Estimates and standard errors are very similar across the two multilevel models. However, although the individual-level comparisons are similar for the non-multilevel model, the county-level comparisons tend to be smaller in magnitude and have smaller standard errors.*

Predictive summaries in practice

Predictive comparisons, like any other automatic summary of a model, cannot be universally applicable, because the best approach in any problem must be tailored to the specifics of the application. We agree with this point but note that the overwhelming current practice in applied statistics of regression models is simply to report coefficient estimates (and standard errors), with no sense of their implications on the original scale of the data. Average predictive comparisons are not intended to be a replacement for regression coefficients; rather, they summarize a model in a way that can complement the coefficient estimates in order to make their scale more interpretable. Thus, we agree that there is no such thing as a "one size fits all" method—but that is what the current standard approach implicitly assumes.

The example illustrates the effectiveness and convenience of predictive comparisons. In this multilevel dataset with a binary outcome measure, the comparisons clarify the overall role of each individual- and group-level predictor in the presence of multiple interactions, as well as illustrate the relative size of the varying coefficients. They can also be used to understand and compare models directly, in a way that is difficult to do using logistic regression coefficients.

21.5 R^2 and explained variance

As discussed in the context of the examples in Chapters 3 and 4, it can be helpful to understand a model through R^2, the proportion of variance explained by its linear predictors. Although explained variance can be misleading (as illustrated in Figure 3.9 on page 42), it can be a useful measure of the relative importance of different sources of variation in a particular dataset. Here we discuss how to generalize R^2 to multilevel models.

Explained variation in linear models

Consider a linear regression written as $y_i = X_i\beta + \epsilon_i$, $i = 1, \ldots, n$. One way to summarize the fit of the regression is by the proportion of variance explained:

$$R^2 = 1 - \frac{\overset{n}{\underset{i=1}{V}} \epsilon_i}{\overset{n}{\underset{i=1}{V}} y_i}, \tag{21.6}$$

where V represents the finite-sample variance operator, $\overset{n}{\underset{i=1}{V}} x_i = \frac{1}{n-1} \sum_{i=1}^{n} (x_i - \bar{x})^2$. In a multilevel model, the predictors "explain" the data at different levels, and R^2 can be generalized in a variety of ways.

One sort of definition of "explained variance," which we do *not* want to use, is $R^2 = 1 - \frac{\text{residual variance under the larger model}}{\text{residual variance under the null model}}$, with various choices of the null model corresponding to predictions at different levels. This definition is not so helpful to us because our goal is to understand a particular model being fit on its own terms.

Here we present a slightly different approach, computing (21.6) at each level of the model and thus coming up with several R^2 values for any particular multilevel model. This approach has the virtue of summarizing the fit at each level and requiring no additional null models to be fit. In defining this summary, our goal is not to dismiss other definitions of R^2 but rather to add another tool to the understanding of multilevel models.

We introduce notation (to which we shall return in the next chapter in the context of the analysis of variance) for a multilevel model with M error terms. (For example, $M = 2$ in the varying-intercept model of radon in houses within counties. For the varying-intercept, varying-slope model, $M = 3$, with regression models for house radon levels, county intercepts, and county slopes.) For convenience, in this section and the next, we shall refer to each error term—that is, the data and each batch of modeled parameters—as a "level" of the model.

For each level m, we write the model as

$$\theta_k^{(m)} = \hat{\theta}_k^{(m)} + \epsilon_k^{(m)}, \quad \text{for } k = 1, \ldots, K^{(m)}, \tag{21.7}$$

where the $\hat{\theta}_k^{(m)}$'s are the linear predictors (that is, the linear combination of coefficients and predictors) at that level of the model, and the errors $\epsilon_k^{(m)}$ come from a distribution with mean zero and standard deviation $\sigma^{(m)}$. At the lowest (data) level of the model, the $\theta_k^{(m)}$'s correspond to the individual data points (the y_i's in the radon model). At higher levels of the model, the $\theta_k^{(m)}$'s represent batches of comparisons or regression coefficients (county intercepts α_j and slopes β_j in the radon model). Because we work with each batch of parameters separately, we shall suppress the superscripts (m) for the rest of this section.

Proportion of variance explained at each level

For each level (21.7) of the model, we first consider the variance explained by the linear predictors $\hat{\theta}_k$. Generalizing from the classical expression (21.6), we define

$$R^2 = 1 - \frac{\text{E}\left(\overset{K}{\underset{k=1}{V}} \epsilon_k\right)}{\text{E}\left(\overset{K}{\underset{k=1}{V}} \theta_k\right)}, \tag{21.8}$$

where the expectations average over the uncertainty in the fitted model (using the posterior simulations). Any particular multilevel model will then have more than one R^2: one for the data level and one for each batch of modeled parameters. In our simulation-based approach to inference, the expectations in the numerator and denominator of (21.8) can be evaluated by averaging over posterior simulation draws, as we discuss later in this section.

R^2 will be close to 0 when the average residual error variance is approximately equal to the average variance of the θ_k's. R^2 will be close to 1 when the residual errors ϵ_k are close to zero for each posterior sample. Thus, R^2 is larger when the $\hat{\theta}_k$'s more closely approximate the θ_k's.

In classical least squares regression, (21.8) reduces to the usual definition of R^2: the numerator of the ratio becomes the residual variance, and the denominator is simply the variance of the data. Averaging over uncertainty in the regression coefficients leads to a lower value for R^2, as with the classical "adjusted R^2" measure. We shall discuss this connection further. It is possible for our measure (21.8) to be negative, much like adjusted R^2, if a model predicts so poorly that, on average, the residual error variance is larger than the variance of the data.

As a model improves (by adding better predictors and thus improving the $\hat{\theta}_k$'s), we would generally expect both R^2 and the amount of pooling (discussed more thoroughly in the next section) to increase for all levels of the model. Increasing R^2 corresponds to more of the variation being explained at that level of the regression model, and a high level of pooling implies that the model is pulling the ϵ_k's strongly toward the population mean for that level.

Adding a predictor at one level does *not* necessarily increase R^2 and the amount of pooling at other levels of the model, however. In fact, it is possible for an individual-level predictor to improve prediction at the data level but decrease R^2 at the group level (see Section 21.7). Here we merely note that a model can have different explanatory power at different levels.

Connections to classical definitions

Our general expression for explained variance reduces to classical R^2 for simple linear regression with the least squares estimate for the vector of coefficients. We present this correspondence here, together with the less frequently encountered explained variance for the basic multilevel model.

The classical normal linear regression model can be written as $y_i = X_i\beta + \epsilon_i, i = 1, \ldots, n$, with a $n \times p$ matrix X of predictors and errors ϵ_i that are normal with zero mean and constant variance σ^2.

If we plug in the least squares estimate, $\hat{\beta} = (X^tX)^{-1}X^ty$, then the proportion of variance explained (21.8) simply reduces to the classical definition

$$R^2 = 1 - \frac{\mathrm{E}\left(\overset{n}{\underset{i=1}{\mathrm{V}}}\,\epsilon_i\right)}{\mathrm{E}\left(\overset{n}{\underset{i=1}{\mathrm{V}}}\,y_i\right)} = 1 - \frac{y^t(I-H)\,y}{y^t I_c\,y},$$

where I is the $n \times n$ identity matrix, $H = X(X^tX)^{-1}X^t$, and I_c is the $n \times n$ matrix with $1 - 1/n$ along the diagonal and $1/n$ off the diagonal.

In a simulation-based context, to fully evaluate our expression (21.8) for R^2, one would also average over posterior uncertainty in β and σ. Under the standard noninformative prior density that is uniform on $(\beta, \log \sigma)$, the proportion of variance explained (21.8) becomes

$$R^2 = 1 - \left(\frac{n-3}{n-p-2}\right) \frac{y^t(I-H)\,y}{y^t I_c\, y},$$

where p is the number of columns of X.

This is remarkably similar to the classical adjusted R^2. In fact, if we plug in the classical estimate, $\hat{\sigma}^2 = y^t(I-H)\,y/(n-p)$, rather than averaging over the marginal posterior distribution for σ^2, then (21.8) becomes

$$R^2 = 1 - \left(\frac{n-1}{n-p}\right) \frac{y^t(I-H)\,y}{y^t I_c\, y},$$

which is exactly classical adjusted R^2. Because $\frac{n-3}{n-p-2} > \frac{n-1}{n-p}$ for $p>1$, our Bayesian adjusted R^2 leads to a lower measure of explained variance than the classical adjusted R^2. This makes sense, since the classical adjusted R^2 could be considered too high because it does not account for uncertainty in σ.

Setting up the computations in R and Bugs

We can add the R^2 computation to a multilevel model in three steps:

1. For each level of the model, add one line in the Bugs code to define the residual error: `e.y[i]<-y[i]-y.hat[i]`, and so forth.

2. Add the errors to the list of parameters saved in the `bugs()` call from R.

3. For each level of the model, compute R^2 as defined in (21.8) in R.

We illustrate with the varying-intercept, varying-slope model for the radon data. We take the Bugs model from Sections 17.1–17.2 and add lines at each level to compute the error terms:

Bugs code
```
model {
    for (i in 1:n){
        y[i] ~ dnorm (y.hat[i], tau.y)
        y.hat[i] <- a[county[i]] + b[county[i]]*x[i]
        e.y[i] <- y[i] - y.hat[i]                      # data-level errors
    }
    tau.y <- pow(sigma.y, -2)
    sigma.y ~ dunif (0, 100)

    for (j in 1:J){
        a[j] <- B[j,1]
        b[j] <- B[j,2]
        B[j,1:2] ~ dmnorm (B.hat[j,], Tau.B[,])
        B.hat[j,1] <- g.a.0 + g.a.1*u[j]
        B.hat[j,2] <- g.b.0 + g.b.1*u[j]
        for (k in 1:2){                                # group-level errors
            E.B[j,k] <- B[j,k] - B.hat[j,k]
        }
    }
    . . .
```

In the call to `bugs()`, the parameters `e.y` and `E.B` must be saved[4] (along with `a`, `b`, and any other parameters of interest). We can then compute explained variance in R as follows:

---

[4] The label E.B follows our general convention of labeling matrices with capital letters.

```
rsquared.y <- 1 - mean (apply (e.y, 1, var)) / var (y)                    R code
e.a <- E.B[,,1]
e.b <- E.B[,,2]
rsquared.a <- 1 - mean (apply (e.a, 1, var)) / mean (apply (a, 1, var))
rsquared.b <- 1 - mean (apply (e.b, 1, var)) / mean (apply (b, 1, var))
```

Alternatively, we could define `e.a` and `e.b` by name in the Bugs model.

The inner variances (within the above `apply()` calls) represent variation across the K items at each level (in the radon example, these items are the n data points, the J slopes, and the J intercepts). The outer means represent averages over the simulations representing the uncertainty distributions. The denominator of the expression for R_y^2 has a simpler format because y is an observed vector and does not have any simulation uncertainty.

21.6 Summarizing the amount of partial pooling

When fitting a multilevel model, it can be useful to get a sense of how much pooling is being performed for each group of parameters. The amount of pooling depends on the group-level variance and the information available within each group. Here we define a simple numerical summary of the amount of pooling of a set of group effects α_j, $j = 1, \ldots, J$. These could be varying intercepts in a simple nested model or a batch of varying parameters in a more complicated model with varying slopes and non-nested levels.

Consider the basic multilevel model with data $y_i \sim N(\alpha_{j[i]}, \sigma_y^2)$, with population distribution $\alpha_j \sim N(\mu_\alpha, \sigma_\alpha^2)$ and hyperparameters $\mu_\alpha, \sigma_y, \sigma_\alpha$ known. Let n_j be the number of measurements in each group j, and label $\bar{y}_j$ as the average of the y_i's within the group. For each group j, the multilevel estimate of the parameter α_j is

$$\hat{\alpha}_j^{\text{multilevel}} = \omega_j \mu_\alpha + (1 - \omega_j) \bar{y}_j, \tag{21.9}$$

where

$$\omega_j = 1 - \frac{\sigma_\alpha^2}{\sigma_\alpha^2 + \sigma_y^2 / n_j} \tag{21.10}$$

is a "pooling factor" that represents the degree to which the estimates are pooled together (that is, based on μ_α) rather than estimated separately (based on the raw data $\bar{y}_j$). The extreme possibilities, $\omega = 0$ and 1, correspond to no pooling ($\hat{\alpha}_j = \bar{y}_j$) and complete pooling ($\hat{\alpha}_j = \mu_\alpha$), respectively. The (posterior) variance of the parameter α_j is

$$\text{var}(\alpha_j) = (1 - \omega_j)\sigma_y^2 / n_j. \tag{21.11}$$

The statistical literature sometimes labels $1 - \omega$ as the "shrinkage" factor, a notation we find confusing since a shrinkage factor of zero corresponds to complete shrinkage toward the population mean. To avoid ambiguity, we use the "pooling factor" terminology instead. The form of expression (21.10) matches the form of the definition (21.6) of R^2 in Section 21.5.

The concept of pooling is used to help understand multilevel models in two distinct ways: comparing the estimates of different parameters in a group, and summarizing the pooling of the model as a whole. When comparing, it is usual to consider several parameters α_j with a common population (prior) distribution but different data variances; thus, $\bar{y}_j \sim N(\alpha_j, \sigma_{yj}^2)$. Then ω_j can be defined as in (21.10), with σ_{yj} in place of σ_y. Then each group is associated with a different pooling factor that is larger if the amount of information $(1/\sigma_{yj}^2)$ in the data for that group is small.

Pooling factors for individual coefficients in a multilevel model

The pooling factor for an individual coefficient captures the weighted averaging between the within-group data and the group-level model:

$$\hat{\alpha}_j^{\text{multilevel}} = \omega_j \hat{\alpha}_j^{\text{complete pooling}} + (1 - \omega_j)\hat{\alpha}_j^{\text{no pooling}}. \qquad (21.12)$$

The pooling factor ω_j can range from 0 (no pooling) to 1 (complete pooling). Coefficients with higher values of ω are estimated more from the group-level model and less from the within-group data.

Computing the pooling factor.   It can be helpful to use ω_j to summarize the amount of pooling of a multilevel parameter; however, it is generally impractical to compute it using the implicit definition in (21.12). The difficulty is that additional computations would be required to determine the "no pooling" and "complete pooling" estimates.

Instead, we make use of the following formula from the linear model for a parameter $\alpha_j = \hat{\alpha}_j + \epsilon_j$, where ϵ_j has a $N(0, \sigma_\alpha^2)$ prior distribution:

$$\text{pooling factor } \lambda_j = \frac{(\text{standard error of } \epsilon_j)^2}{\sigma_\alpha^2}. \qquad (21.13)$$

Computation in Bugs.   To apply formula (21.13), we must perform inferences about the group-level errors ϵ_j, which can be defined easily enough in the Bugs model. For example, the varying intercepts for the counties j in the radon example have the model, $\alpha_j \sim N(\hat{\alpha}_j, \sigma_\alpha^2)$, where the $\hat{\alpha}_j$'s are the linear predictors at the group level. In the Bugs code, we simply add this sort of line inside the "`for (j in 1:J)`" loop:

Bugs code
```
e.a[j] <- a[j] - a.hat[j]
```

In the R code, we must then add `e.a` to the list of parameters to be saved, and then we can calculate the vector of J pooling factors. For example,

R code
```
omega <- (sd(e.a)/sigma.a)^2
omega <- pmin (omega, 1)
```

The second line above[5] is needed to keep the estimated pooling factors below 1, which occasionally happens due to simulation variability (and can also happen with non-normal models).

Figure 21.8a illustrates with a plot of the pooling factors versus sample sizes for the varying intercepts for the 85 counties in the radon model with the floor predictor and county effects but no county-level predictors. The counties with small sample sizes have pooling factors ω_j near 1 (that is, nearly complete pooling), whereas the larger counties have pooling factors near 0, close to the no-pooling estimates.

Figure 21.8b displays pooling factors for the radon model with floor of measurement as an individual-level predictor and uranium as a county-level predictor. Adding the county-level predictor increases the pooling, which is appropriate since the new model has a lower group-level error. (As discussed in Section 12.6, σ_α declines from 0.34 to 0.13 when county uranium is added to the model.) There is more pooling to this better-fitting model, yielding more precise estimates of the county effects.

---

[5] The R function `pmin()` performs parallel minimization, in this case computing, for each element of ω, the minimum of itself and 1.

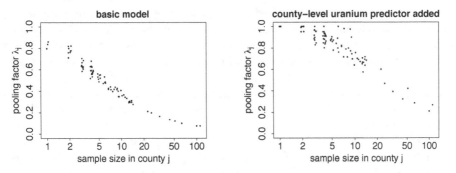

Figure 21.8 *Pooling factors for county intercepts in two versions of the radon model, (a) with no county-level predictors and (b) including the county-level uranium predictor. For each model, ω_j decreases with sample size—that is, the most pooling occurs with the small counties. There is more pooling in the second model, which makes sense—adding the county-level predictor reduces the variance of the group-level errors.*

Summary pooling factor for each batch of parameters

We can define a summary measure, λ, for the average amount of pooling at each level of a multilevel model, summarizing the extent to which the variance of the residuals ϵ_k is reduced by the pooling of the multilevel model:

$$\lambda = 1 - \frac{\overset{K}{\underset{k=1}{V}} \, E(\epsilon_k)}{E\left(\overset{K}{\underset{k=1}{V}} \, \epsilon_k\right)}. \qquad (21.14)$$

The denominator in this expression is the numerator in expression (21.8)—the average variance in the ϵ_k's, that is, the unexplained component of the variance of the θ_k's. The numerator in the ratio term of (21.14) is the variance among the point estimates (the partial-pooling estimators) of the ϵ_k's. If this variance is high (close to the average variance in the ϵ_k's), then λ will be close to 0 and there is little pooling. If this variance is low, then the estimated ϵ_k's are pooled closely together, and the pooling factor λ will be close to 1.

The striking similarity of expressions (21.6) and (21.14), which define R^2 and λ, respectively, suggests that the two concepts can be understood in a common framework. We consider each to represent the fraction of variance explained, first by the linear predictor μ and then by the multilevel model for ϵ.

Setting up the computation in R and Bugs

We can compute λ at each level using the errors e.y, e.a, e.b defined at the end of Section 21.5 for computing R^2. Once the Bugs model has been fit, the pooling factors can be computed in R as follows:

```
lambda.y <- 1 - var (apply (e.y, 2, mean)) / mean (apply (e.y, 1, var))      R code
lambda.a <- 1 - var (apply (e.a, 2, mean)) / mean (apply (e.a, 1, var))
lambda.b <- 1 - var (apply (e.b, 2, mean)) / mean (apply (e.b, 1, var))
```

Discussion

The proportion of variance explained (21.8) and the pooling factor (21.14) can be easily calculated at each stage of a multilevel model. In general, R^2 will be informative wherever regression predictors (including group indicators) are present, and λ will be relevant at the hierarchical stages of the model. The measures of explained variance and partial pooling conveniently summarize the fit at each level of the model and the degree to which estimates are pooled toward their population models. Together, they clarify the role of predictors at different levels of a multilevel model. They can be derived from a common framework of comparing variances at each level of the model, which also means that they do not require the fitting of additional null models.

Expressions (21.8) and (21.14) are closely related to the usual definitions of adjusted R^2 in simple linear regression and pooling in balanced one-way hierarchical models. From this perspective, they unify the data-level concept of R^2 and the group-level concept of pooling or shrinkage, and also generalize these concepts to account for uncertainty in the variance components. Further, as illustrated for the radon application, they can help us understand more complex multilevel models.

Other challenges include defining explained variance and partial pooling factors for generalized linear models, either on the scale of the data or of the latent parameters.

21.7 Adding a predictor can *increase* the residual variance!

Multilevel models can behave in ways that are unexpected from the perspective of classical statistics. We illustrate with the radon model from Chapter 12. We first fit a stripped-down multilevel model for the home radon levels, including the county-level uranium predictor but no individual-level predictors (not even the floor of measurement); thus,

$$\text{model 1:} \quad \begin{aligned} y_i &\sim \text{N}(\alpha_{j[i]}, \sigma_y^2) \\ \alpha_j &\sim \text{N}(\gamma_0 + \gamma_1 u_j, \sigma_\alpha^2). \end{aligned}$$

Fitting this model to the Minnesota radon data yields estimated variance components $\sigma_y = 0.80, \sigma_\alpha = 0.12$.

We then add the house-level floor indicator x_i:

$$\text{model 2:} \quad \begin{aligned} y_i &\sim \text{N}(\alpha_{j[i]} + \beta x_i, \sigma_y^2) \\ \alpha_j &\sim \text{N}(\gamma_0 + \gamma_1 u_j, \sigma_\alpha^2), \end{aligned}$$

yielding new estimates of $\sigma_y = 0.76, \sigma_\alpha = 0.16$. The house-level standard deviation has decreased—which makes sense since we have added a predictor at that level—but the variation at the county level has *increased*, which is a surprise. In classical regression models, the residual variance can only go down, not up, when a predictor is added.[6]

What is going on? After some thought, we realized that in model 2, the counties with more basements happened to have higher county coefficients α_j. In model 1, some of the variation in county radon levels was canceled by an opposite variation in the proportion of basements. The increased between-county variance in model 2 indicates true variation among counties that happened to be masked by the first model.

Figure 21.9 shows a hypothetical extreme version of this situation: the three counties have identical average radon levels (thus, $\sigma_\alpha = 0$ for model 1, which has no basement predictor), but only because the naturally low-radon county has many basements and the naturally high-radon county has few basements. (Such a pattern

---

[6] We ignore the minor increase in the variance estimate that corresponds to reducing the degrees of freedom by 1 and thus dividing by $n - k - 1$ instead of $n - k$ in the variance calculation.

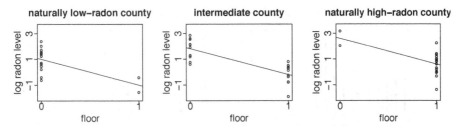

Figure 21.9 *Hypothetical data from three counties illustrating how adding an individual-level predictor can decrease the group-level variance. In each county, radon levels are higher in homes with basements. The county with low natural radon levels has more homes with basements, and the county with high natural radon levels has fewer homes with basements. As a result, the average radon level in the three counties is identical, but when floor of measurement is (appropriately) included as a predictor, the counties appear more different.*

can happen, for example, if low-radon areas have sandy soil in which basements are easy to dig, with high-radon areas having rocky soil where basements are less commonly built.) Model 2, which controls for basements, reveals the true underlying variation among the counties, and thus σ_α increases.

This pattern, caused by correlation between individual-level variables and group-level errors, does not occur in classical regression. When it occurs in multilevel regression, the model fit can be improved by including the average of x as a group-level predictor (in this example, the proportion of houses in the county that have basements). When the county-level basement proportion is added as a group-level predictor, its coefficient is estimated at -0.41 (with a standard error of 0.2), and the estimated residual standard deviations at the data and county levels are 0.76 and 0.14.

For the radon problem, the county-level basement proportion is difficult to interpret directly but rather serves as a proxy for underlying variables (for example, the type of soil that is prevalent in the county).

In other settings, especially in social science, individual averages that are used as group-level predictors are often interpreted as "contextual effects." For example, in the police stops example in Section 15.1, one might suspect that police behavior in a precinct is influenced by the ethnic composition of the local residents. However, we must be suspicious of this sort of conclusion without further information. As the radon example illustrates, it is possible to have between-level correlations without the need for a "contextual" story. As usual in these models, we try to be careful to state the regression results in a predictive rather than causal manner ("the counties with more basements tend to have lower radon levels, after controlling for the basement statuses of the individual houses").

21.8 Multiple comparisons and statistical significance

A meta-analysis of a set of randomized experiments

In the wake of concerns about the health effects of low-frequency electric and magnetic fields, an experiment was performed to measure the effect of electromagnetic fields at various frequencies on the functioning of chick brains. At each of several frequencies of electromagnetic fields (1 Hz, 15 Hz, 30 Hz, ..., 510 Hz), a randomized experiment was performed to estimate the effect of exposure, compared to a control condition of no electromagnetic field. The researchers reported, for each fre-

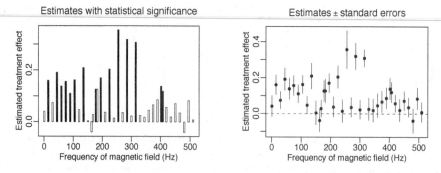

Figure 21.10 *(a) Estimated effects of electromagnetic fields on calcium efflux from chick brains, shaded to indicate different levels of statistical significance, adapted from Blackman et al. (1988). A separate experiment was performed at each frequency. (b) Same results presented as estimates ± standard errors. As discussed in the text, the first plot, with its emphasis on statistical significance, is misleading.*

quency, the estimated treatment effect (the average difference between treatment and control measurements) and the standard error (that is, $\sqrt{\sigma_T^2/n_T + \sigma_C^2/n_C}$; see Section 2.3).

In the article reporting this study, the estimates at the different frequencies were summarized by their statistical significance, as we illustrate in Figure 21.10a by using different shading for results that are more than 2.3 standard errors from zero (that is, statistically significant at the 99% level), between 2.0 and 2.3 standard errors from zero (statistically significant at the 95% level), and so forth. The researchers used this sort of display to hypothesize that one process was occurring at 255, 285, and 315 Hz (where effects were highly significant), another at 135 and 225 Hz (where effects were only moderately significant), and so forth. The estimates are all of relative calcium efflux, so that an effect of 0.1, for example, corresponds to a 10% increase compared to the control condition.

In the chick brain experiment, the researchers made the common mistake of using statistical significance as a criterion for separating the estimates of different effects. As we discuss in Section 2.5, this approach does not make sense. At the very least, it is more informative to show the estimated treatment effect and standard error at each frequency, as in Figure 21.10b.

Multilevel model

The confidence intervals at different frequencies in Figure 21.10b overlap substantially, which implies that the estimates could be usefully pooled using a multilevel model. The raw data from these experiments were not available,[7] so we analyzed the estimates, which we shall label as y_j for each subexperiment j. Because the chicken brains were randomly assigned to the treatment and control groups, we can assume the y_j's are unbiased estimates of the treatment effects θ_j; and because the sample sizes were not tiny, it is reasonable to assume that the estimation errors are approximately normally distributed; thus,

$$y_j \sim \mathrm{N}(\theta_j, \sigma_j^2).$$

---

[7] When we asked the experimenter to share the data with us, he refused. This was disturbing considering this was a government-funded study of public health interest, but we did not feel like putting in the effort to wrest the data from him.

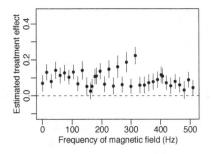

Figure 21.11 *Multilevel estimates and standard errors for the effects of magnetic fields, partially pooled from the separate estimates displayed in Figure 21.10. The standard errors of the original estimates were large, and so the multilevel estimates are pooled strongly toward the common mean estimate of 0.1.*

Our default model for the treatment effects is simply

$$\theta_j \sim \mathrm{N}(\mu_\theta, \sigma_\theta^2).$$

If we assume each σ_j is known and equal to the standard error of the estimate y_j, we can easily perform inference about the θ_j's (as well as the hyperparameters $\mu_\theta, \sigma_\theta$) using Bugs, supplying y, σ, and J as data:

```
model {
  for (j in 1:J){
    y[j] ~ dnorm (theta[j], tau.y[j])
    tau.y[j] <- pow(sigma.y[j], -2)
  }
  for (j in 1:J){
    theta[j] ~ dnorm (mu.theta, tau.theta)
    e.theta[j] <- theta[j] - mu.theta
  }
  tau.theta <- pow(sigma.theta, -2)
  mu.theta ~ dnorm (0, .0001)
  sigma.theta ~ dunif (0, 100)
}
```
                                                                    Bugs code

(The line defining `e.theta` is there to allow the computation of the partial pooling factor λ, which turns out to be 0.49, implying that the estimates are pooled, on average, halfway toward the group-level model, which in this case is simply the average of all the treatment effects.)

Figure 21.11 displays the inferences for the treatment effects θ_j, as estimated from the multilevel model. The inferences shown here represent partial pooling of the separate estimates y_j toward the grand mean μ_θ.

More generally, the multilevel model can be seen as a way to estimate the effects at each frequency j, without setting "nonsignificant" results to zero. Some of the apparently dramatic features of the original data as plotted in Figure 21.10a—for example, the negative estimate at 480 Hz and the pair of statistically significant estimates at 405 Hz—do not stand out so much in the multilevel estimates, indicating that these features could be easily explained by sampling variability and do not necessarily represent real features of the underlying parameters.

Potential criticisms of the multilevel model

The above multilevel model can be criticized because it pools all the estimates toward their common mean of 0.1 with an assumed normal distribution for the true θ_j's. This would not be appropriate if, for example, the treatment had a positive effect at some frequencies and a zero effect at others. Or, as hypothesized by the authors of the original study, that the true treatment effects could fall into three groups: a set of large effects near 0.3, a set of moderate effects near 0.15, and a set of zero effects.

We could explore this possibility by fitting a mixture model for the θ_j's. But the resulting inferences would not differ much from our multilevel analysis that used a normal distribution. The reason that changing the model would not do much is that the uncertainty bounds for the individual estimates are so high. Even if, for example, we fit a model with three clusters of effects, it would not be so clear which points correspond to which cluster. The estimates at 255, 285, and 315 Hz appear to be one cluster, but in fact the point at 255 Hz has a high standard error (see Figure 21.10b) and could very well belong to a lower cluster, whereas the estimates at 45, 135, 225, and other points are consistent with being in the higher group. Similarly, several of the estimates are not statistically significant (see Figure 21.10a) but they are almost all positive, and in aggregate they do not appear to be zero. There is certainly no sharp dividing line between the "low" and the "moderate" estimates.

To put it another way: we do not "believe" the estimates in Figure 21.11 in the same way as we trust the estimated county radon levels in Chapters 12 and 13. The difference is that the lognormal distribution for county radon levels seems reasonable enough (in that radon in a house is affected by many small multiplicative factors), whereas it seems more plausible that the effect of electromagnetic fields on calcium efflux could be concentrated at a few frequencies. However, given the variability in the parameter estimates, we believe the unpooled estimates in Figure 21.10 even less, and we would recommend using the multilevel model and estimates as a starting point for further analysis of these data.

Relevance to multiple comparisons

One of the risks in statistical analysis of complex data is overinterpretation of patterns that could be explained by random variation. Multilevel modeling is sometimes viewed with skepticism as just one more kind of model that can be fit without the evidence of the data. Actually, though, multilevel modeling can reduce overinterpretation. For example, Figure 21.10 shows a dramatic pattern of three points that stand apart from all the others, but the multilevel estimates in Figure 21.11 show these to be part of a larger group of relatively large effects for frequencies up to 300 Hz or so. The partial pooling has revealed the fragility of the patterns in the raw data (or in the no-pooling estimates), which can be explained by sampling variability.

21.9 Bibliographic note

Achen (1982), Aitkin and Longford (1986), and Kiss (2003) present some methods for understanding multilevel models. Textbooks such as Ramsey and Schafer (2001) and Fox (2002) are useful places to start.

The distinction between finite-population and superpopulation variances is fundamental in multilevel models and has been considered by many researchers, including Searle, Casella, and McCulloch (1992) and Gelman (2005). The different definitions of "fixed" and "random" effects, with references, appear in Gelman (2005). See also Kreft and De Leeuw (1998, section 1.3.3) for a discussion of the multiplicity of definitions of fixed and random effects and coefficients, and Robinson (1998) for a historical overview. See Rosenthal, Rosnow, and Rubin (2000) for more on contrasts in linear models.

The material on average predictive comparisons is from Gelman and Pardoe (2007); the prison example appears in Pardoe and Weidner (2006) along with a lively discussion. See also Graubard and Korn (1999), King, Tomz, and Wittenberg (2000), Pardoe (2001), and Pardoe and Cook (2002) for related work on numerical and graphical summaries of marginal predictive comparisons. The measures of R^2 and partial pooling for multilevel models come from Gelman and Pardoe (2006); see Wherry (1931), Pratt (1987), Bentler and Raykov (2000), Goldstein, Browne, and Rasbash (2002), Afshartous and De Leeuw (2002), Xu (2003), Gustafson and Clarke (2004), and Browne et al. (2005) for related ideas.

Kane, Rockoff, and Staiger (2006) describe the study of value added in teaching; some important earlier work in this area includes Hanushek (1971), Summers and Wolfe (1977), and Ehrenberg and Brewer (1994). The Advanced Placement study is described by Wainer, Gelman, and Wang (2000). The example of the increasing residual variance comes from Gelman and Price (1998). The chick brain studies, along with the inappropriate analysis based on statistical significance, come from Blackman et al. (1988). For more on multiple comparisons in hierarchical models, see Gelman and Tuerlinckx (2000).

21.10 Exercises

1. Uncertainty and variability: for the radon model in Section 21.1, give examples of how the parameter estimates and standard deviations might look if the sample size is increased in the following ways:

   (a) 4 times as many houses measured within each existing county.

   (b) 4 times as many counties, but the same number of houses measured in each county.

   (c) 4 times as many counties, with 4 times the number of houses measured in each county (thus, 16 times as many houses in total).

2. Superpopulation and finite-population standard deviations: fit a non-nested multilevel model to the Winter Olympics data (see Exercises 11.3 and 13.3).

   (a) Fit the model using `lmer()` to get quick estimates of the standard-deviation parameters (at the levels of data, judge, and skater).

   (b) Fit the model in Bugs and get point estimates and 50% intervals for the superpopulation standard-deviation parameters and the corresponding finite-sample standard deviations.

3. Superpopulation and finite-population standard deviations:

   (a) Fit a varying-intercept, varying-slope model to the data in the folder **radon**. Get point estimates and 50% intervals for the superpopulation and finite-population standard deviations.

   (b) Repeat step (a) but just using a sample of 10 counties. The inference for the superpopulation standard deviation should now be much more uncertain than the finite-population standard deviation.

(c) Repeat but just using a sample of 5 counties. The inferences should be even more different now.

4. Contrasts: fit in Bugs a varying-intercept model to the radon data with log uranium as a group-level predictor. You will compare inferences for the superpopulation contrast (that is, the slope for log uranium in the county-level model) and the corresponding finite-population contrast (that is, the coefficient of log uranium for the intercepts for the particular counties in the data). You will need to postprocess the simulations in R in order to get simulations for the finite-population contrast.

(a) Compare the inferences (estimates and standard errors) for the superpopulation and finite-population contrasts.

(b) Repeat part (a), but fitting the model just to the first three counties in the dataset.

5. Average predictive comparisons:

(a) Take the model of well switching from Exercise 14.2 and estimate average predictive comparisons for the input variables in the model (including the index variable for villages).

(b) Take the model of rodent infestation from Exercise 14.3 and estimate average predictive comparisons for the input variables in the model (including the index variables for buildings and for community districts).

6. Explained variance and partial pooling: go to the model you fit to the one-fifth sample of the radon data in Exercise 12.9. Fit the model in Bugs and compute the percentage of explained variance and partial pooling factor at each of the two levels of the model.

Analysis of variance

Analysis of variance (ANOVA) refers to a specific set of methods for data analysis and to a way of summarizing multilevel models:

- As a tool for data analysis, ANOVA is typically used to learn the relative importance of different sources of variation in a dataset. For example, Figure 13.8 displays success rates of pilots at a flight simulator under five different treatments at eight different airports. How much of the variation in the data is explained by treatments, how much by airports, and how much remains after these factors have been included in a linear model?

- If a multilevel model has already been fit, it can be summarized by the variation in each of its batches of coefficients. For example, in the radon modeling in Chapter 12, how much variation in radon levels is explained by floor of measurement and how much by geographical variation? Or, in the analysis of public opinion by state in Section 14.1, how much of the variation is explained by demographic factors (sex, age, ethnicity, education), and how much by states and regions?

These "analysis of variance" questions can be of interest even for models that are primarily intended for prediction, or for estimating particular regression coefficients.

The sections of this chapter address the different roles of ANOVA in multilevel data analysis. We begin in Section 22.1 with a brief review of the goals and methods of classical analysis of variance, outlining how they fit into our general multilevel modeling approach. Sections 22.2 and 22.3 explain how ANOVA (or, more precisely, a set of computations that are inspired by classical ANOVA) can be used to summarize inferences from multilevel models.

Having showed how to compute the ANOVA corresponding to a given multilevel model, we discuss the converse in Section 22.4: if an ANOVA decomposition is desired, how to set up and interpret the corresponding model. We start with simple one-way and two-way structures and move to more complicated designs such as latin squares and split plots.

In Section 22.5, we discuss two statistical methods associated with ANOVA—the analysis of covariance and contrast analysis—and interpret them as multilevel models with individual-level and group-level predictors, respectively. We cannot cover all the topics of the analysis of variance in this chapter, but we hope to show the connections with multilevel modeling to make it clear how to construct appropriate models for the highly structured data that arise in many application areas, especially those with designed experiments.

22.1 Classical analysis of variance

In classical statistics, ANOVA refers either to a family of additive data decompositions, or to a method of testing the statistical significance of added predictors in a linear model. We shall discuss each of these interpretations in turn and then explain

```
> summary (aov (y ~ factor (treatment) + factor(airport)))
```

	Df	Sum Sq	Mean Sq	F value	Pr(>F)	
factor(treatment)	4	0.0783	0.0196	0.3867	0.8163	
factor(airport)	7	3.9437	0.5634	11.1299	1.187e-06	***
Residuals	28	1.4173	0.0506			

Figure 22.1 *Classical analysis of variance (as computed in R) for the flight simulator data. The usual focus of this sort of analysis is on the p-values, which indicate that the variation among treatments is* not *statistically significant (that is, it could be explained by chance alone), but the airport variation cannot be plausibly attributed to chance. Compare to the multilevel ANOVA display in Figure 22.5 on page 495.*

what parts of classical ANOVA we will keep and what parts we will discard when moving to multilevel analysis.

Classical ANOVA as additive data decomposition

In many examples with multilevel structure, it is helpful to perform a simple "data decomposition." For the flight simulator data indexed by treatments i and airports j, we can write

$$y_i = \mu + \gamma_{j[i]} + \delta_{k[i]} + \epsilon_i,$$

or, equivalently,

$$y_{jk} = \mu + \gamma_j + \delta_k + \epsilon_{jk},$$

in either case decomposing the data y into treatment effects, airport effects, and residuals. In this case, with one observation i per cell (j, k), the residuals are equivalent to treatment $\times$ airport interactions.

In general, additive decompositions are equivalent to regressions on index variables and their interactions, and the classical analysis of variance can be viewed as a summary of an additive decomposition. In classical ANOVA, the model is estimated using least squares, with the estimate of each batch of coefficients (except for the mean level μ) constrained to sum to 0. However, the focus of interest is typically not the coefficient estimates but rather their variances, as we discuss next, in the context of the ANOVA table.

Sources of variation and degrees of freedom. Figure 22.1 illustrates for the flight simulator data. Each row of the table represents a set of index variables: the treatments j, airports k, and residuals i, with *degrees of freedom* (Df) defined as the number of coefficients in that group, minus the number of constraints required for the coefficients to be identifiable in a classical regression. Thus,

- 5 treatment effects minus 1 constraint = 4 degrees of freedom

- 8 airport effects minus 1 constraint = 7 degrees of freedom

- 40 residuals minus 12 constraints (1 mean, 4 treatment effects, and 7 airport effects) = 28 degrees of freedom.

The degrees of freedom can be more formally defined in the language of matrix algebra, but we shall not go into such details here.

Sums of squares. To continue with the description of Figure 22.1, the *sum of squares* for each row of the table is derived from the classical coefficient estimates (recall that these are constrained to sum to 0 in the least squares estimate). Thus, the sums of squares for treatments, airports, and residuals are $\sum_{i=1}^{40} \hat{\gamma}_{j[i]}^2$, $\sum_{i=1}^{40} \hat{\delta}_{k[i]}^2$, and $\sum_{i=1}^{40} \hat{\epsilon}_i^2$, respectively.

Balance. If the data are *balanced*, then the sums of squares in the table add up to the "total sum of squares" of the data, $\sum_{i=1}^{40}(y_i - \hat{\mu})^2$. Roughly speaking, in a balanced design there are the same number of observations in each row and each column of the data. The flight simulator example is trivially balanced because the data are complete, with 8 observations for each treatment, 5 for each airport, and 1 for each treatment × airport interaction. The data would be unbalanced if, for example, we were missing the data from the YN condition for Nagoya (recall Figure 13.8 on page 290), or if we had replications for some cells and not for others. Balance is much less important in multilevel modeling than in classical ANOVA, and we do not discuss it further here.

Mean squares, F ratios, and p-values. For each source of variation in the ANOVA table, the *mean square* is defined as the sum of squares divided by the degrees of freedom. The mean square is important in classical ANOVA because it can be used for hypothesis testing. If a given row in the table actually has zero variation, then its mean square will, on average, equal the mean square of the residuals in the model. The ratios of mean squares are called the *F statistics*, and the usual goal of classical ANOVA is to find F-ratios that are significantly greater than 1.

In the flight simulator example, the treatment mean square is less than the residual mean square, indicating no evidence for variation between treatments. Airport effects, however, seem to be present—their mean square is 11.1 times greater than the residual mean square. The discrepancy of a mean square from the "null hypothesis" value of 1 is tested using the F_{ν_1, ν_2}, where ν_1 and ν_2 are the degrees of freedom of the numerator and denominator, respectively. The p-values in the table indicate the statistical significance of the F-tests. As usual, the p-values are only considered significant if they are close to 0 or 1 (typically, if less than 0.05 or greater than 0.95). Thus, that the treatment mean square is *not* statistically significantly less than what would be expected under the null hypothesis, but the airport mean square is statistically significant, indicating that we can reject the hypothesis that the "airport" factor has no effect on the outcome. Equivalently, we can say that the between-airport variation is greater than what could be expected by chance, given the variation in the data.

Classical ANOVA for model comparison

We have just illustrated how ANOVA is used to summarize data, and how this ANOVA corresponds to a "default" linear model with indicator variables. We now discuss the other classical role of the analysis of variance, which is to summarize hypothesis tests within an existing family of models. The basic idea is that an analyst has fit a linear regression model and is considering fitting a larger model formed by adding predictors to the first model. (The two models are said to be *nested*, with this term having a different meaning from its use in multilevel models as in Chapters 12 and 13.)

When comparing nested models, ANOVA is related to the classical test of the hypothesis that the smaller model is true, which is equivalent to the hypothesis that the additional predictors all have coefficients of zero when included in the larger model. Suppose the smaller model has k_1 predictors, the larger model has $k_1 + k_2$ predictors, and each model is fit to n data points. If the predictions from the smaller and larger regressions are $X_1 \hat{\beta}_1$ and $X_2 \hat{\beta}_2$, respectively, then the classical ANOVA for testing the model expansion can be written as

	Df	SS	MS
Model expansion	k_2	$\sum_i (y_i - X_1\hat{\beta}_1)^2 - \sum_i (y_i - X_2\hat{\beta}_2)^2$	SS/k_2
Residuals	$n-k_1-k_2$	$\sum_i (y_i - X_2\hat{\beta}_2)^2$	$SS/(n-k_1-k_2)$

If the ratio of these sums of squares is statistically significantly greater than 1 (as compared to the $F_{k_2,n-k_1-k_2}$ distribution), then the improvement in fit from the model expansion cannot be reasonably explained by chance.

22.2 ANOVA and multilevel linear and generalized linear models

When moving to multilevel modeling, the key idea we want to take from the analysis of variance is the estimation of the importance of different batches of predictors ("components of variation" in ANOVA terminology). As usual, we focus on estimation rather than testing: instead of testing the null hypothesis that a variance component is zero, we estimate the standard deviation of the corresponding batch of coefficients. If this standard deviation is estimated to be small, then the source of variation is minor—we do not worry about whether it is exactly zero. In the social science and public health examples that we focus on, it can be a useful research goal to identify important sources of variation, but it is rare that anything is truly zero.

Notation

As always, any multilevel model can be expressed in several ways. For ANOVA, we write the models to emphasize the grouping of regression coefficients into "sources of variation," with each batch corresponding to one row of the ANOVA table. We use the notation $m = 1, \ldots, M$ for the rows of the table. Each row m represents a batch of J_m regression coefficients $\beta_j^{(m)}$, $j = 1, \ldots, J_m$. We denote the mth subvector of coefficients as $\beta^{(m)} = (\beta_1^{(m)}, \ldots, \beta_{J_m}^{(m)})$ and the corresponding classical least squares estimate as $\hat{\beta}^{(m)}$. These estimates are subject to c_m linear constraints, yielding $(df)_m = J_m - c_m$ degrees of freedom. We label the constraint matrix as $C^{(m)}$, so that $C^{(m)}\hat{\beta}^{(m)} = 0$ for all m. For notational convenience, we label the grand mean as $\beta_1^{(0)}$, corresponding to the (invisible) zeroth row of the ANOVA table and estimated with no linear constraints.

In classical ANOVA, a linear model is fit to the data points y_i, $i = 1, \ldots, n$, and can be written as

$$y_i = \sum_{m=0}^{M} \beta_{j_i^m}^{(m)} \qquad (22.1)$$

where j_i^m indexes the appropriate coefficient j in batch m corresponding to data point i. Thus, each data point pulls one coefficient from each row in the ANOVA table. Equation (22.1) could also be expressed as a linear regression model with a design matrix composed entirely of 0's and 1's. The coefficients β_j^M of the last row of the table correspond to the residuals or error term of the model.

ANOVA can also be applied more generally to regression models, in which case we can have any design matrix X, and (22.1) would be generalized to

$$y_i = \sum_{m=0}^{M} \sum_{j=1}^{J_m} X_{ij}^{(m)} \beta_j^{(m)}. \qquad (22.2)$$

The essence of analysis of variance is in the structuring of the coefficients into batches—hence the notation $\beta_j^{(m)}$—going beyond the usual linear model formulation that has a single indexing of coefficients β_j. We assume that the structure (22.1), or the more general regression parameterization (22.2), has already been constructed using knowledge of the data structure. To use ANOVA terminology, we assume the sources of variation have already been set, and our goal is to perform inference for each variance component.

We shall follow our usual practice and model each batch of regression coefficients as a sample from a normal distribution with mean 0 and superpopulation standard deviation σ_m:

$$\beta_j^{(m)} \sim N(0, \sigma_m^2), \quad \text{for } j = 1, \ldots, J_m, \quad \text{for each batch } m = 1, \ldots, M. \quad (22.3)$$

With factors that have only a finite number of levels (for example, 50 states), the superpopulation is difficult to intepret in itself except as a tool that allows better inferences for the individual coefficients.

We model the underlying coefficients β as unconstrained (unlike the least squares estimates) but in many cases will summarize them by subtracting off their averages, as in Section 19.4. The mean of 0 for each batch in (22.3) comes naturally from the ANOVA decomposition structure (pulling out the grand mean, main effects, interactions, and so forth), and the standard deviations represent the magnitudes of the variance components corresponding to each row of the table.

The finite-population standard deviation

One measure of the importance of each row or "source" in the ANOVA table is the standard deviation of its constrained regression coefficients, the finite-population standard deviation

$$s_m = \sqrt{\frac{1}{J_m - 1} \sum_{j=1}^{J_m} \left(\beta_j^{(m)} - \bar{\beta}^{(m)} \right)^2}, \quad (22.4)$$

where $\bar{\beta}^{(m)} = \sum_{j=1}^{J_m} \beta_j^{(m)} / J_m$. As discussed in the context of definitions (21.1) on page 460, s_m captures the variation in the existing J_m levels of factor m in the data, in comparison to σ_m, which reflects the potential uncertainty in the superpopulation.

Variance estimation is often presented in terms of the superpopulation standard deviations σ_m, but in our ANOVA summaries, we focus on the finite-population quantities s_m. However, for computational reasons, the parameters σ_m are sometimes useful intermediate quantities to estimate.

Generalized linear models

The multilevel ANOVA framework does not require the normal distribution or, for that matter, even linearity. All that is needed is that the parameters β can be grouped into reasonable batches, with the magnitude of each batch summarized by a standard deviation. We illustrate this process in the next section with a logistic regression. For generalized linear models, coefficients and variance parameters on the logarithmic or logit scales can be interpreted as discussed in Chapters 5–6 and 14–15.

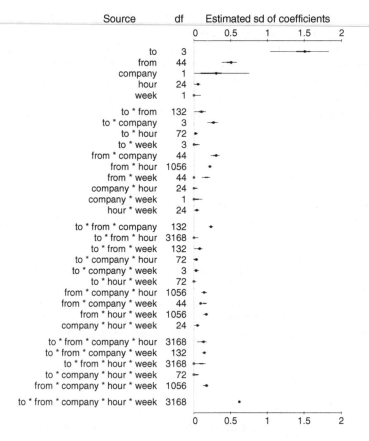

Figure 22.2 *ANOVA display for the World Wide Web data. The bars indicate 50% and 95% intervals for the finite-population standard deviations s_m. The display makes apparent the magnitudes and uncertainties of the different components of variation. Since the data are on the logarithmic scale, the standard-deviation parameters can be interpreted directly. For example, $s_m = 0.20$ corresponds to a coefficient of variation of $\exp(0.2) - 1 \approx 0.2$ on the original scale, and so the exponentiated coefficients $\exp(\beta_j^{(m)})$ in this batch correspond to multiplicative increases or decreases in the range of 20%. (The dots on the bars show simple classical estimates of the variance components that were used as starting points in the multilevel analysis.)*

22.3 Summarizing multilevel models using ANOVA

It can be helpful to graph the estimates of variance components, especially for complex data structures with many levels of variation. In basic multilevel models (that is, the models covered in this book), each variance parameter corresponds to a set of coefficients—for example, $y \sim N(X\beta, \sigma_y^2)$, or $\alpha \sim N(\mu_\alpha, \sigma_\alpha^2)$. As discussed in Section 21.2, the standard deviation of a set of coefficients gives a sense of their predictive importance in the model. An analysis-of-variance plot, which shows the relative scale of different variance components, can be a useful tool in understanding a model.

A five-way factorial analysis: internet connect times

We illustrate the analysis of variance with an example of a linear model fitted for exploratory purposes to a highly structured dataset.

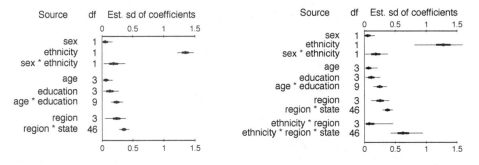

Figure 22.3 *Analysis of variance (ANOVA) display for two logistic regression models of the probability that a survey respondent prefers the Republican candidate for the 1988 U.S. presidential election, based on data from seven CBS News polls. Point estimates and error bars show median estimates, 50% intervals, and 95% intervals of the finite-population standard deviations s_m. The demographic factors are those used by CBS to perform non-response adjustments, and states and regions are included because we were interested in estimating average opinions by state. The large coefficients for ethnicity, region, and state suggest that it might make sense to include interactions, hence the inclusion of the ethnicity $\times$ region and ethnicity $\times$ state coefficients in the second model.*

Data were collected by an internet infrastructure provider on connect times for messages processed by two different companies. Messages were sent every hour for 25 consecutive hours, from each of 45 locations to 4 different destinations, and the study was repeated one week later. It was desired to quickly summarize these data to learn about the importance of different sources of variation in connect times.

Figure 22.2 shows the Bayesian ANOVA display for an analysis of logarithms of connect times on the five factors: destination ("to"), source ("from"), service provider ("company"), time of day ("hour"), and week. The data have a full factorial structure with no replication, so the full five-way interaction at the bottom represents the "error" or lowest-level variability.

Each row of the plot shows the estimated finite-population standard deviation of the corresponding group of parameters, along with 50% and 95% uncertainty intervals. We can immediately see that the lowest-level variation is more important in variance than any of the factors except for the main effect of the destination. Company has a large effect on its own and, perhaps more interestingly, in interaction with to, from, and in the three-way interaction. (By comparison, a classical analysis of variance reveals that all the main effects and almost all the interactions are statistically significant, but it does not give a good sense of the relative magnitudes of the different variance components.)

Figure 22.2 would not normally represent the final statistical analysis for this sort of problem. The ANOVA plot represents a default model and is a tool for data exploration—for learning about which factors are important in predicting the variation in the data—and can be used to construct more focused models or design future data collection.

A multilevel logistic regression: vote preference broken down by state and demographics

We illustrate the use of ANOVA for understanding an existing model with the vote preference study described in Section 14.1. There, we focused on hierarchical modeling as a tool for estimating state opinions; here, we examine the fitted models

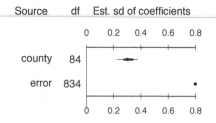

Figure 22.4 *One-way analysis of variance for the radon data. The dots, thick lines, and thin lines represent medians, 50% intervals, and 95% intervals for the finite-population standard deviation s, for each source of variation.*

themselves to see the relative importance of different inputs in predicting vote preferences. The left plot of Figure 22.3 displays the analysis of variance from the basic model, which shows that ethnicity is by far the most important demographic factor, with state also explaining quite a bit of variation.

The natural next step is to consider interactions among the most important factors, as shown in the plot on the right side of Figure 22.3. The ethnicity × state × region interactions are surprisingly large: the differences between African Americans and others vary dramatically by state.

22.4 Doing ANOVA using multilevel models

Our general solution to the ANOVA problem is simple: we treat *every* row in the table as a batch of "varying coefficients"—that is, a set of regression coefficients drawn from a distribution with mean 0 and some standard deviation to be estimated from the data. We illustrate in the rest of this chapter with several examples of basic data structures to which the analysis of variance is often applied.

One-way ANOVA: radon measurements within counties

Some of the essential elements of multilevel analysis of variance can be seen in the simplest case of a one-way structure. We illustrate with the radon example from Chapter 12, simplified to measurements within counties, ignoring all individual-level and group-level predictors. The model is then

$$y_i \sim \mathrm{N}(\alpha_{j[i]}, \sigma_y^2), \text{ for } i = 1, \ldots, n$$
$$\alpha_j \sim \mathrm{N}(\mu_\alpha, \sigma_\alpha^2), \text{ for } j = 1, \ldots, J.$$

As discussed in Section 19.4, we use redundant parameters to speed the computation. We also add lines in the Bugs model to define the finite-population standard deviations, s_ϵ and s_α:

Bugs code
```
for (i in 1:n){
   e.y[i] ~ y[i] - y.hat[i]
}
s.y <- sd(e[])
s.a <- sd(a[])
```

Figure 22.4 shows the result of fitting this model to the Minnesota radon data. The variation among houses within counties is much larger than the variation of county mean radon levels. The between-county variation is estimated to be about 0.3; the analysis is on the log scale, so this corresponds to a multiplicative variation

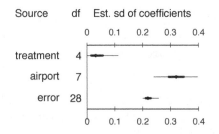

Figure 22.5 *Two-way analysis of variance for the flight simulator data. The dots, thick lines, and thin lines represent medians, 50% intervals, and 95% intervals for the finite-population standard deviation s, for each source of variation.*

of $\exp(0.3) \approx 1.3$. That is, average radon levels vary by about 30% among Minnesota counties. Although this variation is small, it is clearly "statistically significant": its 95% confidence interval in Figure 22.4 shows that we are fairly certain that the between-county standard deviation falls between 0.2 and 0.4. (The within-county standard deviation is estimated much more accurately, which makes sense given its larger degrees of freedom.)

Two-way ANOVA: flight simulator data

We illustrate two-way analysis of variance with the flight simulator experiment described in Section 13.5. The data, displayed in Figure 13.8, are the success rates of pilots in flight simulators, under five different experimental treatments at eight different airports. Figure 22.1 on page 488 displays the classical two-way ANOVA for these data. The corresponding multilevel model is (13.9) on page 289, a model we can also write as

$$\begin{aligned}
y_i &= \mu + \gamma_{j[i]} + \delta_{k[i]} + \epsilon_i, \text{ for } i = 1, \dots, n \\
\gamma_j &\sim \text{N}(0, \sigma_\gamma^2), \text{ for } j = 1, \dots, J \\
\delta_k &\sim \text{N}(0, \sigma_\delta^2), \text{ for } k = 1, \dots, K \\
\epsilon_i &\sim \text{N}(0, \sigma_\epsilon^2), \text{ for } i = 1, \dots, n.
\end{aligned} \quad (22.5)$$

In the terminology of the analysis of variance, the data have $n - 1 = 39$ degrees of freedom, which can be decomposed into:

- $J - 1 = 4$ degrees of freedom for γ

- $K - 1 = 7$ degrees of freedom for δ

- The remaining 28 degrees of freedom for ϵ.

The multilevel ANOVA is performed by fitting model (22.5) and summarizing by the estimated variance components. We could use the superpopulation variance parameters $\sigma_y, \sigma_\gamma, \sigma_\delta$, but, for reasons described in Section 21.2, we prefer to work with the finite-population variance parameters s_y, s_γ, s_δ defined in (21.1) on page 460. Each of these standard deviations is defined in terms of the model parameters and thus varies across the simulations produced by the multilevel inference in Bugs.

Figure 22.5 summarizes the inference for the variance components for the flight simulator data. Treatment effects are small (with coefficients estimated to be less than 0.05 in absolute value), airport effects are on the order of 0.3 (which is large, considering that the outcomes y are proportions and thus fall between 0 and 1), and the scale of the errors is moderately large, at about 0.2.

Two-way ANOVA with replication

When replications are present, it is possible to estimate the two-way interactions separately from the measurement error. The additive model can be expanded to

$$y_i = \mu + \gamma_{j[i]} + \delta_{k[i]} + \eta_{j[i],k[i]} + \epsilon_i,$$

with separate variance components for the γ_j's, δ_k's, $\eta_{j,k}$'s, and ϵ_i's. For example, if the flight simulator data had four replications per cell (thus, $5 \times 8 \times 4 = 160$ observations), the ANOVA would look like

Source	df
treatment	4
airport	7
treatment $\times$ airport	28
error	120

The 120 degrees of freedom for error correspond to the 160 data points, minus the 40 cell means. Each row of the table would correspond to a single variance component.

Unbalanced designs

Multilevel analysis of variance works the same for unbalanced as for balanced designs. For example, in the flight simulator analysis, if we had 160 observations in the 40 cells, but not necessarily evenly distributed at 4 per cell, then the ANOVA table would have the same structure.

Nested designs

Consider an experiment on 4 treatments for an industrial process applied to 20 machines (randomly divided into 4 groups of 5), with each treatment applied 6 times independently on each of its 5 machines. For simplicity, we assume no systematic time effects, so that the 6 measurements on each machine are replications. The ANOVA table is then

Source	df
treatment	3
machine	16
error	100

Because the design is nested, it does not make sense to consider treatment $\times$ machine interactions. If expressed in terms of measurements i, machines j, and treatments k, the nested model can be written identically as the two-way non-nested model (22.5). The multilevel analysis automatically accounts for the nesting.

22.5 Adding predictors: analysis of covariance and contrast analysis

Individual-level predictors and analysis of covariance

Analysis of covariance is a decomposition of the sources of variation of a dataset (as in ANOVA), after adjusting for a predictor or set of predictors. In the multilevel modeling context, analysis of covariance corresponds to an ANOVA-like decomposition of a model that includes individual-level predictors.

For example, Figure 22.4 on page 494 displays the one-way analysis of variance for the Minnesota radon data, showing the estimated county-level and house-level

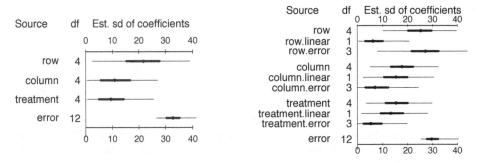

Figure 22.6 *ANOVA displays for a 5×5 latin square experiment (an example of a crossed three-way structure) for the data shown on page 292: (a) with no group-level predictors, (b) contrast analysis including linear trends for rows, columns, and treatments. See also the plots of coefficient estimates and trends on page 293.*

variation. Suppose we are interested in estimating these sources of variation, after controlling for the floor of measurement (recall from Chapter 12 that radon levels tends to be higher in houses with basements). We then fit the model

$$y_i \quad \sim \quad N(\alpha_{j[i]} + x_i\beta, \sigma_y^2), \text{ for } i = 1, \ldots, n$$
$$\alpha_j \quad \sim \quad N(\mu_\alpha, \sigma_\alpha^2), \text{ for } j = 1, \ldots, J,$$

and display the estimated variance components σ_β, σ_y (or their finite-sample counterparts, s_β, s_y). In this particular example, the estimated variance parameters are not much changed from the simple ANOVA, and so we do not display the analysis of covariance table here.

Group-level predictors and contrast analysis

Adding predictors at the group level in a multilevel model corresponds to the classical method of *contrasts* in the analysis of variance. We illustrate with the latin square data displayed in Figure 13.11 on page 292. First we perform the analysis of variance for the three-level model with no additional predictors; then we add group-level predictors and show the contrast analysis.

Multilevel ANOVA with no contrasts.   In Section 13.5, we fit a model including row effects, column effects, treatment effects, and linear trends for each of these factors. We shall first fit the model without the linear trends and simply estimate the magnitudes of the row, column, and treatment effects. The model is the following stripped-down version of (13.10):

$$y_i \quad \sim \quad N(\mu + \beta_{j[i]}^{\text{row}} + \beta_{k[i]}^{\text{column}} + \beta_{l[i]}^{\text{treat}}, \sigma_y^2), \text{ for } i = 1, \ldots, 25$$
$$\beta_j^{\text{row}} \quad \sim \quad N(0, \sigma_{\beta,\text{row}}^2), \text{ for } j = 1, \ldots, 5$$
$$\beta_k^{\text{column}} \sim \quad N(0, \sigma_{\beta,\text{column}}^2), \text{ for } k = 1, \ldots, 5$$
$$\beta_l^{\text{treat}} \sim \quad N(0, \sigma_{\beta,\text{treat}}^2), \text{ for } l = 1, \ldots, 5, \tag{22.6}$$

and we summarize by the finite-sample standard deviations. Figure 22.6a shows the results: none of the effects are large compared to residual variation, and the sample size is small enough that it is difficult to distinguish column and treatment effects from zero.

Multilevel ANOVA with contrasts.   We next add *linear contrasts* for the rows, columns, and treatments, expanding the model by replacing the group-level models in (22.6) with

$$
\begin{aligned}
\beta_j^{\text{row}} &\sim \text{N}(\gamma^{\text{row}} \cdot (j - 3), \sigma_{\beta\,\text{row}}^2) \\
\beta_k^{\text{column}} &\sim \text{N}(\gamma^{\text{column}} \cdot (k - 3), \sigma_{\beta\,\text{column}}^2) \\
\beta_l^{\text{treat}} &\sim \text{N}(\gamma^{\text{treat}} \cdot (l - 3), \sigma_{\beta\,\text{treat}}^2).
\end{aligned}
$$

Figure 22.6b shows the new ANOVA, with each factor decomposed into a linear contrast and residuals from the contrast. The column and treatment effects are mostly captured by the linear contrasts, whereas the variation in the row effects does not follow a linear trend.

These ANOVA displays give a reasonable quick summary, but in this particular application it is probably more useful to simply display each group of parameter estimates, along with the estimated linear contrasts, directly, as in Figure 13.12 on page 293. We have shown the ANOVA here to connect to classical contrast analysis.

22.6 Modeling the variance parameters: a split-plot latin square

Multilevel data structures can be characterized by the number of grouping factors (that is, rows of the ANOVA table) and the number of groups in each. For example, the radon example has one grouping factor—counties—which takes on 85 values. The latin square example of the previous section had three factors—row, column, and treatment—each of which took on five levels. Designed experiments sometimes go even further in this direction: for example, a so-called 2^6 design has six factors, each of which can take on two different values (thus a total of 64 data points in a complete design, or more if there is replication).

Different models are appropriate for differently shaped data structures. With few grouping factors and many levels per factor, the models discussed so far in this book—modeling the coefficients but leaving the variance components unmodeled (in Bayesian terms, to be estimated using noninformative prior distributions)—are appropriate. At the other extreme, when the number of levels per grouping factor is small, it can be helpful to model the variance parameters, and when there are many factors, it is possible to use partial pooling to do this estimation. We illustrate with a hierarchical data structure with two treatment factors and nine grouping factors (including interactions).

Crossed and nested ANOVA: a split-plot design

In a *split-plot* design, units are clustered, and there are two treatment factors, with one factor applied to groups and the other to individual units. For example, in an educational experiment with students within classrooms, several teaching methods might be applied at the classroom level, with individual interventions applied to individual students. In this sort of design, the individual-level treatments are typically estimated with higher precision than the group-level treatments, and part of the goal of ANOVA is to assess the importance of both factors. (The term "split-plot" refers to agricultural experiments, where the groups are large plots that are split into subplots, which play the role of individual units in our analysis.)

For example, here are the variance components for a $5 \times 5 \times 2$ split-plot latin square:

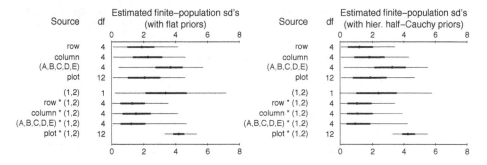

Figure 22.7 *ANOVA display for a split-plot latin square experiment: posterior medians, 50%, and 95% intervals for finite-population standard deviations s_k. (a) The left plot shows inferences given uniform prior distributions on the σ_k's; (b) the right plot shows inferences given a hierarchical half-Cauchy model with scale fit to the data. The half-Cauchy model gives sharper estimates, indicating the power of hierarchical modeling for these highly uncertain quantities.*

Source	df
row	4
column	4
(A,B,C,D,E)	4
plot	12
(1,2)	1
row $\times$ (1,2)	4
column $\times$ (1,2)	4
(A,B,C,D,E) $\times$ (1,2)	4
plot $\times$ (1,2)	12

In this example, there are 25 plots with five full-plot treatments (labeled A, B, C, D, E), and each plot is divided into two subplots with subplot varieties (labeled 1 and 2). The horizontal line in the table separates the main-plot from the subplot effects. In a classical analysis, it is easy enough to decompose the 49 degrees of freedom to the rows in the ANOVA table; the tricky part of the analysis is to know which residuals are to be used for which comparisons.

The analysis is straightforward using multilevel models. We first present the results (using data from an agricultural experiment) with a simple multilevel model that leaves the variance parameters unmodeled, then with an expanded model that includes a hierarchical model for the variance parameters themselves.

Multilevel model with noninformative prior distributions for the variance parameters

To perform ANOVA using multilevel analysis, we simply set up a linear model with a batch of coefficients corresponding to each source of variation; thus,

$$y_i = \alpha^0 + \alpha_{j[i]}^{\text{row}} + \alpha_{k[i]}^{\text{col}} + \alpha_{l[i]}^{\text{ABCDE}} + \alpha_{m[i]}^{\text{plot}} + \alpha_{n[i]}^{12} + \alpha_{j[i],n[i]}^{\text{col}\times 12} + \alpha_{k[i],n[i]}^{\text{ABCDE}\times 12} + \alpha_{l[i],n[i]}^{\text{plot}\times 12}.$$

Here the data i run from 1 to 50, and we have used the variables j, k, l, m, n to index rows, columns, group-level treatments, plots, and individual-level treatments. There is no replication in the study (that is, there is only one measurement per subplot), and so the last term in the additive model, $\alpha_{l,n}^{\text{plot}\times 12}$, corresponds to data-level errors.

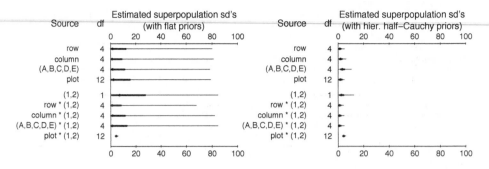

Figure 22.8 *Posterior medians, 50%, and 95% intervals for standard-deviation parameters σ_k estimated from a split-plot latin square experiment. (a) The left plot shows inferences given uniform prior distributions on the σ_k's; and (b) the right plot shows inferences given a hierarchical half-Cauchy model with scale fit to the data. The half-Cauchy model gives much sharper inferences, using the partial pooling that comes with fitting a hierarchical model. Compare to Figure 22.7 (which is on a different scale).*

Each batch of coefficients α is then assigned its own normal distribution with mean 0 and standard deviation estimated from the data: $\alpha_j^{\text{row}} \sim \text{N}(0, (\sigma^{\text{row}})^2)$, and so forth. We can then fit the model and summarize using finite-population standard deviations. Figure 22.7a illustrates.

Multilevel model with noninformative prior distributions for the variance parameters

Each row of the ANOVA table corresponds to a different variance component, and the split-plot ANOVA can be understood as a linear model with nine variance components, $\sigma_1, \ldots, \sigma_9$—one for each row of the table. The default performed earlier assigns a uniform prior distribution to each of the parameters σ_k.

More generally, we can set up a hierarchical model, where the variance parameters have a common distribution with hyperparameters estimated from the data. We consider a half-Cauchy prior distribution with peak 0 and scale A, and with a uniform prior distribution on A. The hierarchical half-Cauchy model allows most of the variance parameters to be small but with the occasionally large σ_α, which seems reasonable in the typical settings of analysis of variance, in which most sources of variation are small but some are large. See Section 19.6 for further discussion of the half-Cauchy model for multilevel variance parameters.

Figure 22.7b shows inferences for the finite-population standard-deviation parameters s_α for each row of the latin square split-plot ANOVA under this new model. The inferences from the half-Cauchy prior distribution are slightly more precise than with the uniform, with the most pooling occurring for the variance component that has just one degree of freedom. The Cauchy scale parameter A was estimated at 1.8, with a 95% interval of $[0.5, 5.1]$.

Superpopulation and finite-population standard deviations

As discussed in Section 21.2, finite-population inferences can be much more precise than superpopulation inferences when the number of groups is small. We illustrate here by displaying, in Figure 22.8, the inferences for the superpopulation standard deviations in the split-plot latin square example, again separately for the uniform and hierarchical half-Cauchy prior distributions for the standard-deviation parameters σ_k.

As the left plot shows, the uniform prior distribution does not rule out the potential for some extremely high values of the variance components—the degrees of freedom are low, and the interlocking of the linear parameters in the latin square model results in difficulty in estimating any single variance parameter. In contrast, the hierarchical half-Cauchy model performs a great deal of shrinkage, especially of the high ranges of the intervals. (For most of the variance parameters, the posterior medians are similar under the two models, but the 75th and 97.5^{th} percentiles are shrunk by the hierarchical model.) This is an ideal setting for hierarchical modeling of variance parameters in that it combines separately imprecise estimates of each of the individual σ_k's.

22.7 Bibliographic note

Searle, Casella, and McCulloch (1992) review classical analysis of variance and variance-component models. Kirk (1995) provides an introductory treatment from the perspective of experimental psychology.

The multilevel ANOVA approach described in this chapter comes from Gelman (2005), and is based on earlier work of Green and Tukey (1960), Nelder (1965a, b), Yates (1967), and Lane and Nelder (1982). The hierarchical model for variance parameters appears in Gelman (2006). See also many of the references in Section 21.9 for related ideas.

McCullagh (2005) points out that ANOVA can be applied more generally to nonexchangeable models such as arise in genetics or, more generally, in models with structured interactions.

22.8 Exercises

1. Take a varying-intercept model from one of the exercises in Part 2 of this book and construct the corresponding ANOVA plot as in Section 22.3.

2. Analysis of variance for meta-analysis: consider the magnetic-fields experiments from Section 21.8 (data in the folder chicks) as an analysis-of-variance problem. Identify the sources of variation and construct the ANOVA plot, making whatever assumptions are necessary to do this using the available data.

3. Three-way designs: perform the analysis of variance as described in Section 22.4 for the data in the folder olympics, which are figure-skating ratings classified by judges, skaters, and measurement criterion (see Exercise 11.3). You will need to identify the sources of variation, then fit the model and display the estimated standard-deviation parameters and their uncertainties.

4. Hierarchical modeling of variance parameters: consider the model for height and earnings shown in Figure 13.10 on page 291.

   (a) Start with a varying-intercept model (with intercepts varying by ethnicity, age, and ethnicity × age), with a hierarchical model for the variance parameters as in Section 22.6. Make plots similar to Figures 22.7 and 22.8 to compare the inferences under flat and hierarchical prior distributions.

   (b) Repeat (a) for a varying-intercept, varying slope model (with both intercepts and slopes varying by ethnicity, age, and their interaction).

5. Multivariate analysis of variance for multilevel models: consider how the models

and displays of this chapter could be generalized to varying-intercept, varying-slope models.

## CHAPTER 23

Causal inference using multilevel models

Causal inference using regression has an inherent multilevel structure—the data give comparisons between units, but the desired causal inferences are within units. Experimental designs such as pairing and blocking assign different treatments to different units within a group. Observational analyses such as pairing or panel study attempt to capture groups of similar observations with variation in treatment assignment within groups.

23.1 Multilevel aspects of data collection

Hierarchical analysis of a paired design

Section 9.3 describes an experiment applied to school classrooms with a paired design: within each grade, two classes were chosen within each of several schools, and each pair was randomized, with the treatment assigned to one class and the control assigned to the other. The appropriate analysis then controls for grade and pair.

Including pair indicators in the Electric Company experiment. As in Section 9.3, we perform a separate analysis for each grade, which could be thought of as a model including interactions of treatment with grade indicators. Within any grade, let n be the number of classes (recall that the treatment and measurements are at the classroom, not the student, level) and J be the number of pairs, which is $n/2$ in this case. (We use the general notation n, J rather than simply "hard-coding" $J = n/2$ so that our analysis can also be used for more general randomized block designs with arbitrary numbers of units within each block.)

The basic analysis has the form

$$y_i \sim \mathrm{N}(\alpha_{j[i]} + T_i\theta, \sigma_y^2), \text{ for } i = 1, \ldots, n$$
$$\alpha_j \sim \mathrm{N}(\mu_\alpha, \sigma_\alpha^2), \text{ for } j = 1, \ldots, J.$$

By including the pair indicators, this model controls for all information used in the design.

Here is the Bugs code for this model, as fit to classrooms from a single grade, and using `pair[i]` to index the pair to which classroom i belongs:

```
model {
  for (i in 1:n){
    y[i] ~ dnorm (y.hat[i], tau.y)
    y.hat[i] <- a[pair[i]] + theta*treatment[i]
  }
  for (j in 1:n.pair){
    a[j] ~ dnorm (mu.a, tau.a)
  }
  theta ~ dnorm (0, .0001)
  tau.y <- pow(sigma.y, -2)
  sigma.y ~ dunif (0, 100)
```

Bugs code

```
      mu.a ~ dnorm (0, .0001)
      tau.a <- pow(sigma.a, -2)
      sigma.a ~ dunif (0, 100)
      }
   }
```

Fitting all four grades at once. We can fit the above model to each grade separately, or we can expand it by allowing each of the parameters to vary by grade.
For convenience, we use two index variables: `grade`, which indexes the grades of the
classrooms, and `grade.pair`, which indexes the grades of the pairs. (This particular
experiment involves $n = 192$ classrooms clustered into $J = 96$ pairs, so `grade` is a
vector of length 192 and `grade.pair` has length 96. The entries of both vectors are
1's, 2's, 3's, and 4's.) Here is the Bugs model:

Bugs code
```
      model {
        for (i in 1:n){
          y[i] ~ dnorm (y.hat[i], tau.y[grade[i]])
          y.hat[i] <- a[pair[i]] + theta[grade[i]]*treatment[i]
        }
        for (j in 1:n.pair){
          a[j] ~ dnorm (mu.a[grade.pair[j]], tau.a[grade.pair[j]])
        }
        for (k in 1:n.grade){
          theta[k] ~ dnorm (0, .0001)
          tau.y[k] <- pow(sigma.y[k], -2)
          sigma.y[k] ~ dunif (0, 100)
          mu.a[k] ~ dnorm (0, .0001)
          tau.a[k] <- pow(sigma.a[k], -2)
          sigma.a[k] ~ dunif (0, 100)
        }
      }
```

Writing the model this way has the advantage that the Bugs output (not shown
here) displays inferences for all the parameters at once. The treatment effects θ are
large for the first two grades and closer to zero for grades 3 and 4; the intercepts
α_j are highly variable for the first 11 pairs (which correspond to classes in grade
1), vary somewhat for the next bunch of pairs (which are grade 2 classes), and vary
little for the classes in higher grades. The residual data variance and the between-
pair variance both decrease for the higher grades, all of which are consistent with
the compression of the scores for higher grades at the upper end of the range of
data (see Figure 9.4 on page 174).

Controlling for pair indicators and pre-test score. As discussed in Section 9.3,
the next step is to include pre-test class scores as an input in the regression. The
treatments are assigned randomly within each grade and pair, so it is not necessary
to include other pre-treatment information, but adding pre-test to the analysis can
improve the precision of the estimated treatment effects.

For simplicity, we return to the model on page 503 that fits one grade at a
time. We need to alter this Bugs model only slightly, by simply adding the term
`+ b.pre.test*pre.test[i]` to the expression for `y.hat[i]`, and then at the end
of the model, placing the prior distribution, `b.pre.test ~ dnorm(0,.0001)`.

Figure 23.1 displays the estimated treatment effects for the models controlling
for pairing, with and without pre-test score. The final analysis, including pairing
and pre-test as inputs, clearly shows a positive treatment effect in all grades, with
narrower intervals than the corresponding estimates without including the pairing
information (see Figure 9.5 on page 176).

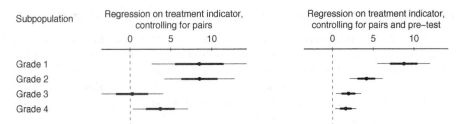

Figure 23.1 *Estimates, 50%, and 95% intervals for the effect of the Electric Company television show in each grade as estimated from hierarchical models that account for the pairing of classrooms in the experimental design. Displayed are regression coefficients for the treatment indicator: (a) also controlling for pair indicators, (b) also controlling for pair indicators and pre-test scores. Compare to Figure 9.5 on page 176, which shows estimates not including the pairing information. The most precise estimates are those that control for both pairing and pre-test.*

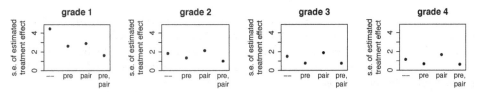

Figure 23.2 *Standard errors for estimated treatment effects in each grade for each of four models: (−−) with no other predictors in the model, (pre) controlling for pre-test, (pair) controlling for indicators for pairing, (pre, pair) controlling for pre-test and pair indicators. Unsurprisingly, controlling for more pre-test information tends to reduce estimation errors. Figures 9.5 and 23.1 display the estimated treatment effects and uncertainty bounds for the four models in each grade.*

To show more clearly the improvements from including more pre-treatment data in the model, Figure 23.2 displays the standard deviations of the estimated treatment effect in each grade as estimated from models excluding or including the pairing and pre-test information. In this case, the pre-test appears to contain more information than the pairing, and it is most effective to include both inputs, especially in grade 1, where there appears to be wide variation among classes.

Hierarchical analysis of randomized-block and other structured designs

More generally, pre-treatment information used in the design should be included as inputs so that the coefficients for the treatment indicators correspond to causal effects. For example, consider designs such as randomized blocks (in which data are partitioned into groups or "blocks," with random treatment assignment within each block) or latin squares (in which experimental units in a row × column data structure are assigned treatments randomly with constraints on the randomization so that treatment assignments are balanced within each row and column), the blocks, or the rows and columns, represent pre-treatment factors that can be accounted for using a multilevel model. This combines our general advice for causal inference in Chapter 9 with the idea from Chapter 12 of multilevel modeling as a general approach for handling categorical variables as regression predictors.

23.2 Estimating treatment effects in a multilevel observational study

In Section 10.1 we introduced the example of estimating the effect of the Infant Health and Development Program in the context of a (constructed) observational study. In that classical regression analysis, we included indicator variables for the 8 sites in order to control for unobserved site-specific characteristics that might be associated with both selection into treatment as well as the outcome (test scores). Here we extend to a multilevel model.

Varying intercepts

The simplest extension is just to allow for varying intercepts across site. If we denote post-treatment outcomes by y, treatment assignment by T, and the vector of confounding covariates for person i as X_i, we can write

$$y_i \sim N(\alpha_{j[i]} + T_i\theta + X_i\beta, \sigma_y^2), \text{ for } i = 1, \ldots, n,$$
$$\alpha_j \sim N(\mu_\alpha, \sigma_\alpha^2), \text{ for } j = 1, \ldots, J.$$

Here, the matrix of predictors X does not include a constant term, since we have included a constant in the model for the α_j's. The treatment effect θ is estimated at 9.1 with a standard error of 2.1. (By comparison, our estimate from the classical regression with site indicators was 8.8 with standard error of 2.1, and the experimental benchmark is 7.4.)

Assumptions satisfied? The original justification for including site information when estimating the causal effect of the treatment was that it is plausible that unobserved site characteristics may be associated with both selection into treatment and subsequent test scores. However, the model above implicitly assumes that the intercepts (which capture unobserved characteristics of the sites) are independent of the other predictors in the model.

How can we resolve this conceptual inconsistency? One approach is to allow the intercepts to be correlated with the treatment variable. We can accomplish this by creating an aggregated version T^{agg} of the treatment variable, defined so that T_j^{agg} is the average value of T_i for the members of group j—in this case, the proportion who received the treatment, among the people in the dataset in site j. We then add this measure as a group-level predictor:

$$\alpha_j \sim N(\gamma_0 + \gamma_1 T_j^{\text{agg}}, \sigma_\alpha^2).$$

In our example, this changes the estimate of the treatment effect to 9.0, with the standard error virtually unchanged at 2.1. Addition of aggregated measures of the other predictors brings the estimate to 8.9.

Varying treatment effects

The demographic composition of the sample varied across sites, as did treatment implementation to some degree. Therefore we might expect treatment effects to vary also. A perhaps more interesting use of multilevel models in this example is to investigate variation in treatment effects across sites: we fit the model

$$y_i \sim N(\alpha_{j(i)} + T_i\theta_{j(i)} + X_i\beta, \sigma_y^2), \text{ for } i = 1, \ldots, n,$$
$$\begin{pmatrix} \alpha_j \\ \theta_j \end{pmatrix} \sim N\left(\begin{pmatrix} \mu_\alpha \\ \mu_\theta \end{pmatrix}, \begin{pmatrix} \sigma_\alpha^2 & \rho\sigma_\alpha\sigma_\theta \\ \rho\sigma_\alpha\sigma_\theta & \sigma_\theta^2 \end{pmatrix} \right), \text{ for } j = 1, \ldots, J.$$

Figure 23.3 displays the inferences for the treatment effects θ_j for each site from

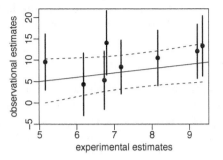

Figure 23.3 *Comparison of observational treatment effects (dark dots) and 95% intervals (dark vertical lines) across sites from multilevel models against the experimental benchmark (solid line with slope 1) and corresponding 95% confidence band (dotted curves). All observational intervals cover the corresponding experimental estimates; they all reflect greater uncertainty as compared to the experimental intervals as well.*

the multilevel model fit to observational data. The experimental estimates for each site (also calculated using a multilevel model) are referenced by the line with slope 1. A 95% interval for each observational estimate has been plotted (dark vertical line) for each site. The dotted curves display an approximate 95% confidence band for the experimental estimates. The observational intervals all cover the corresponding experimental estimate (our best comparison point for this example), but with greater uncertainty than the experimental estimates.[1]

23.3 Treatments applied at different levels

As noted in Section 20.1, treatments are sometimes implemented on *groups* of individuals (experimental units) rather than the individual units themselves. The choice of experimental design may be motivated by several concerns.

Suppose, for instance, that we want to evaluate a new method for teaching long division by teaching a random sample of third-grade students and then evaluating outcomes on math tests six months later. Sampling and logistical considerations would motivate a nested design such as including students in 40 classrooms across 20 schools in the study. It is easier and cheaper to train half the teachers (20 rather than 40) to implement the new method. Moreover, even if all teachers were trained, it would be inconvenient to divide each classroom into two parts, teaching the old method to half the students and the new method to the others. Finally if some of the students in a classroom were taught the new method and others the old, it is possible that the students would share information about the methods they were taught. This could influence their subsequent test scores and violate the assumption of independent outcomes which is standard in regression modeling (for further discussion of this principle of stable unit treatment values, see the end of Section 9.3). These are all motivations for randomizing classrooms rather than the students within classrooms (they might even motivate randomizing schools rather than classrooms). This is called a group- or cluster-randomized experimental design.

As with many of the other sampling and experimental designs discussed in this book, it is appropriate to analyze data from a grouped experiment using multilevel

---

[1] The displayed inferences actually come from a simpler version of the model in which the correlation ρ of the group-level errors was set to zero. For our example, the estimation with the full model including ρ was unstable, and there was no evidence that the correlation differed from zero.

models. In an extreme sense, it is as if we have only randomized J experimental units (where J is the number of groups) rather than the n individual units.

Analysis of an educational-subsidy program

In 1997 the Mexican federal government implemented Progresa (Programa de Educacion, Salud y Alimentacion), a program that provides cash subsidies to low-income families if they send their children to school (rather than, for instance, having them leave school to go to work at an early age) and visit health clinics. This program was randomly assigned to 506 eligible localities.[2]

Here we analyze a convenience subsample of the Progresa data that includes 81 localities and approximately 7000 households. The primary goal for this analysis is to determine if there was an effect of the program on enrollment in school.

A standard analysis that ignores the grouping might be to simply run a logistic regression of post-program enrollment on the treatment indicator and possibly some additional predictors to increase efficiency (baseline enrollment, work status, age, sex, and poverty status). This analysis yields an estimated treatment effect of 0.51 with standard deviation 0.09—a highly statistically significant result. This estimate implies that program availability increased the probability of enrollment by (at most) about 13%.

In contrast, we can build a multilevel logistic regression model of the form,

$$\Pr(y_i = 1) = \text{logit}^{-1}(\alpha_{j[i]} + X_i\beta), \text{ for } i = 1, \ldots, n$$
$$\alpha_j \sim \text{N}(U_j\gamma, \sigma_\alpha^2), \text{ for } j = 1, \ldots, J,$$

where X is the matrix of individual-level predictors just described, and U is the matrix of group-level predictors, in this case simply a constant term and the treatment indicator.

Here is the corresponding Bugs model:[3]

Bugs code
```
model {
  for (i in 1:n){
    y[i] ~ dbin (p.bound[i], 1)
    p.bound[i] <- max(0, min(1, p[i]))
    logit(p[i]) <- Xbeta[i]
    Xbeta[i] <- a[village[i]] + b.1*enroll97[i] + b.2*work97[i] +
                b.3*poor[i] + b.4*male[i] + b.5*age97[i]
  }
  b.1 ~ dnorm (0, .0001)
  b.2 ~ dnorm (0, .0001)
  b.3 ~ dnorm (0, .0001)
  b.4 ~ dnorm (0, .0001)
  b.5 ~ dnorm (0, .0001)
  for (j in 1:J){
    a[j] ~ dnorm (a.hat[j], tau.a)
    a.hat[j] <- g.0 + g.1*program[j]
  }
  g.0 ~ dnorm (0, .0001)
  g.1 ~ dnorm (0, .0001)
```

[2] Localities not assigned to receive the program immediately were given the program a few years later.

[3] An alternative parameterization would store X as a matrix and express the linear predictor as Xbeta[i] <- a[village[i]] + inprod(b[],X[i,]), which would allow us to model the coefficients b more conveniently.

```
    tau.a <- pow(sigma.a, -2)
    sigma.a ~ dunif (0, 100)
}
```

The multilevel analysis yields a treatment coefficient estimate of 0.17 with a standard error of 0.20. In this case, correctly accounting for our uncertainty has a substantial impact on the results.[4]

Unmodeled varying intercepts? One strategy sometimes used (inappropriately) to account for the type of clustering observed in this experiment is to include indicator variables for each group—here each village. In this example such a model would yield a treatment effect estimate of 0.6 with standard error 1.4. While this estimate is also statistically insignificant, the uncertainty here is more than six times the uncertainty in the varying-intercept multilevel model.

A split-plot experiment

Figure 22.7 on page 499 shows the analysis of variance for a split-plot experiment: a design in which one set of treatments is applied at the individual level, and another set of treatments is applied at the group level. This is a randomized experiment with the probabilities of treatment assignment based on observed pre-treatment variables, and so the analysis is straightforward:

- Regress the outcome on the treatment indicators (these are treatments 1, 2 at the subplot (individual) level, A, B, C, D, E at the main-plot (group) level, and the 10 interactions A*1, A*2, ..., E*1, E*2.

- Also include in the regression the indicators for the 5 rows, the 5 columns, and the 25 main plots (groups).

For causal inference we should look at the coefficient estimates, which we display in Figure 23.4.

23.4 Instrumental variables and multilevel modeling

Section 10.5 discussed how instrumental variables can be used to identify causal effects in certain prescribed situations (when a valid instrument exists). We return to this same example, the randomized Sesame Street experiment, to illustrate extensions of the standard model to a multilevel framework.

The basic model can be written as

$$\begin{pmatrix} y_i \\ T_i \end{pmatrix} \sim N\left(\begin{pmatrix} \alpha + \beta T_i \\ \gamma + \delta z_i \end{pmatrix}, \begin{pmatrix} \sigma_y^2 & \rho\sigma_y\sigma_T \\ \rho\sigma_y\sigma_T & \sigma_T^2 \end{pmatrix} \right), \text{ for } i = 1, \dots, n,$$

where in this example z represents the randomized encouragement to watch Sesame Street, T represents whether the child subsequently watched or not (the desired "treatment" variable which in other contexts might be called the "compliance" variable), and y is the outcome measure, a post-treatment score on a letter recognition test. In this model, β is the causal effect of watching the show.

Recall that this experiment was randomized within combinations of site and setting. Therefore we first extend to include varying intercepts for each equation,

---

[4] We are analyzing here a nonrandom subset of the data and do not intend the results of this analysis to represent effects of the treatment over the entire experiment.

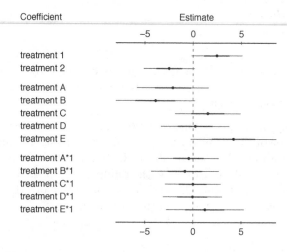

Figure 23.4 *Coefficient estimates for a split-plot experiment: an example of causal inference for a design with treatments at individual and group levels. The ANOVA table summarizing this analysis appears in Figure 22.7 on page 499. (This plot omits the five interactions A*2, ..., E*2, since, with only two levels of the numerical factor, these are simply the opposites of A*1, ..., E*1.)*

and we allow those intercepts to be correlated:

$$\begin{pmatrix} y_i \\ T_i \end{pmatrix} \sim \mathrm{N}\left(\begin{pmatrix} \alpha_{j[i]} + \beta T_i \\ \gamma_{j[i]} + \delta z_i \end{pmatrix}, \begin{pmatrix} \sigma_y^2 & \rho_{yT}\sigma_y\sigma_T \\ \rho\sigma_y\sigma_T & \sigma_T^2 \end{pmatrix}\right), \text{ for } i = 1, \dots, n,$$

$$\begin{pmatrix} \alpha_j \\ \gamma_j \end{pmatrix} \sim \mathrm{N}\left(\begin{pmatrix} \mu_\alpha \\ \mu_\gamma \end{pmatrix}, \begin{pmatrix} \sigma_\alpha^2 & \rho_{\alpha\gamma}\sigma_\alpha\sigma_\gamma \\ \rho\sigma_\alpha\sigma_\gamma & \sigma_\gamma^2 \end{pmatrix}\right), \text{ for } j = 1, \dots, J. \qquad (23.1)$$

After creating a variable called `siteset` that represents the 9 existing combinations of site and setting (school or home), and bundling the outcome y and encouragement T into a single $n \times 2$ data matrix `yt`, we fit the following Bugs model:

Bugs code
```
model {
    for (i in 1:n){
        yt[i,1:2] ~ dmnorm (yt.hat[i,], Tau.yt[,])        # data model
        yt.hat[i,1] <- a[siteset[i]] + b*yt[i,2]
        yt.hat[i,2] <- g[siteset[i]] + d*z[i]
    }
    for (j in 1:J){
        ag[j,1:2] ~ dmnorm (mu.ag[1:2], Tau.ag[1:2,1:2])
        a[j] <- ag[j,1]
        g[j] <- ag[j,2]
    }

    # data level
    Tau.yt[1:2,1:2] <- inverse(Sigma.yt[,])
    Sigma.yt[1,1] <- pow(sigma.y,2)
    sigma.y ~ dunif (0, 100)                # noninformative prior on sigma.a
    Sigma.yt[2,2] <- pow(sigma.t,2)
    sigma.t ~ dunif (0, 100)                # noninformative prior on sigma.b
    Sigma.yt[1,2] <- rho.yt*sigma.y*sigma.t
    Sigma.yt[2,1] <- Sigma[1,2]             # noninformative prior on rho
    rho.yt ~ dunif (-1,1)
```

```
d ~ dnorm (0, .0001)
b ~ dnorm (0, .0001)

# group level
  Tau.ag[1:2,1:2] <- inverse(Sigma.ag[,])
  Sigma[1,1] <- pow(sigma.a,2)
  sigma.a ~ dunif (0, 100)
  Sigma[2,2] <- pow(sigma.g,2)
  sigma.g ~ dunif (0, 100)
  Sigma[1,2] <- rho.ag*sigma.a*sigma.g
  Sigma[2,1] <- Sigma[1,2]
  rho.ag ~ dunif(-1,1)

  mu.ag[1] ~ dnorm(0, .0001)
  mu.ag[2] ~ dnorm(0, .0001)
}
```

The causal parameter of interest is **b**.

One advantage of Bugs is that it allows us to model the standard form of this model directly, but it turns out that modeling the reduced form of the model is more efficient and the algorithm will converge much more quickly. This just requires a simple change to the fourth line of the above model:

```
yt.hat[i,1] <- a[siteset[i]] + b*d*z[i]                          Bugs code
```

Conditioning on pre-treatment variables

The instrumental variables model can be augmented by conditioning on pre-treatment scores as well, changing the fourth and fifth lines of the model to:

```
yt.hat[i,1] <- a[siteset[i]] + b*d*z[i] + phi.y*pretest[i]       Bugs code
yt.hat[i,2] <- g[siteset[i]] + d*z[i] + phi.t*pretest[i]
```

and specifying prior distributions for `phi.y` and `phi.t`.

The results from this model indicate a treatment effect distribution centered at 13.5 with a standard error of about 3.8. Results from two-stage least squares were similar with an estimate of 14.1 and standard error of 3.9, although in general the two approaches can give different answers.

Varying treatment effects

Because randomization occurred at the individual level, we could also extend this model to include varying treatment effects (similar to the coding for the varying intercepts). In this example, however, the sample sizes in each group were too small to estimate varying treatment effects reliably.

Group-level randomization

It is common to see an instrument that was assigned at the group level. This occurs in the case of a group-randomized experiment or, for example, when state policies are used as instruments in an analysis of individual outcomes. In these settings, the varying intercept model presented here is appropriate; however, varying treatment effects cannot be identified.

23.5 Bibliographic note

The books and articles referred to in Sections 9.9 and 10.8 include many examples with treatments at different levels, but we have few specific references for causal inference and multilevel models. Oakes (2004) and the accompanying discussion consider the challenges of interpreting multilevel coefficients causally, and Sobel (2006) considers the assumptions involved in estimating treatment effects in the presence of interactions between individuals and groups.

For discussion of prior distributions for Bayesian instrumental variables models, see, for instance, Dreze (1976), Maddala (1976), Kleibergen and Zivot (2003), and Hoogerheide, Kleibergen, and van Dijk (2006).

23.6 Exercises

1. Fit a varying-intercept model (without varying slopes) to the Sesame Street data (in folder `sesame`) and compare to the results in Section 23.4.

2. Simulate data from a group randomized experiment and then fit using a classical and then a multilevel model. The confidence intervals from the classical fit should be too narrow.

3. Generate data from an instrumental variables model randomized at the group level where differences exist between groups. Fit a varying-intercept model to these data. Compare these results to results from a classical two-stage least squares fit as described in Section 10.5.

4. Generate data from an instrumental variables model randomized at the individual level where treatment effects vary across groups. Fit a varying-intercept, varying-slope model to these data. Compare these results to results from a classical two-stage least squares model.

5. Explain why varying treatment effects can be identified when the instrument is randomized at the individual level but not when the instrument is randomized at the group level.

## CHAPTER 24

Model checking and comparison

There are generally many options available when modeling a data structure, and once we have successfully fit a model, it is important to check its fit to data. It is also often necessary to compare the fits of different models.

Our basic approach for checking model fit is—as we have described in Sections 8.3–8.4 for simple regression models—to simulate replicated datasets from the fitted model and compare these to the observed data. We discuss the general approach in Section 24.1 and illustrate in Section 24.2 with an extended example of a set of models fit to an experiment in animal learning. The methods we demonstrate are not specific to multilevel models but become particularly important as models become more complicated.

Although the methods described here are quite simple, we believe that they are not used as often as they could be, possibly because standard statistical techniques were developed before the use of computer simulation. In addition, fitting multilevel models is a challenge, and users are often so relieved to have successfully fit a model with convergence that there is a temptation to stop and rest rather than check the model fit. Section 24.3 discusses some tools for comparing different models fit to the same data.

Posterior predictive checking is a useful direct way of assessing the fit of the model to various aspects of the data. Our goal here is not to compare or choose among models but rather to explore the ways in which any of the models being considered might be lacking.

24.1 Principles of predictive checking

Monitoring the quality of a statistical model implies the detection of systematic differences between the model and observed data. Posterior predictive checks set this up by generating replicated datasets from the predictive distribution of the fitted model; these replicated datasets are then compared to the observed dataset with respect to any features of interest. The functions of data and model parameters that we use to compare to the model are called *test summaries* or *discrepancy variables*; we also consider the special case of *test statistics*, which depend on the (replicated) data only. This is the formal treatment of the simulation-based model-checking introduced in Section 8.3.

We use the notation $y = (y_1, \ldots, y_n)$ for discrete observed data, X for the matrix of predictor variables, and θ for the vector of all parameters. We assume that a model has been fit and that we have a set of simulations, $\theta^{(s)}$, $s = 1, \ldots, n_{\text{sims}}$.

We further assume that, for each of these draws, a replicated dataset, $y^{\text{rep}(s)}$, has been simulated from the predictive distribution of the data $p(y^{\text{rep}}|X, \theta = \theta^{(s)})$; the ensemble of simulated datasets $(y^{\text{rep}(s)}, \ldots, y^{\text{rep}(n_{\text{sims}})})$ thus represents the posterior predictive distribution, $p(y^{\text{rep}}|X, y)$. For simplicity, we suppress the conditioning on X in the notation that follows, but in some examples we shall allow X to vary and simulate X^{rep} as well (see Section 24.2). Predictive simulation of the replicated datasets y^{rep}, conditional on θ, is usually extremely easy—typically requiring nothing more than simulation from known independent distributions—even though obtaining posterior simulations of θ usually requires complicated Markov chain simulation methods.

We check the model by means of discrepancy variables $T(y, \theta)$. If θ were known, one could perform a goodness-of-fit test by comparing the observed $T(y, \theta)$ to the distribution of the discrepancy variables in the replications, $T(y^{\text{rep}}, \theta)$, with the statistical significance of the test summarized by a p-value, $p = \Pr(T(y^{\text{rep}}, \theta) > T(y, \theta)|y, \theta)$. (Here, we consider only one-sided tests, with the understanding that the corresponding two-sided p-value is $2 \cdot \min(p, 1-p)$.) In the more usual case of unknown θ, the test comparison is averaged over the uncertainty in θ (that is, the posterior distribution), with a posterior predictive p-value, $\Pr(T(y^{\text{rep}}, \theta) > T(y, \theta)|y) = \int \Pr(T(y^{\text{rep}}, \theta) > T(y, \theta)|y, \theta)p(\theta|y)d\theta$, which can be estimated from the simulations by $\sum_{s=1}^{n_{\text{sims}}} 1_{T(y^{\text{rep}(s)}, \theta^{(s)}) > T(y, \theta^{(s)})}/n_{\text{sims}}$, where 1_A is the indicator function that is 1 if the condition A is true and 0 otherwise.

As to the choice of discrepancy variables, we focus here on methods for detecting systematic discrepancies between model and data, not on the related problem of discovering outliers in otherwise-reasonable mdels. Some of the discrepancy variables we develop have been used in Bayesian methods for outlier detection but there with a focus on individual observations rather than on larger patterns. By comparison, the discrepancy variables we consider often average over sections of the data. In addition, we seek discrepancy variables that are easy to interpret and are also generally applicable to a wide range of problems. In many cases, this means that we would like to check qualitative features of the model (for example, independence, monotonicity, and unimodality) to give a better understanding of directions of model improvement.

Classical residual plots and binned residual plots can be viewed as particular examples of graphical discrepancy variables. If $T(y, \theta)$ is a graph, rather than a number, it would not make sense to compute a p-value, but we still could compare the plot, explicitly or implicitly, with what would be obtained under replications from the model. This is the approach illustrated in Sections 8.3–8.4.

The application of the predictive check method goes as follows. Several discrepancy variables are chosen to reveal interesting features of the data or discrepancies between the model and the data. For each discrepancy variable, each simulated *realized value* $T(y, \theta^{(s)})$ is compared to the corresponding simulated *replicated value* $T(y^{\text{rep}(s)}, \theta^{(s)})$. Large and systematic differences between realized and replicated values indicate a misfit of the model to the data, in the sense that the observed data do not look typical, in this respect, of the data predicted under the model. In some cases, differences between the realized data and replications are apparent visually; other times, it can be useful to compute the p-value of a realized discrepancy to see whether it could plausibly have arisen by chance under the model.

In any applied problem, it is appropriate to check aspects of the data and model that are of particular substantive interest. By their nature, such diagnostics can never be routine or automatic, but we can give some general suggestions. First, it is often useful to display the entire dataset (or, if that is not possible for a highly multivariate problem, various data-rich summaries) and compare to some predictive replications of the data to get an idea of what would be expected under the model. Patterns seen in this sort of exploratory check can be used as the basis for more systematic model checks. As in exploratory data analysis in general, the reference distribution of replicated datasets provides a standard of comparison by which the observed discrepancies can be measured—the goal is not to find statistical significance but rather to reveal areas where the data look different from what would have been expected under the model. Second, one can directly compute the predictive distribution of any function of data and parameters using the posterior simulations of (θ, y^{rep})—and thus directly check the fit with respect to any easily computable discrepancy variable of interest. Third, it often makes sense to set up discrepancy variables with an eye toward how the model might be improved— for example, summarizing between-group variability if one is considering fitting a random effects model.

24.2 Example: a behavioral learning experiment

Experimental data and historical context

We investigate the effectiveness of various model checks for an analysis of a logistic regression model applied to data from a well-known experiment on behavioral learning conducted around 1950. In this experiment, each of 30 dogs was given a sequence of 25 trials; in each trial, a light was switched on for ten seconds and then an electric shock was applied to the metal cage in which the dog was sitting. In each trial, the dog had an opportunity, once the light went on, to jump into an adjoining cage and thus avoid the shock. In the initial trial, all the dogs received the shock (since they did not know the meaning of the signal), and in the succeeding 24 trials they learned to avoid it. The left side of Figure 24.1 displays the experimental data for the 30 dogs, ordered by the time of the last trial in which they were shocked. (This ordering has nothing to do with the order in which the experiment was performed on the dogs; we choose it simply to make the between-dog variation in the data more visible.) Interest lies in the factors that affected the dogs' learning; in particular, did they learn more from successful avoidances than from shocks? Another question is: can the variation in responses among the 30 dogs be explained by a single stochastic learning model, or is there evidence in the data for underlying between-dog variation?

We choose this example to study model-checking methods because the data and the associated stochastic learning model have an interesting structure, with replications over dogs and the probability of avoidance of an electric shock dependent on previous outcomes. As we shall see, the sequential nature of the model has important implications for some of the predictive distributions used to calibrate the model checks. Specifically, the logistic regression model fits these data reasonably well but has interesting systematic patterns of misfit. These data are also of historical interest because the analysis by Bush and Mosteller in 1955 includes an early example of simulation-based model checking: they compared the observed data to simulations of 30 dogs from their model (with parameters fixed at their maximum likelihood estimates, which was reasonable in this case since they are accurately estimated from this dataset).

Setting up the logistic regression model and estimating its parameters

We use the notation $y_{jt} = 1$ or 0 to indicate a shock or avoidance for trial t on dog j, for $t = 0, \ldots, 24$ and $j = 1, \ldots, 30$. We fit a logistic regression model,

$$\Pr(y_{jt} = 1) = \text{logit}^{-1}(\beta_0 + \beta_1 X_{1jt} + \beta_2 X_{2jt}), \qquad (24.1)$$

```
   REAL DOGS              FAKE DOGS (logit model)      FAKE DOGS (log model)

 trial number              trial number                 trial number

0   5  10  15  20  24     0   5  10  15  20  24        0   5  10  15  20  24
SS.S.S..................  ..SS.SSS................  S.SS....................
SSSS.S..................  SSSSS.SS................  SSS.S...................
SSSSSSS.................  S.SSS...S...............  SSSSS...................
SSSSSSS.................  SSSS.SS.S...............  SS.SS.S.................
SSSSSSSS................  .SSSS.SSS.S.............  SSS.S.S.................
SSS.S..S................  S...SS..S.S.............  SSSSSSS.................
SSSS.S.S................  SSSSSS.SSSS.............  SSSS.S.S................
SSS..S.SS...............  SSS..SSS.S.S............  SSS.SSS.S...............
SSSS.S..S...............  SSSSSS..SSSS............  SSSS..S.S...............
SSS.....S...............  SSS.S.SS..SS............  SSSSSSSSS...............
S.SSSS.SSS..............  SSSSSSS...SS............  SSSSSSSS.S..............
SSSSSSS..SS.............  SSSSSSS.SS.SS...........  SS...S...S.S............
SS.S...S..S.S...........  SSSS.S.S....S...........  SSSSSSSSS..SS...........
SSSSS.SSSSSS............  .SSSSSSS.S.S............  SSSSS.SSS..S............
SSSS.SS..S.S............  SSSSSSSS.SSSS...........  SSS..S....S.............
SSSSS.....SS............  SS.SSSSSSSS.SS..........  SS.S.SSS.S.S............
SSS.S.S...S.S...........  SSS.S.SS..SS....S.......  SSSSSS.SSS..S...........
SSSS......S.S...........  SSSSS...SS.SS.SSS.......  SSSS..S.S...S...........
SSSSSSS.SSSSSS..........  SSSSS..S..SS...S........  SSSSS....S...S..........
S..SS....S.S.S..........  SSSSS.SS.S......S.......  SSSSSSSSS...SSS.........
SSSSS....SS.S...........  SSSSSSSS......S..S.......  SSSSSSSSS..S..S.........
S.S.SSS.S....S..........  SSSSSSS..SSSSS...S.......  SS.SSS..S......S........
SSSS..S.S..S.S..........  .SSSSSSSSS...S....S......  SSSSSS.S.SS.S.S........
SSSSSS.S...S.S..........  SSSSSSS.S...S....S......  SSSSS.S.S.S.S..S........
SSSSS..S..SS..S.S.......  SSSS.SSSS...SS..S.S......  SSSSSSSSSS......S.......
SSSSSSSSSS.....S........  SSSSS.....S..S.....S....  SSSSSS..S...S...SS......
SS.S.S........SS.......  SSSSSSS.SSSSSSS..S.SS.......  SSSSSSS.S.S.S.SS.S......
SSSS..SS...S.S.S.S......  SSSSSSSS.SSS.S.S...S.S...  SSSSSS.S.S.............S.
SSSSSSSS...S.SSS..S.....  SS.SSSS..S.............S.  SSSSSSSSSSS.SS.....S.....
SSSSS.S.S..S.SSS.....S..S  SSSSSS.SSS...S.........S.  SSSSS.....S............S
```

Figure 24.1 *On the left, sequence of shocks ("S") and avoidances (".") for 25 trials on each of 30 dogs. The dogs here are ordered by the time of the last shock, with ties broken randomly. In the middle and right, similar displays for 30 dogs simulated from the (classical) logistic and logarithmic regression models conditional on the estimated parameters for each model. See also Figure 24.5.*

where

$$X_{1jt} = \sum_{k=0}^{t-1}(1 - y_{jk}) = \text{number of previous avoidances}$$

$$X_{2jt} = \sum_{k=0}^{t-1} y_{jk} = \text{number of previous shocks.} \qquad (24.2)$$

Upon reflection, we can realize that this model is not ideal for these data. In particular, the experiment is designed so that all dogs get shocked on trial 0 (which in fact happens, as is shown in the leftmost column of Figure 24.1), whereas the logistic regression model is structured to always have a nonzero probability of both outcomes. This problem could be addressed in many ways, for example, by fitting a logarithmic instead of a logistic link or simply by fitting the model to the data excluding trial 0.

To start with, however, we fit the logistic regression model to the entire dataset and examine what aspects of model misfit are uncovered by various predictive checks. We believe this is an interesting question because it is standard practice to fit a logistic regression model to binary data without seriously considering its appropriateness. (One reason for this is that for many problems there is no clearly preferable alternative to logistic regression, and simple tricks like changing the link function or discarding noninformative data are not generally sufficient to allow a simple model to fit.) In these cases of routine use, we would like to have routine model checks (by analogy to residual plots in normal regressions) that would give the user some idea of the model's problems. The goal of such methods is not to "accept" or "reject" a model, but rather to highlight important areas where it does not fit the data.

The model is easy to fit in Bugs, and the posterior medians of β_0, β_1, and β_2 are 1.80, -0.35, and -0.21, respectively. The negative coefficients β_1, β_2 imply that the probability of shock declines after either an avoidance or a shock, with $|\beta_1| > |\beta_2|$ implying that avoidances have a larger effect.

We can write the Bugs model in three parts: first, calculation of the number of previous shocks and avoidances recursively in terms of y[j,t], the indicator for a shock of dog j at time t; second, the classical logistic regression model; and, third, the noninformative prior distributions: For convenience, we enclose both of the first two parts in the same for (j in 1:n.dogs) loop:

```
model {                                                              Bugs code
  for (j in 1:n.dogs){
    n.avoid[j,1] <- 0
    n.shock[j,1] <- 0
    for (t in 2:n.trials){
      n.avoid[j,t] <- n.avoid[j,t-1] + 1 - y[j,t-1]
      n.shock[j,t] <- n.shock[j,t-1] + y[j,t-1]
    }
    for (t in 1:n.trials){
      y[j,t] ~ dbin (p[j,t], 1)
      logit(p[j,t]) <- b.0 + b.1*n.avoid[j,t] + b.2*n.shock[j,t]
    }
  }
  b.0 ~ dnorm (0, .0001)
  b.1 ~ dnorm (0, .0001)
  b.2 ~ dnorm (0, .0001)
}
```

Defining predictive replications for the dog example

To perform model checks, the data must be compared to a reference distribution of possible replicated datasets. In a usual logistic regression model, this would be performed by fixing the matrix X of predictors and then, for each simulated parameter vector β^l, drawing the 25×30 responses $y_{jt}^{\mathrm{rep}\,(s)}$ independently,

$$\Pr(y_{jt}^{\mathrm{rep}\,(s)}) = \mathrm{logit}^{-1}(X_{jt}\beta^{(s)}), \qquad (24.3)$$

to yield a simulated dataset $y^{\mathrm{rep}\,(s)}$. (The notation X_{jt} indicates the vector of predictors, $(1, X_{1jt}, X_{2jt})$ defined in (24.2).) Computing this for n_{sims} parameter vectors yields n_{sims} simulated datasets.

A stochastic learning model is more complicated, however, because the predictor variables X depend on previous outcomes y. Simulation of replicated data for a new dog must thus be performed sequentially. For each simulated parameter vector $\beta^{(s)}$:

- For each dog, $j = 1, \ldots, 30$:

  - For trial $t = 0, \ldots, 24$:

1. Compute the vector of predictors, $X_{jt}^{\text{rep}\,(s)}$, based on the previous t trials for dog j.
2. Simulate $y_{jt}^{\text{rep}\,(s)}$ as in (24.3).

Predictive replications in Bugs. We can create the simulations in Bugs or directly in R. The Bugs model has two parts: recursive definition of the number of shocks/avoidances, and the probability model for the individual outcomes.

Bugs code
```
for (j in 1:n.dogs){
  n.avoid.rep[j,1] <- 0
  n.shock.rep[j,1] <- 0
  for (t in 2:n.trials){
    n.avoid.rep[j,t] <- n.avoid.rep[j,t-1] + 1 - y.rep[j,t-1]
    n.shock.rep[j,t] <- n.shock.rep[j,t-1] + y.rep[j,t-1]
  }
  for (t in 1:n.trials){
    y.rep[j,t] ~ dbin (p.rep[j,t], 1)
    logit(p.rep[j,t]) <- b.0+b.1*n.avoid.rep[j,t]+b.2*n.shock.rep[j,t]
  }
}
```

Once this is set up in Bugs, we can save y^{rep} along with everything else. The only difficulty is that we are now saving 750 additional parameters, and if we save thousands of iterations, we will run into computer storage problems. (Nowadays, 750,000 bytes are nothing on the computer, but in R, computations with a matrix of this size can be slow.) We set up and call the Bugs model as follows, using the n.thin option in Bugs to save only every 100th iteration:

R code
```
n.dogs <- nrow(y)
n.trials <- ncol(y)
data <- list ("y", "n.dogs", "n.trials")
inits <- function (){
  list (b.0=rnorm(1), b.1=rnorm(1), b.2=rnorm(1))
}
parameters <- c ("b.0", "b.1", "b.2", "y.rep", "y.mean.rep")
fit.logit.1 <- bugs (data, inits, parameters, "dogs.logit.1.bug",
  n.chains=3, n.iter=2000, n.thin=100)
```

With 3 chains of 2000 iterations each, saving every 100th iteration of the last half of each chain, we are left with 30 simulation draws, which are enough for reasonable estimates, standard errors, and predictive checks.

Predictive replications in R. Alternatively, we can write an R function to simulate predictive replications, given the simulations of $\beta_0, \beta_1, \beta_2$ from the Bugs model. We first set up an empty array for the n_{sims} replicated datasets and then fill it up, one dog at a time and one trial at a time. All the computations are done in vector form, simulating all n_{sims} random replications at once.

R code
```
y.rep <- array (NA, c(n.sims, n.dogs, n.trials))
for (j in 1:n.dogs){
  n.avoid.rep <- rep (0, n.sims)
  n.shock.rep <- rep (0, n.sims)
  for (t in 1:n.trials){
    p.rep <- invlogit (b.0 + b.1*n.avoid.rep + b.2*n.shock.rep)
    y.rep[,j,t] <- rbinom (n.sims, 1, p.rep)
    n.avoid.rep <- n.shock.rep + 1 - y.rep[,j,t]
    n.shock.rep <- n.shock.rep + y.rep[,j,t]
  }
}
```

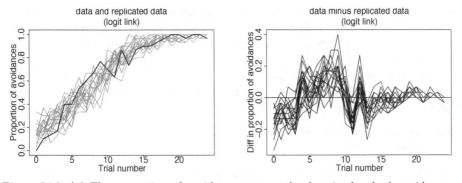

Figure 24.2 *(a) The proportion of avoidances among the dogs in the shock-avoidance experiment, as a function of the trial number. The solid line shows the data, and the light lines represent 20 simulated replications from the model. This plot can be seen as a predictive check of a vector test statistic, $T(y)$, compared to replications $T(y^{\text{rep}})$.*
(b) Plots of $T(y^{\text{rep}}) - T(y)$ for the 20 simulations of y^{rep}. The systematic differences from the horizontal line represent aspects of the data that are not captured by the model.

Direct comparison of simulated to real data

The most basic predictive check is a visual comparison of the observed data to a replication under the assumed model. Figure 24.1 shows the observed data (in the left part of the figure), along with a single replicated dataset, y^{rep} (in the center part; ignore the right part of the figure for now). The visual display shows some interesting differences between the real and simulated dogs.

The visual comparison is aided by ordering the 30 dogs in each dataset in order of the time of their last shock. We program this as a function in R:

```
dogsort <- function (y){
  n.dogs <- nrow(y)
  n.trials <- ncol(y)
  last.shock <- rep (NA, n.dogs)
  for (j in 1:n.dogs){
    last.shock[j] <- max ((1:n.trials)[y[j,]==1])}
  y[order(last.shock),]
}
```
R code

To make the left and center displays in Figure 24.1, we then simply print y and y.rep[1,,], with the latter being the first of the n_{sims} random replications of the data. (Since the simulations from the Bugs output are randomly ordered before being returned to R, the first simulated replication matrix is as good as any other.)

Strictly speaking, a posterior predictive check should compare y to several draws of y^{rep}, but in this case a single draw is informative because of the internal replication of 30 independent dogs in a single dataset.

More focused model checks

The graphical comparison in Figure 24.1 can be used to suggest more focused diagnostics. For example, the replicated dogs appear to have too few shocks in the early trials, compared to the real dogs. This is substantively relevant because the purpose of the model is to understand the learning behavior of the dogs. To check this pattern more formally, we display in Figure 24.2a the proportion of avoidances among the 30 dogs (that is, $1 - \bar{y}_{.t}$) versus time t. Overlain on the graph are the corresponding time series for 20 random draws $y^{\text{rep}\,(s)}$ from the predictive distribution under the estimated model. Compared to the data, the model predicts too many avoidances in the first two trials and too slow an improvement in the first five trials. The model thus does not capture the rate of learning at the beginning of the experiment.

We programmed this model check in R by first writing a function to generate the test variable (in this case, the mean number of avoidances over time) and then displaying it for the replications and observed data:

R code
```
test <- function (data){
  colMeans (1-data)
}
.plot (c(0,n.trials-1), c(0,1), xlab="Time",
  ylab="Proportion of avoidances", type="n")
mtext ("data and replicated data\n(logit link)", 3)
for (s in 1:20){
  lines (0:(n.trials-1), test (y.sim[s,,]), lwd=.5, col="gray")
}
lines (0:(n.trials-1), test (y), lwd=3)
```

To sharpen the contrast between replications and data, we display in Figure 24.2b the difference between the replicated and observed proportions of avoidances over time. In R, we first create the function for the difference of the test variables,

R code
```
test.diff <- function (data, data.rep){
  test (data) - test (data.rep)
}
```

then determine the range of these differences (to use in scaling the plot),

R code
```
diff.range <- NULL
for (s in 1:20){
  diff.range <- range (diff.range, test.diff (y, y.rep[s,,]))
}
```

and then set up the plot, graph the differences between the data and each of 20 simulations, and graph the zero line, which shows the standard of comparison.

R code
```
plot (c(0,24), diff.range, xlab="Time", ylab="Proportion of avoidances",
  type="n")
mtext ("data minus replicated data\n(logit link)", 3)
for (s in 1:20){
  lines (0:(n.trials-1), test.diff (y, y.sim[s,,]), lwd=.5, col="gray")
}
abline (0, 0, lwd=3)
```

Figure 24.2b shows the data to be consistently lower than the simulations at some points and higher at others, indicating a statistically significant discrepancy between model and data.

Numerical test statistics. For another example, Figure 24.3 displays predictive checks for two simple test statistics: the mean and standard deviation of the number of shocks per dog. In each plot, the observed value of $T(y)$ is shown as a vertical bar in a histogram representing 1000 draws of $T(y^{\text{rep}})$ from the posterior distribution. From Figure 24.3a, we see that the mean number of shocks is fit well by the model. Figure 24.3b shows that the observed standard deviation is a bit higher than expected under the model, but the discrepancy is not statistically significant; that is, we could expect to see such discrepancies occasionally just by chance, even if the model were correct.

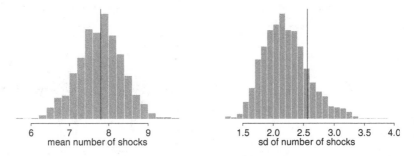

Figure 24.3 *Predictive checks for the (a) mean and (b) standard deviation of the number of shocks among the 30 dogs. The vertical bars indicate the observed values of the test statistics $T(y)$, and the histograms display $T(y^{\text{rep}})$ from 1000 draws of y^{rep} under the logistic model.*

Thus, these two aspects of the data are fit reasonably well; however, the systematic problem we have found in the early trials indicates a problem with the model.

Fitting and checking a logarithmic regression model

We now move to a more reasonable logarithmic regression model for the same data (which was in fact fit by the psychologists who performed the early data analyses):

$$\Pr(y_{jt} = 1) = \exp(\beta_1 X_{1jt} + \beta_2 X_{2jt}), \tag{24.4}$$

with X_{1jt} and X_{2jt} the number of previous avoidances and shocks, respectively, as defined in (24.2). Unlike the logistic model (24.1), this model has no constant term because the probability of shock is fixed at 1 at the beginning of the experiment. In addition, β_1 and β_2 are restricted to be negative.

The Bugs model is similar to the logistic regression on page 517 except with the `logit(p[j,t])` line changed to:

```
log(p[j,t]) <- b.1*n.avoid[j,t] + b.2*n.shock[j,t]
```
Bugs code

The log model omits the intercept term, b_0, so that the probability of a shock is 1 for the first trial. In addition, the coefficients β_1, β_2 must be constrained to be negative, so we give them the following noninformative distributions:

```
b.1 ~ dunif (-100, 0)
b.2 ~ dunif (-100, 0)
```
Bugs code

The median estimates of the parameters β_1 and β_2 in the logarithmic model are -0.24 and -0.08, with standard errors of 0.02 and 0.01, respectively. The coefficient for avoidances, β_1, is estimated to be more negative than β_2, indicating that avoidances have a larger effect than shocks in reducing the probability of future shocks. Transforming back to the probability scale, the median estimates for $(e^{\beta_1}, e^{\beta_2})$ are $(0.79, 0.92)$, indicating that an avoidance or a shock multiplies the predicted probability of shock by an estimated factor of 0.79 or 0.92, respectively.

Having fit this improved model, we check its fit using predictive replications, which we simulate from the model as described earlier in this section (except using the logarithmic rather than the logistic link). A single random draw from the predictive distribution of 30 new dogs is displayed on the right side of Figure 24.1. A check of the average number of avoidances over time—Figure 24.4, replicating Figure 24.2—shows no apparent discrepancies with this aspect of the data: the logarithmic link has fixed the problem with the early trials.

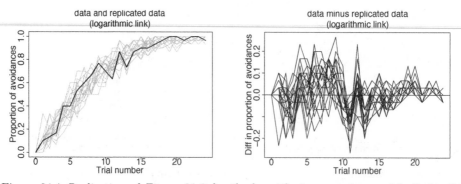

Figure 24.4 *Replication of Figure 24.2 for the logarithmic regression model. Under this estimated model, the simulated patterns of the rate of avoidances over time fit reasonably well to the data.*

Fitting and checking a multilevel model with no additional learning from avoidances

The logarithmic regression model fits the data reasonably well, but a key issue not addressed so far is separating between-dog variability from stochastic learning. The data from the 30 dogs vary quite a bit. In the stochastic learning model this is explained by the fact that dogs learn more from avoidances than from shocks (that is, $\beta_1 < \beta_2 < 0$) so that a dog that gets lucky at the beginning of the experiment is likely to perform well throughout, even in the absence of any real differences between dogs. However, one may consider an alternative explanation that the dogs indeed differ and, perhaps, there may be no additional learning associated with avoidances.

These two hypotheses—extra learning from avoidances, or between-dog variability—can be distinguished by a multilevel model of the form (24.4) but with parameters that vary by dog:

$$\Pr(y_{jt} = 1) = \exp(\beta_{1j}X_{1jt} + \beta_{2j}X_{2jt}), \qquad (24.5)$$

with the parameters β_1, β_2 both constrained to be negative for each dog (to ensure that $\Pr(y_{jt} = 1)$ is always less than 1).

We can easily fit this model all at once, but we fit it in stages in order to understand the workings of each part. The model we have already fit corresponds to extra learning from avoidances. The other extreme is to allow no extra learning from avoidances but to allow dogs to vary in their ability to perform the task. This model could be written as

$$\Pr(y_{jt} = 1) = \exp(\beta_j(X_{1jt} + X_{2jt})), \qquad (24.6)$$

where $X_{1jt} + X_{2jt}$ is the number of previous trials, or simply $t - 1$.

In Bugs, we write the model as before, but applying the same coefficient β to both shocks and avoidances, and allowing it to vary by dog. The parameters β_j are restricted to be negative (so that the probabilities p_{jt} fall below 1). To form a multilevel model, we work with the negative of the β_j's, which we assign a log-normal distribution (that is, the values of $\log(-\beta_j)$'s are modeled with a normal distribution) with hyperparameters μ_β and σ_β. Here is the rewritten fragment of the Bugs model:

Bugs code
```
log(p[j,t]) <- b[j]*(n.avoid[j,t]+n.shock[j,t])
}
```

```
       REAL DOGS              FAKE (simple multilevel model)   FAKE (full multilevel model)

  trial number                trial number                     trial number

0   5  10  15  20  24        0   5  10  15  20  24             0   5  10  15  20  24
SS.S.S..................     S.S.....................          SS.S....................
SSSS.S..................     SSSSS...................          SSSS....................
SSSSSSS.................     SSSSS...................          SSSSS...................
SSSSSSS.................     S.S....S................          SSSSS...................
SSSSSSSS................     SSS.S..S................          SSS.S...................
SSS.S..S................     SSSSS...S...............          SSS.S.S.................
SSSS.S.S................     SSSSS.SS................          SSS.SSS.................
SSS..S.SS...............     SSSSSS..S...............          SSSSSS.S................
SSSS.S...S..............     S.SS.....S..............          SSS...S.S...............
SSS.....S...............     SS.SSSS...S.............          SSS.....S...............
S.SSSS.SSS..............     SS.SS.S.S..S............          SSSSS...S...............
SSSSSSS..SS.............     SS..S.S....S............          SSSSSSSSSS..............
SS.S...S..S.............     SSSS......S.............          S.SSSS..S.S.............
SSSSS.SSSSSS............     SSSSS.S.SS..S...........          SSS.S..S..S.............
SSSS.SS..S.S............     SSS....S...S............          SSS........S............
SSSSS.....SS............     SSS.S.SS.S.S.S..........          SSSSS.SS.SSS............
SSS.S.S...S.S...........     SS..SS...S.....S........          SSS.SSS.....S...........
SSSS......S.S...........     SSS.SS.S.......S........          SSSSSSSSSSS.S...........
SSSSSSS.SSSSSS..........     S.S...SSS....S..........          SSSSSS.......S..........
S..SS....S.S.S..........     SS..S....S......S.......          SSS..SSSS.....S.........
SSSSS....SS.S...........     SS..SSS.SS......S.......          SSSSS..S.S.S..S.........
S.S.SSS.S....S..........     SSSSSS.SS..SS..SS.......          SSSS.SSSSS.S....S.......
SSSS..S.S..S...........     SS.SS..S...SS.S..S.......          SSSSSSSSS.SS.SSSS.......
SSSSSS.S...S.S..........     SSS.S........S...S.......         SSS.....S......SS.S......
SSSSS..S..SS..S.S.......     SSSS.....S..S...S.......          SSS..SS.S.S.S....S......
SSSSSSSSSS......S........    SSSS..SSSS.S......S......          SSSS.SS.S........S......
SS.S.S.......SS.........     SSSSS.....SSS....S.S.....          SSSSS.S............S.....
SSSS..SS...S.S.S.......      SSSS..S..S...........S...         SSSSSSS.S.S..S.....S.....
SSSSSSSS...S.SSS..S......     SSS.SS................S..         SSSS.................S....
SSSSS.S.S..S.SSS.....S..S    S.SSS..S..........S....S.         SSSSSSSS.........S.SS....
```

Figure 24.5 *On the left, sequence of shocks ("S") and avoidances (".") for 25 trials on each of 30 dogs, copied from the left panel of Figure 24.1. The dogs here are ordered by the time of the last shock, with ties broken randomly. In the middle and right, similar displays for 30 dogs simulated from the multilevel model, with and without no extra learning from avoidances. Compare to Figure 24.1 on page 516.*

```
    b[i] <- -b.neg[i]
    b.neg[i] ~ dlnorm (mu.b, tau.b)
  }
  mu.b ~ dnorm (0, .0001)
  tau.b <- pow(sigma.b, -2)
  sigma.b ~ dunif (0, 100)
```

When this model is fit to the dog data, the average of the 30 estimated values of e^{β_j} is 0.86—thus, the probability of a shock is multiplied by about 0.86 after each trial. This estimate is broadly consistent with our previous results: it falls between the estimates of e^{β_1} and e^{β_2} in the classical logarithmic regression model with differential learning. The standard deviation of the 30 dog parameters e^{β_j} in the multilevel model is estimated to be 0.04, indicating a fairly small range of variation in the "abilities" of the dogs.

The left and middle panels of Figure 24.5 show the data y (repeated from the left panel of Figure 24.1) and a simulated replicated dataset y^{rep} from the multilevel model (24.6) with no differential learning. The model clearly does not fit the data—the actual data show many cases of a dense series of shocks with no further avoidances, whereas the fake data shows many dogs with a long series of avoidances followed by a single late shock.

Fitting and checking the full multilevel model

Thus, it appears that the variation between dogs in the data is not simply explained by varying abilities; rather, there is evidence that the experience of avoidances introduces additional learning. We combine the features of between-dog variation and differential learning with the multilevel model (24.5), in which both the learning-after-avoidance and learning-after-shock parameters vary by dog. The regression part of the Bugs model becomes

Bugs code

```
log(p[j,t]) <- b.1[j]*n.avoid[j,t] + b.2[j]*n.shock[j,t])
```

with a normal distribution for $(\log(-\beta_{1j}), \log(-\beta_{2j}))$, a bivariate version of the lognormal distribution in the Bugs model on page 522.

The average of the 30 estimated values of $e^{\beta_{1j}}$ and $e^{\beta_{2j}}$ are 0.78 and 0.91 (with estimated between-dog standard deviations of 0.05 and 0.03, respectively). Hence, as with the non-multilevel log model, the probability of a shock is multiplied by about 0.8 following each avoidance and 0.9 following each shock, with relatively little between-dog variation in the parameters.

The group pooling factor (21.14) from page 479 is 0.73 for the parameters β_{1j} and 0.77 for the parameters β_{2j}—thus, the estimates are shrunk approximately three-fourths toward the group-level model, or to put it another way, the estimates are close to complete pooling.

Having fit the model, we can as usual check its fit by displaying a replicated dataset, which is displayed in the rightmost panel of Figure 24.5. The replications look generally comparable to the real dogs, and, as with Figure 24.4, plots of average avoidances over time show no systematic discrepancies between model and data. Given the relatively small size of this dataset, it is difficult to say much more about this comparison.

Review of example

We have fit several models to the dog data and found that the variation between the response patterns of the dogs can largely be explained by learning more from avoidances than from shocks. This path-dependence or positive-feedback pattern implies that when a dog gets lucky and avoids the shock early, learning, and hence further avoidances, will proceed faster. There is evidence for a small amount of between-dog variation, but less than the difference between the effects of shocks and avoidances.

We used predictive simulations to check the fit of the different models to data, using raw data displays and also summaries of the rate of avoidances over time. The predictive checks do not automatically lead us to an appropriate model and will not necessarily "reject" a poorly fitting model—for example, the mean and standard-deviation summaries in Figure 24.3 on page 521 do not point out the serious failings of the logistic model. Rather, predictive checking can be a useful tool in revealing aspects of disagreement between model and data, which can motivate a search for more reasonable models.

24.3 Model comparison and deviance

When fitting several models to the same dataset, it can be helpful to compare them using summary measures of fit. A standard summary that is programmed in Bugs is the *deviance*, which is -2 times the log-likelihood; that is, -2 times the logarithm of the probability of the data given the estimated model parameters.

Deviance and AIC in classical generalized linear models

In classical generalized linear models, adding a parameter to a model is expected to increase the fit—even if the new parameter represents pure noise. Adding a noise predictor is expected to reduce the deviance by 1, and adding k predictors that are pure noise is expected to reduce the deviance by k. More precisely, adding k noise predictors will reduce the deviance by an amount corresponding to a χ^2 distribution with k degrees of freedom.

Thus, if k predictors are added and the deviance declines by significantly *more* than k, then we can conclude that the observed improvement in predictive power is statistically significant. Thus,

$$\text{adjusted deviance} = \text{deviance} + \text{number of predictors} \qquad (24.7)$$

can be used as an adjusted measure that approximately accounts for the increase in fit attained simply by adding predictors to a model. (The analogy in simple linear regression is the adjusted R^2.)

The next step, beyond checking whether an improvement in deviance is statistically significant, is to see if it is estimated to increase out-of-sample predictive power. On average, a predictor needs to reduce the deviance by 2 in order to improve the fit to new data. The *Akaike information criterion* is defined as

$$
\begin{aligned}
\text{AIC} &= \text{deviance} + 2 \cdot (\text{number of predictors}) \\
&= \text{adjusted deviance} + \text{number of predictors}. \qquad (24.8)
\end{aligned}
$$

In classical regression or generalized linear modeling, a new model is estimated to reduce out-of-sample prediction error if the AIC decreases.

Deviance and DIC in multilevel models

The ideas of deviance and AIC apply to multilevel models also, but with the difficulty that the "number of parameters" is not so clearly defined. Roughly speaking, the number of parameters in a multilevel model depends on the amount of pooling—a batch of J parameters corresponds to one parameter if there is complete pooling, J independent parameters if there is no pooling, and something in between with partial pooling. For example, with the varying-intercept radon models, the coefficients for the 85 county indicators represent something fewer than 85 independent parameters. Especially for the counties with small sample sizes, the group-level regression explains much of the variation in the intercepts, so that in the multilevel model they are not estimated independently. When the model is improved and the group-level variance decreases, the effective number of independent parameters also decreases.

In multilevel models, the *mean deviance* (that is, the deviance averaged over all the n_{sims} simulated parameter vectors) plays the role of the adjusted deviance (24.7). The effective number of parameters is called p_D, and the measure of out-of-sample predictive error is the *deviance information criterion*,

$$\text{DIC} = \text{mean deviance} + 2p_D, \qquad (24.9)$$

which plays the role of the Akaike information criterion in (24.8). We shall not further discuss the computation of p_D here, except to say that it is unstable to estimate—even from Bugs simulations that have otherwise converged—and so we currently use it only in a provisional sense.

We illustrate the use of deviance and DIC by comparing the fit of the models fit to the dog data in Section 24.2. Figure 24.6 shows the mean deviance, effective

Model	Mean deviance	Effective # params, p_D	DIC
Classical logistic	570	4	574
Classical logarithmic	550	5	555
Simple multilevel	544	21	565
Full multilevel	533	29	562

Figure 24.6 *Average deviance, estimated effective number of parameters, and deviance information criterion (DIC) for each of four models fit to the dog data in Section 24.2. Model fit improves as we go down the table, as can be seen from the decreasing values of mean deviance. However, the effective number of parameters increases for the larger models. The best model for out-of-sample predictions, as measured by DIC, is the classical logarithmic model.*
We would still prefer the full multilevel model here, since we expect the dogs to vary in their underlying parameters. But the improvement in fit, compared to the classical logarithmic regression, is not estimated to result in more accurate predictions.

number of parameters, and DIC for each. As the models get more complicated, the mean deviance decreases, which makes sense—with more structure, we can fit the data better. The largest jump is from the logit to the log model, which makes sense, since as we saw in Section 24.2, the logarithmic link fit the data much better.

However, the improvement in fit when moving to the multilevel models is counterbalanced by the increase in p_D, the effective number of parameters. As a result, the estimated out-of-sample prediction error, DIC, actually increases slightly for these models. The multilevel models do fit the data better, an improvement greater than would be expected by chance—as we can see from the mean deviances—but the DIC values suggest that they would not actually do as well in predicting for new dogs.

In summary, we are *not* saying that the classical logarithmic model is "best" here. We prefer the full multilevel model as a more complete description of the data. However, the DIC results are interesting in suggesting that further improvement is possible, perhaps constraining the β_{1j} and β_{2j} parameters more than is done by the multilevel model we have fit so far.

Finally, the values of p_D in Figure 24.6 illustrate the instability of the estimates of this quantity: the classical models are estimated to have 4 and 5 "effective parameters" each, even though they only have 3 and 2 parameters, respectively. The multilevel models are estimated to have 21 (out of a possible 30) and 29 (out of a possible 60) parameters each, even though we have seen that their pooling factors are close to 1 (that is, the models are close to complete pooling). So we treat these p_D and DIC results as suggestive rather than definitive.

24.4 Bibliographic note

The simulation-based model checking approach presented in Chapter 8 and here is based on ideas of Box (1980), Rubin (1984), Gelman, Meng, and Stern (1996), Gelman et al. (2003, chapter 6), and Gelman (2003); see also Stone (1974) and Gelfand, Dey, and Chang (1992) for related ideas. Sinharay and Stern (2003) apply predictive checks to hierarchical models. See Cook and Weisberg (1999) for a comprehensive overview of classical regression diagnostics. Pardoe (2001) and Pardoe and Cook (2002) extend these ideas to simulation-based inference.

The logarithmic model for the dog data comes from Bush and Mosteller (1955), who also performed simulation-based model checking. Further analyses of these data were considered by Sternberg (1963) and Gelman et al. (2000).

AIC comes from Akaike (1973) and is related to C_p (Mallows, 1973); DIC comes from Spiegelhalter et al. (2002); see also Hodges and Sargent (2001), Vaida and Blanchard (2002), and Spiegelhalter (2006) for related work and discussion.

24.5 Exercises

1. Download the data in the folder **dogs** and fit some other models, for example using as a predictor the result from the previous trial, or the previous two trials, rather than the total number of shocks and avoidances.

   (a) Fit this model, as usual building up from simpler versions (first a single-level model, then varying intercepts, then varying slopes, then adding other predictors as appropriate). Plot the data and fitted model to make sure that your model makes sense.

   (b) Use Bugs to simulate replicated datasets from your model, and make various plots to compare the replicated with the actual data.

2. Model checking with non-nested levels: the folder **supreme.court** contains data regarding U.S. Supreme Court votes for all justices across several issues.

   (a) Fit an ideal-point model to these data (see Section 14.3) and then use replicated data and graphical displays to check the model fit. (This is a huge dataset, and so the model will certainly not fit in many ways. Your goal here is not simply to "reject" the model but rather to understand the ways in which it does not fit.)

   (b) How might you expand the model to fix these problems?

3. Model checking for multilevel logistic regression:

   (a) Do some simulation-based graphical checking for the logistic regression model that you fit in Exercises 14.5–14.6 to the data from the speed-dating experiment.

   (b) How might you expand the model to fix the problems you have found?

4. Model checking for ordered categorical regression:

   (a) Do some simulation-based graphical checking for the ordered logistic regression model that you fit in Exercise 17.11 to the data from the storable-voting experiment.

   (b) How might you expand the model to fix the problems you have found?

## CHAPTER 25

Missing-data imputation

Missing data arise in almost all serious statistical analyses. In this chapter we discuss a variety of methods to handle missing data, including some relatively simple approaches that can often yield reasonable results. We use as a running example the Social Indicators Survey, a telephone survey of New York City families conducted every two years by the Columbia University School of Social Work. Nonresponse in this survey is a distraction to our main goal of studying trends in attitudes and economic conditions, and we would like to simply clean the dataset so it could be analyzed as if there were no missingness. After some background in Sections 25.1–25.3, we discuss in Sections 25.4–25.5 our general approach of random imputation. Section 25.6 discusses situations where the missing-data process must be modeled (this can be done in Bugs) in order to perform imputations correctly.

Missing data in R and Bugs

In R, missing values are indicated by NA's. For example, to see some of the data from five respondents in the data file for the Social Indicators Survey (arbitrarily picking rows 91–95), we type

```
cbind (sex, race, educ_r, r_age, earnings, police)[91:95,]
```
R code

and get

	sex	race	educ_r	r_age	earnings	police
[91,]	1	3	3	31	NA	0
[92,]	2	1	2	37	135.00	1
[93,]	2	3	2	40	NA	1
[94,]	1	1	3	42	3.00	1
[95,]	1	3	1	24	0.00	NA

R output

In classical regression (as well as most other models), R automatically excludes all cases in which any of the inputs are missing; this can limit the amount of information available in the analysis, especially if the model includes many inputs with potential missingness. This approach is called a complete-case analysis, and we discuss some of its weaknesses below.

In Bugs, missing *outcomes* in a regression can be handled easily by simply including the data vector, NA's and all. Bugs explicitly models the outcome variable, and so it is trivial to use this model to, in effect, impute missing values at each iteration.

Things become more difficult when predictors have missing values. For example, if we wanted to model attitudes toward the police, given earnings and demographic predictors, then the model would *not* automatically account for the missing values of earnings. We would have to remove the missing values, impute them, or model them. In Bugs, regression predictors are typically unmodeled and so Bugs does not know how to draw from a predictive distribution for them. To handle missing data in the predictors, Bugs regression models such as those in Part IIB need to be extended by modeling (that is, supplying distributions for) the input variables.

25.1 Missing-data mechanisms

To decide how to handle missing data, it is helpful to know why they are missing. We consider four general "missingness mechanisms," moving from the simplest to the most general.

1. *Missingness completely at random.* A variable is *missing completely at random* if the probability of missingness is the same for all units, for example, if each survey respondent decides whether to answer the "earnings" question by rolling a die and refusing to answer if a "6" shows up. If data are missing completely at random, then throwing out cases with missing data does not bias your inferences.

2. *Missingness at random.* Most missingness is *not* completely at random, as can be seen from the data themselves. For example, the different nonresponse rates for whites and blacks (see Exercise 25.1) indicate that the "earnings" question in the Social Indicators Survey is not missing completely at random.

   A more general assumption, *missing at random*, is that the probability a variable is missing depends only on available information. Thus, if sex, race, education, and age are recorded for all the people in the survey, then "earnings" is missing at random if the probability of nonresponse to this question depends only on these other, fully recorded variables. It is often reasonable to model this process as a logistic regression, where the outcome variable equals 1 for observed cases and 0 for missing.

   When an outcome variable is missing at random, it is acceptable to exclude the missing cases (that is, to treat them as NA's), as long as the regression controls for all the variables that affect the probability of missingness. Thus, any model for earnings would have to include predictors for ethnicity, to avoid nonresponse bias.

   This missing-at-random assumption (a more formal version of which is sometimes called the ignorability assumption) in the missing-data framework is the basically same sort of assumption as ignorability in the causal framework. Both require that sufficient information has been collected that we can "ignore" the assignment mechanism (assignment to treatment, assignment to nonresponse).

3. *Missingness that depends on unobserved predictors.* Missingness is no longer "at random" if it depends on information that has not been recorded and this information also predicts the missing values. For example, suppose that "surly" people are less likely to respond to the earnings question, surliness is predictive of earnings, and "surliness" is unobserved. Or, suppose that people with college degrees are less likely to reveal their earnings, having a college degree is predictive of earnings, and there is also some nonresponse to the education question. Then, once again, earnings are not missing at random.

   A familiar example from medical studies is that if a particular treatment causes discomfort, a patient is more likely to drop out of the study. This missingness is not at random (unless "discomfort" is measured and observed for all patients).

   If missingness is not at random, it must be explicitly modeled, or else you must accept some bias in your inferences.

4. *Missingness that depends on the missing value itself.* Finally, a particularly difficult situation arises when the probability of missingness depends on the (potentially missing) variable itself. For example, suppose that people with higher earnings are less likely to reveal them. In the extreme case (for example, all persons earning more than \$100,000 refuse to respond), this is called *censoring*, but even the probabilistic case causes difficulty.

Censoring and related missing-data mechanisms can be modeled (as discussed in Section 18.5) or else mitigated by including more predictors in the missing-data model and thus bringing it closer to missing at random. For example, whites and persons with college degrees tend to have higher-than-average incomes, so controlling for these predictors will somewhat—but probably only somewhat—correct for the higher rate of nonresponse among higher-income people. More generally, while it can be possible to predict missing values based on the other variables in your dataset, just as with other missing-data mechanisms, this situation can be more complicated in that the nature of the missing-data mechanism may force these predictive models to extrapolate beyond the range of the observed data.

General impossibility of proving that data are missing at random

As discussed above, missingness at random is relatively easy to handle—simply include as regression inputs all variables that affect the probability of missingness. Unfortunately, we generally cannot be sure whether data really are missing at random, or whether the missingness depends on unobserved predictors or the missing data themselves. The fundamental difficulty is that these potential "lurking variables" are unobserved—by definition—and so we can never rule them out. We generally must make assumptions, or check with reference to other studies (for example, surveys in which extensive follow-ups are done in order to ascertain the earnings of nonrespondents).

In practice, we typically try to include as many predictors as possible in a model so that the "missing at random" assumption is reasonable. For example, it may be a strong assumption that nonresponse to the earnings question depends only on sex, race, and education—but this is a lot more plausible than assuming that the probability of nonresponse is constant, or that it depends only on one of these predictors.

25.2 Missing-data methods that discard data

Many missing data approaches simplify the problem by throwing away data. We discuss in this section how these approaches may lead to biased estimates (one of these methods tries to directly address this issue). In addition, throwing away data can lead to estimates with larger standard errors due to reduced sample size.

Complete-case analysis

A direct approach to missing data is to exclude them. In the regression context, this usually means *complete-case analysis*: excluding all units for which the outcome or any of the inputs are missing. In R, this is done automatically for classical regressions (data points with any missingness in the predictors or outcome are ignored by the regression). In Bugs, missing values in unmodeled data are not allowed, so these cases must be excluded in R before sending the data to Bugs, or else the variables with missingness must be explicitly modeled (see Section 25.6).

Two problems arise with complete-case analysis:

1. If the units with missing values differ systematically from the completely observed cases, this could bias the complete-case analysis.

2. If many variables are included in a model, there may be very few complete cases, so that most of the data would be discarded for the sake of a simple analysis.

Another simple approach is *available-case analysis*, where different aspects of a problem are studied with different subsets of the data. For example, in the 2001 Social Indicators Survey, all 1501 respondents stated their education level, but 16% refused to state their earnings. We could thus summarize the distribution of education levels of New Yorkers using all the responses and the distribution of earnings using the 84% of respondents who answered that question. This approach has the problem that different analyses will be based on different subsets of the data and thus will not necessarily be consistent with each other. In addition, as with complete-case analysis, if the nonrespondents differ systematically from the respondents, this will bias the available-case summaries. For example in the Social Indicators Survey, 90% of African Americans but only 81% of whites report their earnings, so the "earnings" summary represents a different population than the "education" summary.

Available-case analysis also arises when a researcher simply excludes a variable or set of variables from the analysis because of their missing-data rates (sometimes called "complete-variables analyses"). In a causal inference context (as with many prediction contexts), this may lead to omission of a variable that is necessary to satisfy the assumptions necessary for desired (causal) interpretations.

Nonresponse weighting

As discussed previously, complete-case analysis can yield biased estimates because the sample of observations that have no missing data might not be representative of the full sample. Is there a way of reweighting this sample so that representativeness is restored?

Suppose, for instance, that only one variable has missing data. We could build a model to predict the nonresponse in that variable using all the other variables. The inverse of predicted probabilities of response from this model could then be used as survey weights to make the complete-case sample representative (along the dimensions measured by the other predictors) of the full sample. This method becomes more complicated when there is more than one variable with missing data. Moreover, as with any weighting scheme, there is the potential that standard errors will become erratic if predicted probabilities are close to 0 or 1.

25.3 Simple missing-data approaches that retain all the data

Rather than removing variables or observations with missing data, another approach is to fill in or "impute" missing values. A variety of imputation approaches can be used that range from extremely simple to rather complex. These methods keep the full sample size, which can be advantageous for bias and precision; however, they can yield different kinds of bias, as detailed in this section.

Whenever a single imputation strategy is used, the standard errors of estimates tend to be too low. The intuition here is that we have substantial uncertainty about the missing values, but by choosing a single imputation we in essence pretend that we know the true value with certainty.

Mean imputation. Perhaps the easiest way to impute is to replace each missing value with the mean of the observed values for that variable. Unfortunately, this strategy can severely distort the distribution for this variable, leading to complications with summary measures including, notably, underestimates of the standard deviation. Moreover, mean imputation distorts relationships between variables by "pulling" estimates of the correlation toward zero.

Last value carried forward. In evaluations of interventions where pre-treatment measures of the outcome variable are also recorded, a strategy that is sometimes used is to replace missing outcome values with the pre-treatment measure. This is often thought to be a conservative approach (that is, one that would lead to underestimates of the true treatment effect). However, there are situations in which this strategy can be *anti*conservative. For instance, consider a randomized evaluation of an intervention that targets couples at high risk of HIV infection. From the regression-to-the-mean phenomenon (see Section 4.3), we might expect a reduction in risky behavior even in the absence of the randomized experiment; therefore, carrying the last value forward will result in values that look worse than they truly are. Differential rates of missing data across the treatment and control groups will result in biased treatment effect estimates that are anticonservative.

Using information from related observations. Suppose we are missing data regarding the income of fathers of children in a dataset. Why not fill these values in with mother's report of the values? This is a plausible strategy, although these imputations may propagate measurement error. Also we must consider whether there is any incentive for the reporting person to misrepresent the measurement for the person about whom he or she is providing information.

Indicator variables for missingness of categorical predictors. For unordered categorical predictors, a simple and often useful approach to imputation is to add an extra category for the variable indicating missingness.

Indicator variables for missingness of continuous predictors. A popular approach in the social sciences is to include for each continuous predictor variable with missingness an extra indicator identifying which observations on that variable have missing data. Then the missing values in the partially observed predictor are replaced by zeroes or by the mean (this choice is essentially irrelevant). This strategy is prone to yield biased coefficient estimates for the other predictors included in the model because it forces the slope to be the same across both missing-data groups. Adding interactions between an indicator for response and these predictors can help to alleviate this bias (this leads to estimates similar to complete-case estimates).

Imputation based on logical rules. Sometimes we can impute using logical rules: for example, the Social Indicators Survey includes a question on "number of months worked in the previous year," which all 1501 respondents answered. Of the persons who refused to answer the earnings question, 10 reported working zero months during the previous year, and thus we could impute zero earnings to them. This type of imputation strategy does not rely on particularly strong assumptions since, in effect, the missing-data mechanism is known.

25.4 Random imputation of a single variable

When more than a trivial fraction of data are missing, however, we prefer to perform imputations more formally. In order to understand missing-data imputation, we start with the relatively simple setting in which missingness is confined to a single variable, y, with a set of variables X that are observed on all units. We shall consider the case of imputing missing earnings in the Social Indicators Survey.

Figure 25.1 *Histogram of earnings (in thousands of dollars) in the Social Indicators Survey: (a) for the 988 respondents who answered the question and had positive earnings, (b) deterministic imputations for the 241 missing values from a regression model, (c) random imputations from that mode. All values are topcoded at 100, with zero values excluded.*

Simple random imputation

The simplest approach is to impute missing values of earnings based on the observed data for this variable. We can write this as an R function:

R code
```
random.imp <- function (a){
   missing <- is.na(a)
   n.missing <- sum(missing)
   a.obs <- a[!missing]
   imputed <- a
   imputed[missing] <- sample (a.obs, n.missing, replace=TRUE)
   return (imputed)
}
```

(To see how this function works, take a small dataset and evaluate the function line by line.) We use `random.imp` to create a *completed data* vector of earnings:

R code
```
earnings.imp <- random.imp (earnings)
```

imputing into the missing values of the original `earnings` variable. This approach does not make much sense—it ignores the useful information from all the other questions asked of these survey responses—but these simple random imputations can be a convenient starting point. A better approach is to fit a regression to the observed cases and then use that to predict the missing cases, as we show next.

Zero coding and topcoding

We begin with some practicalities of the measurement scale. We shall fit the regression model to those respondents whose earnings were observed and positive (since, as noted earlier, the respondents with zero earnings can be identified from their zero responses to the "months worked" question). In addition, we shall "topcode" all earnings at $100,000—that is, all responses above this value will be set to $100,000—before running the regression. Figure 25.1a shows the distribution of positive earnings after topcoding.

R code
```
topcode <- function (a, top){
   return (ifelse (a>top, top, a))
}
earnings.top <- topcode (earnings, 100)    # earnings are in $thousands
hist (earnings.top[earnings>0])
```

The topcoding reduces the sensitivity of the results to the highest values, which in this survey go up to the millions. By topcoding we lose information, but the main use of earnings in this survey is to categorize families into income quantiles, for which purpose topcoding at \$100,000 has no effect.

Similarly, we topcoded number of hours worked per week at 40 hours. The purpose of topcoding was not to correct the data—we have no particular reason to disbelieve the high responses—but rather to perform a simple transformation to improve the predictive power of the regression model.

Using regression predictions to perform deterministic imputation

A simple and general imputation procedure that uses individual-level information uses a regression to the nonzero values of earnings. We begin by setting up a data frame with all the variables we shall use in our analysis:

```
sis <- data.frame (cbind (earnings, earnings.top, male, over65, white,       R code
    immig, educ_r, workmos, workhrs.top, any.ssi, any.welfare, any.charity))
```

and then fit a regression to positive values of earnings:

```
lm.imp.1 <- lm (earnings ~ male + over65 + white + immig + educ_r +          R code
    workmos + workhrs.top + any.ssi + any.welfare + any.charity,
    data=SIS, subset=earnings>0)
```

We shall describe these predictors shortly, but first we go through the steps needed to create deterministic and then random imputations. We first get predictions for all the data:

```
pred.1 <- predict (lm.imp.1, SIS)                                            R code
```

To get predictions for the entire data vector, we must include the data frame, `sis`, in the `predict()` call. Simply writing `predict(lm.imp.1)` would give predictions only for the data used in the fitting, which in this case are the subset of cases for which earnings are positive and for which none of the variables used in the regression are missing.

Next we write a little function to create a completed dataset by imputing the predictions into the missing values:

```
impute <- function (a, a.impute){                                           R code
    ifelse (is.na(a), a.impute, a)
}
```

and use this to impute missing earnings:

```
earnings.imp.1 <- impute (earnings, pred.1)                                  R code
```

Transforming and topcoding.   For the purpose of predicting incomes in the low and middle range (where we are most interested), we can do better by working on the square root scale of income, topcoded to 100 (in thousands of dollars):

```
lm.imp.2.sqrt <- lm (I(sqrt(earnings.top)) ~ male + over65 + white +        R code
    immig + educ_r + workmos + workhrs.top + any.ssi + any.welfare +
    any.charity, data=SIS, subset=earnings>0)
display (lm.imp.2.sqrt)
pred.2.sqrt <- predict (lm.imp.2.sqrt, SIS)
pred.2 <- topcode (pred.2.sqrt^2, 100)
earnings.imp.2 <- impute (earnings.top, pred.2)
```

Here is the fitted model:

R output

```
                      coef.est coef.se
(Intercept)            -1.67     0.44
male                    0.32     0.13
over65                 -1.44     0.58
white                   0.96     0.15
immig                  -0.62     0.14
educ_r                  0.79     0.07
workmos                 0.33     0.03
workhrs.top             0.06     0.01
any.ssi                -0.97     0.55
any.welfare            -1.35     0.37
any.charity            -1.17     0.60
  n = 988, k = 11
  residual sd = 1.96, R-Squared = 0.44
```

Figure 25.1b shows the deterministic imputations:

R code
```
hist (earnings.imp.2[is.na(earnings)])
```

From this graph, it appears that most of the nonrespondents have incomes in the middle range (compare to Figure 25.1a). Actually, the central tendency of Figure 25.1b is an artifact of the deterministic imputation procedure. One way to see this is through the regression model: its R^2 is 0.44, which means that the explained variance from the regression is only 44% of the total variance. Equivalently, the explained standard deviation is $\sqrt{0.44} = 0.66 = 66\%$ of the data standard deviation. Hence, the predicted values from the regression will tend to be less variable than the original data. If we were to use the resulting deterministic imputations, we would be falsely implying that most of these nonrespondents had incomes in the middle of the scale.

Random regression imputation

We can put the uncertainty back into the imputations by adding the prediction error into the regression, as discussed in Section 7.2. For this example, this involves creating a vector of random predicted values for the 241 missing cases using the normal distribution, and then squaring, as before, to return to the original dollar scale:

R code
```
pred.4.sqrt <- rnorm (n, predict (lm.imp.2.sqrt, SIS),
    sigma.hat (lm.imp.2.sqrt))
pred.4 <- topcode (pred.4.sqrt^2, 100)
earnings.imp.4 <- impute (earnings.top, pred.4)
```

Figure 25.1c shows the resulting imputed values from a single simulation draw. Compared to Figure 25.1b, these random imputations are more appropriately spread across the range of the population.

The new imputations certainly do not look perfect—in particular, there still seem to be too few imputations at the topcoded value of $100,000—suggesting that the linear model on the square root scale, with normal errors, is not quite appropriate for these data. (This makes sense given the spike in the data from the topcoding.) The results look much better than the deterministic imputations, however.

Figure 25.2 illustrates the deterministic and random imputations in another way. The left plot in the figure shows the deterministic imputations as a function of the predicted earnings from the regression model. By the definition of the imputation procedure, the values are identical and so the points fall along the identity line. The right plot shows the random imputations, which follow a generally increasing

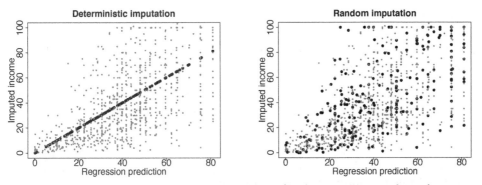

Figure 25.2 *Deterministic and random imputations for the 241 missing values of earnings in the Social Indicators Survey. The deterministic imputations are exactly at the regression predictions and ignore predictive uncertainty. In contrast, the random imputations are more variable and better capture the range of earnings in the data. See also Figure 25.1.*

pattern but with scatter derived from the unexplained variance in the model. (The increase in variance as a function of predicted value arises from fitting the model on the square root scale and squaring at the end.)

Predictors used in the imputation model

We fit a regression of earnings on sex, age, ethnicity, nationality, education, the number of months worked in the previous year and hours worked per week, and indicators for whether the respondent's family receives each of three forms of income support (from disability payments, welfare, and private charities).

It might seem strange to model earnings given information on income support—which is, in part, a consequence of earnings—but for the purposes of imputation this is acceptable. The goal here is not causal inference but simply accurate prediction, and it is acceptable to use any inputs in the imputation model to achieve this goal.

Two-stage modeling to impute a variable that can be positive or zero

In the Social Indicators Survey, we only need to impute the positive values of earnings: the "hours worked" and "months worked" questions were answered by everyone in the survey, and these variables are a perfect predictor of whether the value of earnings (more precisely, employment income) is positive. For the missing cases of earnings, we can impute 0 if workhrs = 0 and workmos = 0, and impute a continuous positive value when either of these is positive. This imputation process is what was described above, with the regression based on $n = 988$ data points and displayed in Figure 25.2. The survey as a whole included 1501 families, of whom 272 reported working zero hours and months and were thus known to have zero earnings. Of the 1229 persons reporting positive working hours or months, 988 responded to the earnings question and 241 did not.

Now suppose that the workhrs and workmos variables were *not* available, so that we could not immediately identify the cases with zero earnings. We would then impute missing responses to the earnings question in two steps: first, imputing an indicator for whether earnings are positive, and, second, imputing the continuous positive values of earnings.

Mathematically, we would impute earnings y given regression predictors X in a two-step process, defining

$$y = I^y y^{\mathrm{pos}},$$

where $I^y = 1$ if $y > 0$ and 0 otherwise, and $y^{\mathrm{pos}} = y$ if $y > 0$. The first model is a logistic regression for I^y:

$$\Pr(I_i^y = 1) = \mathrm{logit}^{-1}(X_i \alpha),$$

and the second part is a linear regression for the square root of y^{pos}:

$$\sqrt{y_i^{\mathrm{pos}}} \sim \mathrm{N}(X_i \beta, \sigma^2).$$

The first model is fit to all the data for which y is observed, and the second model is fit to all the data for which y is observed and positive.

We illustrate with the earnings example. First we fit the two models:

R code

```
glm.sign <- glm (I(earnings>0) ~ male + over65 + white +
    immig + educ_r + any.ssi + any.welfare + any.charity,
    data=SIS, family=binomial(link=logit))
display (glm.sign)
lm.ifpos.sqrt <- lm (I(sqrt(earnings.top)) ~ male + over65 + white +
    immig + educ_r + any.ssi + any.welfare + any.charity,
    data=SIS, subset=earnings>0)        # (same as lm.imp.2 from above)
display (lm.ifpos.sqrt)
```

Then we impute whether missing earnings are positive:

R code

```
pred.sign <- rbinom (n, 1, predict (glm.sign, data, type="response"))
pred.pos.sqrt <- rnorm (n, predict (lm.ifpos.sqrt, SIS),
    sigma.hat(lm.ifpos.sqrt))
```

and then impute the earnings themselves:

R code

```
pred.pos <- topcode (pred.pos.sqrt^2, 100)
earnings.imp <- impute (earnings, pred.sign*pred.pos)
```

Matching and hot-deck imputation

A different way to impute is through *matching*: for each unit with a missing y, find a unit with similar values of X in the observed data and take its y value. This approach is also sometimes called "hot-deck" imputation (in contrast to "cold deck" methods, where the imputations come from a previously collected data source). Matching imputation can be combined with regression by defining "similarity" as closeness in the regression predictor (for example, $0.32 \cdot \mathtt{male} - 1.44 \cdot \mathtt{over65} + 0.96 \cdot \mathtt{white} + \cdots$ for the model on page 536). Matching can be viewed as a nonparametric or local version of regression and can also be useful in some settings where setting up a regression model can be challenging.

For example, the New York City Department of Health has the task of assigning risk factors to all new HIV cases. The risk factors are assessed from a reading of each patient's medical file, but for a large fraction of the cases, not enough information is available to determine the risk factors. For each of these "unresolved" cases, we proposed taking a random imputation from the risk factors of the five closest resolved cases, where "closest" is defined based on a scoring function that penalizes differences in sex, age, the clinic where the HIV test was conducted, and other information that is available on all or most cases.

More generally, one could estimate a propensity score that predicts the missingness of a variable conditional on several other variables that are fully observed, and then match on this propensity score to impute missing values.

25.5 Imputation of several missing variables

It is common to have missing data in several variables in an analysis, in which case one cannot simply set up a model for a single partially observed variable y given a set of fully observed X variables. In fact, even in the Social Indicators Survey example, some of the predictor variables (ethnicity, interest income, and the indicators for income supplements) had missing values in the data, which we crudely imputed before running the regression for the imputations. More generally, we must think of the dataset as a multivariate outcome, any components of which can be missing.

Routine multivariate imputation

The direct approach to imputing missing data in several variables is to fit a multivariate model to all the variables that have missingness, thus generalizing the approach of Section 25.4 to allow the outcome Y as well as the predictors X to be vectors. The difficulty of this approach is that it requires a lot of effort to set up a reasonable multivariate regression model, and so in practice an off-the-shelf model is typically used, most commonly the multivariate normal or t distribution for continuous outcomes, and a multinomial distribution for discrete outcomes. Software exists to fit such models automatically, so that one can conceivably "press a button" and impute missing data. These imputations are only as good as the model, and so they need to be checked in some way—but this automatic approach is easy enough that it is a good place to start, in any case.

Iterative regression imputation

A different way to generalize the univariate methods of the previous section is to apply them iteratively to the variables with missingness in the data. If the variables with missingness are a matrix Y with columns $Y_{(1)}, \ldots, Y_{(K)}$ and the fully observed predictors are X, this entails first imputing all the missing Y values using some crude approach (for example, choosing imputed values for each variable by randomly selecting from the observed outcomes of that variable); and then imputing $Y_{(1)}$ given $Y_{(2)}, \ldots, Y_{(K)}$ and X; imputing $Y_{(2)}$ given $Y_{(1)}, Y_{(3)}, \ldots, Y_{(K)}$ and X (using the newly imputed values for $Y_{(1)}$), and so forth, randomly imputing each variable and looping through until approximate convergence.

For example, the Social Indicators Survey asks about several sources of income. It would be helpful to use these to help impute each other since they have non-overlapping patterns of missingness. We illustrate for the simple case of imputing missing data for two variables—interest income and earnings—using the same fully observed predictors used to impute earnings in the previous section.

We create random imputations to get the process started:

```
interest.imp <- random.imp (interest)                               R code
earnings.imp <- random.imp (earnings)
```

and then we write a loop to iteratively impute. For simplicity in demonstrating the programming, we set up the function on the original (non-square-root) scale of the data:

```
n.sims <- 10                                                        R code
for (s in 1:n.sims){
  lm.1 <- lm (earnings ~ interest.imp + male + over65 + white +
    immig + educ_r + workmos + workhrs.top + any.ssi + any.welfare +
```

```
                   any.charity)
      pred.1 <- rnorm (n, predict(lm.1), sigma.hat(lm.1))
      earnings.imp <- impute (earnings, pred.1)

      lm.2 <- lm (interest ~ earnings.imp + male + over65 + white +
        immig + educ_r + workmos + workhrs.top + any.ssi + any.welfare +
        any.charity)
      pred.2 <- rnorm (n, predict(lm.2), sigma.hat(lm.2))
      interest.imp <- impute (interest, pred.2)
    }
```

This code could be easily elaborated to handle topcoding, transformations, and two-stage modeling for variables that could be zero or positive (see Exercise 25.4). These operations should be done within the imputation loop, not merely tacked on at the end.

Iterative regression imputation has the advantage that, compared to the full multivariate model, the set of separate regression models (one for each variable, $Y_{(k)}$) is easier to understand, thus allowing the imputer to potentially fit a reasonable model at each step. Moreover, it is easier in this setting to allow for interactions (difficult to do using most joint model specifications).

The disadvantage of the iterative approach is that the researcher has to be more careful in this setting to ensure that the separate regression models are consistent with each other. For instance, it would not make sense to impute age based on income but then to later ignore age when imputing income.

Moreover, even if such inconsistencies are avoided, the resulting specification will not in general correspond to any joint probability model for all of the variables being imputed. It is an open research project to develop methods to diagnose problems with multivariate imputations, by analogy to the existing methods such as residual plots for finding problems in regressions. In the meantime, it makes sense to examine histograms and scatterplots of observed and imputed data to check that the imputations are reasonable.

25.6 Model-based imputation

Missing data can be handled in Bugs by modeling the input variables that have missingness. This requires some work, however: with multiple missing input variables, a multivariate model is required, and this can be particularly tricky when some of the variables are discrete. So in practice it can be helpful to do some simple imputation in R, as we have described, before then analyzing completed datasets. When more is known about the missing-data mechanism (for example, with censored or truncated data; see the model on page 405), it can make more sense to explicitly model the missingness in Bugs.

Nonignorable missing-data models

Realistic censored-data problems often have particular complications. For example, in the study of death penalty appeals described in Section 6.3, we are interested in the duration of the appeals process for individual cases. For example, if a death sentence is imposed in 1983 and its final appeal is decided in 1994, then the process lasted 11 years. It is challenging to estimate the distribution of these waiting times, and to model them based on case-level predictors, because our dataset includes appeals only up to the year 1995. Figure 25.3 illustrates. The censoring model, by analogy to model (18.17) on page 404, looks like:

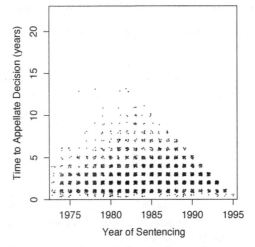

Figure 25.3 *Delays in state appeals court for death penalty cases, plotted versus year of sentencing (jittered to allow individual data points to be visible). We only have results up to the year 1995. The data show a steady increase in delay times for the first decade, but after that, the censoring makes the graph difficult to interpret directly.*

$$y_i = \begin{cases} z_i & \text{if } z_i \leq 1995 - t_i \\ \text{censored} & \text{otherwise,} \end{cases}$$

where y_i is the observed waiting time for case i, z_i is the ultimate waiting time, and t_i is the year of sentencing. We shall not analyze these data further here; we have introduced this example just to illustrate the complexities that arise in realistic censoring situations. The actual analysis for this problem is more complicated because death sentences have three stages of review, and cases can be waiting at any of these stages.

Imputation in multilevel data structures

Imputing becomes more complicated with clustered data. Suppose, for instance, that we have individual-level observations on children grouped within schools (for instance, test scores and demographics), and then measurements pertaining to the schools themselves (for instance, school policies and characteristics such as public versus private). We would not want to impute on a standard individual-level dataset where the school-level measurements are just repeated over each individual in the same school because, if a given school measurement is missing, such an approach would not be likely to impute the same value of this variable for each member of the group (as it should).

Our general advice in this situation is to create two datasets, as in Figure 11.3 on page 239, one with only individual-level data, and one with group-level data and do separate imputations within each dataset while using results from one in the other (perhaps iterating back and forth). For instance, one could first impute individual-level variables using individual-level data and observed group-level measurement. Then in the group-level dataset one could include aggregated forms of the individual-level measurements when imputing missingness at this level.

25.7 Combining inferences from multiple imputations

Rather than replacing each missing value in a dataset with one randomly imputed value, it may make sense to replace each with several imputed values that reflect our uncertainty about our imputation model. For example, if we impute using a regression model we may want our imputations to reflect not only sampling variability (as random imputation should) but also our uncertainty about the regression coefficients in the model. If these coefficients themselves are modeled, we can draw a new set of missing value imputations for each draw from the distribution of the coefficients.

Multiple imputation does this by creating several (say, five) imputed values for each missing value, each of which is predicted from a slightly different model and each of which also reflects sampling variability. How do we analyze these data? The simple idea is to use each set of imputed values to form (along with the observed data) a *completed* dataset. Within each completed dataset a standard analysis can be run. Then inferences can be combined across datasets.

For instance, suppose we want to make inferences about a regression coefficient, β. We obtain estimates $\hat{\beta}_m$ in each of the M datasets as well as standard errors, $s_1, \ldots, s_M$. To obtain an overall point estimate, we then simply average over the estimates from the separate imputed datasets; thus, $\hat{\beta} = \frac{1}{m} \sum_{m=1}^{M} \hat{\beta}_m$. A final variance estimate V_β reflects variation within and between imputations:

$$V_\beta = W + \left(1 + \frac{1}{m}\right) B,$$

where $W = \frac{1}{m} \sum_{m=1}^{M} s_m^2$, and $B = \frac{1}{m-1} \sum_{m=1}^{M} (\hat{\beta}_m - \hat{\beta})^2$.

If missing data have been included in the main data analysis (as when variables X and y are given distributions in a Bugs model), the uncertainty about the missing-data imputations is automatically included in the Bayesian inference, and the above steps are not needed.

25.8 Bibliographic note

Little and Rubin (2002) provide an overview of methods for analysis with missing data. For more on multiple imputation in particular, see Rubin (1987, 1996). "Missing at random" and related concepts were formalized by Rubin (1976). A simple discrete-data example appears in Rubin, Stern, and Vehovar (1995). King et al. (2001) review many of the practical costs and benefits of multiple imputation.

For routine imputation of missing data, Schafer (1997) presents a method based on the multivariate normal distribution, Liu (1995) uses the t distribution, and Van Buuren, Boshuizen, and Knook (1999) use interlocking regressions. Abayomi, Gelman, and Levy (2005) discuss methods for checking the fit of imputation models, and Troxel, Ma, and Heitjan (2004) present a method to assess sensitivity of inferences to missing-data assumptions.

Software for routine imputation in R and SAS has been developed by Van Buuren and Oudshoom (2000), Raghunathan, Van Hoewyk, and Solenberger (2001), and Raghunathan, Solenberger, and Van Hoewyk (2002). An overview of some imputation software is at www.missing-data.com.

Specialized imputation models have been developed for particular problems, with multilevel models used to adjust for discrete predictors. Some examples include Clogg et al. (1991), Belin et al. (1993), and Gelman, King, and Liu (1998). See also David et al. (1986).

Meng (1994), Fay (1996), Rubin (1996), Clayton et al. (1998), and Robins and

Wang (2000) discuss situations in which the standard rules for combining multiple imputations have problems. Barnard and Meng (1994) and Robins and Wang (2000) propose alternative variance estimators and reference distributions.

For more on the Social Indicators Survey, see Garfinkel and Meyers (1999). The death-sentencing example is discussed by Gelman, Liebman, et al. (2004) and Gelman (2004a); see also Finkelstein et al. (2006).

25.9 Exercises

1. Based on the summaries at the very end of Section 25.2, show that the response rates for the "earnings" question in the Social Indicators Survey are statistically significantly different for whites and blacks.

2. Take a complete dataset (with no missingness) of interest to you with two variables, x and y. Call this the "full data."

   (a) Write a program in R to cause approximately half of the values of x to be missing. Design this missingness mechanism to be at random but *not* completely at random; that is, the probability that x is missing should depend on y. Call this new dataset, with missingness in x, the "available data."

   (b) Perform the regression of x on y (that is, with y as predictor and x as outcome) using complete-case analysis (that is, using only the data for which both variables are observed) and show that it is consistent with the regression on the full data.

   (c) Perform the complete-case regression of y on x and show that it is *not* consistent with the corresponding regression on the full data.

   (d) Using just the available data, fit a model in R for x given y, and use this model to randomly impute the missing x data. Perform the regression of y on x using this imputed dataset and compare to your results from (c).

3. Nonignorable missing data: in Exercise 9.13, you estimated the effects of incumbency in U.S. congressional elections, discarding uncontested elections.

   (a) Construct three "bad" imputation procedures and one "good" imputation procedure for these uncontested elections.

   (b) Define clearly how to interpret these imputations. (These election outcomes are not actually "missing"—it is known that they were uncontested.)

   (c) Fit the model to the completed dataset under each of the imputation procedures from (a) and compare the results.

4. Use iterative regression to impute missing data for all the income components in the Social Indicators Survey (data at folder sis).

Appendixes

Six quick tips to improve your regression modeling

A.1 Fit many models

Think of a series of models, starting with the too-simple and continuing through to the hopelessly messy. Generally it's a good idea to start simple. Or start complex if you'd like, but prepare to quickly drop things out and move to the simpler model to help understand what's going on. Working with simple models is not a research goal—in the problems we work on, we usually find complicated models more believable—but rather a technique to help understand the fitting process.

A corollary of this principle is the need to be able to fit models relatively quickly. Realistically, you don't know what model you want to be fitting, so it's rarely a good idea to run the computer overnight fitting a single model. At least, wait until you've developed some understanding by fitting many models.

A.2 Do a little work to make your computations faster and more reliable

This sounds like computational advice but is really about statistics: if you can fit models faster, you can fit more models and better understand both data and model. But getting the model to run faster often has some startup cost, either in data preparation or in model complexity.

Data subsetting

Related to the "multiple model" approach are simple approximations that speed the computations. Computers are getting faster and faster—but models are getting more and more complicated! And so these general tricks might remain important. A simple and general trick is to break the data into subsets and analyze each subset separately. For example, break the 85 counties of radon data randomly into three sets of 30, 30, and 25 counties, and analyze each set separately.

The *advantage* of working with data subsets is that computation is faster on data subsets, for two reasons: first, the total data size n is smaller, so each regression computation is faster; and, second, the number of groups J is smaller, so there are fewer parameters, and the Gibbs sampling requires fewer updates per iteration.

The two *disadvantages* of working with data subsets are: first, the simple inconvenience of subsetting and performing separate analyses; and, second, the separate analyses are not as accurate as would be obtained by putting all the data together in a single analysis. If computation were not an issue, we would like to include all the data, not just a subset, in our fitting.

In practice, when the number of groups is large, it can be reasonable to perform an analysis on just one random subset, for example one-tenth of the data, and inferences about the quantities of interest might be precise enough for practical purposes.

Redundant parameterization

Sections 19.4–19.5 discuss redundant additive and multiplicative parameterizations. These steps add extra parameters to a Bugs model, and can be confusing at first, but can really pay off in speed of computation. In addition, the recentering and scaling required in defining the adjusted parameters can have a convenient statistical interpretation in terms of finite-population inference for the groups in the dataset.

Fake-data and predictive simulation

When computations get stuck, or a model does not fit the data, it is usually not clear at first if this is a problem with the data, the model, or the computation. Fake-data and predictive simulation (discussed in general in Chapter 8 and for multilevel models in Sections 16.7 and 24.1–24.2) are effective ways of diagnosing problems. First use fake-data simulation to check that your computer program does what it is supposed to do, then use predictive simulation to compare the data to the fitted model's predictions.

A.3 Graphing the relevant and not the irrelevant

Graphing the fitted model

Graphing the data is fine (see Appendix B) but it is also useful to graph the estimated model itself (see lots of examples of regression lines and curves throughout this book). A table of regression coefficients does not give you the same sense as graphs of the model. This point should seem obvious but can be obscured in statistical textbooks that focus so strongly on plots for raw data and for regression diagnostics, forgetting the simple plots that help us understand a model.

Don't graph the irrelevant

Are you sure you really want to make those quantile-quantile plots, influence diagrams, and all the other things that spew out of a statistical regression package? What are you going to do with all that? Just forget about it and focus on something more important. A quick rule: any graph you show, be prepared to explain.

A.4 Transformations

Consider transforming every variable in sight:

- Logarithms of all-positive variables (primarily because this leads to multiplicative models on the original scale, which often makes sense)

- Standardizing based on the scale or potential range of the data (so that coefficients can be more directly interpreted and scaled); an alternative is to present coefficients in scaled and unscaled forms

- Transforming before multilevel modeling (thus attempting to make coefficients more comparable, thus allowing more effective second-level regressions, which in turn improve partial pooling).

Plots of raw data and residuals can also be informative when considering transformations (as with the log transformation for arsenic levels in Section 5.6).

In addition to univariate transformations, consider interactions and predictors created by combining inputs (for example, adding several related survey responses to create a "total score"). The goal is to create models that *could* make sense (and can then be fit and compared to data) and that include all relevant information.

A.5 Consider all coefficients as potentially varying

Don't get hung up on whether a coefficient "should" vary by group. Just allow it to vary in the model, and then, if the estimated scale of variation is small (as with the varying slopes for the radon model in Section 13.1), maybe you can ignore it if that would be more convenient.

Practical concerns sometimes limit the feasible complexity of a model—for example, we might fit a varying-intercept model first, then allow slopes to vary, then add group-level predictors, and so forth. Generally, however, it is only the difficulties of fitting and, especially, understanding the models that keeps us from adding even more complexity, more varying coefficients, and more interactions.

A.6 Estimate causal inferences in a targeted way, not as a byproduct of a large regression

Don't assume that a regression coefficient can be interpreted causally. If you are interested in causal inference, consider your treatment variable carefully and use the tools of Chapters 9, 10, and 23 to address the difficulties of comparing comparable units to estimate a treatment effect and its variation across the population. It can be tempting to set up a single large regression to answer several causal questions at once; however, in observational settings (including experiments in which certain conditions of interest are observational), this is not appropriate, as we discuss at the end of Chapter 9.

APPENDIX B

Statistical graphics for research and presentation

Statistical graphics are sometimes summarized as "exploratory data analysis" or "presentation" or "data display." But these only capture part of the story. Graphs are a way to communicate graphical and spatial information to ourselves and others. Long before worrying about how to convince others, you first have to understand what's happening yourself.

Why to graph

Going back through the dozens of examples in this book, what are our motivations for graphing data and fitted models? Ultimately, the goal is communication (to self or others). More immediately, graphs are comparisons (to zero, to other graphs, to horizontal lines, and so forth). We "read" a graph both by pulling out the expected (for example, the slope of a fitted regression line, the comparisons of a series of confidence intervals to zero and each other) and the unexpected.

In our experience, the unexpected is usually not an "outlier" or aberrant point but rather a systematic pattern in some part of the data. For example, consider the binned residual plots in Section 5.6 for the well-switching models. There was an unexpectedly low rate of switching from wells that were just barely over the dangerous level for arsenic, possibly suggesting that people were moderating their decisions when in this ambiguous zone, or that there was other information not included in the model that could explain these decisions.

Often the most effective graphs simply show us what a fitted model is doing. Consider, for example, the graphs in Section 6.5 of the ordered regression and the data for the storable voting experiment or in Section 14.1 of the data-level logistic model and state-level linear model for political opinions.

We consider three uses of graphics in statistical analysis:

1. Displays of raw data, often called "exploratory analysis." These don't have to look pretty; the goal is to see things you did not expect or even know to look for.

2. Graphs of fitted models and inferences, sometimes overlaying data plots in order to understand model fit, sometimes structuring or summarizing inference for many parameters to see a larger pattern. In addition, we can plot simulations of replicated data from fitted models and compare them to comparable plots of raw data.

3. Graphs presenting your final results—a communication tool. Often your most important audience here is yourself—in presenting all of your results clearly on the page, you'll suddenly understand the big picture.

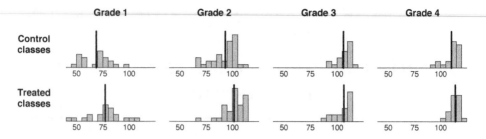

Figure B.1 *Data from the Electric Company experiment, from Figure 9.4 on page 174, displayed in a different orientation to allow easier comparison between treated and control groups in each grade. For each histogram, the average is indicated by a vertical line.*

B.1 Reformulating a graph by focusing on comparisons

Creative thinking might be needed to display numerical data effectively, but your creativity can sometimes be enhanced by carefully considering your goals. Just as in writing, you have to rearrange your sentences sometimes to make yourself clear. For example, consider the graph of the Electric Company data in Figure 9.4 on page 174. Rather than try to cleverly put all the points on a single plot, we arrange them on a 4 × 2 grid, using a common scale for all the graphs to facilitate comparisons among grades and between treatment and control. We also extend the axis all the way to zero, which is not strictly necessary, in the interest of clarity of presentation. In the Electric Company example, as in many others, we are not concerned with the exact counts in the histogram; thus, we simplify the display by eliminating y-axes, and we similarly clarify the x-axis by removing tick marks and using minimal labeling.

Graphs as comparisons

All graphical displays can be considered as comparisons. When making a graph, line things up so that the most important comparisons are clearest. Comparisons are clearest when scales are lined up. Considering Figure 9.4: for each of the two treatments, the histograms for the four grades are lined up and can be directly compared.

In Figure 9.4, we primarily want to compare treatment to control. The comparison of grades is useful—if for no other reason than to ground ourselves and confirm that scores are higher in the higher grades—but we are really more interested in the comparison of treatment to control within each grade.

Thus, it might be more helpful to arrange the histograms as shown in Figure B.1, with treatment and control aligned for each grade. With four histograms arranged horizontally on a page, we need to save some space and so we restrict the x-axes to the combined range of the data. We also indicate the average value in each group with a vertical line to allow easier comparisons of control to treatment in each grade.

No single graph does it all

Sometimes it makes sense to withhold information in order to present a clearer picture. Figure 9.4 (or Figure B.1) shows the outcomes for each classroom in the Electric Company experiment. The scatterplots in Figure 9.6 show pre-test data as well, revealing a high correlation between pre-test and post-test in each grade. The scatterplots certainly show important information, and we are glad to be able to show them, but we prefer the histograms as a starting point for seeing the comparison between treatment and control—at least for this randomized experiment in which the two groups are well balanced.

Graphs of fitted models

It can be helpful to graph a fitted model and data on the same plot, as we have done throughout the book. See Chapters 3–5 for many simple examples, Figure 6.3 on page 120 for a more elaborate example, and Chapters 12–13 for similar plots of multilevel models.

We also like to graph sets of estimated parameters (see, for example, in Figure 4.6 on page 74). Graphs of parameter estimates can be thought of as proto-multilevel models in that the graph suggests a relation between the y-axis (the parameter estimates being displayed) and the x-axis (often time, or some other index of the different data subsets being fit by a model). These graphs contain an implicit model, or a comparison to an implicit model, the same way that any scatterplot contains the seed of a regression or correlation model.

Another use of graphics with fitted models is to plot predicted datasets and compare them visually to actual data, as discussed in Sections 8.3–8.4. For data structures more complicated than simple exchangeable batches or time series, plots can be tailored to specific aspects of the models being checked, as in Section 24.2. As a special case, plots of residuals and binned residuals can be seen as visual comparisons to the hypothesis that the errors from a model are independent with zero mean.

B.2 Scatterplots

Units

When describing or designing a scatterplot, the first thing to decide is the unit of analysis. That is "each dot represents a student" or "each dot represents a county" or whatever. The x and y values have no interpretation until you define the units.

The x and y axes

To get yourself up to speed, start by applying to scatterplots everything you know about linear regression. There's an x variable and a y variable defined on a bunch of units, and you're trying to summarize the average relation between x and y or alternatively to predict y from x where "prediction" includes uncertainty as well as point estimation. This issue is well covered in many recent introductory textbooks which introduce scatterplots first and then move to regression.

Let's start with some bad ideas. First, there is something called a scatterplot matrix for multivariate data, which is a set of scatterplots of all pairs of variables. This can be informative, but it's like regressing every variable versus every other variable. As with regression, we often learn more from scatterplots that are more carefully chosen. For example, if two variables have a time or causal order, we usually prefer to put "before" on the x-axis and "after" on the y-axis.

A common strategy that particularly disturbs us is plotting by index number, for example, plotting data from the 50 states in alphabetical order. In this case the x variable contains little or no information, and the plot is comparable to running

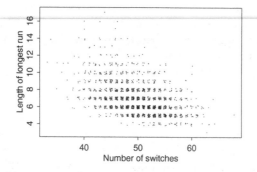

Figure B.2 *Length of longest run (sequence of successive heads or successive tails) versus number of runs (sequences of heads or tails) in each of 2000 independent simulations of 100 coin flips. Each dot on the graph represents a sequence of 100 coin flips; the points are jittered so they do not overlap. When plotted on this graph, the results from an actual sequence of 100 coin flips will most likely fall on a square with a large number of dots. In contrast, a sequence of heads and tails that is artificially created to look "random" will probably have too many runs that are not long enough, and hence will fall on the lower right of this graph.*

a regression on random numbers. An example that is not *necessarily* bad is using, as the x variable, the order of entry of units into the study. This can make sense if one expects or fears time trends (but it would probably be better to plot versus time itself rather than merely order). If there are no major time patterns, however, the choice of x variable might better be spent elsewhere.

You can make as many plots as you want (or as your paper budget allows), but it is useful to think a bit about each plot, just as it is useful to think a bit about each regression you run. This is as good a time as any to recommend that along with every regression you run, you should make a scatterplot. And, in addition, you should be making residual plots where necessary. We'll get to that later.

Jittering

If several data points have the same data values, add a small random number to each so that they do not fall on top of each other. This is called jittering. Jitter just enough so that the discrete nature of the data is still clear. For example, if data points are integers, we might add a random uniform number between -0.3 and $+0.3$ to each x and y value (see Figure B.2). Methods such as plotting 2's, 3's, or cute symbols for multiple data points can be misleading visually, and from a theoretical perspective are unsatisfying in that the display of any unit then depends too strongly on the other data values.

Symbols and auxiliary lines

The symbols of a scatterplot are important because they correspond to the units of analysis in your studies. It can be appropriate to use more than one scatterplot for multilevel data structures. At least in theory you can display five variables easily with a scatterplot: x, y, symbol, symbol size, and symbol color.

Symbols are best for discrete variables, and it's worth putting a little effort into making these symbols distinguishable and also appropriate. For example, we used open circles to indicate open seats in Figure 7.4. In plotting data from an experiment or observational study, you can use different large symbols for treated units and

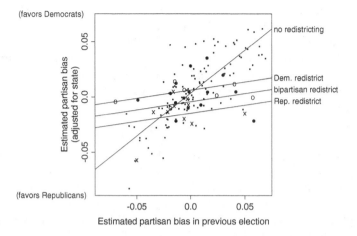

Figure B.3 *Effect of redistricting on partisan bias. Each symbol represents a state and election year, solid circles, open circles, and crosses representing Democratic, bipartisan, and Republican redistricting. The small dots are the control cases—state-years that did not immediately follow a redistricting. Lines show fit from a regression model.*

dots for controls (see Figure B.3). Symbol size can be useful, but it is not always as flexible as one might hope, and we have not had much success in using symbol size for continuous variables.

Color is just great and you should use it as much as possible, even though for printing reasons we do not use color in this book.

We sometimes have had success using descriptive symbol names (for example, two-letter state abbreviations). But if there are only two or three categories, we're happier with visually distinct symbols. For example, to distinguish men and women, we would not use M and W or even M and F. In genealogical charts, men and women are often indicated by open squares and open circles, respectively, but even these symbols are hard to tell apart in a group. We prefer clearly distinguishable symbols—for example, in Figure B.5, open circles for men and solid circles for women.

These suggestions are all based on our subjective experience and attempts at logical reasoning; as far as we know, they have not been validated (or disproved) in any systematic study. We think such a study would be a good idea.

Figure B.3 shows an example of one of the most common regressions: a comparison of treatments to control with a before and after measurement. In this case, the units are state legislative elections, and the plot displays a measure of "partisan bias" in two successive election years. The "treatments" are different kinds of redistricting plans, and the "control" points (indicated by dots on the figure) indicate pairs of elections with no intervening redistricting. We display all the data and also show the regression lines on the same scale. As a matter of fact, we did not at first think of fitting nonparallel regression lines; it was only after making the figure and displaying parallel lines that we realized that nonparallel lines (that is, an interaction between the treatment and the "before" measurement) are appropriate. The interaction is, in fact, crucial to the interpretation of these data: (1) when there is no redistricting, partisan bias is not systematically changed; (2) the largest effect of any kind of redistricting is to bring partisan bias, on average, to near zero. The lines and points together show this much more clearly than any numerical summary.

Another useful kind of line to display is a "default line," which is usually a horizontal line at 0 or a 45-degree line indicating equality of x and y.

When a graph has multiple lines, label them directly, not using symbol codes and a key (which requires the reader—and you—to go back and forth between graph and key). Examples of our recommended approach include Figure 5.11 on page 91, Figure 14.11 on page 313, and Figure 15.2 on page 328.

Shape of the plotting region

The shape of a plot conveys information implicitly. When x and y are the same units on the same scale, we use a square plot with the same scale on the two axes even if that means that large parts of the plot are blank (see Figure B.3). Conversely, if x and y are not the same variable, we are careful *not* to use a square plot so as not to implicitly send the wrong message. When we are presenting several plots of different variables, we sometimes use dimensions for the different plots as a visual cue that they have different meanings.

Displaying the results of model fitting

In a regression with one or two inputs, it is possible to display essentially all the information (all the information if one of the variables is discrete) in a single plot. When additional predictors are present, we have to summarize the data in some way. Ideally, the outcome variable is displayed on the y-axis, symbols indicate the input variable of interest (think of treatments and control here), and the x-axis displays predicted values or some other combination of all the variables that are being controlled for.

When there is more than one control variable, one approach is to plot on the x-axis the linear predictor created from all the control variables with coefficients estimated from their regression models. For example, with a regression model of the form $y_i = \beta_0 + \beta_1 X_{i1} + \beta_2 X_{i2} + \beta_3 X_{i3} + \epsilon_i$, one can plot y_i versus $\beta_0 + \beta_2 X_{i2} + \beta_3 X_{i3}$ with different symbols for different values of X_{i1}. In that plot one would plot dotted lines of $y = c + x$, for $c = \beta_1 x_1$ for the different values for x_1, to illustrate the expected relationship. Figure B.3 shows an example with one predictor that plays the role of "treatment" and other "background" predictors which are combined in the x-axis.

More generally we can overlay the model on a plot of data (conversely when plotting a modeled relationship, we try to include data on this plot appropriately), even if it takes a bit of work to figure out how to do this reasonably. In our own work such plots have been crucial to our understanding, as illustrated by Figure B.3.

Maps

Often when you have a map, you're better off with a scatterplot (but of course there's no reason to throw away the map). For example, if you have data on the occurrence of some medical condition by location and you map it to see whether it's clustered in low-income areas, it might make more sense to plot rates versus income. But the map might be useful in suggesting which variables to consider plotting.

With this use of maps as an explanatory tool in mind, we focus on mapping methods that will reveal unexpected patterns but only when something real is

Figure B.4 *Summary of a forecast of the 1992 U.S. presidential election performed one month before the election. (a) States that Bill Clinton was forecasted to win are shaded. (b) For each state, the proportion of the box that is shaded represents the probability of Clinton winning the state; the width of the box is proportional to the number of electoral votes for the state. The second map conveys more information and is also less misleading.*

going on. Maps are often tricky to read because they can show spurious patterns. For example, a map of the United States shading in different counties with different colors inevitably draws attention to the counties that are geographically larger and perhaps also those that are unusually shaped. At the very least one could replace the shading by a small colored circle in each county, perhaps with larger circles for more populous counties. (However, this would not be appropriate for a geological map of oil reserves: we are usually thinking about social statistics here.) Another approach is to plot "thermometers" within a geographic unit (see Figure B.4).

The problem of unequal population density is sometimes attacked by distorted maps that approximately preserve the shapes of, for example, states, while making their areas proportional to population. We find these maps more distracting than useful because they draw attention to the shapes, which are usually nothing that anybody cares about.

In addition to any possible distorted geographical effects, there are more subtle difficulties in mapping which relate to problems of summarizing inferences with point estimates (see, for example, Gelman and Price, 1999).

Calibration plots

A calibration plot is a plot of observed values on the y-axis versus expected (forecasted) values on the x-axis. If all is well, the expected value of y given x in such a plot is just x. So we make this a square plot with identical axes and a comparison line at $y = x$. See, for example, Figure B.5, which evaluates the calibration of students' guesses of their exam scores.

In general, a forecast supplies a distribution, not just a point estimate, for each data point. In this case, the "expected" or "forecasted" value for any datum is just the mean (or expectation) of the forecast distribution for the datum. The desired relation is $\mathrm{E}(y|x) = x$.

When forecasting discrete outcomes, however, the problem gets more complicated: the expected values are continuous but the observed values are discrete (for example, for binary data, the observed values are 0's and 1's, and the expected values are proportions between 0 and 1). The calibration plot is then virtually unreadable, as the points cluster in discrete values on the y-axis. (See Figure B.6a for an example.) So instead, it is standard practice to order the x values and then divide them into categories or bins $j = 1, \ldots, J$. In each category we compute the averages $\bar{x}_j$ and $\bar{y}_j$ and then plot the J values of $(\bar{x}_j, \bar{y}_j)$. Figure B.6b shows an example in which the data can take on 5 possible outcomes.

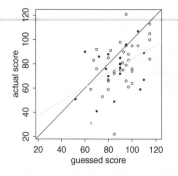

Figure B.5 *Actual versus guessed midterm exam scores for a class of 53 students. Each symbol represents a student; empty circles are men, solid circles are women, and ? has unknown sex. The 45° line represents perfect guessing, and the dotted line is the linear regression of actual score on guessed score. (The separate regression lines for men and women were similar.) Both men and women tended to perform worse than their guesses. That the slope of the regression line is less than 1 is an instance of the "regression effect" (see Section 4.3): if a student's guessed score is x points higher than the mean guess, then his or her actual score is, on average, only about 0.6x higher than the mean score. A square scatterplot is used because the horizontal and vertical axes are on the same scale.*

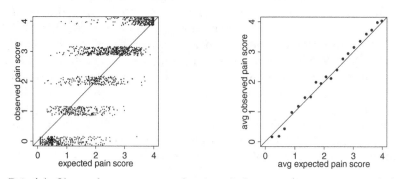

Figure B.6 *(a) Observed versus expected pain relief scores (0 = no pain relief, ..., 5 = complete pain relief) for data from the analysis of Sheiner, Beal, and Dunne (1997). Observed pain relief scores are jittered. (b) Average observed versus averaged expected pain relief scores, with data divided into 20 equally sized bins defined by ranges of expected pain relief scores.*

Whether in the continuous or discrete case, we prefer to put "observed" on the y-axis and "expected" on the x-axis (rather than the reverse), because in the calibration context, the expected value is the predictor and the observed value is the outcome. See Section 8.2 for related discussion of residual plots.

Residual plots

If all is going well, the points on the calibration plot will mostly fall near the 45-degree line, meaning there will be much empty space on the plot. A natural next step is to plot $y - x$ versus x; that is, "deviation from predicted" versus "predicted." This is the residual plot. In fact "deviation from predicted" can be plotted versus just about anything, not just predicted values (see Figure B.7). Residual plots should not be square and should have a dotted line at $y = 0$ rather than $y = x$.

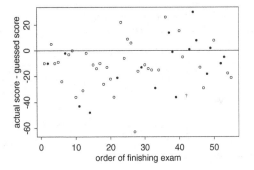

Figure B.7 *Difference between actual and guessed midterm exam scores, plotted against the order of finishing the exam. The exact order is only relevant for the first 20 or 25 students, who finished early; the others all finished within five minutes of each other at the end of the class period. Each symbol represents a student; empty circles are men, solid circles are women, and ? has unknown sex. The horizontal line represents perfect guessing. The students who finished early were highly overconfident, whereas the other students were less biased in their predictions.*

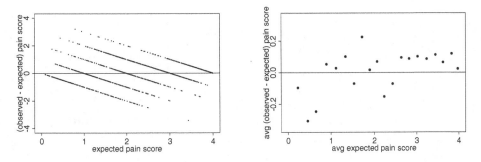

Figure B.8 *(a) Residuals (observed - expected) of pain relief scores versus expected pain relief scores from Figure B.6. (b) Average residuals versus expected pain scores, with measurements divided into 20 equally sized bins defined by ranges of expected pain scores. The average prediction errors are relatively small (as can be seen from the scale of the y-axis), but with a consistent pattern that low predictions are too low and high predictions are too high.*

As with calibration plots, it is generally a good idea to bin the points in a residual plot if the outcomes are discrete (see Figure B.8).

B.3 Miscellaneous tips

We conclude with some suggestions derived from our experiences using graphs in data analysis, first presenting a few ideas that have proved generally useful, then going through a variety of specific techniques through a series of examples.

A display of several time series of opinion polls

Each subgraph of Figure B.9 shows a time series of the support in the polls for the Republican candidate for U.S. president, as a proportion of the two-party support, for a given election year, in the months leading up to the election.

Tip: Put many little graphs on the same page. Do it with a slick graphics package if possible; otherwise, use scissors, tape, and a reducing copy.

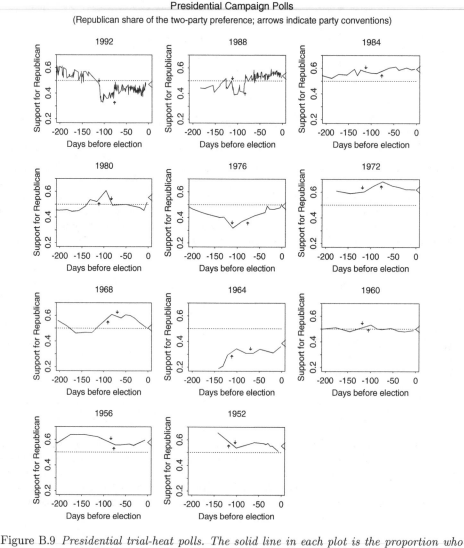

Figure B.9 *Presidential trial-heat polls. The solid line in each plot is the proportion who would vote for the Republican candidate for president, among those who report a preference for the Democratic or Republican candidates. The 1992 and 1998 graphs include data from all available nationwide polls; plots for the other years are from the Gallup Report. The upward arrow marks the time of the Republican convention, and the downward arrow marks the time of the Democratic convention. The triangle at the end of each time series indicates what actually happened in the election.*

Tip: When you have multiple graphs, use a common scale.

Tip: Put a light line to indicate what "no effect" would be. (There is a dotted line at 50% in each graph.)

Tip: It's worth putting in little details and doing it right. For example, each graph also indicates, with arrows, the times of the political conventions. The Republican conventions are shown with up arrows (because the Republicans improve in the polls then), and the Democratic conventions are indicated with down arrows (corresponding to the drop in the Republican poll numbers).

Tip: Keep the lines on a graph thin, even if each plot has only one line. A fat line conveys no more information and just makes the information harder to see.

By comparison, we got the data from printed reports from Gallup that had graphs like ours for each election year, but with two thick lines on each graph displaying the Democratic and the Republican shares of the polls. For our purposes, we didn't care about undecideds and third parties, so we just display the Republican proportion of the two-party support.

Tip: Repeat axis labels as necessary to make mini-graphs easier to read. Once you know what they say, your eye easily ignores the labels.

We originally created this graph to help us understand the history of the pre-election polls at a glance—exploratory data analysis—and later we fixed it up for final presentation. (In the original, exploratory, stage, we wrote in the arrows by hand.)

Significant digits and uncertainty

When reporting the output from a statistical analysis, you should always imagine yourself in the position of the reader of the report. It is important not to overwhelm the reader with irrelevant material. For the simplest (but still important) example, consider the reporting of numerical results (either alone or in tables).

Do not include too many significant digits in numbers you report. The relevant comparison is not to an absolute number of decimal places but to the uncertainty and variability in the numbers being presented. For example, the confidence interval [3.276, 6.410] would be more clearly written as [3.3, 6.4]. (An exception is that it makes sense to save lots of extra digits for intermediate steps in computations. For example, $51.7643 - 51.7581$.) A related issue is that you can often make a list or table of numbers more clear by first subtracting out the average (or for a table, row and column averages). The appropriate number of significant digits depends on the uncertainty. But in practice, three digits are usually enough because if more were necessary, we would subtract out the mean first.

Maybe the biggest source of too many significant digits is from computer output. One solution is to set the rounding in the computer program (for example in R, `options(digits=2)`).

Titles and captions

All titles and axis labels should be meaningful. In addition, each figure should be accompanied by a caption so that it makes sense even for the reader who skips the rest of the article.

Histograms

Histograms are for plotting values of a single variable. Whenever possible, use a scatterplot, but sometimes it is convenient look at just one variable, especially when arranged in a grid such as in Figure B.1 on page 552. When looking at one variable, we prefer histograms to snazzier methods such as density estimation because we feel more connected to the actual numerical values this way.

There's some confusion on this point. The purpose of a histogram is to display a set of numbers, not to approximate an underlying distribution function. It's a good idea to divide your histogram into more bins than "necessary" so that you can get

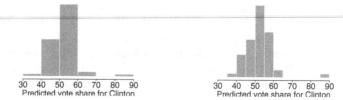

Figure B.10 *Histograms of the forecast proportion of the two-party vote for Bill Clinton in 1992 in each of the 50 states and the District of Columbia, displayed with two different choices of bin width: (a) the bin width automatically assigned by R, (b) the bin width set manually with the R command* `hist(y,breaks=seq(30,90,5))`.

an idea of the variability in the histogram itself. Do not use the default bin width in R (see Figure B.10).

General advice

Plot numerical data and inferences as graphs, not as tables. A good example is the multilevel logistic regression of vote preference on demographic and geographic predictors, with graphs on pages 306–307 that show coefficient estimates and standard errors, along with curves of the fitted model and data. Or, for a simpler example, Figure 15.9 on page 337 graphs the inference from a simple regression.

Multiple plots per page. A graph can almost always be made smaller than you think and still be readable. This then leaves room for more plots on a grid, which then allows more patterns to be seen at once and compared.

Don't plot the index numbers. For example, Figure 14.9 on 312 plots estimates for the 50 states versus average state income, rather than simply listing the states in alphabetical order. For another example, the dogs in Figure 24.1 are ordered by the time of their last shock, rather than by their ID numbers, which turn out to have no meaning in this problem.

Never display a graph you can't explain. Give a full caption for every graph (as we try to do in this book). This explains to yourself and others what you are trying to show and what you have learned from each plot. Avoid displaying graphs that have been made simply because they are conventional. For example, regressions are commonly equipped with quantile-quantile plots of residuals, but for most applications the information in such a plot is irrelevant, and a distraction from the more relevant results that could be presented.

B.4 Bibliographic note

For statistical graphics in R, the book by Murrell (2005) is an excellent overview and starting point. Fox (2002) is also helpful in that it focuses on regression models.

On the topic of statistical graphics more generally, much of the most important and influential work has appeared in books, including Bertin (1967, 1983), Chambers et al. (1983), Cleveland (1985, 1993), Tufte (1983, 1990), and Wainer (1984, 1997).

There are various systematic ways of studying statistical graphics. One useful approach is to interpret graphs as model checking (for example, if residuals are not independent of x, then there is some model violation), as we have discussed in Chapter 24. Another approach is to perform experiments to find out how well people can gather information from various graphical displays (for example, are line plots easier to read than histograms). This is discussed by Cleveland (1985). More research is needed on both these approaches: relating to probability models is important for allowing us to understand graphs and devise graphs for new problems; and effective display is important for communicating to ourselves as well as others.

For some ideas on the connections between statistical theory, modeling, and graphics, see Tukey (1977), Wilkinson (2005), and (for our own perspective) Gelman (2004a).

Some of the ideas considered in this chapter are explored by Gelman, Pasarica, and Dodhia (2002), Wand (1997), Wainer (2001), and Friendly and Kwan (2003). Ehrenberg (1978) and Tukey (1977) discuss tabular displays in detail. An important topic not discussed in the present book is dynamic graphics; see Buja et al. (1988) and Buja, Cook, and Swayne (1999).

B.5 Exercises

1. Find an example of a published article in a statistics or social science journal in which too many significant digits are used.

2. Find an example of a published article in a statistics or social science journal in which there is *not* a problem with too many significant digits being used.

3. Take any data analysis exercise from this book and present the *raw data* in several different ways. Discuss the advantages and disadvantages of each presentation.

4. Take any data analysis exercise from this book and present the *fitted model* in several different ways. Discuss the advantages and disadvantages of each presentation.

APPENDIX C

Software

C.1 Getting started with R, Bugs, and a text editor

Follow the instructions at www.stat.columbia.edu/~gelman/arm/software/ to download, install, and set up R and Bugs on your Windows computer. The webpage is occasionally updated as the software improves, so we recommend checking back occasionally. R, OpenBugs, and WinBugs have online help with more information available at www.r-project.org, www.math.helsinki.fi/openbugs/, and www.mrc-bsu.cam.ac.uk/bugs/.

Set up a *working directory* on your computer for your R work. Every time you enter R, your working directory will automatically be set, and the necessary functions will be loaded in.

Configuring your computer display for efficient data analysis

We recommend working with three nonoverlapping open windows, as pictured in Figure C.1: an R console, the R graphics window, and a text editor (ideally a program such as Emacs or WinEdt that allows split windows, or the script window in the Windows version of R). When programming in Bugs, the text editor will have two windows open: a file (for example, project.R) with R commands, and a file (for example, project.bug) with the Bugs model. It is simplest to type commands into the text file with R commands and then cut and paste them into the R console. This is preferable to typing in the R console directly because copying and altering the commands is easier in the text editor. To run Bugs, there is no need to open a Bugs window; R will do this automatically when the function bugs() is called (assuming you have set up your computer as just described, which includes loading the R2WinBUGS package in R). The only reason to manually open a Bugs window is to access the manuals and examples in its Help menu.

Software updates

Here we discuss how to set up and run the statistical packages that we use to fit regressions and multilevel models. All this software is under development, so some of the details of the code and computer output in the book may change along with the programs. We recommend periodically checking the websites for R, Bugs, and other software and updating as necessary.

C.2 Fitting classical and multilevel regressions in R

Using R for classical regression and miscellaneous statistical operations

The lm() and glm() functions fit linear and generalized linear models in R. Many examples appear in Part 1 of this book; you can see the R documentation and other references given at the end of this chapter for instructions and further examples.

We have prepared several functions including display(), sim(), se.coef(), and

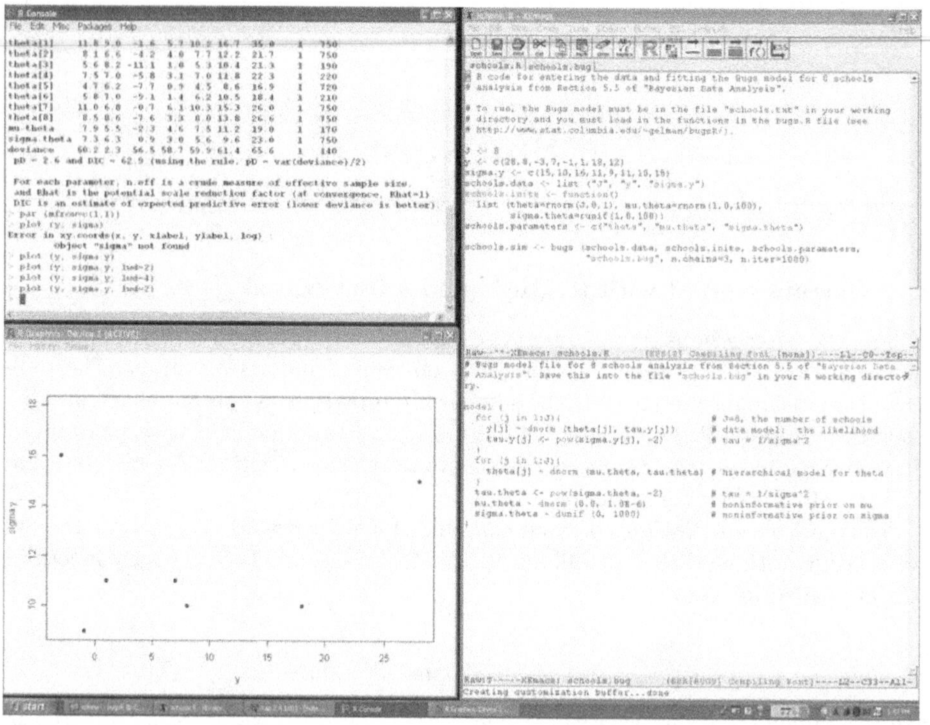

Figure C.1 *Configuration of a computer screen with R console, R graphics window, and a text editor (in this case, Xemacs) with two windows, one for R script and one for a Bugs model. We call Bugs from R, so there is no need to have an open Bugs window on the screen.*

`sigma.hat()`, for displaying, accessing, and generating simulations summarizing the inferences from linear and generalized linear models in R, as well as functions such as `bayesglm()` and `bayespolr()` for fitting Bayesian generalized linear models and ordered logistic regressions. All these functions, and a few others, are in the R package `arm` (*a*pplied *r*egression and *m*ultilevel modeling) and are loaded in automatically if you have followed the instructions in Section C.1. Online help is available for these as for all R functions.

If you are having trouble with any of these functions, we suggest going to the website, `www.stat.columbia.edu/~gelman/arm/software/` and downloading the latest versions of everything.

The `lmer()` function for multilevel modeling

Our starting point for fitting multilevel models is `lmer()` ("linear mixed effects," but it also fits nonlinear models), a function that is currently part of the `lme4` package in R and can fit a variety of multilevel models using point estimation of variance parameters. We use `lmer()` for most of the examples in Part 2A of this book; as discussed in Section 16.1, `lmer()` is a good way to get quick approximate estimates before full multilevel modeling using Bugs. Various generalizations of `lmer()` are under development that would generalize it to perform fully Bayesian simulation-based inference; check the webpage `www.stat.columbia.edu/~gelman/arm/software/` for our links to the latest updates.

R packages

Go to the R webpage for information on R packages. To use a package, you must first install it (which can be done from the R console), then in any session you load in the package as needed using the `library()` function. Installation needs to be done only once, but you must load in the package with every R session. (You can load in our most frequently used packages automatically by putting lines into the `Rprofile.site` file, which is set up in the R directory on your computer if you follow the instructions in Section C.1.)

The most important packages for our purposes are `arm` (which has our own functions), `Matrix` and `lme4` (which include `lmer()` in its current form) and `R2WinBUGS` (which allows us to run Bugs from R, as described in the next section).

Other packages are helpful for specific purposes. For example, `hett` is a package that fits robust regression using the *t* model (see Section 6.6). To install, use the `install.packages()` function in R to download packages from the web. Then, in any session where we want to fit *t* regressions, for example, we type `library("hett")` (or include this line in any function call) and we are ready to go. To get help, we can click on Help at the top of the R window, then on "Html help," then on Packages, then on the package name (in this case, `hett`), then on the name of the function of interest (in this case, `tlm`). Alternatively, we can simply type `help(tlm)` or `?tlm` directly from the console.

Other R packages are available, and continue to be developed, to fit various complex models. The `MASS` package (which is automatically loaded if you follow the instructions in Section C.1) includes tools for fitting a variety of models. The `GAMM` package fits *generalized additive mixed models*, an adaptation of regression and generalized linear models that allows arbitrary nonlinear transformations of the input variables to be fit by the data (with "mixed" referring to the possibility of varying coefficients, that is, multilevel models); `sem` fits models for structural equations and instrumental variables (as shown in Section 10.6); and `MCMCpack` fits a variety of models, including multilevel linear regression for panel data.

C.3 Fitting models in Bugs and R

Calling Bugs from R

Currently, our main tool for fitting multilevel models is Bugs, which can be called from R using the `bugs()` function. (Type `?bugs` from the R console for more information.) As described in Part 2B of this book, we can use Bugs to fit models of essentially arbitrary complexity, but for large datasets or models with many parameters, Bugs becomes slow to converge.

Programming in R

The flexibility of Bugs makes it the preferred choice for now. If Bugs is too slow, or if it does not work for a particular model (yes, this happens!), then we program the Gibbs sampler and Metropolis algorithms directly in R. This will typically be faster than Bugs in computation time, and also it can converge in fewer iterations, because in programming the algorithm ourselves we have direct control and can use updating rules that are tailored to the particular model being fit. See Gelman et al. (2003, appendix C) for an example and Section 18.7 for an example using the Umacs package in R. (See `www.stat.columbia.edu/~gelman/arm/software/`.) If even R is too slow, the Gibbs and Metropolis algorithms can be programmed in Fortran or C. Researchers are also developing compiled libraries for fast computation of multilevel models, linkable from R.

C.4 Fitting multilevel models using R, Stata, SAS, and other software

Several other programs are available to fit multilevel models. We shall briefly consider several popular packages, showing how they can be used to fit six prototype models.

We prefer R and Bugs for their flexibility, both in model fitting and in processing the resulting inferences, but we recognize that it is helpful to know how to fit multilevel models in software with which you are already familiar.

In addition to differences in syntax, the different packages display output differently. For example, we prefer to present estimated variance components in terms of standard deviations and (for varying-slope models) correlations, but some programs report variances and covariances. We shall assume that as a user of these other packages, you will be able to interpret the output and understand its relation to our notation in Section 2A of this book.

Six prototype models; fitting in R

We briefly present six example models along with the code needed to fit them in R using `lmer()`. The models can be fit in Bugs as described in Part 2B of this book. We follow with code in other packages. These examples do not come close to exhausting the kinds of multilevel models that we are fitting—but we hope they will be enough to get you started if you are using software other than R and Bugs.

1. Varying-intercept linear regression with data y, predictor x, and grouping variable `group`: $y_i = \alpha_{\text{group}[i]} + \beta x_i + \epsilon_i$ (that is, `group` is an index variable taking on integer values 1 through J, where J is the number of groups):

R code
```
lmer (y ~ x + (1 | group))
```

2. Same as example 1, but with a group-level predictor u (a vector of length J):

R code
```
u.full <- u[group]
lmer (y ~ x + u.full + (1 | group))
```

(We need to define `u.full` to make a predictor that is the same length as the data; currently, the `lmer()` function in R does not take group-level predictors.)

3. Same as example 2, but with varying intercepts and varying slopes: $y_i = \alpha_{\text{group}[i]} + \beta_{\text{group}[i]} x_i + \epsilon_i$, where the J pairs (α_j, β_j) follow a bivariate normal distribution with mean vector $(\gamma_0^\alpha + \gamma_1^\alpha u_j, \gamma_0^\beta + \gamma_1^\beta u_j)$ and unknown 2×2 covariance matrix, with all parameters estimated from the data:

R code
```
lmer (y ~ x + u.full + x:u.full + (1 + x | group))
```

4. Go back to example 1, but with binary data and logistic regression: $\Pr(y_i = 1) = \text{logit}^{-1}(\alpha_{\text{group}[i]} + \beta x_i)$:

R code
```
lmer (y ~ x + (1 | group), family=binomial(link="logit"))
```

5. Go back to example 1, but with count data and overdispersed Poisson regression with offset $\log(z)$: $y_i \sim$ overdispersed $\text{Poisson}(z_i \exp(\alpha_{\text{group}[i]} + \beta x_i))$. This example includes overdispersion and an offset because both are important components to realistic count-data models. To fit quickly in R:

```
log.z <- log(z)
lmer (y ~ x + (1 | group),offset=log.z,family=quasipoisson(link="log"))
```
R code

6. A two-way data structure with replication: for convenience, label the index variables for the groupings as state and occupation, so that the model is $y_i = \mu + \alpha_{\text{state}[i]} + \beta_{\text{occupation}[i]} + \gamma_{\text{state}[i], \text{occupation}[i]} + \epsilon_i$. We want the α's, the β's, and the γ's to be modeled (each with their own normal distribution); for simplicity, we assume no other predictors in the model. To fit:

```
state.occupation <- max(occupation)*(state - 1) + occupation
lmer (y ~ 1 + (1 | state) + (1 | occupation) + (1 | state.occupation))
```
R code

(The first line was needed to define an index variable that sweeps over all the states and occupations.)

Fitting in Stata

Stata (www.stata.com) is a statistical package that is particularly popular in social science and survey research. A wide range of multilevel models can be fit in Stata as extensions of the basic regression framework.

1. Varying-intercept linear regression:

```
xtmixed y x || group:
```
Stata code

or

```
xtreg y x, i(group)
```
Stata code

or

```
gllamm y x, i(group) adapt
```
Stata code

2. Varying-intercept linear regression with a group-level predictor:

Stata has no concept of a vector of length shorter than the current dataset, so we have to create ufull and merge it with the dataset that includes x and y.

```
xtmixed y x ufull || group:
```
Stata code

or

```
xtreg y x ufull, i(group) re
```
Stata code

or

```
gen cons = 1
eq grp_c: cons
gllamm y x u, i(group) nrf(1) eqs(grp_c) adapt
```
Stata code

3. Varying-intercept, varying-slope linear regression with a group-level predictor:

```
xtmixed y x ufull || group: x, cov(unstruct)
```
Stata code

or

Stata code
```
gen cons = 1
eq grp_c: cons
eq grp_u: u
gllamm y x u, i(group) nrf(2) eqs(grp_c grp_u) adapt
```

4. Varying-intercept logistic regression:

Stata code
```
xtlogit y x, i(group)
```

or

Stata code
```
gllamm y x, i(group)family(binom)link(logit)
```

5. Varying-intercept overdispersed Poisson regression:

Stata code
```
xtnbreg y x, exposure(z) i(group)
```

Alternatively,

Stata code
```
xtnbreg y x, i(group) re offset(log.z)
```

or

Stata code
```
gllamm y x, i(group)offset(log.z) family(poi)link(log)
```

6. Varying-intercept linear regression with nested and non-nested groupings:

Stata code
```
egen state_occup = group(state occup)
xtmixed y || _all: R.state || _all: R.occup || _all: R.state_occup
```

Fitting in SAS

SAS (www.sas.com) is a statistical package that is particularly popular in biomedical research. As with Stata, many multilevel models can be fit in SAS by specifying grouping of the data.

1. Varying-intercept linear regression:

SAS code
```
proc mixed;
  class group;
  model y = x;
  random intercept / subject=group;
run;
```

2. Varying-intercept linear regression with a group-level predictor:

SAS has no concept of a vector of length shorter than the current dataset, so we have to create ufull and merge it with the dataset that includes x and y.

SAS code
```
proc mixed;
  class group;
  model y = x ufull;
  random intercept / subject=group;
run;
```

3. Varying-intercept, varying-slope linear regression with a group-level predictor:

```
proc mixed;                                                   SAS code
  class GROUP;
  model y = x ufull x*ufull;
  random intercept x / subject=group type=un;
run;
```

4. Varying-intercept logistic regression:

```
proc nlmixed;                                                 SAS code
  parms b0 b1 s2;
  xbeta = b0 + b1*x + a;
  p = exp(xbeta)/(1+exp(xbeta));
  model y ~ binary(p);
  random a ~ normal(0,s2) subject = group;
run;
```

5. Varying-intercept overdispersed Poisson regression:

```
proc nlmixed;                                                 SAS code
  parms b0=6 b1=-4 k=0.5 s2=1;
  xbeta = b0 + b1*x + a;
  mu = exp (logz + xbeta);
  p = mu/(mu+1/k);
  loglik = lgamma(y+1/k) - lgamma(y+1) - lgamma(1/k) +
    y*log(p) + (1/k)*log(1-p)
  model y ~ general(loglik);
  random a ~ normal(0,s2) subject = group;
run;
```

This nlmixed code specifies the negative binomial likelihood function; the parameters are then estimated by numerical integration. The parms statement names the parameters that will be estimated and gives starting values for them. The four lines that follow code the loglikelihood function that is going to be maximized. The model statement gives the response and the loglikelihood function, and the random statement defines the random intercept.

An alternative approach uses a preprogrammed negative binomial model:

```
proc glimmix;                                                 SAS code
  class group;
  model y = x / solution dist = negbin offset = logz;
  random intercept / subject=group;
run;
```

The output of this run has a scale parameter (which equals k in the nlmixed code) to capture the overdispersion.

6. Varying-intercept linear regression with nested and non-nested groupings:

```
proc mixed;                                                   SAS code
  class state occupation
  model y = ;
  random state occupation state*occupation;
run;
```

Fitting in SPSS

SPSS (www.spss.com) is a statistical package that is particularly popular in psychology and experimental social science. Some multilevel models can be fit in SPSS by specifying grouping in data.

1. Varying-intercept linear regression:

SPSS code
```
mixed
  y with x
  /fixed = x
  /print = solution testcov
  /random intercept | subject(group)
```

2. Varying-intercept linear regression with a group-level predictor:

SPSS has no concept of a vector of length shorter than the current dataset, so we have to create ufull and merge it with the dataset that includes x and y.

SPSS code
```
mixed
  y with x ufull
  /fixed = x ufull
  /print = solution testcov
  /random intercept | subject(group)
```

3. Varying-intercept, varying-slope linear regression with a group-level predictor:

SPSS code
```
mixed
  y with x ufull
  /fixed = x ufull x*ufull
  /print = solution testcov
  /random intercept | subject(group) covtype(un)
```

We are not aware how to fit the other three examples (multilevel logistic regression, multilevel Poisson regression, and non-nested linear regression) in SPSS.

Fitting in AD Model Builder

AD Model Builder (otter-rsch.com/admodel.htm) is a package based on C++ that performs maximum likelihood or posterior simulation given the likelihood or posterior density function, a flexibility that is particularly helpful for nonlinear models.

1. Varying-intercept linear regression:

ADMB code
```
g = -0.5*norm2(z);
alpha = gamma_a + s(1)*z;
for (i=1;i<=n;i++)
  mu(i) = alpha(group(i)) + beta*x(i);
g += -n*log(s(0)) - 0.5*norm2((y-mu)/s(0));
```

Here, g is the log-likelihood.

2. Varying-intercept linear regression with a group-level predictor:

ADMB code
```
g = -0.5*norm2(z);
alpha = gamma_a(0) + gamma_a(1)*u + s(1)*z;
for (i=1;i<=n;i++)
  mu(i) = alpha(group(i)) + beta*x(i);
g += -n*log(s(0)) - 0.5*norm2((y-mu)/s(0));
```

3. Varying-intercept, varying-slope linear regression with a group-level predictor:

ADMB code

```
g = -0.5*(norm2(z1)+norm2(z2));
alpha = gamma_a(0) + gamma_a(1)*u + s(1)*z1;
w = sqrt(1.0-square(rho));
beta = gamma_b(0) + gamma_b(1)*u + s(2)*(rho*z1 + w*z2);
for (i=1;i<=n;i++)
  mu(i) = alpha(group(i)) + beta(group(i))*x(i);
g += -n*log(s(0)) - 0.5*norm2((y-mu)/s(0));
```

4. Varying-intercept logistic regression:

ADMB code

```
g = -0.5*norm2(z);
alpha = gamma_a + s*z;
for (i=1;i<=n;i++)
  eta(i) = alpha(group(i)) + beta*x(i);
g += y*eta - sum(log(1+exp(eta)));
```

5. Varying-intercept overdispersed Poisson regression:

ADMB code

```
g = -0.5*norm2(z);
alpha = gamma_a + s*z;
for (i=1;i<=n;i++)
{
  lambda = offset(i)*exp(alpha(group(i)) + beta*x(i));
  omega = 1.0+lambda/kappa;
  g += log_negbinomial_density(y(i),lambda,omega);
}
```

The negative binomial distribution may be viewed as an overdispersed Poisson distribution (with omega being the overdispersion coefficient).

6. Varying-intercept linear regression with nested and non-nested groupings:

ADMB code

```
g = -0.5*(norm2(z_a)+norm2(z_b)+norm2(z_g));
alpha = s(1)*z_a;
beta = s(2)*z_b;
gamma = s(3)*z_g;
for (i=1;i<=n;i++)
  eta(i) = mu + alpha(state(i)) + beta(occupation(i)) +
    gamma(state_occupation(i));
g += -n*log(s(0)) - 0.5*norm2((y-eta)/s(0));
```

Fitting in HLM, MLWin, and other software

HLM and MLWin are statistical programs specifically designed to fit multilevel models. They can fit models such as those in the preceding examples using a menu-based point-and-click approach.

One can also fit some or all of the models using other statistical packages, with varying degrees of difficulty. See here for an overview of many packages: www.mlwin.com/softrev/index.html; the descriptions there are not all up to date but they should provide a good starting point.

It is also possible to call Bugs using other software, including Stata, SAS, Python, Excel, and Matlab; go to the link at the Bugs homepage for "running from other software," currently at www.mrc-bsu.cam.ac.uk/bugs/winbugs/remote14.shtml

C.5 Bibliographic note

R (R Project, 2000) and Bugs (Spiegelhalter et al., 1994, 2002) have online help. In addition, Fox (2002) describes how to implement regressions in R, and Murrell (2005) shows R graphics. Becker, Chambers, and Wilks (1988) describes S, the predecessor to R. Venables and Ripley (2002) discuss statistical methods in R (or, essentially equivalently, S), focusing on nonparametric methods that are not covered here; the functions and examples used in that book are in the MASS package.

The lmer() function for fitting multilevel models is described by Bates (2005a, b), continuing on earlier work of Pinheiro and Bates (2000). Other R packages have been written for specific multilevel models; for example, MCMCpack (Martin and Quinn, 2002b).

For Bugs code, the books by Congdon (2001, 2003) present a series of examples. Kerman (2006) presents Umacs, and appendix C of Gelman et al. (2003) has examples of direct coding of Bayesian inference in R. The implementation of Bugs using R, as done in this book, is described by Sturtz, Ligges, and Gelman (2004). An open-source version of Bugs called OpenBugs (Thomas and O'Hara, 2005) is also under development.

Several software packages for multilevel models are reviewed by Centre for Multilevel Modelling (2005), including Stata, SAS, MLWin, and HLM. Rabe-Hesketh and Everitt (2003) is a good introduction to Stata, and Rabe-Hesketh and Skrondal (2005) describe how to fit multilevel models in Stata. The methods used by AD Model Builder are described by Fournier (2001) and Skaug and Fournier (2006).

Finally, various open-source software has been written and is under development for Bayesian inference and multilevel modeling; see, for example, Graves (2003), Plummer (2003), and Warnes (2003).

References

Aaronson, D. (1998). Using sibling data to estimate the impact of neighborhoods on children's educational outcomes. *Journal of Human Resources* **33**, 915–946.

Abayomi, K., Gelman, A., and Levy, M. (2005). Diagnostics for multivariate imputations. Technical report, Department of Statistics, Columbia University.

Achen, C. (1982). *Interpreting and Using Regression.* Newbury Park, Calif.: Sage.

Achen, C. (1986). *Statistical Analysis of Quasi-Experiments.* Berkeley: University of California Press.

Afshartous, D., and De Leeuw, J. (2002). Decomposition of prediction error in multi-level models. Technical report, Department of Statistics, University of California, Los Angeles.

Agodini, R., and Dynarski, M. (2004). Are experiments the only option? A look at dropout prevention programs. *Review of Economics and Statistics* **86**, 180–194.

Agresti, A. (2002). *Categorical Data Analysis,* second edition. New York: Wiley.

Agresti, A., and Coull, B. A. (1998). Approximate is better than exact for interval estimation of binomial proportions. *American Statistician* **52**, 119–126.

Ainsley, A. E., Dyke, G. V., and Jenkyn, J. F. (1995). Inter-plot interference and nearest-neighbour analysis of field experiments. *Journal of Agricultural Science* **125**, 1–9.

Aitkin, M., and Longford, N. (1986). Statistical modelling issues in school effectiveness studies (with discussion). *Journal of the Royal Statistical Society A* **149**, 1–43.

Akaike, H. (1973). Information theory and an extension of the maximum likelihood principle. In *Proceedings of the Second International Symposium on Information Theory,* ed. B. N. Petrov and F. Csaki, 267–281. Budapest: Akademiai Kiado. Reprinted in *Breakthroughs in Statistics,* ed. S. Kotz, 610–624. New York: Springer-Verlag, 1992.

Albert, A., and Anderson, J. A. (1984). On the existence of maximum likelihood estimates in logistic regression models. *Biometrika* **71**, 1–10.

Albert, J. H., and Chib, S. (1993). Bayesian analysis of binary and polychotomous response data. *Journal of the American Statistical Association* **88**, 669–679.

Almond, D., Chay, K. Y., and Lee, D. S. (2005). The costs of low birth weight. *Quarterly Journal of Economics* **120**, 1031–1083.

Amemiya, T. (1981). Qualitative response models: a survey. *Journal of Economic Literature* **19**, 481–536.

Anderson, D. A. (1988). Some models for overdispersed binomial data. *Australian Journal of Statistics* **30**, 125–148.

Angrist, J. D. (1990). Lifetime earnings and the Vietnam era draft lottery: evidence from Social Security administrative records. *American Economic Review* **80**, 313–336.

Angrist J. D., and Evans, W. N. (1998). Children and their parents' labor supply:

evidence from exogenous variation in family size. *American Economic Review* **88**, 450–477.

Angrist, J. D., Graddy, K., and Imbens, G. W. (2000). The interpretation of instrumental variables estimators in simultaneous equations models with an application to the demand for fish. *Review of Economic Studies* **67**, 499–527.

Angrist, J. D., Imbens, G. W., and Rubin, D. B. (1996). Identification of causal effects using instrumental variables. *Journal of the American Statistical Association* **91**, 444–455.

Angrist, J. D., and Krueger, A. (1999). Empirical strategies in labor economics. In *Handbook of Labor Economics*, volume 3A, ed. O. Ashenfelter and D. Card, 1278–1366. Amsterdam: North-Holland.

Angrist, J. D., and Krueger, A. (2001). Instrumental variables and the search for identification: from supply and demand to natural experiments. *Journal of Economic Perspectives* **15**, 69–85.

Ansolablehere, S., Rodden, J., and Snyder, J. M. (2005). Purple America. Technical report, Department of Political Science, Massachusetts Institute of Technology.

Ansolabehere, S., and Snyder, J. M. (2002). Using term limits to estimate incumbency advantages when officeholders retire strategically. Technical report, Department of Political Science, Massachusetts Institute of Technology.

Ansolabehere, S., Snyder, J. M., and Stewart, C. (2000). Old voters, new voters, and the personal vote: using redistricting to measure the incumbency advantage. *American Journal of Political Science* **44**, 17–34.

Ashenfelter, O., and Krueger, A. (1994). Estimates of the economic return to schooling from a new sample of twins. *American Economic Review* **84**, 1157–1173.

Ashenfelter, O., Zimmerman, P., and Levine, D. (2003). *Statistics and Econometrics: Methods and Applications*. New York: Wiley.

Atkinson, A. C. (1985). *Plots, Transformations, and Regression*. Oxford University Press.

Bafumi, J. (2005). The stubborn American voter. Technical report, Department of Political Science, Columbia University.

Bafumi, J., Gelman, A., and Park, D. K. (2005). Issues in ideal point estimation. *Political Analysis* **13**, 171–187 .

Baker, S. G., and Kramer, B. S. (2001). Good for women, good for men, bad for people: Simpson's paradox and the importance of sex-specific analysis in observational studies. *Journal of Women's Health and Gender-Based Medicine* **10**, 867–872.

Balke, A., and Pearl, J. (1997). Bounds on treatment effects from studies with imperfect compliance. *Journal of the American Statistical Association* **92**, 1172–1176.

Ball, S., and Bogatz, G. A. (1972). Reading with television: an evaluation of the Electric Company. Report PR-72-2. Princeton, N.J.: Educational Testing Service.

Ball, S., Bogatz, G. A., Kazarow, K. M., and Rubin, D. B. (1972). Reading with television: a follow-up evaluation of the Electric Company. Report PR-74-15. Princeton, N.J.: Educational Testing Service.

Bannerjee, S., Gelfand, A. E., and Carlin, B. P. (2003). *Hierarchical Modeling and Analysis for Spatial Data*. London: CRC Press.

Barnard, J., Frangakis, C., Hill, J. L., and Rubin, D. B. (2003). A principal stratification approach to broken randomized experiments: a case study of vouchers in

New York City (with discussion). *Journal of the American Statistical Association* **98**, 299–311.

Barnard, J., McCulloch, R. E., and Meng, X. L. (1996). Modeling covariance matrices in terms of standard deviations and correlations, with application to shrinkage. *Statistica Sinica* **10**, 1281–1311.

Barnard, J., and Meng, X. L. (1994). Exploring cross-match estimators with multiply-imputed data sets. *Proceedings of the American Statistical Association, Section on Survey Research Methods.* Alexandria, Va.: American Statistical Association.

Barry, S. C., Brooks, S. P., Catchpole, E. A., and Morgan, B. J. T. (2003). The analysis of ring-recovery data using random effects. *Biometrics* **59**, 54–65.

Bates, D. (2005a). Fitting linear models in R using the lme4 package. *R News* **5** (1), 27–30. cran.r-project.org/doc/Rnews/Rnews_2005-1.pdf

Bates, D. (2005b). The mlmRev package.
cran.r-project.org/doc/Rnews/Rnews_2005-1.pdfpackages/mlmRev.pdf

Beck, N., and Katz, J. (1995). What to do (and not to do) with time-series cross-section data. *American Political Science Review* **89**, 634–647.

Beck, N., and Katz, J. (1996). Nuisance vs. substance: specifying and estimating time-series cross-section models. *Political Analysis* **6**, 1–36.

Becker, R. A., Chambers, J. M., and Wilks, A. R. (1988). *The New S Language: A Programming Environment for Data Analysis and Graphics.* Pacific Grove, Calif.: Wadsworth.

Belin, T. R., Diffendal, G. J., Mack, S., Rubin, D. B., Schafer, J. L., and Zaslavsky, A. M. (1993). Hierarchical logistic regression models for imputation of unresolved enumeration status in undercount estimation (with discussion). *Journal of the American Statistical Association* **88**, 1149–1166.

Bentler, P. M., and Raykov, T. (2000). On measures of explained variance in nonrecursive structural equation models. *Journal of Applied Psychology* **85**, 125–131.

Berger, J. O. (1985). *Statistical Decision Theory and Bayesian Analysis*, second edition. New York: Springer-Verlag.

Berk, R. A. (2004). *Regression Analysis: A Constructive Critique.* Thousand Oaks, Calif.: Sage.

Bernardo, J. M. (1979). Reference prior distributions for Bayesian inference (with discussion). *Journal of the Royal Statistical Society B* **41**, 113–147.

Bernardo, J. M., and Smith, A. F. M. (1994). *Bayesian Theory.* New York: Wiley.

Bertin, J. (1967, 1983). *Semiology of Graphics.* Translated by W. J. Berg. Madison: University of Wisconsin Press.

Besag, J., and Higdon, D. (1999). Bayesian analysis of agricultural field experiments (with discussion). *Journal of the Royal Statistical Society B* **61**, 691–746.

Besag, J., York, J., and Mollie, A. (1991). Bayesian image restoration, with two applications in spatial statistics (with discussion). *Annals of the Institute of Statistical Mathematics* **43**, 1–59.

Blackman, C. F., Benane, S. G., Elliott, D. J., House, D. E., and Pollock, M. M. (1988). Influence of electromagnetic fields on the efflux of calcium ions from brain tissue in vitro: a three-model analysis consistent with the frequency response up to 510 Hz. *Bioelectromagnetics* **9**, 215–227.

Blalock, H. M. (1961). Evaluating the relative importance of variables. *American Sociological Review* **26**, 866–874.

Bock, R. D., ed. (1989). *Multilevel Analysis of Educational Data.* New York: Academic Press.

Bogart, W. T., and Cromwell, B. A. (2000). How much is a neighborhood school worth? *Journal of Urban Economics* **47**, 280–306.

Bogatz, G. A., and Ball, S. (1971). *The Second Year of Sesame Street: A Continuing Evaluation*, two volumes. Princeton, N.J.: Educational Testing Service.

Boscardin, W. J. (1996). Bayesian analysis for some hierarchical linear models. Ph.D. thesis, Department of Statistics, University of California, Berkeley.

Boscardin, W. J., and Gelman, A. (1996). Bayesian regression with parametric models for heteroscedasticity. *Advances in Econometrics* **11A**, 87–109.

Box, G. E. P. (1980). Sampling and Bayes inference in scientific modelling and robustness. *Journal of the Royal Statistical Society A* **143**, 383–430.

Box, G. E. P., and Cox, D. R. (1964). An analysis of transformations (with discussion). *Journal of the Royal Statistical Society B* **26**, 211–252.

Box, G. E. P., Hunter, J. S., and Hunter, W. G. (2005). *Statistics for Experimenters*, second edition. New York: Wiley.

Box, G. E. P., and Tiao, G. C. (1973). *Bayesian Inference in Statistical Analysis*. New York: Wiley Classics.

Bradley, R. A., and Terry, M. E. (1952). The rank analysis of incomplete block designs: I. The method of paired comparison. *Biometrika* **39**, 324–345.

Brainard, J., and Burmaster, D. E. (1992). Bivariate distributions for height and weight of men and women in the United States. *Risk Analysis* **12**, 267–275.

Bring, J. (1994). How to standardize regression coefficients. *American Statistician* **48**, 209–213.

Brooks, S., and Gelman, A. (1998). General methods for monitoring convergence of iterative simulations. *Journal of Computational and Graphical Statistics* **7**, 434–455.

Brooks-Gunn, J., Liaw, F. R., and Klebanov, P. K. (1992). Effects of early intervention on cognitive function of low birth weight preterm infants. *Journal of Pediatrics* **120**, 350–359.

Browne, W. J., Subramanian, S. V., Jones, K., and Goldstein, H. (2005). Variance partitioning in multilevel logistic models that exhibit over-dispersion. *Journal of the Royal Statistical Society B* **168**, 599–613.

Browner, W. S., and Newman, T. B. (1987). Are all significant P values created equal? *Journal of the American Medical Association* **257**, 2459–2463.

Buja, A., Asimov, D., Hurley, C., and McDonald, J. A. (1988). Elements of a viewing pipeline for data analysis. In *Dynamic Graphics for Statistics*, ed. W. S. Cleveland and M. E. McGill, 277–308. Belmont, Calif.: Wadsworth.

Buja, A., Cook, D., and Swayne, D. (1999). Inference for data visualization. Talk given at Joint Statistical Meetings. www.research.att.com/~andreas/#dataviz

Bush, R. R., and Mosteller, F. (1955). *Stochastic Models for Learning*. New York: Wiley.

Campbell, D. T., and Stanley, J. C. (1963). *Experimental and Quasi-Experimental Designs for Research*. Chicago: Rand McNally.

Campbell, J. E. (1992). Forecasting the presidential vote in the states. *American Journal of Political Science* **36**, 386–407.

Campbell, J. E. (2002). Is the House incumbency advantage mostly a campaign finance advantage? Department of Political Science, State University of New York at Buffalo.

Card, D., and Krueger, A. (1994). Minimum wages and employment: a case study of the fast-food industry in New Jersey and Pennsylvania. *American Economic Review* **84** 772–784.

Carlin, B. P., and Louis, T. A. (2001). *Bayes and Empirical Bayes Methods for Data Analysis*, second edition. London: CRC Press.

Carlin, J. B. (1992). Meta-analysis for 2×2 tables: a Bayesian approach. *Statistics in Medicine* **11**, 141–158.

Carlin, J. B., and Forbes, A. (2004). *Linear Models and Regression*. Melbourne: Biostatistics Collaboration of Australia.

Carlin, J. B., Wolfe, R., Brown, C. H., and Gelman, A. (2001). A case study on the choice, interpretation, and checking of multilevel models for longitudinal binary outcomes. *Biostatistics* **2**, 397–416.

Carroll, R. J., and Ruppert, D. (1981). On prediction and the power transformation family. *Biometrika* **68**, 609–615.

Carroll, R. J., Ruppert, D., Crainiceanu, C. M., Tosteson, T. D., and Karagas, M. R. (2004). Nonlinear and nonparametric regression and instrumental variables. *Journal of the American Statistical Association* **99**, 736–750.

Casella, A., Gelman, A., and Palfrey, T. (2006). An experimental study of storable votes. *Games and Economic Behavior*.

Centre for Multilevel Modelling (2005). Software reviews of multilevel analysis packages. `multilevel.ioe.ac.uk/softrev`

Chambers, J. M., Cleveland, W. S., Kleiner, B., and Tukey, P. A. (1983). *Graphical Methods for Data Analysis*. Pacific Grove, Calif.: Wadsworth.

Chapman, R. (1973). The concept of exposure. *Accident Analysis and Prevention* **5**, 95–110.

Chay, K., and Greenstone, M. (2003). The impact of air pollution on infant mortality: evidence from geographic variation in pollution shocks induced by a recession. *Quarterly Journal of Economics* **118**, 1121–1167.

Chen, M. H., Shao, Q. M., and Ibrahim, J. G. (2000). *Monte Carlo Methods in Bayesian Computation*. New York: Springer-Verlag.

Chipman, H., George, E. I., and McCulloch, R. E. (2001). The practical implementation of Bayesian model selection (with discussion). In *Model Selection*, ed. P. Lahiri, 67–116. Institute of Mathematical Statistics Lecture Notes 38.

Clayton, D. G., and Kaldor, J. M. (1987). Empirical Bayes estimates of age-standardized relative risks for use in disease mapping. *Biometrics* **43**, 671–682.

Clayton, D. G., Dunn, G., Pickles, A., and Spiegelhalter, D. (1998). Analysis of longitudinal binary data from multi-phase sampling. *Journal of the Royal Statistical Society B* **60**, 71–87.

Cleveland, W. S. (1979). Robust locally weighted regression and smoothing scatterplots. *Journal of the American Statistical Association* **74**, 829–836.

Cleveland, W. S. (1985). *The Elements of Graphing Data*. Pacific Grove, Calif.: Wadsworth.

Cleveland, W. S. (1993). *Visualizing Data*. Summit, N.J.: Hobart Press.

Clinton, J., Jackman, S., and Rivers, D. (2004). The statistical analysis of roll call data. *American Political Science Review* **98**, 355–370.

Clogg, C. C., Rubin, D. B., Schenker, N., Schultz, B., and Wideman, L. (1991). Multiple imputation of industry and occupation codes in Census public-use samples using Bayesian logistic regression. *Journal of the American Statistical Association* **86**, 68–78.

Cochran, W. G. (1968). The effectiveness of adjustment by subclassification in removing bias in observational studies. *Biometrics* **24**, 205–213.

Cochran, W. G. (1977). *Sampling Techniques*, third edition. New York: Wiley.

Cochran, W. G., and Rubin, D. B. (1973). Controlling bias in observational studies: a review. *Sankhya A* **35**, 417–446.

Congdon, P. (2001). *Bayesian Statistical Modelling*. London: Wiley.

Congdon, P. (2003). *Applied Bayesian Modelling*. London: Wiley.

Cook, R. D., and Weisberg, S. (1999). *Applied Regression Including Computing and Graphics*. New York: Wiley.

Cook, S. R., Gelman, A., and Rubin, D. B. (2006). Bayesian model validation. *Journal of Computational and Graphical Statistics*.

Costa, D. L., and Kahn, M. E. (2002). Changes in the value of life, 1940–1980. National Bureau of Economic Research Working Paper #W9396.

Cox, D. R. (1958). *Planning of Experiments*. New York: Wiley.

Cox, G. W., and Katz, J. (1996). Why did the incumbency advantage grow? *American Journal of Political Science* **40**, 478–497.

Crainiceanu, C. M., Ruppert, D., and Wand, M. P. (2005). Bayesian analysis for penalized spline regression using WinBUGS. *Journal of Statistical Software* **14**, 1–24.

Cramer, C. S. (2003). *Logit Models from Economics and Other Fields*. Cambridge University Press.

D'Agostino, R. B., Jr. (1998). Propensity score methods for bias reduction in the comparison of a treatment to a non-randomized control group. *Statistics in Medicine* **17**, 2265–2281.

Daniels, M. J., and Kass, R. E. (1999). Nonconjugate Bayesian estimation of covariance matrices and its use in hierarchical models. *Journal of the American Statistical Association* **94**, 1254–1263.

Daniels, M. J., and Kass, R. E. (2001). Shrinkage estimators for covariance matrices. *Biometrics* **57**, 1173–1184.

Daniels, M. J., and Pourahmadi, M. (2002). Bayesian analysis of covariance matrices and dynamic models for longitudinal data. *Biometrika* **89**, 553–566.

Datta, G. S., Lahiri, P., Maiti, T., and Lu, K. L. (1999). Hierarchical Bayes estimation of unemployment rates for the states of the U.S. *Journal of the American Statistical Association* **94**, 1074–1082.

David, M. H., Little, R. J. A., Samuhel, M. E., and Triest, R. K. (1986). Alternative methods for CPS income imputation. *Journal of the American Statistical Association* **81**, 29–41.

Dawid, A. P. (2000). Causal inference without counterfactuals (with discussion). *Journal of the American Statistical Association* **95**, 407–448.

Dehejia, R. (2003). Was there a Riverside miracle? A framework for evaluating multi-site programs. *Journal of Business and Economic Statistics* **21**, 1–11.

Dehejia, R. (2005a). Practical propensity score matching: a reply to Smith and Todd. *Journal of Econometrics* **125**, 355–364.

Dehejia, R. (2005b). Does matching overcome LaLonde's critique of nonexperimental estimators? A postscript. Technical report, Department of Economics, Columbia University.

Dehejia, R., and Wahba, S. (1999). Causal effects in non-experimental studies:

re-evaluating the evaluation of training programs. *Journal of the American Statistical Association* **94**, 1053–1062.

Dellaportas, P., and Smith, A. F. M. (1993). Bayesian inference for generalized linear and proportional hazards models via Gibbs sampling. *Applied Statistics* **42**, 443–459.

Dempster, A. P., Rubin, D. B., and Tsutakawa, R. K. (1981). Estimation in covariance components models. *Journal of the American Statistical Association* **76**, 341–353.

De Veaux, R., Velleman, P., and Bock, D. (2006). *Stats: Data and Models*. Boston: Addison-Wesley.

Diamond, A., and Sekhon, J. S. (2005). Genetic matching for estimating causal effects. Technical report, Department of Government, Harvard University.

Diggle, P., and Kenward, M. G. (1994). Informative drop-out in longitudinal data analysis. *Journal of the Royal Statistical Society C* **43**, 49–73.

Dobson, A. (2001). *An Introduction to Generalized Linear Models*, second edition. London: CRC Press.

Donohue, J. J., and Wolfers, J. (2006). Uses and abuses of empirical evidence in the death penalty debate. *Stanford Law Review* **25**, 791–845.

Dorman, P., and Hagstrom, P. (1998). Wage compensation for dangerous work revisited. *Industrial and Labor Relations Review* **52**, 116–135.

Drake, C. (1993). Effects of misspecification of the propensity score on estimators of treatment effect. *Biometrics* **49**, 1231–1236.

Dreze, J. H. (1976). Bayesian limited information analysis of the simultaneous equations model. *Econometrica* **44**, 1045–1075.

Du, J. (1998). Valid inferences after propensity score subclassification using maximum number of subclasses as building blocks. Ph.D. thesis, Department of Statistics, Harvard University.

DuMouchel, W. M., and Harris, J. E. (1983). Bayes methods for combining the results of cancer studies in humans and other species (with discussion). *Journal of the American Statistical Association* **78**, 293–315.

Dunson, D. B. (2006). Efficient Bayesian model averaging in factor analysis. Technical Report, Biostatistics Branch, National Institute of Environmental Health Sciences.

Efron, B., and Morris, C. (1975). Data analysis using Stein's estimator and its generalizations. *Journal of the American Statistical Association* **70**, 311–319.

Ehrenberg, A. S. C. (1978). *Data Reduction: Analysing and Interpreting Statistical Data*. New York: Wiley.

Ehrenberg, R., and Brewer, D. (1994). Do school and teacher characteristics matter?: evidence from High School and Beyond. *Economics of Education Review* **13**, 1–17.

Eisenhart, C. (1947). The assumptions underlying the analysis of variance. *Biometrics* **3**, 1–21.

Erikson, R. S. (1971). The advantage of incumbency in congressional elections. *Polity* **3**, 395–405.

Erikson, R. S., and Romero, D. W. (1990). Candidate equilibrium and the behavioral model of the vote. *American Political Science Review* **4**, 1103–1126.

Fair, R. C. (1978). The effect of economic events on votes for President. *Review of Economics and Statistics* **60**, 159–173.

582   REFERENCES

Fay, R. E. (1996). Alternative paradigms for the analysis of imputed survey data. *Journal of the American Statistical Association* **91**, 490–498.

Fay, R. E., and Herriot, R. A. (1979). Estimates of income for small places: an application of James-Stein procedures to census data. *Journal of the American Statistical Association* **74**, 269–277.

Felton, J., Mitchell, J., and Stinson, M. (2003). Web-based student evaluations of professors: the relations between perceived quality, easineness, and sexiness. *Assessment and Evaluation in Higher Education* **29**, 91–108.

Fienberg, S. E. (1977). *The Analysis of Cross-Classified Categorical Data*. Cambridge, Mass.: M.I.T. Press.

Finkelstein, M. O., Levin, B., McKeague, I. M., and Tsai, W. Y. (2006). A note on the censoring problem in empirical case-outcome studies. *Journal of Empirical Legal Studies*.

Firth, D. (1993). Bias reduction of maximum likelihood estimates. *Biometrika* **80**, 27–38.

Fisman, R., Iyengar, S. S., Kamenica, E., and Simonson, I. (2006). Gender differences in mate selection: evidence from a speed dating experiment. *Quarterly Journal of Economics*.

Fournier, D. (2001). An introduction to AD Model Builder Version 6.0.2 for use in nonlinear modeling and statistics. `otter-rsch.com/admodel.htm`

Fox, J. (2002). *An R and S-Plus Companion to Applied Regression*. Thousand Oaks, Calif.: Sage.

Frangakis, C. E., Brookmeyer, R. S., Varadhan, R., Mahboobeh, S., Vlahov, D., and Strathdee, S. A. (2003). Methodology for evaluating a partially controlled longitudinal treatment using principal stratification, with application to a needle exchange program. *Journal of the American Statistical Association* **99**, 239–249.

Frangakis, C. E., and Rubin, D. B. (2002). Principal stratification in causal inference. *Biometrics* **58**, 21–29.

Friedlander, D., and Robins, P. K. (1995). Evaluating program evaluations—new evidence on commonly used nonexperimental methods. *American Economic Review* **85**, 923–937.

Friendly, M., and Kwan, E. (2003). Effect ordering for data displays. *Computational Statistics and Data Analysis* **43**, 509–539.

Frolich, M. (2004). Finite-sample properties of propensity-score matching and weighting estimators. *Review of Economics and Statistics* **86**, 77–90.

Garfinkel, I., and Meyers, M. K. (1999). A tale of many cities: the New York City Social Indicators Survey. School of Social Work, Columbia University.

Gawron, V. J., Berman, B. A., Dismukes, R. K., and Peer, J. H. (2003). New airline pilots may not receive sufficient training to cope with airplane upsets. *Flight Safety Digest* (July–August), 19–32.

Gelfand, A. E., Dey, D. K., and Chang, H. (1992). Model determination using predictive distributions with implementation via sampling-based methods (with discussion). In *Bayesian Statistics 4*, ed. J. M. Bernardo, J. O. Berger, A. P. Dawid, and A. F. M. Smith, 147–167. Oxford University Press.

Gelfand, A. E., Hills, S. E., Racine-Poon, A., and Smith, A. F. M. (1990). Illustration of Bayesian inference in normal data models using Gibbs sampling. *Journal of the American Statistical Association* **85**, 972–985.

Gelfand, A. E., and Sahu, S. K. (1999). Identifiability, improper priors, and Gibbs sampling for generalized linear models. *Journal of the American Statistical Association* **94**, 247–253.

Gelfand, A. E., Sahu, S. K., and Carlin, B. P. (1995). Efficient parametrizations for normal linear mixed models. *Biometrika* **82**, 479–488.

Gelfand, A. E., and Smith, A. F. M. (1990). Sampling-based approaches to calculating marginal densities. *Journal of the American Statistical Association* **85**, 398–409.

Gelman, A. (1997). Using exams for teaching concepts in probability and statistics. *Journal of Educational and Behavioral Statistics* **22**, 237–243.

Gelman, A. (2003). A Bayesian formulation of exploratory data analysis and goodness-of-fit testing. *International Statistical Review* **71**, 369–382.

Gelman, A. (2004a). Exploratory data analysis for complex models (with discussion). *Journal of Computational and Graphical Statistics* **13**, 755–787.

Gelman, A. (2004b). Parameterization and Bayesian modeling. *Journal of the American Statistical Association* **99**, 537–545.

Gelman, A. (2004c). 55,000 residents desperately need your help! *Chance* **17** (2), 28–31.

Gelman, A. (2004d). Treatment effects in before-after data. In *Applied Bayesian Modeling and Causal Inference from Incomplete Data Perspectives*, ed. A. Gelman and X. L. Meng, chapter 18. London: Wiley.

Gelman, A. (2005). Analysis of variance: why it is more important than ever (with discussion). *Annals of Statistics* **33**, 1–53.

Gelman, A. (2006). Prior distributions for variance parameters in hierarchical models. *Bayesian Analysis* **1**, 514–534.

Gelman, A. (2007). Scaling regression inputs by dividing by two standard deviations. Technical report, Department of Statistics, Columbia University.

Gelman, A., Carlin, J. B., Stern, H. S., and Rubin, D. B. (2003). *Bayesian Data Analysis*, second edition. London: CRC Press.

Gelman, A., Fagan, J., and Kiss, A. (2006). An analysis of the NYPD's stop-and-frisk policy in the context of claims of racial bias. *Journal of the American Statistical Association*.

Gelman, A., Goegebeur, Y., Tuerlinckx, F., and Van Mechelen, I. (2000). Diagnostic checks for discrete-data regression models using posterior predictive simulations. *Applied Statistics* **49**, 247–268.

Gelman, A., and Huang, Z. (2006). Estimating incumbency advantage and its variation, as an example of a before/after study (with discussion). *Journal of the American Statistical Association*.

Gelman, A., Huang, Z., van Dyk, D., and Boscardin, W. J. (2006). Transformed and parameter-expanded Gibbs samplers for multilevel linear and generalized linear models. Technical report, Department of Statistics, Columbia University.

Gelman, A., Jakulin, A., Pittau, M. G., and Su, Y. S. (2007). A default prior distribution for logistic and other regression models. Technical report, Department of Statistics, Columbia University.

Gelman, A., and Katz, J. (2005). Moderation in the pursuit of moderation is no vice: the clear but limited advantages to being a moderate for congressional elections. Technical report, Department of Statistics, Columbia University.

Gelman, A., and King, G. (1990). Estimating incumbency advantage without bias. *American Journal of Political Science* **34**, 1142–1164.

Gelman, A., and King, G. (1993). Why are American presidential election campaign polls so variable when votes are so predictable? *British Journal of Political Science* **23**, 409–451.

Gelman, A., and King, G. (1994a). A unified model for evaluating electoral systems and redistricting plans. *American Journal of Political Science* **38**, 514–554.

Gelman, A., and King, G. (1994b). Enhancing democracy through legislative redistricting. *American Political Science Review* **88**, 541–559.

Gelman, A., King, G., and Liu, C. (1998). Multiple imputation for multiple surveys (with discussion). *Journal of the American Statistical Association* **93**, 846–874.

Gelman, A., Liebman, J., West, V., and Kiss, A. (2004). A broken system: the persistent pattern of reversals of death sentences in the United States. *Journal of Empirical Legal Studies* **1**, 209–261.

Gelman, A., and Little, T. C. (1997). Poststratification into many categories using hierarchical logistic regression. *Survey Methodology* **23**, 127–135.

Gelman, A., Meng, X. L., and Stern, H. S. (1996). Posterior predictive assessment of model fitness via realized discrepancies (with discussion). *Statistica Sinica* **6**, 733–807.

Gelman, A., and Nolan, D. (2002). *Teaching Statistics: A Bag of Tricks.* Oxford University Press.

Gelman, A., and Pardoe, I. (2006). Bayesian measures of explained variance and pooling in multilevel (hierarchical) models. *Technometrics* **48**, 241–251.

Gelman, A., and Pardoe, I. (2007). Average predictive comparisons for models with nonlinearity, interactions, and variance components. *Sociological Methodology.*

Gelman, A., Pasarica, C., and Dodhia, R. (2002). Let's practice what we preach: using graphs instead of tables. *American Statistician* **56**, 121–130.

Gelman, A., and Price, P. N. (1998). Discussion of "Some algebra and geometry for hierarchical models, applied to diagnostics," by J. S. Hodges. *Journal of the Royal Statistical Society B* **60**, 532.

Gelman, A., and Price, P. N. (1999). All maps of parameter estimates are misleading. *Statistics in Medicine* **18**, 3221–3234.

Gelman, A., and Rubin, D. B. (1992). Inference from iterative simulation using multiple sequences (with discussion). *Statistical Science* **7**, 457–511.

Gelman, A., Shor, B., Bafumi, J., and Park, D. K. (2005). Rich state, poor state, red state, blue state: what's the matter with Connecticut? Technical report, Department of Political Science, Columbia University.

Gelman, A., and Stern, H. S. (2006). The difference between "statistically significant" and "not significant" is not itself significant. Technical report, Department of Statistics, Columbia University.

Gelman, A., Trevisani, M., Lu, H., and van Geen, A. (2004). Direct data manipulation for local decision analysis, as applied to the problem of arsenic in drinking water from tube wells in Bangladesh. *Risk Analysis* **24**, 1597–1612.

Gelman, A., and Tuerlinckx, F. (2000). Type S error rates for classical and Bayesian single and multiple comparison procedures. *Computational Statistics* **15**, 373–390.

Gerber, A. (2004). Does campaign spending work? *American Behavioral Scientist* **47**, 541–574.

Geweke, J. (2004). Getting it right: joint distribution tests of posterior simulators. *Journal of the American Statistical Association* **99**, 799–804.

Gilks, W. R., Richardson, S., and Spiegelhalter, D., eds. (1996). *Practical Markov Chain Monte Carlo.* London: Chapman and Hall.

Gill, J. (2002). *Bayesian Methods for the Social and Behavioral Sciences*. London: Chapman and Hall.

Gilovich, T., Vallone, R., and Tversky, A. (1985). The hot hand in basketball: on the misperception of random sequences. *Cognitive Psychology* **17**, 295–314.

Giltinan, D., and Davidian, M. (1995). *Nonlinear Models for Repeated Measurement Data*. London: Chapman and Hall.

Girosi, F., and King, G. (2005). Demographic forecasting. Technical report, Department of Government, Harvard University.

Glickman, M. E., and Normand, S. L. (2000). The derivation of a latent threshold instrumental variables model. *Statistica Sinica* **10**, 517–544.

Goldstein, H. (1995). *Multilevel Statistical Models*, second edition. London: Edward Arnold.

Goldstein, H., Browne, W. J., and Rasbash, J. (2002). Partitioning variation in multilevel models. *Understanding Statistics* **1**, 223–232.

Goldstein, H., and Silver, R. (1989). Multilevel and multivariate models in survey analysis. In *Analysis of Complex Surveys*, ed. C. J. Skinner, D. Holt, and T. M. F. Smith, 221–235. New York: Wiley.

Graubard, B. I., and Korn, E. L. (1999). Predictive margins with survey data. *Biometrics* **55**, 652–659.

Graves, T. L. (2003). Yadas: yet another data analysis system. Statistical Sciences Group, Los Alamos National Laboratory.
`www.stat.lanl.gov/yadas/home.html`

Green, B. F., and Tukey, J. W. (1960). Complex analyses of variance: general problems. *Psychometrika* **25**, 127–152.

Greenland, S. (2000). When should epidemiologic regressions use random coefficients? *Biometrics* **56**, 915–921.

Greenland, S. (2005). Multiple bias modelling for analysis of observational data (with discussion). *Journal of the Royal Statistical Society A*.

Greenland, S., Robins, J. M., and Pearl, J. (1999). Confounding and collapsability in causal inference. *Statistical Science* **14**, 29–46.

Greenland, S., Schlessman, J. J., and Criqui, M. H. (1986). The fallacy of employing standardized regression coefficients and correlations as measures of effect. *American Journal of Epidemiology* **123**, 203–208.

Groves, R., Fowler, F. J., Couper, M. P., Lepkowski, J. M., Singer, E., and Tourangeau, R. (2004). *Survey Methodology*. New York: Wiley.

Gustafson, P., and Clarke, B. (2004). Decomposing posterior variance. *Journal of Statistical Planning and Inference* **119**, 311–327.

Gustafson, P., and Greenland, S. (2005). The performance of random coefficient regression in accounting for residual confounding. Technical report, Department of Statistics, University of British Columbia.

Haavelmo, T. (1943). The statistical implications of a system of simultaneous equations. *Econometrica* **11**, 1–12.

Hahn, J. (1998). On the role of the propensity score in efficient semiparametric estimation of average treatment effects. *Econometrica* **66**, 315–331.

Hahn, J., Todd, P., and van der Klaauw, W. (2001). Identification and estimation of treatment effects with a regression-discontinuity design. *Econometrica* **69**, 201–209.

Hamermesh, D. S., and Parker, A. M. (2005). Beauty in the classroom: instructors' pulchritude and putative pedagogical productivity. *Economics of Education Review* **24**, 369–376.

Hansen, B. B. (2004). Full matching in an observational study of coaching for the SAT. *Journal of the American Statistical Association* **99**, 609–619.

Hansen, B. B. (2006). OptMatch package for optimal matching. www.stat.lsa.umich.edu/~bbh/optmatch.html.

Hanushek, E. (1971). Teacher characteristics and gains in student achievement: estimation using micro data. *American Economic Review* **61**, 280–288.

Harrell, F. (2001). *Regression Modeling Strategies*. New York: Springer-Verlag.

Hastie, T. J., and Tibshirani, R. J. (1990). *Generalized Additive Models*. New York: Chapman and Hall.

Hastie, T. J., Tibshirani, R. J., and Friedman, J. (2002). *The Elements of Statistical Learning: Data Mining, Inference, and Prediction*. New York: Springer-Verlag.

Hauer, E., Ng, J. C. N., and Lovell, J. (1988). Estimation of safety at signalized intersections. *Transportation Research Record* **1185**, 48–61. Washington, D.C.: National Research Council.

Healy, M. J. R. (1990). Measuring importance. *Statistics in Medicine* **9**, 633–637.

Heckman, J. (1979). Sample selection bias as a specification error. *Econometrica* **47**, 153–161.

Heckman, J. J., Ichimura, H., and Todd, P. (1997). Matching as an econometric evaluation estimator: evidence from a job training programme. *Review of Economic Studies* **64**, 605–654.

Heckman, J. J., Ichimura, H., and Todd, P. (1998). Matching as an econometric evaluation estimator. *Review of Economic Studies* **65**, 261–294.

Heinze, G., and Schemper, M. (2003). A solution to the problem of separation in logistic regression. *Statistics in Medicine* **12**, 2409–2419.

Heitjan, D. F., Moskowitz, A. J., and Whang, W. (1999). Problems with interval estimates of the incremental cost-effectiveness ratio. *Medical Decision Making* **19**, 9–15.

Henderson, C. R. (1950). Estimation of genetic parameters (abstract). *Annals of Mathematical Statistics* **21**, 309–310.

Henderson, C. R. (1984). *Application of Linear Models in Animal Breeding*. Univerity of Guelph, Ontario, Canada.

Henderson, C. R., Kempthorne, O., Searle, S. R., and Von Krosigk, C. M. (1959). The estimation of environmental and genetic trends from records subject to culling. *Biometrics* **15**, 192–218.

Hill, B. M. (1965). Inference about variance components in the one-way model. *Journal of the American Statistical Association* **60**, 806–825.

Hill, J. L., Brooks-Gunn, J., and Waldfogel, J. (2003). Sustained effects of high participation in an early intervention for low-birth-weight premature infants. *Developmental Pscyhology* **39**, 730–744.

Hill, J. L., and McCulloch, R. E. (2006). Bayesian nonparametric modeling for causal inference. Technical report, School of International and Public Affairs, Columbia University.

Hill, J. L., and Reiter, J. (2006). Interval estimation for treatment effects using propensity score matching. *Statistics in Medicine* **25**, 2230–2256.

Hill, J. L., Reiter, J., and Zanutto, E. (2004). A comparison of experimental and observational data analyses. In *Applied Bayesian and Causal Inference from an Incomplete Data Perspective*, ed. A. Gelman and X. L. Meng. New York: Wiley.

Hill, J. L., Waldfogel, J., and Brooks-Gunn, J. (2002). Assessing the differential impacts of high-quality child care: a new approach for exploiting post-treatment variables. *Journal of Policy Analysis and Management* **21**, 601–627.

Hill, J. L., Waldfogel, J., Brooks-Gunn, J., and Han, W. J. (2005). Maternal employment and child development: a fresh look using newer methods. *Developmental Psychology* **41**, 833–850.

Hirano, K., Imbens, G. W., and Ridder, G. (2003). Efficient estimation of average treatment effects using the estimated propensity score. *Econometrica* **71**, 1161–1189.

Hirano, K., Imbens, G. W., Rubin, D. B., and Zhou, A. (2000). Assessing the effect of an influenza vaccine in an encouragement design. *Biostatistics* **1**, 69–88.

Hodges, J. S. (1998). Some algebra and geometry for hierarchical models, applied to diagnostics (with discussion). *Journal of the Royal Statistical Society B* **60**, 497–536.

Hodges, J. S., Cui, Y., Sargent, D. J., and Carlin, B. P. (2005). Smoothed ANOVA. Technical report, Department of Biostatistics, University of Minnesota.

Hodges, J. S., and Sargent, D. J. (2001). Counting degrees of freedom in hierarchical and other richly parameterized models. *Biometrika* **88**, 367–379.

Hoenig, J. M., and Heisey, D. M. (2001). The abuse of power: the pervasive fallacy of power calculations for data analysis. *American Statistician* **55**, 19–24.

Hoeting, J., Madigan, D., Raftery, A. E., and Volinsky, C. (1999). Bayesian model averaging (with discussion). *Statistical Science* **14**, 382–417.

Hoff, P. D. (2003). Random effects models for network data. In *Dynamic Social Network Modeling and Analysis: Workshop Summary and Papers*, ed. R. Breiger, K. Carley, and P. Pattison, 303–312. Washington, D.C.: National Academies Press.

Hoff, P. D. (2005). Bilinear mixed-effects models for dyadic data. *Journal of the American Statistical Assosciation* **100**, 286–295.

Hoff, P. D., Raftery, A. E., and Handcock, M. S. (2002). Latent space approaches to social network analysis. *Journal of the American Statistical Association* **97**, 1090–1098.

Hoogerheide, L., Kleibergen, F., and van Dijk, H. K. (2006). Natural conjugate priors for the instrumental variables regression model applied to the Angrist-Krueger data. *Journal of Econometrics*.

Hox, J. (2002). *Multilevel Analysis: Techniques and Applications*. Mahwah, N.J.: Lawrence Erlbaum Associates.

Huang, Z., and Gelman, A. (2003). Sampling for Bayesian computation with large datasets. Technical report, Department of Statistics, Columbia University.

Ichimura, H., and Linton, O. (2001). Asymptotic expansions for some semiparametric program evaluation estimators. Institute for Fiscal Studies, Cemmap Working Paper cwp04/01.

Imai, K., and van Dyk, D. A. (2003). A Bayesian analysis of the multinomial probit model using marginal data augmentation. *Journal of Econometrics* **124**, 311–334.

Imai, K., and van Dyk, D. A. (2004). Causal inference with general treatment regimes: generalizing the propensity score. *Journal of the American Statistical Association* **99**, 854–866.

Imbens, G. W. (2000). The role of the propensity score in estimating dose-response functions. *Biometrika* **87**, 706–710.

Imbens, G. W. (2004). Nonparametric estimation of average treatment effects under exogeneity: a review. *Review of Economics and Statistics* **86**, 4–29.

Imbens, G. W., and Angrist, J. (1994). Identification and estimation of local average treatment effects. *Econometrica* **62**, 467–475.

Imbens, G. W., and Rubin, D. B. (1997). Bayesian inference for causal effects in randomized experiments with noncompliance. *Annals of Statistics* **25**, 305–327.

Jackman, S. (2001). Multidimensional analysis of roll call data via Bayesian simulation: identification, estimation, inference and model checking. *Political Analysis* **9**, 227–241.

Jackson, C., Best, N., and Richardson, C. (2006). Improving ecological inference using individual-level data. *Statistics in Medicine* **25**, 2136–2159.

Jacob, B. A., and Lefgren, L. (2004). Remedial education and student achievement: a regression-discontinuity analysis. *Review of Economics and Statistics* **86**, 226–244.

James, W., and Stein, C. (1960). Estimation with quadratic loss. In *Proceedings of the Fourth Berkeley Symposium on Mathematical Statistics and Probability* **1**, ed. J. Neyman, 361–380. Berkeley: University of California Press.

Jaynes, E. T. (1983). *Papers on Probability, Statistics, and Statistical Physics*, ed. R. D. Rosenkrantz. Dordrecht, Netherlands: Reidel.

Jeffreys, H. (1961). *Theory of Probability*, third edition. Oxford University Press.

Joffe, M. M., and Rosenbaum, P. R. (1999). Propensity scores. *American Journal of Epidemiology* **150**, 327–333.

Johnson, V. E. (1996). On Bayesian analysis of multirater ordinal data: an application to automated essay grading. *Journal of the American Statistical Association* **91**, 42–51.

Johnson, V. E. (1997). An alternative to traditional GPA for evaluating student performance (with discussion). *Statistical Science* **12**, 251–278.

Kane, T. J., Rockoff, J. E., and Staiger, D. O. (2006). What does certification tell us about teacher effectiveness? evidence from New York City. Technical Report, School of Business, Columbia University.

Karim, M. R., and Zeger, S. L. (1992). Generalized linear models with random effects; salamander mating revisited. *Biometrics* **48**, 631–644.

Kass, R. E., Carlin, B. P., Gelman, A., and Neal, R. (1998). Markov chain Monte Carlo in practice: a roundtable discussion. *American Statistician* **52**, 93–100.

Kass, R. E., and Wasserman, L. (1996). The selection of prior distributions by formal rules. *Journal of the American Statistical Association* **91**, 1343–1370.

Kedar, O., and Shively, W. P., eds. (2005). Special issue on multilevel modeling. *Political Analysis* **13** (4).

Kenny, D. A., Kashy, D. A., and Bolger, N. (1998). Data analysis in social psychology. In *Handbook of Social Psychology*, ed. D. Gilbert, S. Fiske, and G. Lindzey, 233–265. Boston: McGraw-Hill.

Kerman, J. (2006). Umacs: a universal Markov chain sampler. Technical report, Department of Statistics, Columbia University.

Kerman, J., and Gelman, A. (2006). Fully Bayesian computing. Technical report, Department of Statistics, Columbia University.

Killworth, P. D., Johnsen, E., McCarty, C., Shelley, G. A., and Bernard, H. R. (1998). Estimation of seroprevalence, rape and homelessness in the U.S. using a social network approach. *Social Networks* **20**, 23–50.

King, G. (1986). How not to lie with statistics: avoiding common mistakes in quantitative political science. *American Journal of Political Science* **30**, 666–687.

King, G., Honaker, J., Joseph, A., and Scheve, K. (2001). Analyzing incomplete political science data: an alternative algorithm for multiple imputation. *American Political Science Review* **95**, 49–69.

King, G., Tomz, M., and Wittenberg, J. (2000). Making the most of statistical analyses: improving interpretation and presentation. *American Journal of Political Science* **44**, 341–355.

King, G., and Zeng, L. (2006). The dangers of extreme counterfactuals. *Political Analysis* **14**, 131–159.

Kirk, R. E. (1995). *Experimental Design: Procedures for the Behavioral Sciences*, third edition. Pacific Grove, Calif.: Brooks/Cole.

Kiss, A. (2003). Hierarchical models: what the data are really telling us. Ph.D. thesis, Department of Biostatistics, Columbia University.

Kleibergen, F., and Zivot, E. (2003). Bayesian and classical approaches to instrumental variable regression. *Journal of Econometrics* **114**, 29–72.

Krantz, D. H. (1999). The null hypothesis testing controversy in psychology. *Journal of the American Statistical Association* **94**, 1372–1381.

Kreft, I., and De Leeuw, J. (1998). *Introducing Multilevel Modeling*. London: Sage.

Laird, N. M., and Ware, J. H. (1982). Random-effects models for longitudinal data. *Biometrics* **38**, 963–974.

LaLonde, R. J. (1986). Evaluating the econometric evaluations of training programs using experimental data. *American Economic Review* **76**, 604–620.

LaMotte, L. R. (1983). Fixed-, random-, and mixed-effects models. In *Encyclopedia of Statistical Sciences*, ed. S. Kotz, N. L. Johnson, and C. B. Read, **3**, 137–141. New York: Wiley.

Lancaster, T. (2004). *An Introduction to Modern Bayesian Econometrics*. Oxford: Blackwell.

Landwehr, J. M., Pregibon, D., and Shoemaker, A. C. (1984). Graphical methods for assessing logistic regression models. *Journal of the American Statistical Association* **79**, 61–83.

Lane, P. W., and Nelder, J. A. (1982). Analysis of covariance and standardization as instances of prediction. *Biometrics* **38**, 613–621.

Lange, K. L., Little, R. J. A., and Taylor, J. M. G. (1989). Robust statistical modeling using the *t* distribution. *Journal of the American Statistical Association* **84**, 881–896.

Lavori, P. W., Keller, M. B., and Endicott, J. (1995). Improving the aggregate performance of psychiatric diagnostic methods when not all subjects receive the standard test. *Statistics in Medicine* **14**, 1913–1925.

Leamer, E. (1978). *Specification Searches: Ad Hoc Inference with Nonexperimental Data*. New York: Wiley.

Leamer, E. (1983). Let's take the con out of econometrics. *American Economic Review* **73**, 31–43.

Lechner, M. (1999). Earnings and employment effects of continuous off-the-job training in East Germany after unification. *Journal of Business and Economic Statistics* **17**, 74–90.

Lee, D. S., Moretti, E., and Butler, M. J. (2004). Do voters affect or elect policies? Evidence from the U.S. House. *Quarterly Journal of Economics* **119**, 807–859.

Lenth, R. V. (2001). Some practical guidelines for effective sample size determination. *American Statistician* **55**, 187–193.

Lesaffre, E., and Albert, A. (1989). Partial separation in logistic discrimination. *Journal of the Royal Statistical Society B* **51**, 109–116.

Levitt, S. D., and Wolfram, C. D. (1997). Decomposing the sources of incumbency advantage in the U.S. House. *Legislative Studies Quarterly* **22**, 45–60.

Leyland, A. H., and Goldstein, H., eds. (2001). *Multilevel Modelling of Health Statistics*. Chichester: Wiley.

Liang, K. Y., and McCullagh, P. (1993). Case studies in binary dispersion. *Biometrics* **49**, 623–630.

Lin, C. Y., Gelman, A., Price, P. N., and Krantz, D. H. (1999). Analysis of local decisions using hierarchical modeling, applied to home radon measurement and remediation (with discussion). *Statistical Science* **14**, 305–337.

Linden, L. (2006). Are incumbents really advantaged? The preference for non-incumbents in India. Technical report, Department of Economics, Columbia University.

Lindley, D. V., and Smith, A. F. M. (1972). Bayes estimates for the linear model. *Journal of the Royal Statistical Society B* **34**, 1–41.

Lira, P. I., Ashworth, A., and Morris, S. S. (1998). Effect of zinc supplementation on the morbidity, immune function, and growth of low-birth-weight, full-term infants in northeast Brazil. *American Journal of Clinical Nutrition* **68**, 418S–424S.

Little, R. J. A., and Rubin, D. B. (2002). *Statistical Analysis with Missing Data*, second edition. New York: Wiley.

Liu, C. (1995). Missing data imputation using the multivariate t distribution. *Journal of Multivariate Analysis* **48**, 198–206.

Liu, C. (2004). Robit regression: a simple robust alternative to logistic and probit regression. In *Applied Bayesian Modeling and Causal Inference from Incomplete-Data Perspectives*, ed. A. Gelman and X. L. Meng, 227–238. London: Wiley.

Liu, C., Rubin, D. B., and Wu, Y. N. (1998). Parameter expansion to accelerate EM: the PX-EM algorithm. *Biometrika* **85**, 755–770.

Liu, J. (2002). *Monte Carlo Strategies in Scientific Computing*. New York: Springer-Verlag.

Liu, J., and Wu, Y. N. (1999). Parameter expansion for data augmentation. *Journal of the American Statistical Association* **94**, 1264–1274.

Lohr, S. (1999). *Sampling: Design and Analysis*. Pacific Grove, Calif.: Duxbury.

Loken, E. (2004). Multimodality in mixture models and factor models. In *Applied Bayesian Modeling and Causal Inference from Incomplete-Data Perspectives*, ed. A. Gelman and X. L. Meng, 203–213. London: Wiley.

Longford, N. (1993). *Random Coefficient Models*. Oxford: Clarendon Press.

Longford, N. (2005). *Missing Data and Small-Area Estimation*. New York: Springer-Verlag.

Lord, F. M., and Novick, M. R. (1968). *Statistical Theories of Mental Test Scores*. Reading, Mass.: Addison-Wesley.

Louis, T. A. (1984). Estimating a population of parameter values using Bayes and empirical Bayes methods. *Journal of the American Statistical Association* **78**, 393–398.

Louis, T. A., and Shen, W. (1999). Innovations in Bayes and empirical Bayes methods: estimating parameters, populations and ranks. *Statistics in Medicine* **18**, 2493–2505.

MacLehose, R. F., Dunson, D. B., Herring, A., and Hoppin, J. A. (2006). Bayesian methods for highly correlated exposure data. Technical report, Department of Epidemiology, University of North Carolina.

Maddala, G. S. (1976). Weak priors and sharp posteriors in simultaneous equation models. *Econometrica* **44**, 345–351.

Maddala, G. S. (1983). *Limited Dependent and Qualitative Variables in Econometrics*. Cambridge University Press.

Madigan, D., and Raftery, A. E. (1994). Model selection and accounting for model uncertainty in graphical models using Occam's window. *Journal of the American Statistical Association* **89**, 1535–1546.

Mallows, C. L. (1973). Some comments on C_p. *Technometrics* **15**, 661–675.

Manski, C. F. (1990). Nonparametric bounds on treatment effects. *American Economic Reviews, Papers and Proceedings* **80**, 319–323.

Martin, A. D., and Quinn, K. M. (2002a). Dynamic ideal point estimation via Markov chain Monte Carlo for the U.S. Supreme Court, 1953–1999. *Political Analysis* **10**, 134–153.

Martin, A. D., and Quinn, K. M. (2002b). MCMCpack. scythe.wustl.edu/mcmcpack.html

McCarty, C., Killworth, P. D., Bernard, H. R., Johnsen, E., and Shelley, G. A. (2000). Comparing two methods for estimating network size. *Human Organization* **60**, 28–39.

McCarty, N., Poole, K. T., and Rosenthal, H. (2005). *Polarized America: The Dance of Political Ideology and Unequal Riches*. Cambridge, Mass.: MIT Press.

McCullagh, P. (1980). Regression models for ordinal data (with discussion). *Journal of the Royal Statistical Society B* **42**, 109–142.

McCullagh, P. (2005). Discussion of "Analysis of variance: why it is more important than ever," by A. Gelman. *Annals of Statistics* **33**, 33–38.

McCullagh, P., and Nelder, J. A. (1989). *Generalized Linear Models*, second edition. London: Chapman and Hall.

McDonald, G. C., and Schwing, R. C. (1973). Instabilities of regression estimates relating air pollution to mortality. *Technometrics* **15**, 463–482.

McFadden, D. (1973). Conditional logit analysis of qualitative choice behavior. In *Frontiers in Econometrics*, ed. P. Zarembka, 105–142. New York: Academic Press.

Meng, X. L. (1994). Multiple-imputation inferences with uncongenial sources of input (with discussion). *Statistical Science* **9**, 538–573.

Meng, X. L., and Zaslavsky, A. M. (2002). Single observation unbiased priors. *Annals of Statistics* **30**, 1345–1375.

Metropolis, N., Rosenbluth, A. W., Rosenbluth, M. N., Teller, A. H., and Teller, E. (1953). Equation of state calculations by fast computing machines. *Journal of Chemical Physics* **21**, 1087–1092.

Michalopoulos, C., Bloom, H. S., and Hill, C. J. (2004). Can propensity score methods match the findings from a random assignment evaluation of mandatory welfare-to-work programs? *Review of Economics and Statistics* **86**, 156–179.

Montgomery, D. C. (1986). *Design and Analysis of Experiments*, second edition. New York: Wiley.

Moore, D. S., and McCabe, G. P. (1998). *Introduction to the Practice of Statistics*, third edition. New York: W. H. Freeman.

Morris, C. (1983). Parametric empirical Bayes inference: theory and applications (with discussion). *Journal of the American Statistical Association* **78**, 47–65.

Mosteller, F. (1951). Remarks on the method of paired comparison. *Psychometrika* **16**, 3–9, 203–206, 207–218.

Mosteller, F., and Tukey, J. W. (1977). *Data Analysis and Regression*. Reading, Mass.: Addison-Wesley.

Muller, O., Becher, H., van Zweeden, A. B., Ye, Y., Diallo, D. A., Konate, A. T., Gbangou, A., Kouate, B., and Gareene, M. (2001). Effect of zinc supplementation on malaria and other causes of morbidity in west African children: randomized double blind placebo controlled trial. *British Medical Journal* **322**, 1567–1573.

Murphy, R. T. (1991). Educational effectiveness of Sesame Street: a review of the first twenty years of research, 1969–1989. Report RR-91-55. Princeton, N.J.: Educational Testing Service.

Murrell, P. (2005). *R Graphics*. London: CRC Press.

Nelder, J. A. (1965a). The analysis of randomized experiments with orthogonal block structure, I. Block structure and the null analysis of variance. *Proceedings of the Royal Society A* **273**, 147–162.

Nelder, J. A. (1965b). The analysis of randomized experiments with orthogonal block structure, II. Treatment structure and general analysis of variance. *Proceedings of the Royal Society A* **273**, 163–178.

Nelder, J. A., and Wedderburn, R. W. M. (1972). Generalized linear models. *Journal of the Royal Statistical Society A* **135**, 370–384.

Nepomnyaschy, L., and Garfinkel, I. (2005). The relationship between formal and informal child support: evidence from three-year Fragile Families data. Technical report, School of Social Work, Columbia University.

Neter, J., Kutner, M. H., Nachtsheim, C. J., and Wasserman, W. (1996). *Applied Linear Statistical Models*, fourth edition. Burr Ridge, Ill.: Richard D. Irwin, Inc.

Newman, M. E. J. (2003). The structure and function of complex networks. *SIAM Review* **45**, 167–256.

Newton, M. A., Kendziorski, C. M., Richmond, C. S., Blattner, F. R., and Tsui, K. W. (2001). On differential variability of expression ratios: improving statistical inference about gene expression changes from microarray data. *Journal of Computational Biology* **8**, 37–52.

Neyman, J. (1923). On the application of probability theory to agricultural experiments. Essay on principles. Section 9. Translated and edited by D. M. Dabrowska and T. P. Speed. *Statistical Science* **5**, 463–480 (1990).

Normand, S. L., Glickman, M. E., and Gatsonis, C. A. (1997). Statistical methods for profiling providers of medical care: issues and applications. *Journal of the American Statistical Association* **92**, 803–814.

Normand, S. L., Landrum, M. B., Guadagnoli, E., et al. (2001). Validating recommendations for coronary angiography following acute myocardial infarction in the elderly: a matched analysis using propensity scores. *Journal of Clinical Epidemiology* **54**, 387–398.

Novick, M. R., Jackson, P. H., Thayer, D. T., and Cole, N. S. (1972). Estimating multiple regressions in m-groups: a cross validation study. *British Journal of Mathematical and Statistical Psychology* **25**, 33–50.

Novick, M. R., Lewis, C., and Jackson, P. H. (1973). The estimation of proportions in m groups. *Psychometrika* **38**, 19–46.

Oakes, J. M. (2004). The (mis)estimation of neighborhood effects: causal inference for a practicable social epidemiology (with discussion). *Social Science and Medicine* **58**, 1929–1971.

O'Keefe, S. (2004). Job creation in California's enterprise zones: A comparison using a propensity score matching model. *Journal of Urban Economics* **55**, 131–150.

O'Malley, A. J., and Zaslavsky, A. M. (2005). Cluster-level covariance analysis for survey data with structured nonresponse. Technical report, Department of Health Care Policy, Harvard Medical School.

Pardoe, I. (2001). A Bayesian sampling approach to regression model checking. *Journal of Computational and Graphical Statistics* **10**, 617–627.

Pardoe, I. (2004). Model assessment plots for multilevel logistic regression. *Computational Statistics and Data Analysis* **46**, 295–307.

Pardoe, I. (2006). *Applied Regression Modeling.* New York: Wiley.

Pardoe, I., and Cook, R. D. (2002). A graphical method for assessing the fit of a logistic regression model. *American Statistician* **56**, 263–272.

Pardoe, I., and Simonton, D. K. (2006). Applying discrete choice models to predict Academy Award winners. Technical report, University of Oregon Business School.

Pardoe, I., and Weidner, R. R. (2006). Sentencing convicted felons in the United States: a Bayesian analysis using multilevel covariates (with discussion). *Journal of Statistical Planning and Inference* **136**, 1433–1472.

Park, D. K., Gelman, A., and Bafumi, J. (2004). Bayesian multilevel estimation with poststratification: state-level estimates from national polls. *Political Analysis* **12**, 375–385.

Pauler, D. K., Wakefield, J. C., and Kass, R. E. (1999). Bayes factors for variance component models. *Journal of the American Statistical Association* **94**, 1242–1253.

Pearl, J. (2000). *Causality.* Cambridge University Press.

Persico, N., Postlewaite, A., and Silverman, D. (2004). The effect of adolescent experience on labor market outcomes: the case of height. *Journal of Political Economy* **112**, 1019–1053.

Pinheiro, J. C., and Bates, D. M. (1996). Unconstrained parameterizations for variance-covariance matrices. *Statistics and Computing* **6**, 289–296.

Pinheiro, J. C., and Bates, D. M. (2000). *Mixed-Effects Models in S and S-Plus.* New York: Springer-Verlag.

Plummer, M. (2003). JAGS: a program for analysis of Bayesian graphical models using Gibbs sampling. www-fis.iarc.fr/~martyn/software/jags/

Poole, K. T., and Rosenthal, H. (1997). *Congress: A Political-Economic History of Roll Call Voting.* Oxford University Press.

Pratt, J. W. (1987). Dividing the indivisible: using simple symmetry to partition variance explained. In *Proceedings of the Second International Tampere Conference in Statistics*, ed. T. Pukkila, T., and S. Puntanen, 245–260. Department of Mathematical Sciences/Statistics, University of Tampere, Finland.

Price, P. N., and Gelman, A. (2004). Should you measure the radon concentration in your home? In *Statistics: A Guide to the Unknown*, fourth edition. Pacific Grove, Calif.: Duxbury.

Price, P. N., Nero, A. V., and Gelman, A. (1996). Bayesian prediction of mean indoor radon concentrations for Minnesota counties. *Health Physics* **71**, 922–936.

R Project (2000). The R project for statistical computing. www.r-project.org

Rabe-Hesketh, S., and Everitt, B. S. (2003). *A Handbook of Statistical Analyses using Stata*, third edition. London: CRC Press.

Rabe-Hesketh, S., and Skrondal, A. (2005). *Multilevel and Longitudinal Modeling using Stata*. College Station, Tex.: Stata Press.

Raghunathan, T. E., Solenberger, P. W., and Van Hoewyk, J. (2002). IVEware. www.isr.umich.edu/src/smp/ive/

Raghunathan, T. E., Van Hoewyk, J., and Solenberger, P. W. (2001). A multivariate technique for multiply imputing missing values using a sequence of regression models. *Survey Methodology* **27**, 85–95.

Ramsey, F. L., and Schafer, D. W. (2001). *The Statistical Sleuth*, second edition. Pacific Grove, Calif.: Duxbury.

Rasbash, J., and Browne, W. J. (2003). Non-hierarchical multilevel models. In *Handbook of Quantitative Multilevel Analysis*, ed. J. De Leeuw and I. Kreft. Boston: Kluwer.

Raudenbush, S. W. (1997). Statistical analysis and optimal design for cluster randomized trials. *Psychological Methods* **2**, 173–185.

Raudenbush, S. W., and Bryk, A. S. (2002). *Hierarchical Linear Models*, second edition. Thousand Oaks, Calif.: Sage.

Raudenbush, S. W., and Sampson, R. (1999). Assessing direct and indirect effects in multilevel designs with latent variables. *Sociological Methods and Research* **28**, 123–153.

Raudenbush, S. W., and Xiaofeng, L. (2000). Statistical power and optimal design for multisite randomized trials. *Psychological Methods* **5**, 199–213.

Ripley, B. D. (1988). *Statistical Inference for Spatial Processes*. Cambridge University Press.

Roberts, G. O., and Sahu, S. K. (1997). Updating schemes, correlation structure, blocking and parameterization for the Gibbs sampler. *Journal of the Royal Statistical Society B* **59**, 291–317.

Robins, J. M. (1989). The analysis of randomized and non-randomized AIDS treatment trials using a new approach to causal inference in longitudinal studies. *Health Services Research Methodology: A Focus on AIDS*, ed. L. Sechrest, H. Freeman, and A. Mulley, 113–159. Washington, D.C.: U.S. Public Health Service, National Center for Health Services Research.

Robins, J. M. (1994). Correcting for non-compliance in randomized trials using structural nested mean models. *Communications in Statistics* **23**, 2379–2412.

Robins, J. M. (1998). Confidence intervals for causal parameters. *Statistics in Medicine* **7**, 773–785.

Robins, J. M., and Ritov, Y. (1997). Towards a curse of dimensionality appropriate (CODA) asymptotic theory for semi-parametric models. *Statistics in Medicine* **16**, 285–319.

Robins, J. M., and Rotnitzky, A. (1995). Semiparametric efficiency in multivariate regression models with missing data. *Journal of the American Statistical Association* **90**, 122–129.

Robins, J. M., Rotnitzky, A., and Zhao, L. P. (1995). Analysis of semiparametric regression models for repeated outcomes in the presence of mimssing data. *Journal of the American Statistical Association* **90**, 106–121.

Robins, J. M., and Wang, N. (2000). Inference for imputation estimators. *Biometrika* **87**, 113–124.

Robinson, G. K. (1991). That BLUP is a good thing: the estimation of random effects (with discussion). *Statistical Science* **6**, 15–51.

Robinson, G. K. (1998). Variance components. In *Encyclopedia of Biostatistics*, ed. P. Armitage and T. Colton, **6**, 4713–4719. New York: Wiley.

Rosado, J. L., Lopez, P., Munoz, E., Martinez, H., and Allen, L. H. (1997). Zinc supplementation reduced morbidity, but neither zinc nor iron supplementation affected-growth or body composition of Mexican preschoolers. *American Journal of Clinical Nutrition* **65**, 13–19.

Rosenbaum, P. R. (1984). The consequences of adjustment for a concomitant variable that has been affected by the treatment. *Journal of the Royal Statistical Society A* **147**, 656–666.

Rosenbaum, P. R. (1987). Model-based direct adjustment. *Journal of the American Statistical Association* **82**, 387–394.

Rosenbaum, P. R. (1989). Optimal matching in observational studies. *Journal of the American Statistical Association* **84**, 1024–1032.

Rosenbaum, P. R. (2002a). Covariance adjustment in randomized experiments and observational studies. *Statistical Science* **17**, 286–327.

Rosenbaum, P. R. (2002b). *Observational Studies*. New York: Springer-Verlag.

Rosenbaum, P. R., and Rubin, D. B. (1983a). The central role of the propensity score in observational studies for causal effects. *Biometrika* **70**, 41–55.

Rosenbaum, P. R., and Rubin, D. B. (1983b). Assessing sensitivity to an unobserved binary covariate in an observational study with binary outcome. *Journal of the Royal Statistical Society B* **45**, 212–218.

Rosenbaum, P. R., and Rubin, D. B. (1984). Reducing bias in observational studies using subclassification on the propensity score. *Journal of the American Statistical Association* **79**, 516–524.

Rosenbaum, P. R., and Rubin, D. B. (1985). Constructing a control group using multivariate matched sampling methods that incorporate the propensity score. *American Statistician* **39**, 33–38.

Rosenstone, S. J. (1983). *Forecasting Presidential Elections*. New Haven: Yale University Press.

Rosenthal, R., Rosnow, R. L., and Rubin, D. B. (2000). *Contrasts and Effect Sizes in Behavioral Research*. Cambridge University Press.

Ross, C. E. (1990). Work, family, and well-being in the United States. Survey data available from Inter-university Consortium for Political and Social Research, Ann Arbor, Mich.

Rubin, D. B. (1973). The use of matched sampling and regression adjustment to remove bias in observational studies. *Biometrics* **29**, 185–203.

Rubin, D. B. (1974). Estimating causal effects of treatments in randomized and nonrandomized studies. *Journal of Educational Psychology* **66**, 688–701.

Rubin, D. B. (1976). Inference and missing data. *Biometrika* **63**, 581–592.

Rubin, D. B. (1978). Bayesian inference for causal effects: the role of randomization. *Annals of Statistics* **6**, 34–58.

Rubin, D. B. (1979). Using multivariate matched sampling and regression adjustment to control bias in observational studies. *Journal of the American Statistical Association* **74**, 318–328.

Rubin, D. B. (1980). Using empirical Bayes techniques in the law school validity studies (with discussion). *Journal of the American Statistical Association* **75**, 801–827.

Rubin, D. B. (1984). Bayesianly justifiable and relevant frequency calculations for the applied statistician. *Annals of Statistics* **12**, 1151–1172.

Rubin, D. B. (1987). *Multiple Imputation for Nonresponse in Surveys.* New York: Wiley.

Rubin, D. B. (1996). Multiple imputation after 18+ years (with discussion). *Journal of the American Statistical Association* **91**, 473–520.

Rubin, D. B. (2000). Statistical issues in the estimation of the causal effects of smoking due to the conduct of the tobacco industry. In *Statistical Science in the Courtromm*, ed. J. L. Gastwirth, 322–350. New York: Springer-Verlag.

Rubin, D. B. (2004). Direct and indirect causal effects via potential outcomes (with discussion). *Scandinavian Journal of Statistics* **31**, 161–201.

Rubin, D. B. (2006). *Matched Sampling for Causal Effects.* Cambridge University Press.

Rubin, D. B., Stern, H. S, and Vehovar, V. (1995). Handling "Don't Know" survey responses: the case of the Slovenian plebiscite. *Journal of the American Statistical Association* **90**, 822–828.

Rubin, D. B., and Stuart, E. A. (2005). Matching with multiple control groups and adjusting for group differences. In *Proceedings of the American Statistical Association, Section on Health Policy Statistics.* Alexandria, Va.: American Statistical Association.

Rubin, D. B., and Thomas, N. (2000). Combining propensity score matching with additional adjustments for prognostic covariates. *Journal of the American Statistical Association* **95**, 573–585.

Ruel, M. T., Rivera, J. A., Santizo, M. C., Lonnerdal, B., and Brown, K. H. (1997). Impact of zinc supplementation on morbidity from diarrhea and respiratory infections among rural Guatemalan children. *Pediatrics* **99**, 808–813.

Sacerdote, B. (2004). What happens if we randomly assign children to families? National Bureau of Economic Research Working Paper No. 10894.

Sampson, R. J., Raudenbush, S. W., and Earls, F. (1997). Neighborhoods and violent crime: a multilevel study of collective efficacy. *Science* **277**, 918–924.

Sargent, D. J., Hodges, J. S., and Carlin, B. P. (2000). Structured Markov chain Monte Carlo. *Journal of Computational and Graphical Statistics* **9**, 217–234.

Schafer, J. L. (1997). *Analysis of Incomplete Multivariate Data.* London: Chapman and Hall.

Scott, A., and Smith, T. M. F. (1969). Estimation in multi-stage surveys. *Journal of the American Statistical Association* **64**, 830–840.

Searle, S. R., Casella, G., and McCulloch, C. E. (1992). *Variance Components.* New York: Wiley.

Shadish, W., Cook, T., and Campbell, S. (2002). *Experimental and Quasi-Experimental Designs.* Boston: Houghton Mifflin.

Sheiner, L. B., Beal, S. L., and Dunne, A. (1997). Analysis of non-randomly censored ordered categorical longitudinal data from analgesic trials (with discussion). *Journal of the American Statistical Association* **92**, 1235–1255.

Shen, W., and Louis, T. A. (1998). Triple-goal estimates in two-stage hierarchical models. *Journal of the Royal Statistical Society B* **60**, 455–471.

Sinharay, S., and Stern, H. S. (2003). Posterior predictive model checking in hierarchical models. *Journal of Statistical Planning and Inference* **111**, 209–221.

Skaug, H. J., and Fournier, D. (2006). Automatic approximation of the marginal likelihood in non-Gaussian hierarchical models. *Computational Statistics and Data Analysis.*

Skogan, W. (1990). *Disorder and Decline: Crime and the Spiral of Decay in American Cities.* University of California Press.

Smith, J., and Todd, P. E. (2005). Does matching overcome LaLonde's critique of nonexperimental estimators? (with discussion). *Journal of Econometrics* **120**, 305–375.

Snedecor, G. W., and Cochran, W. G. (1989). *Statistical Methods,* eighth edition. Ames: Iowa State University Press.

Snijders, T. A. B., and Bosker, R. J. (1993). Standard errors and sample sizes for two-level research. *Journal of Educational Statistics* **18**, 237–259.

Snijders, T. A. B., and Bosker, R. J. (1999). *Multilevel Analysis.* London: Sage.

Snijders, T. A. B., Bosker, R., and Guldemond, H. (1999). PINT: Power analysis in two-level designs. `stat.gamma.rug.nl/snijders/multilevel.htm#progPINT`

Sobel, M. E. (1990). Effect analysis and causation in linear structural equation models. *Psychometrika* **55**, 495–515.

Sobel, M. E. (1998). Causal inference in statistical models of the process of socioeconomic achievement: a case study. *Sociological Methods and Research* **27**, 318–348.

Sobel, M. E. (2006). What do randomized studies of housing mobility reveal? Causal inference in the face of interference. Technical report, Department of Sociology, Columbia University.

Speed, T. P. (1990). Introductory remarks on Neyman (1923). *Statistical Science* **5**, 463–464.

Spiegelhalter, D. J. (2006). Two brief topics on modelling with WinBUGS. Presented at IceBUGS conference. `www.math.helsinki.fi/openbugs/IceBUGS/Presentations/SpiegelhalterIceBUGS.pdf`

Spiegelhalter, D. J., Best, N. G., Carlin, B. P., and van der Linde, A. (2002). Bayesian measures of model complexity and fit (with discussion). *Journal of the Royal Statistical Society B* **64**, 583–639.

Spiegelhalter, D., Thomas, A., Best, N., Gilks, W., and Lunn, D. (1994, 2002). BUGS: Bayesian inference using Gibbs sampling. MRC Biostatistics Unit, Cambridge, England. `www.mrc-bsu.cam.ac.uk/bugs/`

Spitzer, E. (1999). The New York City Police Department's "stop and frisk" practices. Office of the New York State Attorney General. `www.oag.state.ny.us/press/reports/stop_frisk/stop_frisk.html`

Sternberg, S. (1963). Stochastic learning theory. In *Handbook of Mathematical Psychology,* volume 2, ed. R. D. Luce, R. R. Bush, and E. Galanter, 1–120. New York: Wiley.

Stigler, S. M. (1977). Do robust estimators work with real data? (with discussion). *Annals of Statistics* **5**, 1055–1098.

Stigler, S. M. (1983). Discussion of "Parametric empirical Bayes inference: theory and applications," by C. Morris. *Journal of the American Statistical Association* **78**, 62–63.

Stigler, S. M. (1986). *The History of Statistics.* Cambridge, Mass.: Harvard University Press.

Stone, M. (1974). Cross-validatory choice and assessment of statistical predictions (with discussion). *Journal of the Royal Statistical Society B* **36**, 111–147.

Sturtz, S., Ligges, U., and Gelman, A. (2005). R2WinBUGS: a package for running WinBUGS from R. *Journal of Statistical Software* **12** (3).

Summers, A., and Wolfe, B. (1977). Do schools make a difference? *American Economic Review* **67**, 639–652.

Tanner, M. A., and Wong, W. H. (1987). The calculation of posterior distributions by data augmentation (with discussion). *Journal of the American Statistical Association* **82**, 528–550.

Thistlethwaite, D., and Campbell, D. (1960). Regression-discontinuity analysis: an alternative to the ex post facto experiment. *Journal of Educational Psychology* **51**, 309–17.

Thomas, A., and O'Hara, B. (2005). OpenBugs. `www.math.helsinki.fi/openbugs`

Thurstone, L. L. (1927a). A law of comparative judgement. *Psychological Review* **34**, 273–286.

Thurstone, L. L. (1927b). The method of paired comparison for social values. *Journal of Abnormal and Social Psychology* **21**, 384–400.

Tiao, G. C., and Box, G. E. P. (1967). Bayesian analysis of a three-component hierarchical design model. *Biometrika* **54**, 109–125.

Tiao, G. C., and Tan, W. Y. (1965). Bayesian analysis of random-effect models in the analysis of variance. I: Posterior distribution of variance components. *Biometrika* **52**, 37–53.

Tiao, G. C., and Tan, W. Y. (1966). Bayesian analysis of random-effect models in the analysis of variance. II: Effect of autocorrelated errors. *Biometrika* **53**, 477–495.

Tinbergen, J. (1930). Determination and interpretation of supply curves: an example. Zeitschrift fur Nationalokonomie. Reprinted in *The Foundations of Econometrics*, eds. D. Hendry and M. Morgan, 233–245. Cambridge University Press.

Tobin, J. (1958). Estimation of relationships for limited dependent variables. *Econometrica* **26**, 24–36.

Trochim, W. M. K. (2001). *The Research Methods Knowledge Base*. Cincinnati: Atomic Dog.

Troxel, A. B., Ma, G., and Heitjan, D. F. (2004). An index of sensitivity to nonignorability. *Statistica Sinica* **14**, 1221–1237.

Tu, W., and Zhou, X. H. (2003). A bootstrap confidence interval procedure for treatment using propensity score subclassification. Technical Report, University of Washington Biostatistics Working Paper Series, Working Paper 200.

Tufte, E. R. (1983). *The Visual Display of Quantitative Information*. Cheshire, Conn.: Graphics Press.

Tufte, E. R. (1990). *Envisioning Information*. Cheshire, Conn.: Graphics Press.

Tukey, J. W. (1977). *Exploratory Data Analysis*. Reading, Mass.: Addison-Wesley.

Vaida, F., and Blanchard, S. (2002). Conditional Akaike informaion for mixed effects models. Technical report, Department of Biostatistics, Harvard University.

Van Buuren, S., Boshuizen, H. C., and Knook, D. L. (1999). Multiple imputation of missing blood pressure covariates in survival analysis. *Statistics in Medicine* **18**, 681–694.

Van Buuren, S., and Oudshoom, C. G. M. (2000). MICE: Multivariate imputation by chained equations (S software for missing-data imputation). `web.inter.nl.net/users/S.van.Buuren/mi/`

van der Linden, W. J., and Hambelton, R. K., eds. (1997). *Handbook of Modern Item Response Theory*. New York: Springer-Verlag.

van Dyk, D. A., and Meng, X. L. (2001). The art of data augmentation (with discussion). *Journal of Computational and Graphical Statistics* **10**, 1–111.

van Geen, A., Zheng, Y., Versteeg, R., et al. (2003). Spatial variability of arsenic in 6000 tube wells in a 25 km^2 area of Bangladesh. *Water Resources Research* **39**, 1140.

Venables, W. N., and Ripley, B. D. (2002). *Modern Applied Statistics with S*, fourth edition. New York: Springer-Verlag.

Verbeke, G., and Molenberghs, G. (2000). *Linear Mixed Models for Longitudinal Data*. New York: Springer-Verlag.

Vikram, H. R., Buenconsejo, J., Hasbun, R., and Quagliarello, V. J. (2003). Impact of valve surgery on 6-month mortality in adults with complicated, left-sided native valve endocarditis: a propensity analysis. *Journal of the American Medical Association* **290**, 3207–3214.

Viscusi, W. K., and Aldy, J. E. (2002). The value of a statistical life: a critical review of market estimates throughout the world. Harvard Law School.

von Mises, R. (1957). *Probability, Statistics, and Truth*, second edition. New York: Dover. Reprint.

Wainer, H. (1984). How to display data badly. *American Statistician* **38**, 137–147.

Wainer, H. (1997). *Visual Revelations*. New York: Springer-Verlag.

Wainer, H. (2001). Order in the court. *Chance* **14** (2), 43–46.

Wainer, H. (2002). The BK-plot: making Simpson's paradox clear to the masses. *Chance* **15** (3), 60–62.

Wainer, H., Gelman, A., and Wang, X. (2000). What happens when advanced exams are used to place students out of introductory courses? Princeton, N.J.: Educational Testing Service.

Wainer, H., Palmer, S., and Bradlow, E. T. (1998). A selection of selection anomalies. *Chance* **11** (2), 3–7.

Walker, S. H., and Duncan, D. B. (1967). Estimation of the probability of an event as a function of several independent variables. *Biometrika* **54**, 167–178.

Waller, L. A., Carlin, B. P., Xia, H., and Gelfand, A. E. (1997). Hierarchical spatio-temporal mapping of disease rates. *Journal of the American Statistical Association* **92**, 607–617.

Wallis, W. A., and Friedman, M. (1942). The empirical derivation of indifference functions. In *Studies in Mathematical Economics and Econometrics in Memory of Henry Schultz*, ed. O. Lange, F. McIntyre, and T. O. Yntema, 267–300. University of Chicago Press.

Wand, M. P. (1997). Data-based choice of histogram bin width. *American Statistician* **51**, 59–64.

Warnes, G. R. (2003). Hydra: a Java library for Markov chain Monte Carlo. `software.biostat.washington.edu/statsoft/MCMC/Hydra`

Weisberg, S. (1985). *Applied Linear Regression*, second edition. New York: Wiley.

West, M. (2003). Bayesian factor regression models in the "large p, small n" paradigm. In *Bayesian Statistics 7*, ed. J. M. Bernardo, M. J. Bayarri, J. O. Berger, A. P. Dawid, D. Heckerman, A. F. M. Smith, and M. West, 733–742. Oxford University Press.

Wherry, R. J. (1931). A new formula for predicting the shrinkage of the coefficient of multiple correlation. *Annals of Mathematical Statistics* **2**, 440–457.

Wiens, B. L. (1999). When log-normal and gamma models give different results: a case study. *American Statistician* **53**, 89–93.

Wikle, C. K., Milliff, R. F., Nychka, D., and Berliner, L. M. (2001). Spatiotemporal hierarchical Bayesian modeling: tropical ocean surface winds. *Journal of the American Statistical Association* **96**, 382–397.

Wilkinson, L. (2005). *The Grammar of Graphics*, second edition. New York: Springer-Verlag.

Wilson, J. Q., and Kelling, G. L. (1982). The police and neigborhood safety: broken windows. *Atlantic Monthly*, March, 29–38.

Witte, J. S., Greenland, S., Hale, R. W., and Bird, C. L. (1994). Hierarchical regression analysis applied to a study of multiple dietary exposures and breast cancer. *Epidemiology* **5**, 612–621.

Wlezien, C., and Erikson, R. S. (2004). The fundamentals, the polls, and the presidential vote. *PS: Political Science and Politics* **37**, 747–751.

Wlezien, C., and Erikson, R. S. (2005). Post-election reflections on our pre-election predictions. *PS: Political Science and Politics* **38**, 25–26.

Woolridge, J. M. (2001). *Econometric Analysis of Cross Section and Panel Data.* Cambridge, Mass.: MIT Press.

Wright, G. C. (1989). Level-of-analysis effects on explanations of voting: the case of the 1982 U.S. Senate elections. *British Journal of Political Science* **19**, 381–398.

Xu, R. (2003). Measuring explained variation in linear mixed effects models. *Statistics in Medicine* **22**, 3527–3541.

Yates, F. (1967). A fresh look at the basic principles of the design and analysis of experiments. *Proceedings of the Fifth Berkeley Symposium on Mathematical Statistics and Probability* **4**, 777–790. Berkeley: University of California Press.

Yusuf, S., Peto, R., Lewis, J., Collins, R., and Sleight, P. (1985). Beta blockade during and after myocardial infarction: an overview of the randomized trials. *Progress in Cardiovascular Diseases* **27**, 335–371.

Zaslavsky, A. M. (1993). Combining census, dual-system, and evaluation study data to estimate population shares. *Journal of the American Statistical Association* **88**, 1092–1105.

Zeger, S. L., and Karim, M. R. (1991). Generalized linear models with random effects; a Gibbs sampling approach. *Journal of the American Statistical Association* **86**, 79–86.

Zellner, A. (1976). Bayesian and non-Bayesian analysis of the regression model with multivariate Student-t error terms. *Journal of the American Statistical Association* **71**, 400–405.

Zheng, T., Salganik, M., and Gelman, A. (2006). How many people do you know in prison?: Using overdispersion in count data to estimate social structure in networks. *Journal of the American Statistical Association.*

Zorn, C. (2005). A solution to separation in binary response models. *Political Analysis* **13**, 157–170.

Author index

Subject index